Lecture Notes in Computer Science　7933

Commenced Publication in 1973
Founding and Former Series Editors:
Gerhard Goos, Juris Hartmanis, and Jan van Leeuwen

Vincenzo Bonifaci   Camil Demetrescu
Alberto Marchetti-Spaccamela (Eds.)

Experimental Algorithms

12th International Symposium, SEA 2013
Rome, Italy, June 5-7, 2013
Proceedings

 Springer

Volume Editors

Vincenzo Bonifaci
Institute for Systems Analysis and Computer Science (IASI)
National Research Council (CNR)
Viale Manzoni 30, 00185 Rome, Italy
E-mail: vincenzo.bonifaci@iasi.cnr.it

Camil Demetrescu
Sapienza University of Rome
Department of Computer and Systems Science
Via Ariosto 25, 00185 Rome, Italy
E-mail: demetres@dis.uniroma1.it

Alberto Marchetti-Spaccamela
Sapienza University of Rome
Department of Computer and Systems Science
Via Ariosto 25, 00185 Rome, Italy
E-mail: alberto@dis.uniroma1.it

ISSN 0302-9743   e-ISSN 1611-3349
ISBN 978-3-642-38526-1   e-ISBN 978-3-642-38527-8
DOI 10.1007/978-3-642-38527-8
Springer Heidelberg Dordrecht London New York

Library of Congress Control Number: 2013938272

CR Subject Classification (1998): F.2, E.1, G.2, G.1.6, G.1, I.2.8

LNCS Sublibrary: SL 2 – Programming and Software Engineering

Typesetting: Camera-ready by author, data conversion by Scientific Publishing Services, Chennai, India

Printed on acid-free paper

Springer is part of Springer Science+Business Media (www.springer.com)

Preface

This volume contains the papers presented at the 12th International Symposium on Experimental Algorithms (SEA 2013) held during June 5–7, 2013, in Rome. The conference was organized under the auspices of the European Association for Theoretical Computer Science (EATCS).

SEA, previously known as WEA (Workshop on Experimental Algorithms), is an international forum for researchers in the area of design, analysis, and experimental evaluation and engineering of algorithms, as well as in various aspects of computational optimization and its applications. The preceding symposia were held in Riga, Monte Verita, Rio de Janeiro, Santorini, Menorca, Rome, Cape Cod, Dortmund, Ischia, Crete, and Bordeaux.

The symposium received 73 submissions. Each submission was reviewed by at least three Program Committee members, and carefully evaluated on originality, quality, and relevance to the conference. Based on extensive electronic discussions, the committee decided to accept 32 papers. In addition to the accepted contributions, the program also included three distinguished plenary lectures by Martin Skutella (Technische Universität Berlin), Andrew V. Goldberg (Microsoft Research Silicon Valley), and Roberto Grossi (Università di Pisa).

We would like to thank all the authors who responded to the call for papers, the invited speakers, the members of the Program Committee, the external referees, and the members of the Organizing Committee. We also gratefully acknowledge the developers and maintainers of the EasyChair conference management system, who provided invaluable support, and Springer for publishing the proceedings of SEA 2013 in their LNCS series and for their support.

April 2013

Vincenzo Bonifaci
Camil Demetrescu
Alberto Marchetti-Spaccamela

Organization

Program Committee

Vincenzo Bonifaci	IASI-CNR, Rome (Co-chair), Italy
Ulrik Brandes	University of Konstanz, Germany
Daniel Delling	Microsoft Research Silicon Valley, USA
Giuseppe Di Battista	Roma Tre University, Italy
Tobias Friedrich	Friedrich-Schiller-Universität Jena, Germany
Ralf Klasing	LaBRI - CNRS, France
Arie Koster	RWTH Aachen University, Germany
Pierre Leone	University of Geneva, Switzerland
Leo Liberti	LIX, Ecole Polytechnique, France
Andrea Lodi	DEIS, University of Bologna, Italy
Alberto Marchetti-Spaccamela	Sapienza Università di Roma (Co-chair), Italy
Nicole Megow	Technische Universität Berlin, Germany
Frank Neumann	University of Adelaide, Australia
Simon Puglisi	University of Helsinki, Finland
Rajeev Raman	University of Leicester, UK
Anita Schöbel	Georg-August University of Goettingen, Germany
Maria Serna	Universitat Politecnica de Catalunya, Spain
Renato Werneck	Microsoft Research Silicon Valley, USA
Christos Zaroliagis	Computer Tech. Institute and University of Patras, Greece
Norbert Zeh	Dalhousie University, Canada

Organizing Committee

Vincenzo Bonifaci	IASI-CNR, Rome
Saverio Caminiti	Sapienza Università di Roma
Camil Demetrescu	Sapienza Università di Roma (Chair)
Alberto Marchetti-Spaccamela	Sapienza Università di Roma

External Reviewers

Alvarez-Miranda, Eduardo	Antoniadis, Antonios
Andoni, Alexandr	Bansal, Nikhil
Angel, Eric	Barbay, Jérémy
Angelini, Patrizio	Bast, Hannah

Belazzougui, Djamal

Bergner, Martin

Berthold, Timo

Blesa, Maria J.

Blum, Christian

Bodlaender, Hans

Bollig, Beate

Bonichon, Nicolas

Bringmann, Karl

Burt, Christina

Butenko, Sergiy

Böckenhauer, Hans-Joachim

Carvalho, Margarida

Chiesa, Marco

Coniglio, Stefano

Cornelsen, Sabine

Costa Fampa, Marcia Helena

Crescenzi, Pierluigi

Crochemore, Maxime

D'Angelo, Gianlorenzo

Da Lozzo, Giordano

Dalkiran, Evrim

Danilewski, Piotr

Di Bartolomeo, Marco

Doerr, Carola

Eisner, Jochen

Fernau, Henning

Fertin, Guillaume

Fischer, Johannes

Fischetti, Matteo

Frangioni, Antonio

Frati, Fabrizio

Fredriksson, Kimmo

Gavoille, Cyril

Goerigk, Marc

Gog, Simon

Grossi, Roberto

Halperin, Dan

Heimel, Max

Hiller, Benjamin

Hoffmann, Frank

Höhn, Wiebke

Hübner, Ruth

Ide, Jonas

Iori, Manuel

Joannou, Stelios

Julstrom, Bryant

Kesselheim, Thomas

Khajavirad, Aida

Kirchler, Dominik

Kontogiannis, Spyros

Krohmer, Anton

Kröller, Alexander

Kötzing, Timo

Larsen, Jesper

Lee, Jon

Ljubic, Ivana

Lucas, Claire

Lulli, Guglielmo

Makris, Christos

Malliaros, F.

Manthey, Bodo

Marinelli, Fabrizio

Molinero, Xavier

Montemanni, Roberto

Moscardelli, Luca

Mucherino, Antonio

Mömke, Tobias

Müller-Hannemann, Matthias

Nallaperuma, Samadhi

Ovsjanikov, Maks

Pacciarelli, Dario

Parriani, Tiziano

Patella, Marco

Petit, Jordi

Radzik, Tomasz

Roselli, Vincenzo

Röttger, Richard

Salvagnin, Domenico

Schieweck, Robert

Schweitzer, Pascal

Shalom, Mordechai

Sutton, Andrew

Tiedemann, Morten

Tollis, Ioannis

Tomescu, Alexandru

Traversi, Emiliano

Tsichlas, Kostas

Vanaret, Charlie

Wagner, Markus

Wieder, Udi

Xhafa, Fatos

Table of Contents

Invited Papers

Transportation Networks and Graph Algorithms I

Combinatorics and Enumeration

Transportation Networks and Graph Algorithms II

Geometry and Optimization

Scheduling and Local Search

Algorithms and Linear Programming Relaxations for Scheduling Unrelated Parallel Machines[*]

Martin Skutella

Fakultät II – Mathematik und Naturwissenschaften,
Institut für Mathematik, Sekr. MA 5-2
Technische Universität Berlin, Straße des 17. Juni 136,
10623 Berlin, Germany
martin.skutella@tu-berlin.de

Since the early days of combinatorial optimization, algorithms and techniques from the closely related area of mathematical programming have played a pivotal role in solving combinatorial optimization problems. This holds both for 'easy' problems that can be solved efficiently in polynomial time, such as, e.g., the weighted matching problem [3], as well as for NP-hard problems whose solution might take exponential time in the worst case, such as, e.g., the traveling salesperson problem [1].

The by far most commonly used and also most successful method for solving NP-hard problems to optimality relies on (mixed) integer linear programming formulations whose linear programming relaxations then yield lower bounds on the value of optimal solutions; see, e.g., [21]. This general approach has also turned out to be extremely useful in the design of approximation algorithms, which compute provably good solutions to NP-hard optimization problems in polynomial time; see, e.g., [19]. Regardless whether one wants to compute a solution that is guaranteed to be optimal or one is satisfied with an approximate solution whose value is provably close to the optimum, in both situations the quality of the linear programming relaxation is critical.

In this talk, we mainly focus on the use of linear programming relaxations in the design of approximation algorithms. The classical problem of scheduling jobs on unrelated parallel machines subject to release dates and with total weighted completion time objective (commonly denoted $R \mid r_{ij} \mid \sum w_j C_j$; see [5]) and special cases of this problem serve as an example for which we discuss the strengths and weaknesses of different types of linear and convex programming relaxations. These formulations and relaxations can mainly be characterized by the different choices of decision variables.

In the following, we only mention some examples and pointer to the literature. We refer to the cited papers and references therein for a more detailed account of work in this area. The work of Hall et al. [6] mainly builds upon linear programming relaxations in which each decision variable corresponds to

---

[*] Supported by the DFG Research Center MATHEON "Mathematics for key technologies" in Berlin and by the DFG Focus Program 1307 within the project "Algorithm Engineering for Real-time Scheduling and Routing".

V. Bonifaci et al. (Eds.): SEA 2013, LNCS 7933, pp. 1–3, 2013.

the completion time of a job. This kind of formulation dates back to Wolsey [20] and Queyranne [10]. For the single-machine sequencing problem Potts [9] gave a formulation in linear-ordering variables. Dyer and Wolsey [2] presented time-indexed relaxations which later turned out to be extremely useful for the design of approximation algorithms; see, e. g., [4,12]. Time-indexed formulations were also extended to unrelated parallel machines [13,14]. In [15], a quadratic programming formulation in assignment variables and a convex quadratic relaxation are presented. Recently, Sviridenko and Wiese [18] study a very strong linear programming relaxation, the configuration LP, whose variables correspond to entire machine configurations.

We finally mention that approximation algorithms based on different types of linear programming relaxations have also been developed for various stochastic machine scheduling problems where processing times of jobs are randomly chosen according to a given distribution; see, e. g., [8,17,7]. Here, mainly linear programming relaxations in completion time variables are used; see, e. g., [11]. Only recently, it has been shown how time-indexed formulations can be put to work in stochastic scheduling [16].

References

1. Applegate, D.L., Bixby, R.E., Chvátal, V., Cook, W.J.: The Traveling Salesman Problem: A Computational Study. Princeton University Press (2006)
2. Dyer, M.E., Wolsey, L.A.: Formulating the single machine sequencing problem with release dates as a mixed integer program. Discrete Applied Mathematics 26, 255–270 (1990)
3. Edmonds, J.: Maximum matching and a polyhedron with 0,1-vertices. Journal of Research National Bureau of Standards Section B 69, 125–130 (1965)
4. Goemans, M.X., Queyranne, M., Schulz, A.S., Skutella, M., Wang, Y.: Single machine scheduling with release dates. SIAM Journal on Discrete Mathematics 15, 165–192 (2002)
5. Graham, R.L., Lawler, E.L., Lenstra, J.K., Rinnooy Kan, A.H.G.: Optimization and approximation in deterministic sequencing and scheduling: A survey. Annals of Discrete Mathematics 5, 287–326 (1979)
6. Hall, L.A., Schulz, A.S., Shmoys, D.B., Wein, J.: Scheduling to minimize average completion time: Off-line and on-line approximation algorithms. Mathematics of Operations Research 22, 513–544 (1997)
7. Megow, N., Uetz, M., Vredeveld, T.: Models and algorithms for stochastic online scheduling. Mathematics of Operations Research 31(3), 513–525 (2006)
8. Möhring, R.H., Schulz, A.S., Uetz, M.: Approximation in stochastic scheduling: The power of LP-based priority policies. Journal of the ACM 46, 924–942 (1999)
9. Potts, C.N.: An algorithm for the single machine sequencing problem with precedence constraints. Mathematical Programming Studies 13, 78–87 (1980)
10. Queyranne, M.: Structure of a simple scheduling polyhedron. Mathematical Programming 58, 263–285 (1993)
11. Schulz, A.S.: Stochastic online scheduling revisited. In: Yang, B., Du, D.-Z., Wang, C.A. (eds.) COCOA 2008. LNCS, vol. 5165, pp. 448–457. Springer, Heidelberg (2008)

12. Schulz, A.S., Skutella, M.: The power of α-points in preemptive single machine scheduling. Journal of Scheduling 5, 121–133 (2002)
13. Schulz, A.S., Skutella, M.: Scheduling unrelated machines by randomized rounding. SIAM Journal on Discrete Mathematics 15, 450–469 (2002)
14. Skutella, M.: Approximation and randomization in scheduling. PhD thesis, Technische Universität Berlin, Germany (1998)
15. Skutella, M.: Convex quadratic and semidefinite programming relaxations in scheduling. Journal of the ACM 48, 206–242 (2001)
16. Skutella, M., Sviridenko, M., Uetz, M.: Stochastic scheduling on unrelated machines (in preparation, 2013)
17. Skutella, M., Uetz, M.: Stochastic machine scheduling with precedence constraints. SIAM Journal on Computing 34, 788–802 (2005)
18. Sviridenko, M., Wiese, A.: Approximating the configuration-LP for minimizing weighted sum of completion times on unrelated machines. In: Goemans, M., Correa, J. (eds.) IPCO 2013. LNCS, vol. 7801, pp. 387–398. Springer, Heidelberg (2013)
19. Williamson, D.P., Shmoys, D.B.: The Design of Approximation Algorithms. Cambridge University Press (2011)
20. Wolsey, L.A.: Mixed integer programming formulations for production planning and scheduling problems. Invited talk at the 12th International Symposium on Mathematical Programming. MIT, Cambridge (1985)
21. Wolsey, L.A.: Integer Programming. Wiley (1998)

The Hub Labeling Algorithm

Andrew V. Goldberg

Microsoft Research Silicon Valley
http://research.microsoft.com/en-us/people/goldberg/

Abstract. Given a weighted graph, a distance oracle takes as an input a pair of vertices and returns the distance between them. The labeling approach to distance oracle design is to precompute a label for every vertex so that the distances can be computed from the corresponding labels, without looking at the graph. In the *hub labeling algorithm (HL)*, a vertex label consists of a set of other vertices (hubs) with distances to the hubs. We survey theoretical and experimental results on HL.

Although computing optimal hub labels is hard, in polynomial time one can approximate them up to a factor of $O(\log n)$. This can be done for the total label size (i.e., memory required to store the labels), the maximum label size (which determines the worst-case query time), and in general for an L_p norm of the vector induced by the vertex label sizes. One can also simultaneously approximate L_p and L_q norms.

Hierarchical labels are a special class of HL for which the relationship "v is a hub of w" is a partial order. For networks with a small highway dimension, one can compute provably small hierarchical labels in polynomial time.

While some graphs admit small labels, there are graphs for which the labels are large. Furthermore, one can prove that for some graphs hierarchical labels are significantly larger than the general ones.

A heuristic for computing hierarchical labels leads to the fastest known implementation of a distance oracle for road networks. One can use label compression to trade off time for space, making the algorithm practical for a wider range of applications. We give experimental results showing that the heuristic hierarchical labels are efficient on road networks as well as on some other graph classes, but not on all graphs.

We also discuss efficient implementations of the provably good approximation algorithms. Although not as fast as the hierarchical heuristic, the algorithms can solve moderate-size problems. For some graph classes, the theoretically justified algorithms compute significantly smaller labels, although for many graphs the label size difference is very small.

V. Bonifaci et al. (Eds.): SEA 2013, LNCS 7933, p. 4, 2013.

Design of Practical Succinct Data Structures
for Large Data Collections

Roberto Grossi and Giuseppe Ottaviano

Dipartimento di Informatica
Università di Pisa
{grossi,ottaviano}@di.unipi.it

Abstract. We describe a set of basic succinct data structures which
have been implemented as part of the *Succinct* library, and applications
on top of the library: an index to speed-up the access to collections of
semi-structured data, a compressed string dictionary, and a compressed
dictionary for scored strings which supports top-k prefix matching.

1 Introduction

Succinct data structures (SDS) encode data in small space and support efficient
operations on them. Encoding is a well studied problem in information theory
and there is a simple lower bound on the required space in bits: if data are
entries from a domain D, encoding each entry with less than $\lceil \log |D| \rceil$ bits cannot
uniquely identify all the entries in D (here logs are to the base 2). Thus any
correct encoding requires at least $\lceil \log |D| \rceil$ bits in the worst case, which is known
as the *information-theoretic lower bound.* Variants of this concept use some form
of entropy of D or other adaptive measures for the space occupancy, in place of
the $\lceil \log |D| \rceil$ term, but the idea is essentially the same.

Going one step beyond encoding data, SDS can also retrieve data in response
to queries. When data are any given subset of elements, an example of query is
asking if an input element belongs to that subset. Without any restriction on the
execution time of the queries, answering them becomes a trivial task to perform:
simply decode the whole encoded data and scan to answer the queries. The
challenge in SDS is to quickly perform its query operations, possibly in *constant
time* per query. To attain this goal, SDS can use extra r bits of *redundancy* in
addition to those indicated by the information-theoretic lower bound.

SDS have been mainly conceived in a theoretical setting. The first results
date back to Elias' papers of 1974 and 1975 on information retrieval [9,10] with
a reference to the Minsky-Papert problem on searching with bounded Hamming
distance. The power of SDS has been extensively discussed in Jacobson's PhD
thesis [17], where he shows how to store data such as binary sequences, trees,
and graphs in small space. The design of SDS is also linked to the bit probe
complexity of data structures, see the literature cited in [22,6], and the time-
space tradeoffs of data structures, e.g. [27]. As of now, there are SDS for sets
of integers, sequences, strings, trees, graphs, relations, functions, permutations,
geometric data, special formats (XML, JSON), and so on [19].

V. Bonifaci et al. (Eds.): SEA 2013, LNCS 7933, pp. 5–17, 2013.

What is interesting for the algorithm engineering community is that, after some software prototype attempts, efficient libraries for SDS are emerging, such as the C++ libraries *libcds* [21], *rsdic* [29], *SDSL* [31], and the Java/C++ library *Sux* [33]. They combine the advantage of data compression with the performance of efficient data structures, so that they can operate directly on compressed data by accessing just a *small* portion of data that needs to be decompressed. The field of applications is vast and the benefit is significant for large data sets that can be kept in main memory when encoded in succinct format.

The preprocessing stage of SDS builds an *index* that occupies r bits (of redundancy). Systematic SDS have a clear separation of the index from the compressed data, with several advantages [3]. When this separation is not obtained, the resulting SDS are called non-systematic because many bits contribute simultaneously both to the index and the compressed data.[1]

Given data chosen from a domain D, the designer of SDS aims at using $r + \lceil \log |D| \rceil$ bits to store data+index, with the main goal of asymptotically minimizing both the space redundancy r and the query time, or at least finding a good trade-off for these two quantities. Optimality is achieved when r and query bounds match the corresponding lower bounds. The problem is not only challenging from a theoretical point of view, where sophisticated upper and lower bounds have been proposed, e.g. [26,35]. Its practical aspects are quite relevant, e.g. in compressed text indexing [12,15], where asymptotically small redundancy and query time quite often do not translate into practical and fast algorithms.

We focus on succinct indexes for semi-structured data and strings but we believe that many SDS should be part of the modern algorithmist's toolbox for all the several data types mentioned so far. We survey a set of fundamental SDS and primitives to represent and index ordered subsets, sequences of bits, sequences of integers, and trees, that can be used as building blocks to build more sophisticated data structures. These SDS have proven to be practical and mature enough to be used as black boxes, without having to understand the inner details. The SDS that we describe are all implemented as part of the *Succinct* C++ library [32]; we give three examples of its applications in the last section.

2   Basic Toolkit for the Designer

In this section we define the most common primitives used when designing SDS, which are also sufficient for the applications of Sect. 4. The details on the algorithm and data structures used to implement them, along with their time and space complexities, are deferred to Sect. 3.

2.1   Subsets and Bitvectors

Bitvectors are the basic building block of SDS. In fact, most constructions consist of a collection of bitvectors and other sequences, that are aligned under some

---

[1] In a certain sense, implicit data structures as the binary heap can be seen as a form of non-systematic SDS where $r = O(\log n)$ bits for n elements.

logic that allows to efficiently translate a position in one sequence to a position in another sequence; this pattern will be used extensively in Sect. 4.

Formally, a *bitvector* is a finite sequence of values from the alphabet $\{0, 1\}$, i.e. *bits*. For a bitvector s we use $|s|$ to denote its length, and s_i to denote its i-th element, where i is 0-based. Bitvectors can be interpreted as *subsets* of an ordered universe U: the ordering induces a numbering from 0 to $|U| - 1$, so a subset X can be encoded as a bitvector s of length $|U|$, where s_i is 1 if the i-th element of U belongs to X, and 0 otherwise. Here the information-theoretic lower bound is $\lceil \log \binom{|U|}{m} \rceil$ where m is the number of 1s [28].

Rank and Select. The Rank and Select primitives form the cornerstone of most SDS, since they are most prominently used to align different sequences under the pattern described above. These operations can be defined on sequences drawn from arbitrary alphabets, but for simplicity we will focus on bitvectors.

- $\text{Rank}_1(i)$ returns the number of occurrences of 1 in $s[0, i)$.
- $\text{Select}_1(i)$ returns the position of the i-th occurrence of 1.

An operation tightly related to Rank/Select is $\text{Predecessor}_1(i)$, which returns the position of the *rightmost* occurrence of 1 preceding or equal to i. Note that $\text{Rank}_1(\text{Select}_1(i)) = i$ and $\text{Select}_1(\text{Rank}_1(i)) = \text{Predecessor}_1(i)$ (Symmetrically, Rank_0, Select_0, and Predecessor_0 can be defined on occurrences of 0s.)

If the bitvector is interpreted as a subset under the correspondence defined above, Rank_1 returns the number of elements in the subset that are strictly smaller than a given element of the universe, while Select_1 returns the elements of the subset in sorted order. Predecessor_1 returns the largest element that is smaller or equal to a given element of the universe.

A basic example of how to use Rank_1 to align different sequences is the *sparse array*. Let A be an array of n elements of k bits each; then, to store it explicitly, kn bits are needed. However, if a significant number of elements is 0 (or any fixed constant), we can use a bitvector z of n bits to encode which elements are zero and which ones are not, and store only the non-zeros in an another array B. To retrieve $A[i]$, we return 0 if $z_i = 0$; otherwise, the value is $B[\text{Rank}_1(i)]$. This allows to reduce the space from kn bits to $n + k|B|$ bits, plus the space taken by the data structure used to efficiently support Rank_1.

2.2   Balanced Parentheses and Trees

Another class of sequences of particular interest is given by sequences of *balanced parentheses* (BP), which can be used to represent arbitrary trees in spaces close to the information-theoretic optimum.

BP sequences are inductively defined as follows: an empty sequence is BP; if α and β are sequences of BP, then also $(\alpha)\beta$ is a sequence of BP, where (and) are called *mates*. For example, $s = (()(()()))$ is a sequence of BP. These sequences are usually represented as bitvectors, where 1 represents (and 0 represents).

Several operations operations can be defined on such sequences; as we will see shortly, the following ones are sufficient for basic tree navigation.

- FindClose(i), for a value i such that $s_i = ($, returns the position $j > i$ such that $s_j =)$ is its mate. (FindOpen(i) is defined analogously.)
- Enclose(i), for a value i such that $s_i = ($, returns the position $j < i$ such that $s_j = ($ and the pair of j and its mate enclose the pair of i and its mate.
- Rank$_($(i) returns the pre-order index of the node corresponding to the parenthesis at position i and its mate; this is just the number of open parentheses preceding i.
- Excess(i) returns the difference between the number of open parentheses and that of close parentheses in the first $i + 1$ positions of s. The sequence of parentheses is balanced if and only if this value is always non-negative, and it is easy to show that it equals $2 \cdot$ Rank$_($(i) $- i$.

Balanced Parentheses Tree Encoding (BP). The BP representation of trees was introduced by Munro and Raman [23]. A sequence of BP implicitly represents an ordinal tree, where each node corresponds to a pair of mates. By identifying each node with the position p of its corresponding open parenthesis, several traversal operations can be reduced to the operations defined above.

Depth-First Unary Degree Sequence (DFUDS). Another tree representation based on balanced parentheses was introduced by Benoit et al. [5]. Called *depth-first unary degree sequence* (DFUDS), it is constructed by concatenating in depth-first order the node degrees encoded in unary, i.e. a degree d is encoded as $(^d)$. It can be shown that by prepending an initial $($, the obtained sequence of parentheses is balanced.

By identifying each node of the tree with the position p of beginning of its degree encoding, traversal operations can be mapped to sequence operations. Compared to the BP representation we loose the Depth operation, but we gain the operation Child which returns the i-th child by performing a single FindClose.

Figure 1 shows an example of a tree represented with BP and DFUDS encodings. Note that both encodings require just 2 bits per node, plus the data structures needed to support the operations. It can be shown that the information-theoretic lower bound to represent an arbitrary tree of n nodes is $2n - o(n)$ bits, so both encodings are asymptotically close to the lower bound.

2.3   Range Minimum Queries

Given a sequence A of elements drawn from a totally ordered universe, a *Range Minimum Query* for the range $[i, j]$, denoted as RMQ(i, j), returns the position of the minimum element in $A[i, j]$ (returning the leftmost in case of ties).

This operation finds application, for example, in suffix arrays, to find the LCP (Longest Common Prefix) of a range of suffixes by using the vector of LCPs of consecutive suffixes: the LCP of the range is just the minimum among the consecutive LCPs. Another application is top-k retrieval (see Sect. 4).

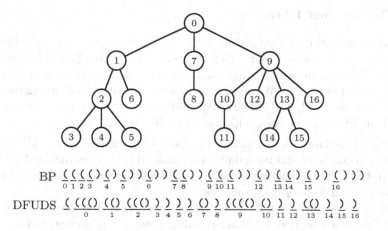

Fig. 1. BP and DFUDS encodings of an ordinal tree

3   Toolkit Implementation

For the operations described in Sect. 2 there exist SDS that can support them in constant time while taking only $r = o(n)$ bits of redundancy, meaning that as the input size n grows to infinity the relative overhead per element of the sequence goes to zero. This fact is the raison d'être for the whole field of SDS: it means that, in a reasonably realistic model of computation, using SDS instead of classical pointer-based ones involves no runtime overhead, and the space overhead needed to support the primitives is negligible.

In practice, however, the constants hidden in the $o(\cdot)$ and $O(\cdot)$ notations are large enough that these data structures become competitive with the classical ones only at unrealistic data sizes. Furthermore, CPU cache and instruction-level parallelism play an important role in the performance of these algorithms, but it is impossible to appreciate their influence in an abstract model of computation.

In recent years, a large effort has been devoted to the *algorithm engineering* of SDS, producing practical data structures that are fast and space-efficient even for small inputs. When compared to their theoretical counterparts, the time bounds often grow from $O(1)$ to $O(\log n)$, and the space bounds from $o(n)$ to $O(n)$, but for all realistic data sizes they are more efficient in both time and space.

To give a sense of how these data structures work, we give a detailed explanation of how Rank can be implemented. The other data structures are only briefly summarized and we refer to the relevant papers for a complete description.

All the SDS described here are *static*, meaning that their contents cannot be changed after the construction. While there has been a significant amount of work in *dynamic* SDS, most results are theoretical, and the practice has not caught up yet. The engineering of dynamic SDS is certainly an interesting research topic, which we believe will receive significant attention in the next few years.

3.1   The *Succinct* Library

The SDS structures described in this section, which to date are among the most efficient, are implemented as part of the *Succinct* library [32]. The library is available with a permissive license, in the hope that it will be useful both in research and applications. While similar in functionality to other existing C++ libraries such as libcds [21], SDSL [31], and Sux [33], we made some radically different architectural choices, which we describe below.

Memory Mapping. As in most static data structures libraries, all the data structures in *Succinct* can be serialized to disk. However, as opposed to libcds, SDSL, and Sux, deserialization is performed by *memory mapping* the data structure, rather than *loading* it into memory.

While being slightly less flexible, memory mapping has several advantages over loading. For example, for short-running processes it is often not necessary to load the full data structure in memory; instead the kernel will load only the relevant pages. If such pages were accessed recently, they are likely to be still in the kernel's page cache, thus making the startup even faster. If several processes access the same structure, the memory pages that hold it are shared among all the processes; with loading, instead, each process keeps its own private copy of the data. Lastly, if the system runs out of memory, it can just un-map unused pages; with loading, it has to *swap* them to disk, thus increasing the I/O pressure.

For convenience we implemented a mini-framework for serialization/memory mapping which uses template metaprogramming to describe recursively a data structure through a single function that lists the members of a class. The mini-framework then automatically implements serialization and mapping functions.

Templates Over Polymorphism. We chose to avoid dynamic polymorphism and make extensive use of C++ templates instead. This allowed us to write idiomatic and modular C++ code without the overhead of virtual functions.

Multi-platform 64-Bit Support. The library is tested under Linux, Mac OS X, and Microsoft Windows, compiled with gcc, clang and MSVC. Like Sux and parts of SDSL, *Succinct* is designed to take advantage of 64-bit architectures, which allow us to use efficient broadword algorithms [20] to speed up several operations on memory words. Another advantage is that the data structures are not limited to 2^{32} elements or less like 32-bit based implementations, a crucial requirement for large datasets, which are the ones that benefit the most from SDS. We also make use of CPU instructions that are not exposed to C++ but are widely available, such as those to retrieve the MSB and LSB of a word, or to reverse its bytes. While all these operations can all be implemented with broadword algorithms, the specialized instructions are faster.

3.2   Rank/Select

Like many SDS, Jacobson's implementation [17] of Rank relies on Four Russians trick by splitting the data into pieces that are small enough that all the answers to the queries for the small pieces can be *tabulated* in small space, and answers to queries on the whole input can be assembled from queries on the small pieces and on a sparse global data structure. Specifically, the input is divided into *super-blocks* of size $\log^2 n$, and the answer to the Rank at the beginning of each super-block is stored in an array. This takes $O(\#\text{blocks} \cdot \log n) = O(n \log n / \log^2 n) = o(n)$.

The super-blocks are then divided into *blocks* of size $\frac{1}{2} \log n$, and the answers to the Rank queries at the beginning of each block are stored *relative to their super-block*: since the superblock is only $\log^2 n$ bits long, the relative ranks cost $O(\log \log n)$ bits each, so overall the block ranks take $O(n \log \log n / \log n) = o(n)$.

The blocks are now small enough that we can *tabulate* all the possible Rank(i) queries: the number of different blocks is at most $2^{\frac{\log n}{2}} = O(\sqrt{n})$, the positions in the block are at most $O(\log n)$, and each answer requires $O(\log \log n)$ bits, so overall the space needed by the table is $O(\sqrt{n} \log n \log \log n) = o(n)$.

To answer Rank(i) on the whole bitvector it is sufficient to sum the rank of its super-block, the relative rank of its block, and the rank inside the block. All three operations take $O(1)$ time, so overall the operation takes constant time.

This data structure can be implemented almost as described, but it is convenient to have fixed-size blocks and super-blocks. This yields an $O(n)$-space data structure, but the time is still $O(1)$. Furthermore, if the blocks are sized as the machine word, the in-block Rank can be efficiently computed with broadword operations, as suggested by Vigna [34], hence avoiding to store the table.

The constant-time data structure for Select is significantly more involved: its practical alternative is to perform a binary search on the super-block ranks, followed by a linear search in the block partial ranks and then inside the blocks. The binary search can be made more efficient by storing the answer to Select$_1$ for every k-th $\mathbf{1}$, so that the binary search can be restricted to a range that contains at most k ones. This algorithm, which is known as *hinted binary search*, can take $O(\log n)$ time but is extremely efficient in practice.

In *Succinct*, Rank and Select are implemented in the `rs_bit_vector` class, which uses the `rank9` data structure [34].

3.3   Elias-Fano Representation of Monotone Sequences

The *Elias-Fano representation of monotone sequences* [9,11] is an encoding scheme to represent a non-decreasing sequence of m integers $\langle x_1, \cdots, x_m \rangle$ from the universe $[0..n)$ occupying $2m + m \lceil \log \frac{n}{m} \rceil + o(m)$ bits, while supporting constant-time access to the i-th integer. It can be effectively used to represent sparse bitvectors (i.e. where the number m of ones is small with respect to the size n of the bitvector), by encoding the sequence of the positions of the ones. Using this representation the retrieval of the the i-th integer can be interpreted as Select$_1(i)$, and similarly it is possible to support Rank$_1$.

The scheme is very simple and elegant. Let $\ell = \lfloor \log(n/m) \rfloor$. Each integer x_i is first encoded in binary into $\lceil \log n \rceil$ bits, and the binary encoding is then split into the first $\lceil \log n \rceil - \ell$ *higher* bits and the last ℓ *lower* bits. The sequence of the higher bits is represented as a bitvector H of $\lceil m + n/2^\ell \rceil$ bits, where for each i, if h_i is the value of the higher bits of x_i, then the position $h_i + i$ of H is set to 1; H is 0 elsewhere. The lower bits of each x_i are just concatenated into a bitvector L of $m\ell$ bits. To retrieve the i-th integer we need to retrieve its higher and lower bits and concatenate them. The lower bits are easily retrieved from L. To retrieve the upper bits it is sufficient to note that $h_i = \text{Select}_H(1, i) - i$. The implementation is straightforward, provided that Select is supported on H.

In *Succinct*, we implement it using the `darray` data structure [24], which supports Select in $O(1)$ time without requiring a data structure to support Rank. The class `elias_fano` implements a sparse bitvector encoded with Elias-Fano.

3.4   Balanced Parentheses

To implement operations on balanced parentheses, variants of a data structure called Range-Min-Max tree [30] have proven the most effective in practice [2], despite their $O(\log n)$ time and $O(n)$ space. The data structure divides the sequence into a hierarchy of blocks, and stores the minimum and maximum Excess value for each block. This yields a tree of height $O(\log n)$ height, that can be traversed to find mate and enclosing parentheses.

In *Succinct* the class `bp_vector` implements the basic operations on balanced parentheses. It uses a variant of the Range-Min-Max tree called *Range-Min tree* [14], which only stores the minimum excess, thus halving the space occupancy with respect to the Range-Min-Max tree. This weakens the range of operations that the data structure can support, but all the important tree navigation operations can be implemented.

3.5   Range Minimum Queries

The RMQ problem is intimately related to the *Least Common Ancestor* (LCA) problem on trees. In fact, the LCA problem can be solved by reducing it to an RMQ on a sequence of integers derived from the tree, while RMQ can be solved by reducing it to an LCA in a tree derived from the sequence, called the *Cartesian Tree* [4]. As shown by Fischer and Heun [13], RMQ on an array A can be reduced to an RMQ on the excess sequence of the DFUDS representation of the *2d-Min-Heap*, which, as noted by Davoodi et al. [7], is an alternative representation of the Cartesian tree. The RMQ operation on an excess sequence can be easily implemented by using the Range-Min tree, so since the DFUDS representation is already equipped with a Range-Min tree to support navigational operations, no extra space is needed.

In *Succinct*, RMQ is implemented in the class `cartesian_tree`. The algorithm is a minor variation, described in [16], of the scheme by Fischer and Heun [13].

4   Applications

We describe some applications of the *Succinct* library that involve handling large collections of semi-structured or textual data. Interestingly, SDS are gaining popularity in other communities such as bioinformatics, Web information retrieval, and networking.

4.1   Semi-indexing Semi-structured Data

With the advent of large-scale distributed processing systems such as MapReduce [8] and Hadoop [1] it has become increasingly common to store large amounts of data in textual semi-structured formats such as JSON and XML, as opposed to the structured databases of classical data warehousing. The flexibility of such formats allows to evolve the data schema without having to migrate the existing data to the new schema; this is particularly important in logging applications, where the set of attributes to be stored usually evolves quickly.

The disadvantage of such formats is that each record, represented by a semi-structured document, must be parsed completely to extract its attributes; in typical applications, the records contain several attributes but each query requires a different small subset of such attributes, thus reading the full document can be highly inefficient. Alternative representations for semi-structured data have been proposed in the literature, many using SDS, that are both compact and support efficient querying of attributes, but they rely on changing the data format, which may be unacceptable in some scenarios where compatibility or interoperability with existing tools is required.

In [25] it was introduced the concept of *semi-index*, which is a systematic SDS that can be used to speed-up the access to an existing semi-structured document without changing its format: the semi-index is stored on a separate file and it takes only a small fraction of the size of the original data.

The semi-index consists of two components: a *positional index*, that is an index of the positions in the document that are starting points of its elements, and a succinct encoding of the parse tree of the document. The positional index can be represented with a bitvector and encoded with Elias-Fano, while the parse tree can be represented with a BP sequence.

For JSON documents [18], the positional index has a 1 in correspondence of the positions where either of the characters {}[],: occur. For each one of these, a pair of parentheses is appended to the BP sequence, specifically ((for { and [,)) for } and], and)(for , and :. An example is shown in Fig. 2. It can be shown that the BP sequence is indeed balanced, and that it is possible to navigate the parse tree of the document by accessing the two index sequences and a minimal part of the original document. An experimental analysis on large-scale document collections has shown speed-ups between 2.5 and 10 times compared to parsing each document, while the space overhead given by the semi-index is at most $\sim 10\%$.

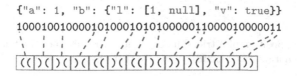

Fig. 2. Semi-index of a JSON document

4.2   Compressed String Dictionaries

A string dictionary is a data structure that maps bijectively a set of n strings to the integer range $[0, n)$. String dictionaries are among the most fundamental data structures, and find application in basically every algorithm that handles strings.

In many scenarios where the string sets are large, the strings are *highly redundant*; a *compressed string dictionary* exploits this redundancy to reduce the space usage of the data structure. Traditionally, string dictionaries are implemented by using *tries*, which are not only fast, but they also offer some degree of compression by collapsing the common prefixes. Tries represented with SDS offer even higher space savings, but the performance suffers a large slow-down because tries can be highly unbalanced, and navigational operations in succinct trees can be costly. Furthermore, prefix compression is not effective for strings that might share other substrings besides the prefix.

In [14] it was introduced a succinct representation for tries that makes use of *path decompositions*, a transformation that can turn a unbalanced tree into a balanced one. The representation also enables compression of the trie *labels*, thus exploiting the redundancy of frequent substrings.

A *path decomposition* of a trie $\mathcal{T}$ is a tree $\mathcal{T}^c$ whose nodes correspond to node-to-leaf paths in $\mathcal{T}$. The tree is built by first choosing a root-to-leaf path π in $\mathcal{T}$ and associating it with the root node u_π of $\mathcal{T}^c$; the children of u_π are defined recursively as the path decompositions of the subtries hanging off the path π, and their edges are labeled with the labels of the edges from the path π to the subtries. An example is shown in Fig. 3. The resulting tree is then encoded using a DFUDS representation, and the sequence of labels is compressed with a simple dictionary compression scheme.

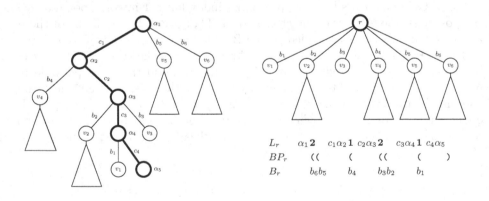

Fig. 3. Path decomposition of a trie. The α_i denote the labels of the trie nodes, c_i and b_i the branching characters (depending on whether they are on the path or not).

Depending on the strategy used to choose the decomposition path π, different properties can be obtained. If we start from the root and recursively choose the child with most descendents, the resulting path is called a *centroid* path, and the resulting *centroid path decomposition* has height $O(\log n)$, regardless of whether the trie is balanced or not. Alternatively, we can choose recursively the leftmost child; the resulting decomposition is called *lexicographic path decomposition* because the numbers associated to the strings of the set respect the lexicographic ordering, but for this decomposition no height guarantees can be given.

In the experiments, large-scale collections of URLs, queries, and web page titles can be compressed down to 32% to 13% of their original size, while maintaining access and lookup times of few microseconds.

4.3   Top-k Completion in Scored String Sets

Virtually every modern application, either desktop, web, or mobile, features some kind of auto-completion of text-entry fields. Specifically, as the user enters a string one character at a time, the system presents k suggestions to speed up text entry, correct spelling mistakes, and help users formulate their intent.

This can be thought of as having a *scored* string set, meaning that each string is assigned a score, and given a prefix p we want to find the k strings prefixed by p that have highest score. We call this problem *top-k completion*. Since the sets of suggestion strings are usually large, space-efficiency is crucial.

A simple solution is to combine the lexicographic trie of Sect. 4.2 with an RMQ data structure. The strings in the string set are stored in the trie, and their scores are stored in lexicographic order in an array R. The set of the indexes of strings that start with a given prefix p is a contiguous range $[a, b]$.

We can then compute $r = \text{RMQ}(a, b)$ to find the index r of the highest-scored string prefixed by p. The second string is either the highest-scored one in $[a, r-1]$ or in $[r+1, b]$, and so on. By using a priority queue it is possible to retrieve the top-k completions one at a time.

It is however possible to do better: we can use again the tries of Sect. 4.2, but with a different path decomposition, where the chosen decomposition path π is the path from the root to the highest-scored leaf. It can be shown that in the resulting *max-score path decomposition* it is possible to find the top-k completions of a given node by visiting just k nodes; since, as noted before, navigational operations are costly in succinct representations, performance is significantly better than the RMQ-based solution.

On large sets of queries and URLs, experiments have shown compression ratios close or better than those of `gzip`, with average times per completion of about one microsecond [16].

References

1. Apache Hadoop, http://hadoop.apache.org/
2. Arroyuelo, D., Cánovas, R., Navarro, G., Sadakane, K.: Succinct trees in practice. In: ALENEX, pp. 84–97 (2010)
3. Barbay, J., He, M., Munro, J.I., Satti, S.R.: Succinct indexes for strings, binary relations and multilabeled trees. ACM Transactions on Algorithms 7(4), 52 (2011)

4. Bender, M.A., Farach-Colton, M.: The LCA problem revisited. In: Gonnet, G.H., Viola, A. (eds.) LATIN 2000. LNCS, vol. 1776, pp. 88–94. Springer, Heidelberg (2000)
5. Benoit, D., Demaine, E.D., Munro, J.I., Raman, R., Raman, V., Rao, S.S.: Representing trees of higher degree. Algorithmica 43(4), 275–292 (2005)
6. Buhrman, H., Miltersen, P.B., Radhakrishnan, J., Venkatesh, S.: Are bitvectors optimal? SIAM Journal on Computing 31(6), 1723–1744 (2002)
7. Davoodi, P., Raman, R., Satti, S.R.: Succinct representations of binary trees for range minimum queries. In: Gudmundsson, J., Mestre, J., Viglas, T. (eds.) COCOON 2012. LNCS, vol. 7434, pp. 396–407. Springer, Heidelberg (2012)
8. Dean, J., Ghemawat, S.: MapReduce: Simplified data processing on large clusters. In: OSDI, pp. 137–150 (2004)
9. Elias, P.: Efficient storage and retrieval by content and address of static files. Journal of the ACM (JACM) 21(2), 246–260 (1974)
10. Elias, P., Flower, R.A.: The complexity of some simple retrieval problems. Journal of the ACM 22(3), 367–379 (1975)
11. Fano, R.: On the number of bits required to implement an associative memory. Memorandum 61. Computer Structures Group, Project MAC. MIT (1971)
12. Ferragina, P., Manzini, G.: Indexing compressed text. J. ACM 52(4), 552–581
13. Fischer, J., Heun, V.: Space-efficient preprocessing schemes for range minimum queries on static arrays. SIAM J. Comput. 40(2), 465–492 (2011)
14. Grossi, R., Ottaviano, G.: Fast compressed tries through path decompositions. In: ALENEX, pp. 65–74 (2012)
15. Grossi, R., Vitter, J.S.: Compressed suffix arrays and suffix trees with applications to text indexing and string matching. SIAM J. Comput. 35(2), 378–407 (2005)
16. Hsu, B.J.P., Ottaviano, G.: Space-efficient data structures for top-k completion. In: Proceedings of the 22st World Wide Web Conference (WWW) (2013)
17. Jacobson, G.: Space-efficient static trees and graphs. In: FOCS, pp. 549–554 (1989)
18. JSON specification, http://json.org/
19. Kao, M.Y. (ed.): Encyclopedia of Algorithms. Springer (2008)
20. Knuth, D.E.: The Art of Computer Programming. Fascicle 1: Bitwise Tricks & Techniques; Binary Decision Diagrams, vol. 4. Addison-Wesley (2009)
21. libcds - Compact Data Structures Library, http://libcds.recoded.cl/
22. Miltersen, P.B.: The bit probe complexity measure revisited. In: Enjalbert, P., Wagner, K.W., Finkel, A. (eds.) STACS 1993. LNCS, vol. 665, pp. 662–671. Springer, Heidelberg (1993)
23. Munro, J.I., Raman, V.: Succinct representation of balanced parentheses and static trees. SIAM Journal on Computing 31(3), 762–776 (2001)
24. Okanohara, D., Sadakane, K.: Practical entropy-compressed rank/select dictionary. In: ALENEX (2007)
25. Ottaviano, G., Grossi, R.: Semi-indexing semi-structured data in tiny space. In: CIKM, pp. 1485–1494 (2011)
26. Pătrașcu, M.: Succincter. In: FOCS 2008, pp. 305–313 (2008)
27. Pătrașcu, M., Thorup, M.: Time-space trade-offs for predecessor search. In: STOC, pp. 232–240 (2006)
28. Raman, R., Raman, V., Satti, S.R.: Succinct indexable dictionaries with applications to encoding k-ary trees, prefix sums and multisets. ACM Trans. Alg. 3(4) (2007)
29. rsdic - Compressed Rank Select Dictionary, http://code.google.com/p/rsdic/
30. Sadakane, K., Navarro, G.: Fully-functional succinct trees. In: SODA 2010, pp. 134–149 (2010)

31. SDSL - Succinct Data Structure Library,
    http://www.uni-ulm.de/in/theo/research/sdsl.html
32. *Succinct* library, http://github.com/ot/succinct
33. Sux - Implementing Succinct Data Structures, http://sux.dsi.unimi.it/
34. Vigna, S.: Broadword implementation of rank/select queries. In: McGeoch, C.C.
    (ed.) WEA 2008. LNCS, vol. 5038, pp. 154–168. Springer, Heidelberg (2008)
35. Viola, E.: Bit-probe lower bounds for succinct data structures. SIAM J. Comput. 41(6), 1593–1604 (2012)

Hub Label Compression

Daniel Delling, Andrew V. Goldberg, and Renato F. Werneck

Microsoft Research Silicon Valley
{dadellin,goldberg,renatow}@microsoft.com

Abstract. The hub labels (HL) algorithm is the fastest known technique for computing driving times on road networks, but its practical applicability can be limited by high space requirements relative to the best competing methods. We develop compression techniques that substantially reduce HL space requirements with a small performance penalty.

1 Introduction

Computing the driving time between two points in a road network is the fundamental building block for location services, which are increasingly important in practice. Dijkstra's algorithm [14] can solve this problem in essentially linear time, but this is too slow for continental road networks. This motivates two-stage algorithms, which use a preprocessing phase to precompute some auxiliary data that is then used to accelerate queries. Several efficient algorithms have recently been developed following this approach, each offering a different tradeoff between preprocessing effort and query times [3–8, 11, 16–19].

This paper focuses on *hub labels* (HL), a labeling algorithm [9, 15] developed by Abraham et al. [3, 4] to work specifically with road networks. For each vertex v in the network, its preprocessing step computes a *label* consisting of a set of hubs (other vertices), together with the distances between v and these hubs. The construction is such that, for any two vertices s and t, there must be at least one hub on the shortest s–t path that belongs to the labels of *both* s and t. Queries can then be answered by simply intersecting the two relevant labels.

The HL algorithm has several attractive properties. First, it is the fastest point-to-point shortest-path algorithm for road networks, for both long-range and (more common) local queries. Second, its query algorithm is by far the simplest: it does not even need a graph data structure, allowing practitioners with no algorithm engineering expertise to implement fast queries. Finally, the concept of labels (and hubs) is intuitive and extremely powerful, naturally lending itself to the implementation of much more sophisticated queries, such as finding nearby points of interest, optimizing ride sharing schemes, or building distance tables [2].

One aspect of HL, however, severely limits its applicability: space usage. Although preprocessing time is in line with most other methods (a few minutes on a modern server [4]), in many settings it produces a prohibitive amount of data. Computing and storing all labels requires up to two orders of magnitude more space than storing the graph itself. For a standard benchmark instance representing Western Europe, labels require roughly 20 GiB of RAM, while a graph-based

V. Bonifaci et al. (Eds.): SEA 2013, LNCS 7933, pp. 18–29, 2013.

algorithm such as contraction hierarchies (CH) [16] requires less than 0.5 GiB. For more realistic representations (with turn costs), space requirements are even higher, rendering HL impractical on most commodity machines.

One could use another algorithm instead, but this would mean sacrificing query speed, ease of use, or flexibility. Storing the labels in external memory is feasible [2], but makes queries orders of magnitude slower. Finally, Abraham et al. [3] propose a (RAM-based) compact label representation that reduces space usage by a factor of roughly 3, but the compression routine itself requires a large amount of time and space (including in-memory access to all labels).

We propose *hub label compression* (HLC), a technique that achieves high compression ratios and works in *on-line* fashion. Compressing labels as they are generated drastically reduces the amount of RAM used during preprocessing, which is only slightly slower than for plain HL. On continental road networks, HLC uses an order of magnitude less space than standard HL (1.8 GiB on Western Europe). Queries are somewhat slower, but still faster than almost all other known algorithms. Crucially, they are still easy to implement (requiring no graph or priority queue), and preserve the full generality of the HL framework.

The remainder of this paper is organized as follows. After a brief overview of the HL algorithm (in Section 2), Section 3 explains the basics of HLC: the data structure, query implementation, and justification for our design decisions. Section 4 proposes optimizations that enable faster queries and better compression. Section 5 explains how the compact data structure can be generated. Section 6 has an experimental analysis of our approach. We conclude in Section 7.

2   Hub Labels

We represent a road network as a directed graph $G = (V, A)$. A vertex $v \in V$ is an intersection, and an arc $(v, w) \in A$ is a road segment with a nonnegative *length* $\ell(v, w)$, typically reflecting driving times. Let $n = |V|$. In the *point-to-point shortest path problem*, we are given a *source* vertex s and a *target* vertex t, and our goal is to find dist(s, t), the total length of the shortest s–t path in G.

The *hub labels* (HL) algorithm [3, 4] solves this problem in two stages. During *preprocessing*, HL creates a forward label $L_f(v)$ and a backward label $L_b(v)$ for each vertex $v \in V$. The forward label $L_f(v)$ consists of a sequence of pairs $(w, \text{dist}(v, w))$, in which $w \in V$ is a *hub* and dist(v, w) is the distance (in G) from v to w. The backward label is similar, with pairs $(u, \text{dist}(u, v))$. By construction, labels obey the *cover property*: for any two vertices s and t, the set $L_f(s) \cap L_b(t)$ must contain at least one hub v that is on the shortest s–t path. Queries are straightforward: to find dist(s, t), simply find the hub $v \in L_f(s) \cap L_b(t)$ that minimizes dist$(s, v) + \text{dist}(v, t)$. By storing the entries in each label sorted by hub ID, this can be done with sequential sweeps over both labels (as in merge sort), which is very simple and cache-friendly. To compute labels for road networks efficiently, Abraham et al. [3, 4] propose a two-step algorithm. First, as in CH [16], one computes the "importance" of each vertex, roughly measuring how many shortest paths it hits. The label for each vertex v is then built in a greedy fashion: the label starts with only the most important vertex as a hub, with more vertices added as needed to ensure every shortest path originated at v is hit. The resulting labels are surprisingly small, with around 100 hubs on average on continental road networks. The resulting queries are fast, but memory requirements are high.

3   Compressed Labels

We now present our basic compression strategy. We describe it in terms of forward labels only, which we denote by $L(\cdot)$ to simplify notation; backward labels can be compressed independently using the same method. As Abraham et al. [4] observe, the forward label $L(u)$ of u can be represented as a tree T_u rooted at u and having the hubs in $L(u)$ as vertices. Given two vertices $v, w \in L(u)$, there is an arc (v, w) in T_u (with length dist(v, w)) if the shortest v–w path in G contains no other vertex of $L(u)$.

Our compression scheme exploits the fact that trees representing labels of nearby vertices in the graph often have many subtrees in common. We assign a unique ID to each distinct subtree and store it only once. Furthermore, each tree is stored using a space-saving recursive representation. More precisely, for any $v \in L(u)$, let $S_u(v)$ be the maximal subtree of T_u rooted at v. This subtree can be described by its root (v itself) together with a list of the IDs of its child subtrees, each paired with an *offset* representing the distance from v to the subtree's root. We call this structure (the root ID together with a list of pairs) a *token*. Common tokens can then be shared by different labels.

The remainder of this section details the actual data structure we use, as well as queries. Section 5 discusses how to actually build the data structure.

Data Structure. Our representation makes standard assumptions [8] for real-world road networks: (1) vertices have integral IDs from 0 to $n-1$ and (2) finite distances in the graph can be represented as unsigned 32-bit integers.

A token is fully defined by the following: (1) the ID r of the *root vertex* of the corresponding subtree; (2) the number k of *child tokens* (representing child subtrees of r); and (3) a list of k pairs (i, δ_i), where i is a token ID and δ_i is the distance from r to the root of the corresponding subtree. We thus represent a token as an array of $2k + 2$ unsigned 32-bit integers. We represent the collection of all subtrees by concatenating all tokens into a single *token array* of unsigned 32-bit integers. In addition, we store an *index*, an array of size n that maps each vertex in V to the ID of its *anchor token*, which represents its full label.

We still have to define how token IDs are chosen. We say that a token is *trivial* if it represents a subtree consisting of a single vertex v, with no child tokens. The ID of such a trivial token is v itself, which is in the range $[0, n)$. *Nontrivial* tokens (those with at least one child token) are assigned unique IDs in the range $[n, 2^{32})$. Such IDs are not necessarily consecutive, however. Instead, they are chosen so as to allow quick access to the corresponding entry in the token array. More precisely, a token that starts at position p in the array has ID $n + p/2$. (This is an integer, since all tokens have an even number of 32-bit integers.) Conversely, the token whose ID is i starts at position $2(i - n)$ in the array. Trivial tokens are not represented in the token array, since the token ID fully defines the root vertex (the ID itself) and the number of children (zero).

Since all IDs must fit in 32 bits, the token array can only represent labelings whose (compressed) size is at most $8(2^{32} - n)$ bytes. For $n \ll 2^{32}$, as is the case in practice, this is slightly less than 32 GiB, and enough to handle all instances we test; bigger inputs could be handled by varying the sizes of each field.

Queries. Since a standard (uncompressed) HL label is stored as an array of hubs (and the corresponding offsets) sorted by ID, a query requires a simple linear scan. With the compact representation, queries require two steps: we first *retrieve* the two labels, then *intersect* them. We discuss each in turn.

Retrieving a label $L(v)$ means transforming its token-based representation T_v into an array of *pairs*, each containing the ID of a hub h and its distance dist(v, h) from v. We can do this by traversing the tree T_v top-down, while keeping track of the appropriate offsets. For efficiency, we avoid recursion and perform a BFS traversal of the tree using the output array itself for temporary storage. More precisely, we do as follows. First, we use the index array to get t_v, the ID of v's anchor token, and initialize the output array with a single element, $(t_v, 0)$. We then process each element of this array in order. Let (t, d) be the element in position p (processed in the p-th step). If $t < n$ (i.e., it is a trivial token), there is nothing to do. Otherwise (if $t \geq n$), we read token t from the token array, starting at position $2(t - i)$. Let w be t's root vertex. We replace (t, d) by (w, d) in the p-th position of the output array and, for each pair (i, δ_i) in the token, append the pair $(i, d + \delta_i)$ to the output array. The algorithm stops when it reaches a position that has not been written to. At this point, each pair in the output array corresponds to a hub together with its distance from v.

The second query step is to *intersect* the two arrays (for source and target) produced by the first step. Since the arrays are *not* sorted by ID, it is not enough to do a linear sweep, as in the standard HL query. We could explicitly sort the labels by ID before sweeping, but this is slow. Instead, we propose using *indexing* to find common hubs without sorting. We first traverse one of the labels to build an index of its hubs (with associated distances), then traverse the second label checking if each hub is already in the index (and adding up the distances). The simplest such index is an array indexed by ID, but this takes $\Theta(n)$ space and may lead to many cache misses. A better alternative is to use a small hash table with a simple hash function (we use ID modulo 1024) and linear probing [10].

Discussion. Our data structure balances space usage, query performance, and simplicity. If compression ratios were our only concern, we could easily reduce space usage with various techniques. We could use fewer bits for some of the fields (notably the number of children). We could use relative (rather than absolute) references and variable-length encoding for the IDs [21]. We could avoid storing the length of each arc (v, w) multiple times in the token array (as offsets in tokens rooted at v) by representing labels as subtrees of the full CH graph [16], possibly using techniques from succinct data structures [20]. Such measures would reduce space usage, but query times could suffer (due to worse locality) and simplicity, arguably the main attraction of HL, would be severely compromised.

4   Variants and Optimizations

We now consider optimizations to our basic compression scheme. They modify the preprocessing stage only, and require *no change* to the query algorithm.

The Token DAG. Conceptually, our compressed representation can be seen as a *token graph*. Each vertex of the graph corresponds to a *nontrivial* token x, and there is an arc (x, y) if and only if y is a child of x in some label. The length of the arc is the offset of y within x. The token graph has some useful properties. By definition, a token x that appears in multiple labels has the same children (in the corresponding trees) in all of them. This means x has the same set of descendants in all labels it belongs to, and by construction these are exactly the vertices in the subgraph reachable from x in the token graph. This implies that this subgraph is a tree, and that the token graph is a DAG. It also implies that the subgraph reachable from x by following only *reverse* arcs is a tree as well: if there were two distinct paths to some ancestor y of x, the direct subgraph reachable from y would not be a tree. We have thus proven the following.

Lemma 1. *The token graph is a DAG in which any two vertices are connected by at most one path.*

Note that all DAG vertices with in-degree zero are anchor tokens, and DAG vertices with out-degree zero (which we call *leaf tokens*) are nontrivial tokens that only have trivial tokens (which are not in the token DAG) as children.

Pruning the DAG. Retrieving a compressed label may require a nonsequential memory access for each internal node in the corresponding tree. To improve locality (and even space usage), we propose two operations. We can eliminate a non-anchor token t (rooted at a vertex v) with a single parent t' in the token DAG as follows. We replace each arc (t, t'') in the DAG by an arc (t', t'') with length equal to the sum of (t', t) and (t, t''). Moreover, in t', we replace the reference to t by a reference to trivial token v. This *1-parent elimination* operation potentially improves query time and space. Similarly, *1-child elimination* applies to a non-anchor token t that has exactly two parents in the DAG, a single nontrivial child t', and no nontrivial children. We can discard t and create direct arcs from each parent of t to t', saving nonsequential accesses with no increase in space.

Flattening. A more aggressive approach to speed up queries is to *flatten* subtrees that occur in many labels. Instead of describing the subtree recursively, we create a single token explicitly listing *all* descendants of its root vertex, with appropriate offsets. We propose a greedy algorithm that in each step flattens the subtree (token) that reduces the expected query time the most, assuming all labels are equally likely to be accessed. Intuitively, our goal is to minimize the average number nonsequential accesses when reading the labels.

Let $\lambda(x)$ be the number of labels containing a nontrivial token x, and let $\alpha(x)$ be the number of proper descendants of x in the token DAG ($\alpha(x)$ is 0 if x is a leaf). The *total access cost* of the DAG is the total number of nonsequential accesses required to access all n labels. (This is n times the expected cost of reading a random label.) If H is the set of all anchor tokens, the total access cost is $\sum_{x \in H}(1 + \alpha(x))$. The share of the access cost attributable to any token x is $\lambda(x) \cdot (1 + \alpha(x))$. Flattening the corresponding subtree would reduce the total access cost by $v(x) = \lambda(x)\alpha(x)$, as a single access would suffice to retrieve x.

Our algorithm starts by traversing the token graph twice in topological order: a direct traversal initializes $\lambda(\cdot)$ and a reverse one initializes $\alpha(\cdot)$. It then keeps the $v(x) = \lambda(x)\alpha(x)$ values in a priority queue. Each step takes the token x with maximum $v(x)$ value, flattens x, then updates all $v(\cdot)$ values that are affected. For every ancestor z of x, we set $\alpha(z) \leftarrow \alpha(z) - \alpha(x)$; for every descendant y of x, we set $\lambda(y) \leftarrow \lambda(y) - \lambda(x)$. If $\lambda(y)$ becomes zero, we simply discard y. Finally, we remove the outgoing arcs from x (making x a leaf) and set $\alpha(x) \leftarrow 0$. We stop when the total size of the token array increases beyond a given threshold. Using Lemma 1, one can show that this algorithm runs in $O(\tau\mu)$ time, where τ is the initial number of tokens in the DAG and μ is the maximum label size.

Discussion. We implemented flattening as described above, but the concept is more general: one could flatten arbitrary subtrees (not just maximal ones), as long as unflattened portions are represented elsewhere with appropriate offsets. The 1-parent and 1-child elimination routines are special cases of this, as is Abraham et al.'s compression technique [3], which splits each label into a subtree containing its root and a (shared) token representing the remaining forest.

With no stopping criterion, the greedy flattening algorithm eventually leads to exactly n (flattened) tokens, each corresponding to a label in its entirety, as in the standard (uncompressed) HL representation. Conversely, we could have a "merge" operation that combines tokens rooted at the same vertex into a single token (not necessarily flattened) representing the union of the corresponding trees. This saves space, but tokens no longer represent minimal labels. Our BFS-based label retrieval technique is still correct (it access all relevant hubs), but it may visit each hub more than once, since Lemma 1 no longer holds. We could fix this by visiting tokens in increasing order of distance (with a heap), as in CH queries [16]. This similarity is not accidental: repeated application of the "merge" operation eventually leads to a single token rooted at each vertex (representing all subtrees rooted at it), with the token graph exactly matching the upward CH graph [16]. In this sense, HLC generalizes both CH and HL.

Reordering. If the tokens corresponding to the endpoints of a DAG arc are not stored consecutively in memory, traversing this arc usually results in a cache miss. Tokens must be stored in increasing order of ID, but these IDs can be chosen to our advantage. By reordering tokens appropriately during preprocessing, we can potentially decrease the number of cache misses during queries.

We propose the following. First, we *mark* a subset M of the DAG arcs such that each token t in the DAG has at most one incoming arc and at most one outgoing arc in M. This creates a collection of vertex-disjoint paths. We then assign consecutive IDs to the vertices along each path (the order among paths is arbitrary). For random queries, this assignment avoids (compared to a random order) approximately $\lambda(t)/n$ cache misses for each marked arc (t, t'). We define the *gain* associated with a set M as the sum of $\lambda(t)$ over all marked arcs (t, t'). One can show that the set M^* with maximum gain can be found using minimum-cost flows. For efficiency, however, we use a greedy heuristic instead. We start with all arcs unmarked, then process the tokens in nonincreasing order of $\lambda(\cdot)$. To process a token t, we mark, among all outgoing arcs (t, t') such that t' has no marked incoming arc, the one maximizing $\lambda(t')$. If no such arc exists, we just skip t. Eventually, this leads to a maximal set M of marked arcs.

5   Label Generation

We now explain how the data structures described in Section 3 can actually be built. To create a compressed representation of an existing set of labels, we start with an empty token array and *tokenize* the labels (i.e., create their token-based representation) one at a time, in any order. To tokenize a label $L(v)$, we traverse the corresponding tree T_v bottom-up. To process a vertex $w \in T_v$, we first build the token t_w that represents it. This can be done because at this point we already know the IDs of the tokens representing the subtrees rooted at w's children. We then pick an ID i to assign to t_w. First, we use hashing to check if t_w already occurs in the token array. If it does, we take its existing ID. Otherwise, we append t_w to the token array and use its position p to compute the ID i, as described in Section 3 ($i = n + p/2$). When the bottom-up traversal of T_v ends, we store the ID of t_v (the anchor token of v) in the index array.

Note that label compression can be implemented in on-line fashion, as labels are generated. Asymptotically, it does not affect the running time: we can compress all labels in linear time. In practice, however, generating a label often requires access to other existing labels [3, 4]. If we are not careful, the extra cost of retrieving existing compressed labels may become the bottleneck.

With that in mind, we modify Abraham et al.'s recursive label generation [4] to compress labels as they are created. Building on the preprocessing algorithm for *contraction hierarchies* (CH) [16], they first find a heuristic *order* among all vertices, then *shortcut* them in this order. To process a vertex v, one (temporarily) deletes v and adds arcs as necessary to preserve distances among the remaining vertices. More precisely, for every pair of incoming and outgoing arcs (u, v) and (v, w) such that $(u, v) \cdot (v, w)$ is the only u–w shortest path, one adds a new shortcut arc (u, w) with $\ell(u, w) = \ell(u, v) + \ell(v, w)$. This procedure outputs the order itself (given by a rank function $r(\cdot)$) and a graph $G^+ = (V, A \cup A^+)$, where A^+ is the set of shortcuts. The number of shortcuts depends on the order; intuitively, it is best to first shortcut vertices that belong to few shortest paths.

Labels are then generated one by one, in reverse contraction (or top-down) order, starting from the last contracted vertex. The first step to process a vertex v is to build an initial label $L(v)$ by combining the labels of v's upward neighbors $U_v = \{u_1, u_2, \ldots, u_k\}$ (u is an upward neighbor of v if $r(u) > r(v)$ and $(u, v) \in A \cup A^+$.) For each $u_i \in U_v$, let T_{u_i} be the (already computed) tree representing its label. We initialize T_v (the tree representing $L(v)$) by taking the first tree (T_{u_1}) in full, and making its root a child of v itself (with an arc of length $\ell(v, u_1)$). We then process the other trees T_{u_i} ($i \geq 2$) in top-down fashion. Consider a vertex $w \in T_{u_i}$ with parent p_w in T_{u_i}. If $w \notin T_v$, we add it—p_w must already be there, since we process vertices top-down. If $w \in T_v$ and its distance label $d_v(w)$ is higher than $\ell(v, u_i) + d_{u_i}(w)$, we update $d_v(w)$ and set w's parent in T_v to p_w.

Once the merged tree T_v is built, we eliminate any vertex $w \in T_v$ such that $d_v(w) > \text{dist}(v, w)$. The actual distance $\text{dist}(v, w)$ can be found by *bootstrapping*, i.e., running a v–w HL query using $L(v)$ itself (unpruned, obtained from T_v) and the label $L(w)$ (which must already exist, since labels are generated top-down).

Our algorithm differs from Abraham et al.'s [4] in that it stores labels in compressed form. To compute $L(v)$, we must retrieve (using the token array) the labels of its upward neighbors, taking care to preserve the parent pointer information that is implicit in the token-based representation. Similarly, bootstrapping requires retrieving the labels of all candidate hubs.

To reduce the cost of retrieving compressed labels during preprocessing, we keep an LRU cache of *uncompressed* labels. Whenever a label is needed, we first look it up in the cache, and only retrieve its compressed version if needed (and add it to the cache). Since labels used for bootstrapping do not need parent pointers and labels used for merging do, we keep an independent cache for each representation. To minimize cache misses, we do not generate labels in strict top-down order; instead, we process vertices in increasing order of ID, deviating from this order as necessary. If we try to process v and realize it has an unprocessed upward neighbor w, we process w first, then come back to v. (We use a stack to keep track of delayed vertices.) The cache hit ratio improves because nearby vertices (with similar labels) tend to have similar IDs in our test instances.

For additional acceleration, we also avoid unnecessary bootstrapping queries. If a vertex v has a single upward neighbor u, there is no need to bootstrap T_v (and u's token can be reused). If v has multiple upward neighbors, we bootstrap T_v in bottom-up order. If we determine that the distance label for a vertex $w \in T_v$ is correct, its ancestors in T_v must be as well, and need not be tested.

6 Experiments

We implemented our HLC algorithm in C++ and compiled it (using full optimization) with Microsoft Visual C++ 2010. We tested it on a machine running Windows Server 2008 R2 with 96 GiB of DDR3-1333 RAM and two 6-core Intel Xeon X5680 CPUs at 3.33 GHz (each CPU has 6×64 KB of L1, 6×256 KB L2,

Table 1. HLC and HL performance, aggregated over forward and backward labels

	PREPROCESSING				QUERIES		
algorithm	time [h:m]	space [MiB]	tokens /vertex	reads /label	hash [ns]	linear [ns]	merge [ns]
basic HLC	00:47	2016.3	4.925	39.46	3338	3832	7035
+1-parents	00:48	1759.0	3.053	36.27	3313	3830	7007
+1-children	00:49	1759.0	2.912	35.78	3304	3809	6996
+5% flat	00:50	1840.1	2.912	12.96	2554	2999	6205
plain HL	00:38	21776.1	2.000	1.00	1208	1315	617

and 12 MB of shared L3 cache). For ease of comparison, all runs were sequential. Our default benchmark instance (made available by PTV AG for the 9th DIMACS Implementation Challenge [13]) represents (Western) Europe; it has $n = 18 \cdot 10^6$ vertices, $m = 42 \cdot 10^6$ arcs, and travel times as the cost function.

Our first experiment examines the effectiveness of all variants of HLC we considered. For reference, we also report the performance of a plain implementation of HL algorithm, where each label is an array of 32-bit integers (hubs and distances), sorted by hub ID. The complicated optimizations proposed by Abraham et al. [3] (such as ID reassignment, 8/24 compression, distance oracles, and index-free queries) will be considered later. We use the default contraction order proposed by Abraham et al. [4] (HL-15), with 78.24 hubs per label on average.

For each method, Table 1 shows the total preprocessing time (including finding the contraction order), the amount of data generated, the average number of (both trivial and nontrivial) tokens per vertex, and the average number of nontrivial tokens read to retrieve a label ("reads/label"). All values are aggregated over both forward and backward labels. We also show average times (over 10^7 random s–t pairs) for three query strategies: hashed index, linear index (an n-sized array), and merging (including a call to the STL sort function for HLC).

The basic version of HLC (described in Section 3) uses an order of magnitude less space than HL, with similar preprocessing time. This makes the label-based approach much more practical: while few current servers have 20 GiB of RAM available, most can easily handle 2 GiB data structures. (For comparison, labels compressed with `gzip` would take 7.7 GiB and would not support fast queries.) Space usage can be reduced by around 10% if 1-parent tokens are eliminated, with little effect on query times; 1-child elimination has a small but positive effect. Greedily flattening tokens until the token array increases by 5% reduces the number of tokens needed to represent each label by more than 70%, but queries are only about 25% faster ("popular" tokens end up in cache anyway, even without flattening). Flattening twice as much (10%) would further reduce query times by only 3%.

Regarding queries, hashing is slightly faster than linear indexing for HLC, and significantly faster than merging (which requires sorting). For HL, whose labels are already sorted by ID, merging is by far the best strategy, due to its favorable access pattern and simplicity. In the end, hashing and worse locality make HLC queries about four times slower than plain HL queries.

Table 2 compares the two versions of HLC (with all optimizations) with other fast algorithms. It includes CH [16] and three variants of HL [3, 4]: *HL-∞ global* is optimized for long-range queries, *HL-15 local* for short-range queries, and *HL prefix* minimizes space usage. HLC-0 and HLC-15 use the same vertex orders as HL-0 and HL-15, respectively [4]. We also include Transit Node Routing (TNR) [5, 6], which uses table lookups for long range queries, CH for local queries, and (optionally) hashing for midrange queries. TNR+AF [8] uses *arc flags* [19] to guide TNR towards the target. We report preprocessing time, space usage, and average time for random queries, considering the best available implementation of each method. All times are sequential and scaled to match our machine [3].

Table 2. Random queries on Europe

method	prepro [h:m]	space [GB]	query [ns]
CH [16]	0:14	0.4	79373
TNR [5]	0:21	2.5	1711
TNR+AF [8]	2:00	5.7	992
HL prefix [3]	≈2:00	5.7	527
HL-15 local [4]	0:37	18.8	556
HL-∞ global [4]	≈60:00	17.7	254
HLC-0	0:30	1.8	2989
HLC-15	0:50	1.8	2554

While standard HL uses much more space than other methods, compression makes the hub labels approach less of an outlier. HLC uses only about 4 times as much space as CH (the most compact method), but random queries are 30 times faster. HLC is comparable to TNR in all three criteria. As we have argued before, however, HLC has advantages that can make it more practical: simplicity (no graphs or priority queues) and flexibility (natural extensions based on hubs).

Moreover, HLC is faster for local queries, which in practice are more common than long-range (or random) ones. Fig. 1 plots the median query times of several algorithms as a function of the *Dijkstra rank* [16]. (Vertex v has Dijkstra rank i with respect to s if v is the i-th vertex scanned by Dijkstra's algorithm from s.) Each point corresponds to 10 000 queries with a given rank. The numbers were taken from [3, 5, 8, 16] and scaled appropriately.

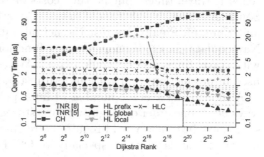

Fig. 1. Local queries

Query times increase with Dijkstra rank for CH (since its search space is bigger), but decreases for TNR (since it can only do fast table lookups for more global queries). Standard HL can reorder vertices to allow long-range queries to skip some unimportant hubs [3]. In contrast, HLC must always visit nearby hubs before getting to more important ones, and its query times are largely independent of the Dijkstra rank. For local queries, HLC is only about three times slower than HL, and much faster than other methods.

Finally, Table 3 reports the performance of HLC on a variety of road networks. We start with Western Europe from PTV, taking both travel times (as before) and travel distances as cost functions. We also test an *expanded* version of this instance with turn costs, which we model by creating a vertex to represent each

Table 3. Results for HLC on various instances: number n of vertices (in millions), average (out-)degree, preprocessing time, space usage, compression ratio (relative to plain HL), average number of hubs per label, and (hash-based) query times

source	input	metric	turns	n [10^6]	deg	time [h:m]	space [MiB]	comp ratio	hubs /label	query [ns]
PTV	W. Europe	time	×	18.0	2.36	0:50	1840.1	11.83	78.24	2554
	W. Europe	time	✓	42.6	2.72	4:10	6363.4	11.02	107.03	3512
	W. Europe	dist	×	18.0	2.36	3:15	2973.9	15.91	171.25	5734
Tiger	US	time	×	23.9	2.41	0:53	2979.6	8.62	69.33	2486
	US	dist	×	23.9	2.41	2:32	4126.2	14.08	157.99	5294
OSM	Australia	time	×	4.9	1.97	0:13	408.8	9.48	51.30	1689
	S. America	time	×	11.3	2.18	0:29	1167.2	8.33	55.25	1865
	N. America	time	×	162.5	2.04	5:52	14560.6	18.25	106.18	3520
	Old World	time	×	188.7	2.02	6:14	17164.4	16.07	94.78	3232
Bing	Europe	default	×	47.9	2.23	1:53	4791.8	15.20	98.53	3264
	Europe	default	✓	107.0	2.26	8:01	13046.6	14.38	113.92	3854
	N. America	default	×	30.3	2.41	1:33	3461.2	14.52	107.74	3437
	N. America	default	✓	72.5	2.61	14:34	13403.8	11.04	132.87	4429

original arc [11]; we follow Delling et al. [11] and set U-turn costs to 100 seconds and all other turn costs to zero. In addition, we test TIGER data [13] representing the USA, as well as four OpenStreetMap (OSM) instances (v. 121812) with realistic travel times (but no turn costs or restrictions), representing Australia, South America, North (and Central) America, and Old World (Africa, Asia, and Europe). Finally, we test the actual data used by Bing Maps (building on Navteq data) in production; the cost function is proprietary, but correlates well with travel times, as one would expect. We consider versions with turn costs (as used in production) and without. All instances are strongly connected.

Although the average number of hubs per label varies significantly, HLC always needs one order of magnitude less storage than plain HL. Without compression, some instances would require more than 200 GiB of RAM, which is hardly practical. This is an issue especially for OSM data, whose vertices represent both topology (intersections) and geometry, leading to sparse but large graphs. With compression, space usage is kept below 20 GiB in every case, which is much more manageable. Queries always remain below 6 μs.

7   Final Remarks

We presented compression techniques that substantially reduce the memory requirements for HL. Not only do they keep queries fast and simple, but they also preserve the flexibility and generality of labels. This makes the label-based approach practical in a wider range of applications. An open problem is to speed up its preprocessing to handle dynamic scenarios (such as real-time traffic) efficiently. Although techniques such as CRP [11, 12] can quickly adapt to changes in the length function, they have more complicated (and much slower) queries.

Acknowledgements. We thank D. Luxen for routable OSM instances [1].

References

1. Project OSRM (2012), http://project-osrm.org/
2. Abraham, I., Delling, D., Fiat, A., Goldberg, A.V., Werneck, R.F.: HLDB: Location-Based Services in Databases. In: Proc. SIGSPATIAL GIS, pp. 339–348. ACM Press (2012)
3. Abraham, I., Delling, D., Goldberg, A.V., Werneck, R.F.: A Hub-Based Labeling Algorithm for Shortest Paths on Road Networks. In: Pardalos, P.M., Rebennack, S. (eds.) SEA 2011. LNCS, vol. 6630, pp. 230–241. Springer, Heidelberg (2011)
4. Abraham, I., Delling, D., Goldberg, A.V., Werneck, R.F.: Hierarchical Hub Labelings for Shortest Paths. In: Epstein, L., Ferragina, P. (eds.) ESA 2012. LNCS, vol. 7501, pp. 24–35. Springer, Heidelberg (2012)
5. Arz, J., Luxen, D., Sanders, P.: Transit Node Routing Reconsidered. In: Bonifaci, V., Demetrescu, C., Marchetti-Spaccamela, A. (eds.) SEA 2013. LNCS, vol. 7933, pp. 55–66. Springer, Heidelberg (2013)
6. Bast, H., Funke, S., Sanders, P., Schultes, D.: Fast Routing in Road Networks with Transit Nodes. Science 316(5824), 566 (2007)
7. Bauer, R., Delling, D.: SHARC: Fast and Robust Unidirectional Routing. ACM JEA 14(2.4), 1–29 (2009)
8. Bauer, R., Delling, D., Sanders, P., Schieferdecker, D., Schultes, D., Wagner, D.: Combining Hierarchical and Goal-Directed Speed-Up Techniques for Dijkstra's Algorithm. ACM JEA 15(2.3), 1–31 (2010)
9. Cohen, E., Halperin, E., Kaplan, H., Zwick, U.: Reachability and Distance Queries via 2-Hop Labels. SIAM J. Computing 32(5), 1338–1355 (2003)
10. Cormen, T., Leiserson, C., Rivest, R., Stein, C.: Introduction to Algorithms, 2nd edn. MIT Press (2001)
11. Delling, D., Goldberg, A.V., Pajor, T., Werneck, R.F.: Customizable Route Planning. In: Pardalos, P.M., Rebennack, S. (eds.) SEA 2011. LNCS, vol. 6630, pp. 376–387. Springer, Heidelberg (2011)
12. Delling, D., Werneck, R.F.: Faster Customization of Road Networks. In: Bonifaci, V., Demetrescu, C., Marchetti-Spaccamela, A. (eds.) SEA 2013. LNCS, vol. 7933, pp. 30–42. Springer, Heidelberg (2013)
13. Demetrescu, C., Goldberg, A.V., Johnson, D.S. (eds.): The Shortest Path Problem: 9th DIMACS Implementation Challenge. DIMACS Book, vol. 74. AMS (2009)
14. Dijkstra, E.W.: A Note on Two Problems in Connexion with Graphs. Numerische Mathematik 1, 269–271 (1959)
15. Gavoille, C., Peleg, D., Pérennes, S., Raz, R.: Distance Labeling in Graphs. Journal of Algorithms 53, 85–112 (2004)
16. Geisberger, R., Sanders, P., Schultes, D., Vetter, C.: Exact Routing in Large Road Networks Using Contraction Hierarchies. Transportation Sci. 46(3), 388–404 (2012)
17. Goldberg, A.V., Harrelson, C.: Computing the Shortest Path: A* Search Meets Graph Theory. In: Proc. SODA 2005, pp. 156–165. SIAM (2005)
18. Goldberg, A.V., Kaplan, H., Werneck, R.F.: Reach for A*: Shortest Path Algorithms with Preprocessing. In: Demetrescu, et al. (eds.) [13], pp. 93–139
19. Lauther, U.: An Experimental Evaluation of Point-To-Point Shortest Path Calculation on Roadnetworks with Precalculated Edge-Flags. In: Demetrescu, et al. (eds.) [13], pp. 19–40
20. Munro, I., Rao, S.: Succinct representation of data structures. In: Mehta, D.P., Sahni, S. (eds.) Handbook of Data Structures and Applications. CRC (2004)
21. Sanders, P., Schultes, D., Vetter, C.: Mobile Route Planning. In: Halperin, D., Mehlhorn, K. (eds.) ESA 2008. LNCS, vol. 5193, pp. 732–743. Springer, Heidelberg (2008)

Faster Customization of Road Networks

Daniel Delling and Renato F. Werneck

Microsoft Research Silicon Valley
{dadellin,renatow}@microsoft.com

Abstract. A wide variety of algorithms can answer exact shortest-path queries in real time on continental road networks, but they typically require significant preprocessing effort. Recently, the customizable route planning (CRP) approach has reduced the time to process a new cost function to a fraction of a minute. We reduce customization time even further, by an order of magnitude. This makes it worthwhile even when a single query is to be run, enabling a host of new applications.

1 Introduction

Computing driving directions in road networks is a fundamental problem of practical importance. Although it can be solved in almost-linear time by Dijkstra's shortest-path algorithm [14], this is not fast enough for interactive queries on continental road networks. This has motivated a wide variety of recent algorithms [2,5,7,8,18,21,22] that rely on a (relatively slow) preprocessing stage to enable much faster queries. Different algorithms offer distinct tradeoffs between preprocessing time, space requirements, and query times. Directly or indirectly, they exploit the fact that road networks have a strong hierarchy when minimizing driving times in free-flow traffic [3]. When other cost functions are optimized, however, their performance can be much worse. Moreover, these approaches should only be used when there are enough queries to amortize the preprocessing cost. This is not true in many practical situations, such as when the cost function changes very frequently (to consider traffic, for example), or when users can choose from several (possibly uncommon) cost functions.

In contrast, the recently proposed *customizable route planning* (CRP) algorithm [11] is lightweight and robust to changes in the cost function (metric). It works in *three* stages. The first, *metric-independent preprocessing*, uses graph partitioning to define the topology of a multilevel overlay graph [23], which is the same regardless of the cost function. The second stage, *customization*, uses the metric to compute the actual costs of the overlay arcs. Finally, the *query stage* uses the output of the first two stages to compute shortest paths in real time (milliseconds). The first stage may take a few minutes (or even hours), but only needs to be run (or updated) when new road segments are built. Metric changes (which are much more frequent) require running only customization, which takes less than a minute. Since it does not rely on strong hierarchies, CRP is robust to metric changes. Unlike most other methods, it can also handle turn costs (and

V. Bonifaci et al. (Eds.): SEA 2013, LNCS 7933, pp. 30–42, 2013.

restrictions) quite naturally, with little effect on performance and space usage. It is thus ideal for a real-world routing engine, and is indeed in use by Bing Maps.

This paper shows how to make customization even faster, enabling a wide range of new applications. To compute overlay arc costs, we propose replacing Dijkstra's algorithm by other approaches, such as contraction and Bellman-Ford. Although they may even *increase* the number of operations (such as arc scans) performed, careful application of algorithm engineering techniques leads to better locality and enables parallelism at instruction and core levels. Remarkably, our new customization routine takes less time (sequentially) than running Dijkstra's algorithm once. Unlike in any other method, *a single query* is enough to amortize the customization cost, making it ideal for highly dynamic applications.

2 Preliminaries

Basics. A road network is usually modeled as a directed graph $G = (V, A)$ with $n = |V|$ vertices and $m = |A|$ arcs. Each vertex $v \in V$ represents an intersection, and each arc $(v, w) \in A$ a road segment. A *metric* (or *cost function*) $\ell : A \to \mathcal{N}$ maps each arc to a positive *length* (or *cost*). In the *point-to-point shortest path problem*, our goal is to compute the minimum-length path in the graph between a *source* s and a *target* t. It can be solved by Dijkstra's algorithm [14], which processes vertices in increasing order of distance from s and stops when t is processed. It runs in essentially linear time in theory and in practice [20].

In this paper, we focus on a more realistic model for road networks, which takes turn costs (and restrictions) into account. We think of each vertex v as having one *entry point* for each of its incoming arcs, and one *exit point* for each outgoing arc. We extend the concept of *metric* by also associating a *turn table* T_v to each vertex v; $T_v[i, j]$ specifies the cost of turning from the i-th incoming arc to the j-th outgoing arc. (The order around each vertex is arbitrary but fixed.)

We can run Dijkstra's algorithm on this *turn-aware* graph by associating distance labels to entry points instead of vertices [11,19]. An alternative approach (often used in practice) is to operate on an *expanded graph* G', where each vertex corresponds to an entry point in G, and each arc represents the concatenation of a turn and an arc in G. This allows standard (non-turn-aware) algorithms to be used, but roughly triples the graph size. In contrast, the turn-aware representation is almost as compact as the simplified one (with no turns at all), since common turn tables can be shared among vertices.

Customizable Route Planning. The *customizable route planning* (CRP) [11] algorithm is a speedup technique that computes shortest paths in three stages: (metric-independent) preprocessing, customization, and queries.

The *preprocessing* stage defines a multilevel overlay [23] of the graph and builds auxiliary data structures. A *partition* of V is a family $\mathcal{C} = \{C_0, \ldots, C_k\}$ of sets $C_i \subseteq V$ such that each $v \in V$ is in exactly one *cell* C_i. A *multilevel partition* of V of L levels is a family of partitions $\{\mathcal{C}^1, \ldots, \mathcal{C}^L\}$, where l denotes the level of a partition $\mathcal{C}^l$. Let U^l be the size of the biggest cell on level l.

We deal only with *nested* multilevel partitions: for each $l \leq L$ and each cell $C_i^l \in \mathcal{C}^l$, there exists a cell $C_j^{l+1} \in \mathcal{C}^{l+1}$ with $C_i^l \subseteq C_j^{l+1}$; we say C_i^l is a *subcell* of C_j^{l+1}. (We assume $\mathcal{C}^0$ consists of singletons and $\mathcal{C}^{L+1} = \{V\}$.) A *boundary arc* on level l is an arc with endpoints in different level-l cells; its endpoints are *boundary vertices*. A boundary arc on level l is also a boundary arc on all levels below.

The preprocessing phase of CRP uses PUNCH [12], a graph-partitioning heuristic tailored to road networks, to create a multilevel partition. Given an unweighted graph and a bound U, PUNCH splits the graph into cells with at most U vertices while minimizing the number of arcs between cells. To find a multilevel partition, one calls PUNCH repeatedly in top-down fashion: after partitioning the full graph, one partitions each subcell independently.

Besides partitioning, the CRP preprocessing phase sets up the topology of the overlay graph. Consider a cell C on any level, as in Fig. 1. Every incoming boundary arc (u, v) (i.e, with $u \notin C$ and $v \in C$) defines an *entry point* for C, and every outgoing arc (v, w) (with $v \in C$ and $w \notin C$) defines an *exit point* for C. The *overlay* for cell C is simply a complete bipartite graph with directed *shortcuts* (gray lines in the figure) between each entry point (filled circle) and each exit point (hollow circle) of C. The overlay of a level l consists of the union of all cell overlays, together with all boundary arcs (black arrows) on this level.

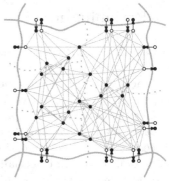

Fig. 1. Overlays of five cells

During the *customization stage*, CRP computes the *lengths* of the shortcuts on the overlay. A shortcut (p, q) within a cell C represents the shortest path (restricted to C) between p and q. Lengths are computed bottom-up, starting at level one. To process a cell, CRP runs Dijkstra's algorithm from each entry point p to find the lengths of all shortcuts starting at p. Processing level-one cells requires running Dijkstra on the original graph, taking turn costs into account. Higher-level cells can use the overlay graph for the level below, which is much smaller and has no explicit turns (turn costs are incorporated into shortcuts).

An s–t CRP *query* runs bidirectional Dijkstra on the overlay graph, but only entering cells containing either s or t. To retrieve the arcs corresponding to each shortcut in the resulting path, one runs bidirectional Dijkstra within the appropriate cell. We accelerate unpacking using an LRU cache to store the unpacking information of a level-i shortcut as a sequence of level-$(i - 1)$ shortcuts.

3   Our Approach

Separating preprocessing from metric customization allows CRP to incorporate a new cost function on a continental road network in less than 30 seconds [11] (sequentially) on a modern server. This is enough for real-time traffic, but too slow to enable on-line personalized cost functions. Accelerating customization even further requires speeding up its basic operation: computing the lengths of the shortcuts within each cell. To do so, we propose different strategies to replace Dijkstra's algorithm.

One strategy is *contraction*, the basic building block of the *contraction hierarchies* (CH) algorithm [18] and an element of many other speedup techniques [2,5,7,8,21]. Instead of computing shortest paths explicitly, we eliminate internal vertices from a cell one by one, adding new arcs as needed to preserve distances; the arcs that eventually remain are the desired shortcuts. For efficiency, not only do we precompute the order in which vertices are contracted, but also abstract away the graph itself. During customization, we simply simulate the actual contraction by following a (precomputed) series of *instructions* describing the basic operations (memory reads and writes) the contraction routine would perform.

Contraction works well on the first overlay level, since it operates directly on the underlying graph, which is sparse. Density quickly increases during contraction, however, making it expensive as cell sizes increase. On higher levels, we compute shortest paths explicitly (as before), but make each computation more efficient. We replace Dijkstra with lightweight algorithms that work better on small graphs, and apply techniques to reduce the size of the search graph.

The next two sections describe each strategy in more detail, including how they can be engineered for better performance in practice.

4 Contraction

The *contraction* approach is based on the *shortcut* operation [18]. To shortcut a vertex v, one removes it from the graph and adds new arcs as needed to preserve shortest paths. For each incoming arc (u, v) and outgoing arc (v, w), one creates a *shortcut* arc (u, w) with $\ell(u, w) = \ell(u, v) + \ell(v, w)$. A shortcut is only added if it represents the only shortest path between its endpoints in the remaining graph (without v), which can be tested by running a *witness search* (local Dijkstra) between its endpoints. CH [18] uses contraction as follows. During preprocessing, it heuristically sorts all vertices in increasing order of importance and shortcuts them in this order; the order and the shortcuts are then used to speed up queries.

We propose using contraction during customization. To process a cell, we can simply contract its internal vertices while preserving its boundary vertices. The arcs (shortcuts) in the final graph are exactly the ones we want. To deal with turn costs appropriately, we run contraction on the expanded graph.

The performance of contraction strongly depends on the cost function. With travel times in free-flow traffic (the most common case), it works very well. Even for continental instances, sparsity is preserved during the contraction process, and the number of arcs less than doubles [18]. Unfortunately, other metrics often need more shortcuts, which leads to denser graphs and makes finding the contraction order much more expensive. Even if a good order is given, simply performing the contraction can still be quite costly [18].

Within the CRP framework, we can deal with these issues by exploiting the separation between metric-independent preprocessing and customization. During preprocessing, we compute a unique contraction order to be used by all metrics. Unlike previous approaches [18], to ensure this order works well even in the worst case, we simply assume that every potential shortcut will be added. Accordingly, we do not perform witness searches during customization. For maximum efficiency, we precompute a sequence of *microinstructions* to describe the entire contraction process in terms of basic operations. We detail each of these improvements next.

Contraction Order. Computing a contraction order that minimizes the number of shortcuts added is NP-hard [6]. In practice, one uses on-line heuristics that pick the next vertex to contract based on a priority function that depends on local properties of the graph [18]. A typical criterion is the difference between the number of arcs added and removed if a vertex v were contracted. We tested similar greedy priority functions to evaluate each vertex v, taking into account parameters such as the number $ia(v)$ of incoming arcs, the number $oa(v)$ of outgoing arcs, and the number $sc(v)$ of shortcuts created or updated if v is contracted (this may be less than $ia(v) \cdot oa(v)$, since self-loops are never needed). We found that picking vertices v that minimize $h(v) = 100sc(v) - ia(v) - oa(v)$ works well. This essentially minimizes the number of shortcuts added, using the current degree as a tiebreaker (the precise coefficients are not important).

This approach gives reasonable orders, but one can do even better by taking the graph topology into account. There exist natural orders that lead to a provably small number of shortcuts for graphs with small separators [10,25] or treewidth [10]. It suffices to find a small separator for the entire graph, recursively contract the two resulting components, then contract the separating vertices themselves. For graphs with $O(\sqrt{n})$-separators (such as planar graphs), such nested dissection leads to $O(n \log n)$ shortcuts. Although real-world road networks are far from planar, they have even smaller separators [12].

This suggests using partitions to guide the contraction order. We create additional *guidance levels* during the preprocessing step, extending our standard CRP multilevel partition downward (to even smaller cells). We subdivide each level-1 cell (of maximum size U) into nested subcells of maximum size U/σ^i, for $i = 1, 2, \dots$ (until cells become too small). Here $\sigma > 1$ is the *guidance step*. For each internal vertex v in a level-1 cell, let $g(v)$ be the smallest i such that v is a boundary vertex on the guidance level with cell size U/σ^i. We use the same contraction order as before, but delay vertices according to $g(\cdot)$. If $g(v) > g(w)$, v is contracted before w; within each guidance level, we use $h(v)$.

Microinstructions. While the contraction order is determined during the metric-independent phase of CRP, we can only *execute* the contraction (follow the order) during customization, once we know the arc lengths. Even with the order given, this execution is expensive [18]. To contract v, we must retrieve the costs (and endpoints) of its incident arcs, then process each potential shortcut (u, w) by either inserting it or updating its current value. This requires data structures supporting arc insertions and deletions, and even checking if a shortcut already exists gets costlier as degrees increase. Each fundamental operation, however, is rather simple: we read the costs of two arcs, add them up, compare the result with the cost of a third arc, and update it if needed. The entire contraction routine can therefore be fully specified by a sequence of triples (a, b, c). Each element in the triple is a memory position holding an arc (or shortcut) length. We must read the values in a and b and write the sum to c if there is an improvement.

Since the sequence of operations is the same for any cost function, we use the metric-independent preprocessing stage to set up, for each cell, an *instruction array* describing the contraction as a list of triples. Each element of a triple represents an offset in a separate *memory array*, which stores the costs of all arcs (temporary or otherwise) touched during the contraction. The preprocessing stage outputs the entire instruction array as well as the size of the memory array.

During customization, entries in the memory array representing input arcs (or shortcuts) are initialized with their costs; the remaining entries (new shortcuts) are set to ∞. We then execute the instructions one by one, and eventually copy the output values (lengths of shortcuts from entry to exit points in the cell) to the overlay graph. With this approach, the graph itself is abstracted away during customization. We do not keep track of arc endpoints, and there is no notion of vertices at all. The code just manipulates numbers (which happen to represent arc lengths). This is cheaper (and simpler) than operating on an actual graph.

Although the space required by the instruction array is metric-independent (shared by all cost functions), it can be quite large. We can keep it manageable by representing each triple with as few bits as necessary to address the memory array. In addition, we use a single *macroinstruction* to represent the contraction of a vertex v whenever the resulting number of writes exceeds an *unrolling threshold* τ. This instruction explicitly lists the addresses of v's c_{in} incoming and c_{out} outgoing arcs, followed by the corresponding $c_{in} \cdot c_{out}$ write positions. The customization phase must explicitly loop over all incoming and outgoing positions, which is slightly slower than reading tuples but saves space.

5 Graph Searches

Although contraction could be used to process the entire hierarchy, it is not as effective at higher levels as it is at level-one cells, since the graphs within each higher-level cell are much denser. In such cases, it is cheaper to actually run graph searches. This section therefore proposes search-based techniques to accelerate higher levels of the hierarchy. Each leads to improvements on its own, and they can be combined in the final algorithm.

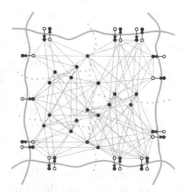

Fig. 2. Pruned overlay

Pruning the Search Graph. To process a cell C, we must compute the distances between its entry and exit points. As shown in Fig. 1, the graph G_C on which we operate within C is the union of subcell overlays (complete bipartite graphs) with some boundary arcs between them. Instead of searching

G_C directly, we first contract its internal exit points (see Fig. 2). Since each such vertex has out-degree one (its outgoing arc is a boundary arc within C), this reduces the number of vertices and edges in the search graph. Note that C's own exit points must be preserved (they are the targets of our searches), but they do not need to be scanned (they have no outgoing arcs).

Improving Locality. Conceptually, to process a cell C we could operate on the full overlay graph, but restricting the searches to vertices inside C. For efficiency, we actually copy the relevant subgraph to a separate memory location, run our searches on it, then copy the results back. This simplifies the searches (there are no special cases), allows us to use sequential local IDs, and improves locality.

Alternative Algorithms. We can further accelerate customization by replacing Dijkstra's algorithm with Bellman-Ford [9,16]. It starts by setting the distance label of the source to 0, and all others to ∞. Each round then scans each vertex once, updating the distance label of its neighbors appropriately. For better performance, we only *active* vertices (i.e., those whose distance improved since the previous round) and stop when there is no active vertex left. While Bellman-Ford cannot scan fewer vertices than Dijkstra, its simplicity and better locality make it competitive. The number of rounds is bounded by the maximum number of arcs on any shortest path, which is small for reasonable metrics but linear in the worst case. One could therefore switch to Dijkstra's algorithm whenever the number of Bellman-Ford rounds reaches a given (constant) threshold.

For completeness, we also tested the Floyd-Warshall algorithm [15]. It computes shortest paths among *all* vertices in the graph, and we just extract the relevant distances. Its running time is cubic, but with its tight inner loop and good locality, it could be competitive with Bellman-Ford on denser graphs.

Multiple-source Executions. Multiple runs of Dijkstra's algorithm (from different sources) can be accelerated if combined into a single execution [22,26]. We apply this idea to Bellman-Ford. Let k be the number of simultaneous executions, from sources $s_1, \ldots, s_k$. For each vertex v, we keep k distance labels: $d_1(v), \ldots, d_k(v)$. All $d_i(s_i)$ values are initialized to zero (each s_i is the source of its own search), and all remaining $d_i(\cdot)$ values to ∞. All k sources s_i are initially marked as active. When Bellman-Ford scans an arc (v, w), we try to update all k distance labels of w at once: for each i, we set $d_i(w) \leftarrow \min\{d_i(w), d_i(v) + \ell(v, w)\}$. If any such distance label actually improves, we mark w as active. This simultaneous execution needs as many rounds as the worst of the k sources, but, by storing the k distances associated with a vertex contiguously in memory, locality is much better. In addition, it enables instruction-level parallelism [26], as discussed next.

Parallelism. Modern CPUs have extended instruction sets with SIMD (single instruction, multiple data) operations, which work on several pieces of data at once. In particular, the SSE instructions available in x86 CPUs can manipulate special 128-bit registers, allowing basic operations (such as additions and comparisons) on four 32-bit words in parallel.

Consider the simultaneous execution of Bellman-Ford from $k = 4$ sources, as above. When scanning v, we first store v's four distance labels in one SSE register. To process an arc (v, w), we store four copies of $\ell(v, w)$ into another register and use a single SSE instruction to add both registers. With an SSE comparison, we check if these tentative distances are smaller than the current distance labels for w (themselves loaded into an SSE register). If so, we take the minimum of both registers (in a single instruction) and mark w as active.

In addition to using SIMD instructions, we can use core-level parallelism by assigning cells to distinct cores. (We also do this for level-1 cells with microinstructions.) In addition, we parallelize the highest overlay levels (where there are few cells per core) by further splitting the sources in each cell into sets of similar size, and allocating them to separate cores (each accessing the entire cell).

6   Experiments

We implemented our algorithm in C++ (using OpenMP for parallelization) and compiled it using Microsoft Visual Studio 2010. Our test machine runs Windows Server 2008 R2 and has 96 GiB of DDR3-1333 RAM and two 6-core Intel Xeon X5680 3.33 GHz CPUs, each with 6×64 KB of L1, 6×256 KB of L2, and 12 MB of shared L3 cache. Unless otherwise mentioned, we run our experiments on a benchmark instance representing the road network of (Western) Europe, made available by PTV AG for the 9th DIMACS Implementation Challenge [13]. The original instance has $n = 18 \cdot 10^6$ vertices, $m = 42 \cdot 10^6$ arcs, and travel times as the cost function. Following Delling et al. [11], we augment it by setting turn costs to 100 s for U-turns (and zero otherwise).

We first evaluate the effectiveness of microinstructions. Each point in Fig. 3 represents the total (sequential) customization time up to a certain cell size, using only microinstructions. Each curve reflects a different guidance step in the contraction order; smaller steps mean heavier use of partition information. Microinstructions use 32-bit addresses, and the unrolling threshold τ is 10.

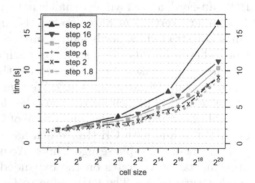

Fig. 3. Customization time up to given cell sizes for various guidance steps

It takes less than 2 s to run the microinstructions on the entire graph if the maximum cell size is 16, and less than 3 s for cells of size 256. Significantly larger cells are much more costly. As expected, smaller guidance steps lead to better contraction orders (smaller instruction arrays), but the effect is not overwhelming: step 4 is about as fast as step 1.8, which is roughly equivalent to nested dissection. Given these results, for the remainder of this paper we only use microinstructions to customize cells of size up to 256 (using 128, 64, 32, 16 and 8 as guidance levels), with 16

Table 1. Time (in milliseconds) spent on each overlay level for different algorithms. The total time includes 1.99 s to process the lowest overlay level (2^8) with microinstructions.

method	2^{11}	2^{14}	2^{17}	2^{20}	total
Dijkstra	1071	710	587	423	4783
4-Dijkstra	1245	850	717	541	5372
4-Dijkstra(SSE)	1036	627	439	327	4425
16-Dijkstra	1184	822	676	486	5161
16-Dijkstra(SSE)	1117	669	464	366	4608
Bellman-Ford	1164	840	753	589	5343
4-Bellman-Ford	930	723	710	603	4962
4-Bellman-Ford(SSE)	693	473	399	295	3852
16-Bellman-Ford	994	797	766	646	5197
16-Bellman-Ford(SSE)	512	335	291	230	3360
Floyd-Warshall	7802	7025	7414	6162	30403

(instead of 32) bits for addresses and $\tau = 50$. To compute all shortcut lengths up to this level, customization takes about 1.99 s to follow the 833 million (write) instructions (which use about 3.1 GB of RAM). In contrast, running Dijkstra-based customization would take 15.23 seconds (even using a so-called phantom level [11]), an order of magnitude slower.

We now consider higher levels. Our second experiment uses 5 levels, with maximum cell sizes 2^8, 2^{11}, 2^{14}, 2^{17}, and 2^{20}. Table 1 reports the total time (on a single core) to compute the shortcut lengths on each of the top 4 levels, as well as the total customization time (including the 1.99 seconds for the lowest level). It shows that individual executions of Dijkstra's algorithm are slightly faster than Bellman-Ford, and Floyd-Warshall is not competitive. Computing distances from 4 boundary vertices at once (prefix "4-" in the table) helps Bellman-Ford, but hurts Dijkstra (which needs more scans). SSE instructions help both algorithms. Due to better locality, the fastest approach is Bellman-Ford with 16 simultaneous searches, which takes 1289 ms to process the top 4 levels, less than half the time taken by plain Dijkstra (2764 ms). We therefore pick 16-Bellman-Ford with SSE as our default approach for the top 4 levels (2^{11}, 2^{14}, 2^{17}, and 2^{20}).

Table 2 compares this default version of CRP with the original implementation of CRP [11] and with CH [18], which has the fastest preprocessing time among state-of-the-art two-phase algorithms. We include two versions of CH: a standard implementation operates on the expanded graph, and a compact version uses explicit turn tables. The latter, due to Geisberger and Vetter [19], was run on a machine comparable to ours, a 2.6 GHz dual 8-core Intel Xeon E5-2670 with 64 GiB of DDR3-1600 RAM. We test our default instance as well as a version with cheaper U-turns (1 s instead of 100 s). For CH, the *customization time* we report corresponds to the entire preprocessing; when a metric changes, CH finds a new order and a different set of shortcuts. The CH space includes shortcuts only, and not original arcs. In every case, (random) query times (and scans) are for finding the distance only. Queries are sequential and customization uses all available cores (12 or 16).

Table 2. Comparison between CRP and two different CH variants on Europe

| | U-TURN: 1 s | | | | U-TURN: 100 s | | | |
| | CUSTOMIZING | | QUERIES | | CUSTOMIZING | | QUERIES | |
algorithm	time [s]	space [MiB]	nmb. scans	time [ms]	time [s]	space [MiB]	nmb. scans	time [ms]
CH compact [19]	410.90	219.42	624	0.27	1753.84	641.95	1998	2.27
CH expanded	1090.52	1442.88	386	0.17	1392.41	1519.48	404	0.19
CRP original [11]	2.10	61.72	3556	1.92	2.44	61.72	3805	1.96
CRP	0.35	70.49	2702	1.60	0.35	70.49	3009	1.64

CH has faster queries (and is more robust to metric changes) on the expanded than on the compact graph, since it can use a more fine-grained contraction order. CRP queries are slower, but still fast enough for interactive applications. More importantly, our CRP customization takes only 0.35 seconds (on 12 cores) and is at least *three orders of magnitude* faster than CH, with much lower metric-dependent space requirements. It is also up to seven times faster than the original CRP customization.

Fig. 3 suggests that a precomputed metric-independent order could lead to faster CH customization times. Indeed, running our microinstruction-based customization up to cells of size $n/2$ (about 9 million) with guidance step 2 takes about 10.7 seconds (sequentially). Using the same same order for CH would lead to comparable customization times. The number of shortcuts generated (231 M) is only twice the number of original arcs in the expanded graph (116 M), but still significant considering that CH (without witness searches) must keep all of them for every metric. CH queries should be comparable to CRP, which performs only 1272 scans and takes 0.89 ms on average in the resulting 21-level setup.

Our last experiment considers other benchmark instances. From the 9th DI-MACS Challenge [13], we take PTV Europe and TIGER USA, each with two cost functions: driving times (with 100 s U-turn costs) and distances. We also consider OpenStreetMap (OSM) data (v. 121812) representing major landmasses and with realistic turn restrictions. Finally, we test the instances used by Bing Maps, which build on Navteq data and include actual turn costs and restrictions; the proprietary "default" metric correlates well with driving times.

For each instance, Table 3 shows the average number of scans and running time (over 100 random queries) of turn-aware Dijkstra, followed by the metric-independent CRP preprocessing time and the amount of metric-independent data generated. It then reports the customization time and the amount of metric-dependent data produced, followed by average statistics about queries (over 100 000 runs): number of scans, time to get the length of the shortest path, and time to get a full description of the path (length and underlying arcs). Queries are sequential and use a (prewarmed) LRU cache for 2^{18} shortcuts; preprocessing and customization run on 12 cores. We use the default CRP settings in every case, with a sixth overlay level (cell size 2^{23}) for the two largest instances.

The table shows that CRP is indeed robust, enabling consistently fast customization and queries. It is slowest for OSM instances, which are very large

Table 3. Performance of CRP on various benchmark instances

| | | | DIJKSTRA | | CRP | | | | | | |
| | | | QUERIES | | PREPRO | | CUSTOM | | QUERIES | | |
source	input	n cost [$\times 10^6$] func	scans [$\times 10^6$]	time [ms]	time [s]	space [MiB]	time [ms]	space [MiB]	nmb. scans	dist [ms]	path [ms]
PTV	Europe	18.0 dist	9.1	3069	796	4151	351	70.5	2916	1.86	2.43
	Europe	18.0 time	15.2	6093	796	4151	347	70.5	3009	1.64	1.81
TIGER	US	23.9 dist	12.1	4790	617	6649	677	111.1	3088	1.84	2.78
	US	23.9 time	13.2	6124	617	6649	664	111.1	2964	1.60	1.89
OSM	Australia	4.9 time	3.4	919	79	531	44	4.6	1108	0.27	0.40
	S. America	11.4 time	9.2	2549	222	2520	256	20.4	1238	0.32	0.61
	N. America	162.7 time	115.8	70542	2752	18675	1202	199.1	2994	1.60	3.63
	Old World	189.4 time	127.0	77121	3650	21538	1234	195.4	2588	1.49	4.20
Bing	N. America	30.3 dflt	28.3	11684	936	8125	762	136.6	3395	1.60	1.91
	Europe	47.9 dflt	37.0	17750	1445	7872	602	120.7	3679	2.10	2.52

because (unlike other inputs) they use vertices to represent both intersections and geometry. Even so, customization takes about a second, and queries take under 2 milliseconds. Preprocessing time is dominated by partitioning. While the amount of metric-dependent data is relatively small, instruction arrays can be quite large. Metric-independent space usage could be reduced using smaller τ or limiting microinstructions to smaller cells (than 256). Curiously, finding the length of a path takes similar time on most instances, but describing the path takes longer on OSM data. For every instance, customization is at least one order of magnitude faster than a single Dijkstra search, and would still be faster *even if run sequentially.* The main reason is the poor locality of Dijkstra's algorithm, whose working set is spread throughout the graph.

7   Final Remarks

We have significantly reduced the time needed to process a cost function to enable interactive queries on road networks. Our customization is an order of magnitude faster than the best previous method [11], and takes less time than a single Dijkstra search. The ability to incorporate a new metric in fractions of a second enables a host of new applications. For example, it allows personalized cost functions: we could store a compact description of the preferences of each user, and run customization on the fly whenever the user accesses the system. Cost functions can even be tuned interactively (in less than a second). Our approach can also be helpful in applications [4,17,24] that repeatedly compute shortest paths on the same underlying graph with a changing cost function. Most importantly, very fast customization has the potential to enable applications that have not even been considered so far.

Acknowledgements. We thank D. Luxen for routable OSM instances [1], C. Vetter for making his code [19] available, and T. Pajor for running it.

References

1. Project OSRM (2012), http://project-osrm.org/
2. Abraham, I., Delling, D., Goldberg, A.V., Werneck, R.F.: Hierarchical Hub Labelings for Shortest Paths. In: Epstein, L., Ferragina, P. (eds.) ESA 2012. LNCS, vol. 7501, pp. 24–35. Springer, Heidelberg (2012)
3. Abraham, I., Fiat, A., Goldberg, A.V., Werneck, R.F.: Highway Dimension, Shortest Paths, and Provably Efficient Algorithms. In: SODA, pp. 782–793 (2010)
4. Bader, R., Dees, J., Geisberger, R., Sanders, P.: Alternative Route Graphs in Road Networks. In: Marchetti-Spaccamela, A., Segal, M. (eds.) TAPAS 2011. LNCS, vol. 6595, pp. 21–32. Springer, Heidelberg (2011)
5. Bast, H., Funke, S., Sanders, P., Schultes, D.: Fast Routing in Road Networks with Transit Nodes. Science 316(5824), 566 (2007)
6. Bauer, R., Columbus, T., Katz, B., Krug, M., Wagner, D.: Preprocessing Speed-Up Techniques is Hard. In: Calamoneri, T., Diaz, J. (eds.) CIAC 2010. LNCS, vol. 6078, pp. 359–370. Springer, Heidelberg (2010)
7. Bauer, R., Delling, D.: SHARC: Fast and Robust Unidirectional Routing. ACM JEA 14(2.4), 1–29 (2009)
8. Bauer, R., Delling, D., Sanders, P., Schieferdecker, D., Schultes, D., Wagner, D.: Combining Hierarchical and Goal-Directed Speed-Up Techniques for Dijkstra's Algorithm. ACM JEA 15(2.3), 1–31 (2010)
9. Bellman, R.: On a routing problem. Q. Appl. Math. 16(1), 87–90 (1958)
10. Columbus, T.: Search space size in contraction hierarchies. Master's thesis, Karlsruhe Institute of Technology (2012)
11. Delling, D., Goldberg, A.V., Pajor, T., Werneck, R.F.: Customizable Route Planning. In: Pardalos, P.M., Rebennack, S. (eds.) SEA 2011. LNCS, vol. 6630, pp. 376–387. Springer, Heidelberg (2011)
12. Delling, D., Goldberg, A.V., Razenshteyn, I., Werneck, R.F.: Graph Partitioning with Natural Cuts. In: Proc. IPDPS, pp. 1135–1146. IEEE Computer Society (2011)
13. Demetrescu, C., Goldberg, A.V., Johnson, D.S. (eds.): The Shortest Path Problem: 9th DIMACS Implementation Challenge. DIMACS Book, vol. 74. AMS (2009)
14. Dijkstra, E.W.: A Note on Two Problems in Connexion with Graphs. Numerische Mathematik 1, 269–271 (1959)
15. Floyd, R.W.: Algorithm 97 (shortest paths). Comm. ACM 5(6), 345 (1962)
16. Ford Jr., L.R., Fulkerson, D.R.: Flows in Networks. Princeton U. Press (1962)
17. Funke, S., Storandt, S.: Polynomial-time Construction of Contraction Hierarchies for Multi-criteria Queries. In: Proc. ALENEX, pp. 41–54. SIAM (2013)
18. Geisberger, R., Sanders, P., Schultes, D., Vetter, C.: Exact Routing in Large Road Networks Using Contraction Hierarchies. Transportation Sci. 46(3), 388–404 (2012)
19. Geisberger, R., Vetter, C.: Efficient Routing in Road Networks with Turn Costs. In: Pardalos, P.M., Rebennack, S. (eds.) SEA 2011. LNCS, vol. 6630, pp. 100–111. Springer, Heidelberg (2011)
20. Goldberg, A.V.: A Practical Shortest Path Algorithm with Linear Expected Time. SIAM J. Comp. 37, 1637–1655 (2008)
21. Goldberg, A.V., Kaplan, H., Werneck, R.F.: Reach for A*: Shortest Path Algorithms with Preprocessing. In: Demetrescu, et al. (eds.) [13], pp. 93–139
22. Hilger, M., Köhler, E., Möhring, R.H., Schilling, H.: Fast Point-to-Point Shortest Path Computations with Arc-Flags. In: Demetrescu, et al. (eds.) [13], pp. 41–72

23. Holzer, M., Schulz, F., Wagner, D.: Engineering Multi-Level Overlay Graphs for Shortest-Path Queries. ACM JEA 13(2.5), 1–26 (2008)
24. Luxen, D., Sanders, P.: Hierarchy Decomposition and Faster User Equilibria on Road Networks. In: Pardalos, P.M., Rebennack, S. (eds.) SEA 2011. LNCS, vol. 6630, pp. 242–253. Springer, Heidelberg (2011)
25. Milosavljevic, N.: On optimal preprocessing for contraction hierarchies. In: Proc. SIGSPATIAL IWCTS, pp. 6:1–6:6 (2012)
26. Yanagisawa, H.: A multi-source label-correcting algorithm for the all-pairs shortest paths problem. In: Proc. IPDPS, pp. 1–10 (2010)

Intriguingly Simple and Fast Transit Routing*

Julian Dibbelt, Thomas Pajor, Ben Strasser, and Dorothea Wagner

Karlsruhe Institute of Technology (KIT), 76128 Karlsruhe, Germany
{dibbelt,pajor,strasser,dorothea.wagner}@kit.edu

Abstract. This paper studies the problem of computing optimal journeys in dynamic public transit networks. We introduce a novel algorithmic framework, called Connection Scan Algorithm (CSA), to compute
journeys. It organizes data as a single array of connections, which it scans
once per query. Despite its simplicity, our algorithm is very versatile. We
use it to solve earliest arrival and multi-criteria profile queries. Moreover, we extend it to handle the minimum expected arrival time (MEAT)
problem, which incorporates stochastic delays on the vehicles and asks
for a set of (alternative) journeys that in its entirety minimizes the
user's expected arrival time at the destination. Our experiments on the
dense metropolitan network of London show that CSA computes MEAT
queries, our most complex scenario, in 272 ms on average.

1 Introduction

Commercial public transit route planning systems are confronted with millions
of queries per hour [12], making fast algorithms a necessity. Preprocessing-based
techniques for computing point-to-point shortest paths have been very successful
on road networks [8,16], but their adaption to public transit networks [2,10] is
harder than expected [1,3,4]. The problem of computing "best" journeys comes
in several variants [14]: The simplest, called *earliest arrival*, takes a departure
time as input, and determines a journey that arrives at the destination as early
as possible. If further criteria, such as the number of transfers, are important,
one may consider *multi-criteria* optimization [7,9]. Finally, a *profile query* [6,7]
computes a set of optimal journeys that depart during a period of time (such
as a day). Traditionally, these problems have been solved by (variants of) Dijkstra's algorithm on an appropriate graph model. Well-known examples are
the time-expanded and time-dependent models [6,10,14,15]. Recently, Delling et
al. [7] introduced RAPTOR. It solves the multi-criteria problem (arrival time and
number of transfers) by using dynamic programming directly on the timetable,
hence, no longer requires a graph or a priority queue.

In this work, we present the *Connection Scan Algorithm* (CSA). In its basic variant, it solves the earliest arrival problem, and is, like RAPTOR, not
graph-based. However, it is not centered around *routes* (as RAPTOR), but elementary *connections*, which are the most basic building block of a timetable.

---

* Partial support by DFG grant WA654/16-1 and EU grant 288094 (eCOMPASS).

V. Bonifaci et al. (Eds.): SEA 2013, LNCS 7933, pp. 43–54, 2013.
© Springer-Verlag Berlin Heidelberg 2013

CSA organizes them as one single array, which it then scans once (linearly) to compute journeys to all stops of the network. The algorithm turns out to be intriguingly simple with excellent spatial data locality. We also extend CSA to handle multi-criteria profile queries: For a full time period, it computes Pareto sets of journeys optimizing arrival time and number of transfers.

Finally, we introduce the *minimum expected arrival time problem* (MEAT). It incorporates uncertainty [5,9,11] by considering stochastic delays on the vehicles. Its goal is to compute a set of journeys that minimizes the user's *expected* arrival time (at the destination). The output can be viewed as a decision graph that provides all relevant alternative journeys at stops where transfers might fail (see Fig. 1). We extend CSA to handle these queries very efficiently. Moreover, we do not make use of heavy preprocessing, thus, enabling

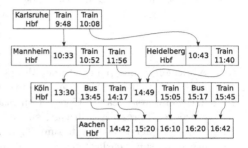

Fig. 1. Delay-robust itinerary from Karlsruhe to Aachen, Germany. A user should try to take the leftmost path. If transfers fail, alternatives are available.

dynamic scenarios including train cancellations, route changes, real-time delays, etc. Our experiments on the dense metropolitan network of London validate our approach. With CSA, we compute earliest arrival queries in under 2 ms, and multi-criteria profile queries for a full period in 221 ms—faster than previous algorithms. Moreover, we solve the most complex of our problems, MEAT, with CSA in 272 ms, fast enough for interactive applications.

This paper is organized as follows. Section 2 sets necessary notion, and Section 3 presents our new algorithm. Section 4 extends it to multi-criteria profile queries, while Section 5 considers MEAT. The experimental evaluation is available in Section 6, while Section 7 contains concluding remarks.

2 Preliminaries

Our public transit networks are defined in terms of their aperiodic *timetable*, consisting of a set of *stops*, a set of *connections*, and a set of *footpaths*. A *stop* p corresponds to a location in the network where a passenger can enter or exit a vehicle (such as a bus stop or train station). Stops may have associated minimum change times, denoted $\tau_{ch}(p)$, which represent the minimum time required to change vehicles at p. A *connection* c models a vehicle departing at a stop $p_{dep}(c)$ at time $\tau_{dep}(c)$ and arriving at stop $p_{arr}(c)$ at time $\tau_{arr}(c)$ without intermediate halt. Connections that are subsequently operated by the same vehicle are grouped into *trips*. We identify them by $t(c)$. We denote by c_{next} the next connection (after c) of the same trip, if available. Trips can be further grouped into *routes*. A route is a set of trips serving the exact same sequence of stops. For correctness, we require trips of the same route to not overtake each other. *Footpaths* enable walking transfers between nearby stops. Each footpath consists of two stops with an associated walking duration. Note that our footpaths are transitively closed. A *journey* is a sequence of connections and footpaths. If two subsequent connections are not part of the same trip, their arrival-departure time-difference must be at least the minimum change time of the stop. Because our footpaths are transitively closed, a journey never contains two subsequent footpaths.

In this paper we consider several well-known problems. In the *earliest arrival problem* we are given a source stop p_s, a target stop p_t, and a departure time τ. It asks for a journey that departs from p_s no earlier than τ and arrives at p_t as early as possible. The *profile problem* asks for the set of all earliest arrival journeys (from p_s to p_t) for every departure at p_s. Besides arrival time, we also consider the number of transfers as criterion: In multi-criteria scenarios one is interested in computing a *Pareto set* of nondominated journeys. Here, a journey J_1 *dominates* a journey J_2 if it is better with respect to every criterion. Nondominated journeys are also called to be *Pareto-optimal*. Finally, the *multi-criteria profile problem* requests a set of Pareto-optimal journeys (from p_s to p_t) for all departures (at p_s).

Usually, these problems have been solved by (variants of) Dijkstra's algorithm on an appropriate graph (representing the timetable). Most relevant to our work is the realistic *time-expanded model* [15]. It expands time in the sense that it creates a vertex for each *event* in the timetable (such as a vehicle departing or arriving at a stop). Then, for every connection it inserts an arc between its respective departure/arrival events, and also arcs that link subsequent connections. Arcs are always weighted by the time difference of their linked events. Special vertices may be added to respect minimum change times at stops. See [14,15] for details.

3   Basic Connection Scan Algorithm

We now introduce the Connection Scan Algorithm (CSA), our approach to public transit route planning. We describe it for the earliest arrival problem and extend it to more complex scenarios in Sections 4 and 5. Our algorithm builds on the following property of public transit networks: We call a connection c *reachable* iff either the user is already traveling on a preceding connection of the same trip $t(c)$, or, he is standing at the connection's departure stop $p_{\mathrm{dep}}(c)$ on time, i.e., before $\tau_{\mathrm{dep}}(c)$. In fact, the time-expanded approach encodes this property into a graph G, and then uses Dijkstra's algorithm to obtain optimal sequences of reachable connections [15]. Unfortunately, Dijkstra's performance is affected by many priority queue operations and suboptimal memory access patterns. However, since our timetables are aperiodic, we observe that G is acyclic. Thus, its arcs may be sorted topologically, e.g., by departure time. Dijkstra's algorithm on G, actually, scans (a subsequence of) them in this order.

Instead of building a graph, our algorithm assembles the timetable's connections into a single array C, sorted by departure time. Given source stop p_s and departure time τ as input, it maintains for each stop p a label $\tau(p)$ representing the earliest arrival time at p. Labels $\tau(\cdot)$ are initialized to all-infinity, except $\tau(p_s)$, which is set to τ. The algorithm scans all connections $c \in C$ (in order), testing if c can be *reached*. If this is the case and if $\tau_{\mathrm{arr}}(c)$ improves $\tau(p_{\mathrm{arr}}(c))$, CSA *relaxes* c by updating $\tau(p_{\mathrm{arr}}(c))$. After scanning the full array, the labels $\tau(\cdot)$ provably hold earliest arrival times for all stops.

Reachability, Minimum Change Times and Footpaths. To account for minimum change times in our data, we check a connection c for reachability by testing if $\tau(p_{\mathrm{dep}}(c)) + \tau_{\mathrm{ch}}(p_{\mathrm{dep}}(c)) \leq \tau_{\mathrm{dep}}(c)$ holds. Additionally, we track whether a preceding connection of the same trip $t(c)$ has been used. We, therefore, maintain for each connection a flag, initially set to 0. Whenever the algorithm identifies a connection c as reachable, it sets the flag of c's subsequent connection c_{next} to 1. Note that for networks with $\tau_{\mathrm{ch}}(\cdot) = 0$, trip tracking can be disabled and testing reachability simplifies to $\tau(p_{\mathrm{dep}}(c)) \leq \tau_{\mathrm{dep}}(c)$. To handle footpaths, each time the algorithm relaxes a connection c, it scans all outgoing footpaths of $p_{\mathrm{arr}}(c)$.

Improvements. Clearly, connections departing before time τ can never be reached and need not be scanned. We do a binary search on C to identify the first relevant connection and start scanning from there (*start criterion*). If we are only interested in *one-to-one queries*, the algorithm may stop as soon as it scans a connection whose departure time exceeds the target stop's earliest arrival time. Also, as soon as one connection of a trip is reachable, so are all subsequent connections of the same trip (and preceding connections of the trip have already been scanned). We may, therefore, keep a flag (indicating reachability) per trip (instead of per connection). The algorithm then operates on these *trip flags* instead. Note that we store all data sequentially in memory, making the scan extremely cache-efficient. Only accesses to stop labels and trip flags are potentially costly, but the number of stops and trips is small in comparison. To further improve spatial locality, we subtract from each connection $c \in C$ the minimum change time of $p_{\mathrm{dep}}(c)$ from $\tau_{\mathrm{dep}}(c)$, but keep the original ordering of C. Hence, CSA requires random access only on small parts of its data, which mostly fits in low-level cache.

4   Extensions

CSA can be extended to profile queries. Given the timetable and a source stop p_s, a profile query computes for every stop p the set of all earliest arrival journeys to p for every departure from p_s, discarding dominated journeys. Such queries are useful for preprocessing techniques, but also for users with flexible departure (or arrival) time. We refer to the solution as a Pareto set of $(\tau_{\mathrm{dep}}(p_s), \tau_{\mathrm{arr}}(p_t))$ pairs.

In the following, we describe the *reverse p–p_t-profile query*, which is needed in Section 5. The forward search works analogously. Our algorithm, pCSA (p for profile), scans once over the array of connections sorted by *decreasing* departure time. For every stop it keeps a partial (tentative) profile. It maintains the property that the partial profiles are correct wrt. the subset of already scanned connections. Every stop is initialized with an empty profile, except p_t, which is set to a constant identity-profile. When scanning a connection c, pCSA *evaluates* the partial profile at the arrival stop $p_{\mathrm{arr}}(c)$: It asks for the earliest arrival time τ^* at p_t over all journeys departing at $p_{\mathrm{arr}}(c)$ at $\tau_{\mathrm{arr}}(c)$ or later. It then *updates* the profile at $p_{\mathrm{dep}}(c)$ by potentially adding the pair $(\tau_{\mathrm{dep}}(c), \tau^*)$ to it, discarding newly dominated pairs, if necessary.

Maintaining Profiles. We describe two variants of maintaining profiles. The first, pCSA-P (P for Pareto), stores them as arrays of Pareto-optimal $(\tau_{\mathrm{dep}}, \tau_{\mathrm{arr}})$ pairs ordered by decreasing arrival (departure) time. Since new candidate entries are generated in order of decreasing departure time, profile updates are a constant-time operation: A candidate entry is either dominated by the last entry or is appended to the array. Profile evaluation is implemented as a linear scan over the array. This is quick in practice, since, compared to the timetable's period, connections usually have a short duration. The identity profile of p_t is handled as a special case. By slightly modifying the data structure, we obtain pCSA-C (C for constant), for which evaluation is also possible in constant time: When updating a profile, pCSA may append a candidate entry, even if it is dominated. To ensure correctness, we set the candidate's arrival time τ^* to that of the dominating entry. We then observe that, independent of the input's source or target stop, profile entries are always generated in the same order. Moreover, each connection is associated with only two such entries, one at its departure stop, relevant for updating, and, one at its arrival stop, relevant for evaluation. For each connection, we precompute *profile indices* pointing to these two entries, keeping them with the connection. Furthermore, its associated departure time and stop may be dropped. Note that the space consumption for keeping all (even suboptimal) profile entries is bounded by the number of connections. Following [6], we also collect—in a quick preprocessing step—at each stop all arrival times (in decreasing order). Then, instead of storing arrival times in the profile entries, we keep *arrival time indices*. For our scenarios, these can be encoded using 16 (or fewer) bits. We call this technique *time indexing*, and the corresponding algorithm pCSA-CT.

Minimum Change Times and Footpaths. We incorporate minimum change times by evaluating the profile at a stop p for time τ at $\tau + \tau_{\mathrm{ch}}(p)$. The trip bit is replaced by a trip arrival time, which represents the earliest arrival time at p_t when continuing with the trip. When scanning a connection c, we take the minimum of the trip arrival time and the evaluated profile at $p_{\mathrm{arr}}(c)$. We update the trip arrival time and the profile at $p_{\mathrm{dep}}(c)$, accordingly. *Footpaths* are handled as follows. Whenever a connection c is relaxed, we scan all incoming footpaths at $p_{\mathrm{dep}}(c)$. However, this no longer guarantees that profile entries are generated by decreasing departure time, making profile updates a non-constant operation for pCSA-P. Also, we can no longer precompute profile indices for pCSA-C. Therefore, we expand footpaths into *pseudoconnections* in our data, as follows. If p_a and p_b are connected by a footpath, we look at all reachable (via the footpath) pairs of incoming connections c_{in} at p_a and outgoing connections c_{out} at p_b. We create a new pseudoconnection (from p_a to p_b, departure time $\tau_{\mathrm{arr}}(c_{\mathrm{in}})$, and arrival time $\tau_{\mathrm{dep}}(c_{\mathrm{out}})$) iff there is no other pseudoconnection with a later or equal departure time and an earlier or equal arrival time. Pseudoconnections can be identified by a simultaneous sweep over the incoming/outgoing connections of p_a and p_b. During query, we handle footpaths toward p_t as a special case of the evaluation procedure. Footpaths at p_s are handled by merging the profiles of stops that are reachable by foot from p_s.

One-to-One Queries. So far we described *all-to-one profile queries*, i.e., from all stops to the target stop p_t. If only the *one-to-one profile* between stops p_s and p_t is of interest, a well-known pruning rule [6,14] can be applied to pCSA-P: Before inserting a new profile entry at any stop, we check whether it is dominated by the last entry in the profile at p_s. If so, the current connection cannot possibly be extended to a Pareto-optimal solution at the source, and, hence, can be pruned. However, we still have to continue scanning the full connection array.

Multi-Criteria. CSA can be extended to compute *multi-criteria* profiles, optimizing triples $(\tau_{\text{dep}}(p_s), \tau_{\text{arr}}(p_t), \#t)$ of departure time, arrival time and number of taken trips. We call this variant mcpCSA-CT. We organize these triples hierarchically by mapping arrival time $\tau_{\text{arr}}(p_t)$ onto *bags* of $(\tau_{\text{dep}}(p_s), \#t)$ pairs. Thus, we follow the general approach of pCSA-CT, but now maintain profiles as $(\tau_{\text{arr}}(p_t), \text{bag})$ pairs. Evaluating a profile, thus, returns a bag. Where pCSA-CT computes the minimum of two departure times, mcpCSA-CT *merges* two bags, i.e., it computes their union and removes dominated entries. When it scans a connection c, $\#t$ is increased by one for each entry of the evaluated bag, unless c is a pseudoconncetion. It then merges the result with the bag of trip $t(c)$, and updates the profile at $p_{\text{dep}}(c)$, accordingly. Exploiting that, in practice, $\#t$ only takes small integral values, we store bags as fixed-length vectors using $\#t$ as index and departure times as values. Merging bags then corresponds to a component-wise minimum, and increasing $\#t$ to shifting the vector's values. A variant, mcpCSA-CT-SSE, uses SIMD-instructions for these operations.

5 Minimum Expected Arrival Time

In this section we aim to provide delay-*robust* journeys that offer sensible backup alternatives at every stop for the case that transfers fail. A tempting approach might be to optimize *reliability*, introduced in [9], possibly together with other criteria. While this produces journeys that have low failure probabilities on their transfers, they are not necessarily robust in our sense: The set of reliable journeys may already diverge at the source stop, and in general, no fall-back alternatives are guaranteed at intermediate stops. On the other hand, on high-frequency urban routes (such as subways) an unreliable transfer might not be a problem, if the next feasible trip is just a few minutes away. To ensure that the user is never left without guidance, we compute a *subset* of connections (rather than journeys) such that at any point along the way, the user is provided with a good (in terms of arrival time) option for continuing his journey toward the destination. We propose to minimize the *expected arrival time* to achieve these goals.

We assume the following simple delay model: A connection c arrives at a random time $\tau_{\text{arr}}^R(c)$ but departs on time at $\tau_{\text{dep}}(c)$. All random arrival times are independent. No connection arrives earlier than its scheduled arrival time $\tau_{\mathrm{arr}}(c)$. To make computations meaningful, we assume an upper bound on all $\tau_{\mathrm{arr}}^{R}(c)$. We further assume that walking is exact. Note that more complex stochastic models have been considered in [5,11], containing dependent random variables to model delays. In this case, however, such models also propagate data errors (besides delays), therefore, requiring precise delay data [5], which is hard to obtain in practice. Also, even basic operations in [11] have super-quadratic running time (in the number of connections), making the approach impractical, already for medium-sized timetables.

For a given target stop p_t, we define for every subset S of connections of the timetable and for every connection c the *expected arrival time* $\hat{\tau}(S, c)$ at p_t, recursively. Let $c_1 \ldots c_n \subseteq S$ be the connections that the user can transfer to at c's arrival stop $p_{\mathrm{arr}}(c)$, ordered by departure time $\tau_{\mathrm{dep}}(c_i)$ (adjusted for footpaths and minimum change times). We define

$$\hat{\tau}(S, c) = \min \left\{ \hat{\tau}(S, c_{\mathrm{next}}), \sum_{i=1}^{n+1} P\left[\tau_{\mathrm{dep}}(c_{i-1}) \leq \tau_{\mathrm{arr}}^{R}(c) < \tau_{\mathrm{dep}}(c_i)\right] \cdot \hat{\tau}(S, c_i) \right\}$$

where $\tau_{\mathrm{dep}}(c_0) = \tau_{\mathrm{arr}}(c)$, $\tau_{\mathrm{dep}}(c_{n+1}) = \infty$, $\hat{\tau}(S, c_{n+1}) = \infty$, and $\hat{\tau}(S, c_{\mathrm{next}}) = \infty$ if c is the last connection of trip $t(c)$. The base of the recursion is defined by the connections c arriving at p_t, for which we define $\hat{\tau}(S, c) = E[\tau_{\mathrm{arr}}^{R}(c)]$. If the possibility of the user not reaching the target is non-zero, the expected arrival time is trivially ∞. Since a connection is assumed to never arrive early, $\hat{\tau}(S, c)$ only depends on connections departing later than c, which guarantees termination. (This is where we require aperiodicity; in periodic networks infinite recursions may occur.) In short, we compute the average over the expected arrival times of each outgoing connection from the stop $p_{\mathrm{arr}}(c)$, weighted by the probability of the user catching it. We define the *minimum expected arrival time* $\hat{\tau}^{*}(c)$ of a connection c as the minimum $\hat{\tau}(S, c)$ over all subsets S. A subset S^{*} minimizes $\hat{\tau}^{*}(c)$, if for every stop p the set of pair $(\tau_{\mathrm{dep}}(c), \hat{\tau}(S^{*}, c))$ induced by those $c \in S^{*}$ that depart at p, does not include dominated connections. (A pair is dominated, if, wrt. another pair, it departs earlier with higher expected arrival time.) Note that removing a dominated pair's connection improves $\hat{\tau}(\cdot)$. Also, all subsets with this property have the same $\hat{\tau}(\cdot)$ and therefore S^{*} is globally optimal. At least one subset S^{*} exists that is optimal for every c, because removing dominated connections is independent of c.

To solve the *minimum expected arrival time problem* (MEAT), we compute a set S^{*}, and output the reachable connections for the desired source stop and departure time. Our algorithm is based directly on pCSA-P, with a different meaning for its stop labels: Instead of mapping a departure time τ_{dep} to the corresponding earliest arrival time τ_{arr} at p_t, the algorithm now maps τ_{dep} to the corresponding minimum expected arrival time $\hat{\tau}^{*}$ at p_t. It does so by maintaining an array of nondominated $(\tau_{\mathrm{dep}}, \hat{\tau}^{*})$ pairs. For a connection c, the label at stop $p_{\mathrm{arr}}(c)$ is evaluated by a linear scan over that array: Following from the recursive definition above, the minimum expected arrival time $\hat{\tau}^{*}(c)$ is computed by a weighted summation of each of the expected arrival times $\hat{\tau}^{*}$ collected during

Table 1. Size figures for our timetables including figures of the time-dependent (TD), colored time-dependent (TD-col), and time-expanded (TE) graph models [6,14,15]

Figures	London		Germany		Europe	
Stops	20 843		6 822		30 517	
Trips	125 537		94 858		463 887	
Connections	4 850 431		976 678		4 654 812	
Routes	2 135		9 055		42 547	
Footpaths	45 652		0		0	
Expanded Footpaths	8 436 763		0		0	
TD Vertices (Arcs)	97 k	(272 k)	114 k	(314 k)	527 k	(1 448 k)
TD-col Vertices (Arcs)	21 k	(71 k)	20 k	(86 k)	79 k	(339 k)
TE Vertices (Arcs)	9 338 k	(34 990 k)	1 809 k	(3 652 k)	8 778 k	(17 557 k)

this scan multiplied with the success probability of the corresponding transfer at $p_{arr}(c)$. An optimization, called *earliest arrival pruning*, first runs an earliest arrival query from the source stop and then only processes connections marked reachable during that query. Note that, since during evaluation we scan over several outgoing connections, pCSA-C is not applicable.

6   Experiments

We ran experiments pinned to one core of a dual 8-core Intel Xeon E5-2670 clocked at 2.6 GHz, with 64 GiB of DDR3-1600 RAM, 20 MiB of L3 and 256 KiB of L2 cache. We compiled our C++ code using g++ 4.7.1 with flags -O3 -mavx.

We consider three realistic inputs whose sizes are reported in Table 1. They are also used in [6,10,7], but we additionally filter them for (obvious) errors, such as duplicated trips and connections with non-positive travel time. Our main instance, London, is available at [13]. It includes tube (subway), bus, tram, Dockland Light Rail (DLR) and is our only instance that also includes footpaths. However, it has no minimum change times. The German and European networks were kindly provided by HaCon [12]. Both have minimum change times. The German network contains long-distance, regional, and commuter trains operated by Deutsche Bahn during the winter schedule of 2001/02. The European network contains long-distance trains, and is based on the winter schedule of 1996/97. To account for overnight trains and long journeys, our (aperiodic) timetables cover one (London), two (Germany), and three (Europe) consecutive days.

We ran for every experiment 10 000 queries with source and target stops chosen uniformly at random. Departure times are chosen at random between 0:00 and 24:00 (of the first day). We report the running time and the number of label comparisons, counting an SSE operation as a single comparison. Note that we disregard comparisons in the priority queue implementation.

Earliest Arrival. In Table 2, we report performance figures for several algorithms on the London instance. Besides CSA, we ran realistic time-expanded

Table 2. Figures for the earliest arrival problem on our London instance. Indicators are: • enabled, ∘ disabled, – not applicable. "Sta." refers to the start criterion. "Trp." indicates the method of trip tracking: connection flag (∘), trip flag (•), none (×). "One." indicates one-to-one queries by either using the stop criterion or pruning.

Alg.	Sta.	Trp.	One.	# Scanned Arcs/Con.	# Reachable Arcs/Con.	# Relaxed Arcs/Con.	# Scanned Footpaths	# L.Cmp. p. Stop	Time [ms]
TE	–	–	∘	20 370 117	—	5 739 046	—	977.3	876.2
TD	–	–	∘	262 080	—	115 588	—	11.9	18.9
TD-col	⌐	–	∘	68 183	—	21 294	—	3.2	7.3
CSA	∘	∘	∘	4 850 431	2 576 355	11 090	11 500	356.9	16.8
CSA	•	∘	∘	2 908 731	2 576 355	11 090	11 500	279.7	12.4
CSA	•	•	∘	2 908 731	2 576 355	11 090	11 500	279.7	9.7
TE	–	–	•	1 391 761	—	385 641	—	66.8	64.4
TD	–	–	•	158 840	—	68 038	—	7.2	10.9
TD-col	–	–	•	43 238	—	11 602	—	2.1	4.1
CSA	•	•	•	420 263	126 983	5 574	7 005	26.6	2.0
CSA	•	×	•	420 263	126 983	5 574	7 005	26.6	1.8

Dijkstra (TE) with two vertices per connection [15] and footpaths [14], realistic time-dependent Dijkstra (TD), and time-dependent Dijkstra using the optimized coloring model [6] (TD-col). For CSA, we distinguish between scanned, reachable and relaxed connections. Algorithms in Table 2 are grouped into blocks.

The first considers one-to-all queries, and we see that basic CSA scans *all* connections (4.8 M), only half of which are reachable. On the other hand, TE scans about half of the graph's arcs (20 M). Still, this is a factor of four more entities due to the modeling overhead of the time-expanded graph. Regarding query time, CSA greatly benefits from its simple data structures and lack of priority queue: It is a factor of 52 faster than TE. Enabling the start criterion reduces the number of scanned connections by 40 %, which also helps query time. Using trip bits increases spatial locality and further reduces query time to 9.7 ms. We observe that just a small fraction of scanned arcs/connections actually improve stop labels. Only then CSA must consider footpaths. The second block considers one-to-one queries. Here, the number of connections scanned by CSA is significantly smaller; journeys in London rarely have long travel times. Since our London instance does not have minimum change times, we may remove trip tracking from the algorithm entirely. This yields the best query time of 1.8 ms on average. Although CSA compares significantly more labels, it outperforms Dijkstra in almost all cases (also see Table 4 for other inputs). Only for one-to-all queries on London TD-col is slightly faster than CSA.

Profile and Multi-Criteria Queries. In Table 3 we report experiments for (multi-criteria) profile queries on London. Other instances are available in Table 4. We compare CSA to SPCS-col [6] (an extension of TD-col to profile queries) and rRAPTOR [7] (an extension of RAPTOR to multi-criteria profile queries).

Table 3. Figures for the (multi-criteria) profile problem on London. "# Tr." is the max. number of trips considered. "Arr." indicates minimizing arrival time, "Tran." transfers. "Prof." indicates profile queries. "# Jn." is the number of Pareto-optimal journeys.

Algorithm	# Tr.	Arr.	Tran.	Prof.	One.	# Jn.	# L.Cmp. p. Stop	Time [ms]
SPCS-col	–	●	○	●	○	98.2	477.7	1 262
SPCS-col	–	●	○	●	●	98.2	372.5	843
pCSA-P	–	●	○	●	○	98.2	567.6	177
pCSA-P	–	●	○	●	●	98.2	436.9	161
pCSA-C	–	●	○	●	–	98.2	1 912.5	134
pCSA-CT	–	●	○	●	–	98.2	1 912.5	104
rRAPTOR	8	●	●	●	○	203.4	1 812.5	1 179
rRAPTOR	8	●	●	●	●	203.4	1 579.6	878
rRAPTOR	16	●	●	●	●	206.4	1 634.0	922
mcpCSA-CT	8	●	●	●	–	203.4	15 299.8	255
mcpCSA-CT-SSE	8	●	●	●	–	203.4	1 912.5	221
mcpCSA-CT-SSE	16	●	●	●	–	206.4	3 824.9	466

Note that in [7] rRAPTOR is evaluated on two-hours range queries, whereas we compute full profile queries. A first observation is that, regarding query time, one-to-all SPCS is outperformed by all other algorithms, even those which additionally minimize the number of transfers. Similarly to our previous experiment, CSA generally does more work than the competing algorithms, but is, again, faster due to its cache-friendlier memory access patterns. We also observe that one-to-all pCSA-C is slightly faster than pCSA-P, even with target pruning enabled, although it scans 2.7 times as many connections because of expanded footpaths. Note, however, that the figure for pCSA-C does not include the post-processing that removes dominated journeys. Time indexing further accelerates pCSA-C, indicating that the algorithm is, indeed, memory-bound. Regarding multi-criteria profile queries, doubling the number of considered trips also doubles both CSA's label comparisons and its running time. For rRAPTOR the difference is less (only 12 %)—most work is spent in the first eight rounds. Indeed, journeys with more than eight trips are very rare. This justifies mcpCSA-CT-SSE with eight trips, which is our fastest algorithm (221 ms on average). Note that using an AVX2 processor (announced for June 2013), one will be able to process 256 bit-vectors in a single instruction. We, therefore, expect mcpCSA-CT-SSE to perform better for greater numbers of trips in the future.

Minimum Expected Arrival Time. In Table 5 we present figures for the MEAT problem on all instances. Besides running time, we also report output complexity in terms of number of stops and arcs of the decision graph (see Fig. 1 for an example). Real world delay data was not available to us. Hence, we follow Disser et al. [9] and assume that the probability of a train being delayed by t minutes (or less) is $0.99 - 0.4 \cdot \exp(-t/8)$. After 30 min (10 min on London) we set this value to 1. Moreover, we also evaluate performance when discretizing

Table 4. Evaluating other instances. Start criterion and trip flags are always used.

Algorithm	#Tr.	Arr.	Tran.	Prof.	One.	Germany #Jn.	#L.Cmp. p. Stop	Time [ms]	Europe #Jn.	#L.Cmp. p. Stop	Time [ms]
TE	−	•	○	○	○	1.0	317.0	117.1	0.9	288.6	624.1
TD-col	−	•	○	○	○	1.0	11.9	3.5	0.9	10.0	21.6
CSA	−	•	○	○	○	1.0	228.7	3.4	0.9	209.5	19.5
TE	−	•	○	○	•	1.0	29.8	11.7	0.9	56.3	129.9
TD-col	−	•	○	○	•	1.0	6.8	2.0	0.9	5.3	11.5
CSA	−	•	○	○	•	1.0	40.8	0.8	0.9	74.2	8.3
pCSA-CT	−	•	○	•	−	20.2	429.5	4.9	11.4	457.6	46.2
rRAPTOR	8	•	•	•	○	29.4	752.1	161.3	17.2	377.5	421.8
rRAPTOR	8	•	•	•	•	29.4	640.1	123.0	17.2	340.8	344.9
mcpCSA-CT-SSE	8	•	•	•	−	29.4	429.5	17.9	17.2	457.6	98.2

Table 5. Evaluating pCSA-P for the MEAT problem on all instances

Network	Max. Delay [min]	Decision Graph # Stops	# Arcs	All-To-One Time [ms]	One-To-One Time [ms]	One-To-One Dis. Time [ms]
Germany	30	8	19	68.1	31.0	24.6
Europe	30	20	46	205.0	169.0	112.0
London	10	2 724	30 243	668.0	491.0	272.0

the probability function at 60 equidistant points [9]. We run pCSA-P on 10 000 random queries and evaluate both the all-to-one and one-to-one (with earliest arrival pruning enabled) setting. Regarding output complexity, on the German and European networks the resulting decision graphs are sufficiently small to be presented to the user. They consist of 8 stops and 19 arcs on average (Germany), roughly doubling on Europe. However, for London these figures are impractically large, increasing to 2 724 (stops) and 30 243 (arcs). Note that in a dense metropolitan network (such as London), trips operate much more frequently, therefore, many more alternate (and fall-back) journeys exist. These must all be captured by the output. Regarding query time, pCSA-P computes solutions in under 205 ms on Germany and Europe for all scenarios. On London, all-to-one queries take 668 ms, whereas one-to-one queries can be computed in 272 ms time. Note that all values are still practical for interactive scenarios.

7 Final Remarks

In this work, we introduced the Connection Scan framework of algorithms (CSA) for several public transit route planning problems. One of its strengths is the conceptual simplicity, allowing easy implementations. Yet, it is sufficiently flexible to handle complex scenarios, such as multi-criteria profile queries. Moreover, we introduced the MEAT problem which considers stochastic delays and asks for a robust set of journeys minimizing (in its entirety) the user's expected arrival time. We extended CSA to MEAT queries in a sound manner. Our experiments on the metropolitan network of London revealed that CSA is faster than existing approaches, and computes solutions to the MEAT problem surprisingly fast in 272 ms time. All scenarios considered are fast enough for interactive applications. For future work, we are interested in investigating network decomposition techniques to make CSA more scalable, as well as more realistic delay models. Also, since CSA does not use a priority queue, parallel extensions seem promising. Regarding multimodal scenarios, we like to combine CSA with existing techniques developed for road networks.

References

1. Bast, H.: Car or Public Transport – Two Worlds. In: Albers, S., Alt, H., Näher, S. (eds.) Efficient Algorithms. LNCS, vol. 5760, pp. 355–367. Springer, Heidelberg (2009)
2. Bast, H., Carlsson, E., Eigenwillig, A., Geisberger, R., Harrelson, C., Raychev, V., Viger, F.: Fast Routing in Very Large Public Transportation Networks using Transfer Patterns. In: de Berg, M., Meyer, U. (eds.) ESA 2010, Part I. LNCS, vol. 6346, pp. 290–301. Springer, Heidelberg (2010)
3. Bauer, R., Delling, D., Wagner, D.: Experimental Study on Speed-Up Techniques for Timetable Information Systems. Networks 57(1), 38–52 (2011)
4. Berger, A., Delling, D., Gebhardt, A., Müller–Hannemann, M.: Accelerating Time-Dependent Multi-Criteria Timetable Information is Harder Than Expected. In: ATMOS. OpenAccess Series in Informatics (OASIcs) (2009)
5. Berger, A., Gebhardt, A., Müller–Hannemann, M., Ostrowski, M.: Stochastic Delay Prediction in Large Train Networks. In: ATMOS, pp. 100–111 (2011)
6. Delling, D., Katz, B., Pajor, T.: Parallel Computation of Best Connections in Public Transportation Networks. ACM JEA (2012) (to appear)
7. Delling, D., Pajor, T., Werneck, R.F.: Round-Based Public Transit Routing. In: ALENEX, pp. 130–140. SIAM (2012)
8. Delling, D., Sanders, P., Schultes, D., Wagner, D.: Engineering Route Planning Algorithms. In: Lerner, J., Wagner, D., Zweig, K.A. (eds.) Algorithmics. LNCS, vol. 5515, pp. 117–139. Springer, Heidelberg (2009)
9. Disser, Y., Müller–Hannemann, M., Schnee, M.: Multi-Criteria Shortest Paths in Time-Dependent Train Networks. In: McGeoch, C.C. (ed.) WEA 2008. LNCS, vol. 5038, pp. 347–361. Springer, Heidelberg (2008)
10. Geisberger, R.: Contraction of Timetable Networks with Realistic Transfers. In: Festa, P. (ed.) SEA 2010. LNCS, vol. 6049, pp. 71–82. Springer, Heidelberg (2010)
11. Goerigk, M., Knoth, M., Müller–Hannemann, M., Schmidt, M., Schöbel, A.: The Price of Robustness in Timetable Information. In: ATMOS, pp. 76–87 (2011)
12. HaCon website (2013), http://www.hacon.de/hafas/
13. London Data Store, http://data.london.gov.uk
14. Müller-Hannemann, M., Schulz, F., Wagner, D., Zaroliagis, C.: Timetable Information: Models and Algorithms. In: Geraets, F., Kroon, L.G., Schoebel, A., Wagner, D., Zaroliagis, C.D. (eds.) Railway Optimization 2004. LNCS, vol. 4359, pp. 67–90. Springer, Heidelberg (2007)
15. Pyrga, E., Schulz, F., Wagner, D., Zaroliagis, C.: Efficient Models for Timetable Information in Public Transportation Systems. ACM JEA 12(2.4), 1–39 (2008)
16. Sommer, C.: Shortest-Path Queries in Static Networks (2012) (submitted), Preprint available at http://www.sommer.jp/spq-survey.html

Transit Node Routing Reconsidered[*]

Julian Arz, Dennis Luxen, and Peter Sanders

Karlsruhe Institute of Technology (KIT), 76128 Karlsruhe, Germany
julian.arz@student.kit.edu, {luxen,sanders}@kit.edu

Abstract. Transit Node Routing (TNR) is a fast and exact distance
oracle for road networks. We show several new results for TNR. First,
we give a surprisingly simple implementation fully based on contraction
hierarchies that speeds up preprocessing by an order of magnitude ap-
proaching the time for just finding a contraction hierarchy (which alone
has two orders of magnitude larger query time). We also develop a very
effective purely graph theoretical locality filter without any compromise
in query times. Finally, we show that a specialization to the online many-
to-one (or one-to-many) shortest path problem.

1   Introduction and Related Work

Route planning in road networks has seen a lot of results from the algorithm
engineering community in recent years. With Dijkstra's seminal algorithm being
the baseline, a number of techniques preprocess the static input graph to achieve
drastic speedups. *Contraction Hierarchies (CH)* [1,2] is a speedup-technique that
has a convenient trade-off between preprocessing effort and query efficiency. Road
networks with millions of nodes and edges can be preprocessed in mere minutes while queries run in about a hundred microseconds. Transit Node Routing
(TNR) [3] is one of the fastest speed-up techniques for shortest path distance
queries in road networks. By preprocessing the input road network even further,
it yields *almost* constant-time queries, in the sense that nearly all queries can
be answered by a small number of table lookups. For a given node the long distance connections almost always enter an arterial network connecting a set of
important nodes – the *access nodes*. The set of these entrances for a particular
node is small on average. The union of all access node set is call the *transit
node set*. Once the this set is identified, a mapping from each node to its access
nodes and pair-wise distances between all transit nodes is stored. Preprocessing
needs to compute a table of distances between the transit nodes, the distances
to the access nodes, and information for a so-called *locality filter*. The filter indicates whether the shortest path might not cross any transit nodes, requiring
an additional local path search.

Many TNR variants have a common drawback that preprocessing time for
TNR is significantly longer than for the underlying speed-up technique. Another
weakness is that the locality filter requires geometric information on the position

---

[*] This work was partially supported by DFG Grant 933/2.

V. Bonifaci et al. (Eds.): SEA 2013, LNCS 7933, pp. 55–66, 2013.
© Springer-Verlag Berlin Heidelberg 2013

of the nodes [4,5,6]. The presence of a geometric component in an otherwise purely graph theoretical method can be viewed as awkward. There are several examples of geometric ingredients in routing techniques being superseded by more elegant and effective graph theoretical ones [7,8] with the locality filter of TNR being the only *survivor* that is still competitive. Geisberger [6] uses CHs to define transit node sets and for local searches, but still uses a geometric locality filter and relies on Highway Hierarchies [9] for preprocessing. In lecture slides [10], Bast describes a simple variant of CH-based preprocessing exploring a larger search space than ours which also computes a super-set of the access nodes as it omits post-search-stalling. No experiments are reported. The geometric locality filter is not touched. In Section 4 we remove all these qualifications and present a simple fully CH-based variant of TNR. It yields surprisingly good preprocessing times and allows for a very effective fully graph-theoretical locality filter. A related technique is *Hub Labeling (HL)* [11] which stores sorted CH search spaces, intersecting them to obtain a distance. A sophisticated implementation can be made significantly faster than TNR since it incurs less cache faults.

In Section 5 we further accelerate TNR queries for the special case that there are many queries with fixed target (or source). This method can be even faster than HL without incurring its space overhead. This can be compared with RPHAST [12] which is also very fast but only works in a *batched* fashion, i.e., RPHAST needs to know all involved nodes in advance whereas our method can decide about the source nodes for every individual s–t query without having to update any precomputed information.

2 Preliminaries

We model the road network as a directed graph $G = (V, E)$, with $|E| = m$ edges and $|V| = n$ nodes. Nodes correspond to locations, e.g. a junction, and edges to the connections between them. Each edge $e \in E$ has an associated cost $c(e)$, where $c : E \to \mathbb{R}^+$. It is called the *weight*. A path $P = \langle s, v_1, v_2, \ldots, t \rangle$ in G is a sequence of nodes such that there exists an edge between each node and the next one in P. The length of a path $c(P)$ is the sum its edge weights. A path with minimum cost between $s, t \in V$ is called a *shortest* path and denoted by $d(s, t)$ with cost $\mu(s, t)$. Note that a shortest path need not be unique. A path $P = \langle v_0, v_1, \ldots, v_p \rangle$ is called *covered* by a node $v \in V$ if and only if $v \in P$.

2.1 Contraction Hierarchies

Contraction Hierarchies heuristically order the nodes by some measure of importance and *contract* them one by one in this order. Contracting means that a node is removed (temporarily) and as few *shortcut* edges as possible are inserted to preserve shortest path distances. The CH search graph is the union of the set of original edges and the set of shortcuts with edges only leading to more important nodes. This graph is a directed acyclic graph (DAG). An important structural property of CHs is that for any two nodes s and t, if there is an s–t-path at all, then there is also a shortest up-down path s–m–t where s–m uses only upward edges and m–t uses only downward edges in the CH. The *meeting node* m is the highest node on this path in the CH. This up-down-path can be found by a variant of bidirectional Dijkstra where only edges to more important nodes are relaxed. The search can be pruned at nodes that are more far away from s or t than the best up-down-path seen so far.

Although the search spaces explored in CH queries are rather small, there is a simple technique called *stall-on-demand* [7] that further prunes search spaces. We use a simplified version of that technique, which leads to queries as fast as those reported in [13]. For every node v that is the end point of a relaxed edge (u, v) it is checked if there exists a downward edge (w, v). The edge is not relaxed if the tentative distance of w plus the edge weight of (w, v) is less than the tentative distance of u plus edge (u, v). If such a node w exists, edge (u, v) can't be part of a shortest path and thus v is not added into the queue. This is done by scanning the edges incident to v. Computing a table of all pair-wise shortest path distances for a set of nodes can be done by running a quadratic number of queries. While this is already significantly faster with CH than with a naive implementation of Dijkstra's algorithm, tables can be computed much more efficiently with the two-phase algorithm of Knopp et al. [14]. Given a set of sources S and targets T, computing large distance tables is a matter of seconds since only $O(|S| + |T|)$ half searches have to be conducted. The quadratic overhead to initialize and update table entries is close to none for $S \cup T = \mathcal{O}(\sqrt{n})$. We refer the interested reader to [2,14] for further details of the method.

3 Transit Node Routing

TNR in itself is not a complete algorithm, but a framework. A concrete instantiation has to find solutions to the following degrees of freedom: It has to identify a set of transit nodes. It has to find access node for all nodes. And it has to deal with the fact that some queries between nearby nodes cannot be answered via the transit nodes set. We now define and introduce the generic TNR framework, conceive a concrete instantiation and then discuss an efficient implementation.

Definition 1. *Formally, the generic TNR framework consists of*

1. *A set $\mathcal{T} \subseteq V$ of transit nodes.*
2. *A distance table $D_{\mathcal{T}} : \mathcal{T} \times \mathcal{T} \to \mathbb{R}_0^+$ of shortest path distances between the transit nodes.*
3. *A forward (backward) access node mapping $A^{\uparrow} : V \to 2^{\mathcal{T}}$ $(A^{\downarrow} : V \to 2^{\mathcal{T}})$. For any shortest s–t-path P containing transit nodes, $A^{\uparrow}(s)$ $(A^{\downarrow}(t))$ must contain the first (last) transit node on P.*
4. *A locality filter $\mathcal{L} : V \times V \to \{true, false\}$. $\mathcal{L}(s, t)$ must be true when no shortest path between s and t is covered by a transit node. False positives are allowed, i.e., $\mathcal{L}(s, t)$ may sometimes be true even when a shortest path contains a transit node.*

Note that we use a simplified version of the generic TNR framework. A more detailed description is in Schultes' Ph.D. dissertation [7]. During preprocessing, $\mathcal{T}$, $D_\mathcal{T}$, $A^\uparrow$, $A^\downarrow$, and some information sufficient to evaluate $\mathcal{L}$ is precomputed. An s–t-query first checks the locality filter. If $\mathcal{L}$ is true, then some fallback algorithm is used to handle the local query. Otherwise,

$$\mu(s,t) = \mu_{min}(s,t) := \min_{\substack{a_s \in A^\uparrow(s) \\ a_t \in A^\downarrow(t)}} \{d_{A^\uparrow}(s,a_s) + D_\mathcal{T}(a_s,a_t) + d_{A^\downarrow}(a_t,t)\}. \quad (1)$$

4   CH Based TNR

Our TNR variant (CH-TNR) is based on CH and does not require any geometric information. We start by selecting a set of transit nodes. Note that local queries are only implicitly defined and we find a locality filter to classify them. For simplicity, we assume that the graph is strongly connected.

Selection of Transit Nodes. CHs order the nodes in such a way that nodes occurring in many shortest paths are moved to the upper part of the hierarchy. Hence, CHs are a natural choice to identify a small node set which covers many shortest paths in the road network. We chose a number of transit nodes $|\mathcal{T}| = k$ and select the highest k nodes from the CH data structure [2]. This choice of $\mathcal{T}$ also allows us to exploit valuable structural properties of CHs. A distance table of pair-wise distances is built on this set with a CH-based implementation of the many-to-many algorithm of Knopp et al. [14].

Finding Access Nodes. We only explain how to find forward access nodes from a node s. The computation of backward access nodes works analogously. We will show that the following simple and fast procedure works: Run a forward CH query from s. Do not relax edges leaving transit nodes. When the search runs out of nodes to settle, report the settled transit nodes.

Lemma 1. *The transit nodes settled by the above procedure find a superset of the access nodes of s together with their shortest path distance.*

Proof. Consider a shortest s–t-path $P := \langle s, \ldots, t \rangle$ that is covered by a node $u \in \mathcal{T}$. Assume that u is the highest transit node on P. A fundamental property of CH is that we can assume P to consist of upward edges leading up to u followed by downward edges to t. Moreover, the forward search of a CH query finds a shortest path to u. Thus, a CH query also finds a shortest path to the first transit node v on P. It remains to show that the pruned forward search of CH-TNR preprocessing does not prune the search before settling v. This is the case since pruning only happens when settling transit nodes and we have defined v to be the first transit node on P.

This superset of access nodes is then reduced using *post-search-stalling* [7]: $\forall t_1, t_2 \in A^\uparrow(v)$: if $d_{A^\uparrow}(v,t_1) + D_\mathcal{T}(t_1,t_2) \le d_{A^\uparrow}(v,t_2)$, discard access node t_2.

Lemma 2. *Post-search-stalling yields a set of access nodes that is minimal in the sense that it only reports nodes that are the first transit node on some shortest path starting on s.*

Proof. Consider a transit node t that is found by our search which is not an access node for s, i.e., there is an access node u on every shortest path from s to t. By Lemma 1, our pruned search found the shortest path to u but did not relax edges out of u. Hence, the only way t can be reported is that it is reported with a distance larger than the shortest path length. Hence, t will be removed by post-search-stalling.

Query. Equation (1) is used to compute an upper bound on the shortest path distance. When the locality filter returns true, an additional local CH query is performed that is pruned at transit nodes.

Search Space Based Locality Filter. Consider a shortest path query from s to t. Let $\bar{S}_\uparrow(s)$ denote the sub-transit-node search space considered by a CH query from s, i.e., those nodes v settled by the forward search from s which are not transit nodes. Analogously, let $\bar{S}_\downarrow(t)$ denote the sub-transit-node CH search space backwards from t. If these two node sets are disjoint, all shortest up-down-paths from s to t must meet in the transit node set and hence, we can safely set $\mathcal{L}(s,t) = $ false. Conversely, if the intersection is non-empty, there might be a meeting node below the transit nodes corresponding to a shortest path not covered by a transit node. A very simple locality filter can be implemented by storing the sub-transit-node search spaces which are computed for finding the access nodes anyway.

Lemma 3. *The locality filter described above fulfils Definition (1).*

Proof. We assume for $s, t \in V \backslash \mathcal{T}$, the distance $\mu(s,t) \neq \mu_{\min}(s,t)$ and thus $\mu(s,t) < \mu_{\min}(s,t)$. Then the meeting node m of a CH query is not a transit node, and it has to be in the forward search space for s, $\bar{S}_\uparrow(s)$ and in the backward search space for t, $\bar{S}_\downarrow(t)$. Hence, $\bar{S}_\uparrow(s) \cap \bar{S}_\downarrow(t) \neq \emptyset$.

The opposite is not true: There can be false positives, when the intersection of two search spaces $S_\uparrow(s)$ and $S_\downarrow(t)$ includes a non-transit-node $\bar{u}$, but the actual meeting node is a transit node. Then, $\mu(s,t) = \mu_{\min}(s,t)$. Preliminary experiments indicate that the average size of these search spaces are much smaller than the full search spaces, e.g. 32 instead of 112 in the main test instance from Section 6. For the locality filter, only node IDs need to be stored. Compared to hub labelling which has to store full search spaces and also distances to nodes this is already a big space saving.

If we are careful to number the nodes in such a way that nearby nodes usually have nearby IDs, the node IDs appearing in a search space will often come from a small range. We compute and store these values in order to facilitate the following *interval check*: When $[\min(\bar{S}_\uparrow(s)), \max(\bar{S}_\uparrow(s))] \cap [\min(\bar{S}_\downarrow(t)), \max(\bar{S}_\downarrow(t))] = \emptyset$, we immediately know that the search spaces are disjoint. As the sole locality filter this would incur too many false positives but it works sufficiently often to drastically reduce the average overhead for the locality filter. We now discuss a much more accurate *lossy* compression of the search spaces.

Graph Voronoi Label Compression. Note that the locality filter remains correct when we add nodes to the search spaces. We do this by partitioning the graph into regions and define the extended search space as the union of all regions that contain a search space node. This helps compression since we can represent a region using a single id, e.g., the number of a node representing the region. This also speeds up the locality filter since instead of intersecting the search spaces explicitly, it now suffices to intersect the (hopefully smaller) sets of block ids. Hence, we want partitions that are large enough to lead to significant compression, yet small and compact enough to keep the false positive rate small. Our solution if a purely graph theoretical adaptation of a geometric concept. Our blocks are *graph Voronoi regions* of the transit nodes. Formally,

$$\text{Vor}(v) := \{u \in V : \forall w \in \mathcal{T} \setminus \{v\} : \mu(u, v) \leq \mu(u, w)\}$$

for $v \in \mathcal{T}$ with ties broken arbitrarily. The intuition behind this is that a positive result of the locality filter means that the search spaces of start and destination come at least close to each other. Computing the Voronoi regions is easy, using a single Dijkstra run with multiple sources on the reversed input graph, as shown by Mehlhorn [15]. We call this filter the *graph Voronoi filter*.

5   TNR Based Many-to-One Computations

Consider a scenario where we have to find many s–t-shortest path distances for a fixed t. The case for fixed s works analogously. For example, this might be interesting for generalizations of A* search to multiple criteria where we can use exact single-criteria searches for pruning the search space. Although we can use Dijkstra's algorithm here (one backward search from t) for precomputing all single-criteria distances, this is expensive when the A* search touches only a small fraction of the nodes.

The idea is to precompute v–t-distances for all transit nodes $v \in \mathcal{T}$ and to store them in a separate array T. This can be done using $|A^{\downarrow}(v)| \cdot |\mathcal{T}|$ table lookups accessing only $|A^{\downarrow}(t)|$ rows of the distance table. Note that T is likely to fit into cache. For a fast locality filter specialized for a particular target node, one can employ highly localized backward search from t, explicitly precomputing the nodes requiring a local query. A non-local query with distance can compute the distance as

$$\mu(s, t) = \min_{a \in A^{\uparrow}(s)} d_{A^{\uparrow}}(s, a) + T[a].$$

Note that this takes time linear rather than quadratic in the number of access nodes and only incurs cache faults for scanning $A^{\uparrow}(s)$. Preliminary experiments indicate that this method can yield an order of magnitude in query time improvement compared to TNR (to around 100ns for the European instance).

Table 1. Scalability experiment with 10 000 transit nodes

Cores	CH			Dist. Table			Exploration			Total		
	[s]	Spdp	Eff.	[s]	Spdp	Eff.	[s]	Spdp	Eff.	[s]	Spdp.	Eff.
1	513	1	1	9.0	1	1	500	1	1	1046	1	1
2	281	1.83	0.91	5.1	1.74	0.88	287	1.74	0.87	596	1.75	0.88
3	203	2.53	0.84	3.9	2.23	0.76	202	2.48	0.83	432	2.42	0.81
4	160	3.20	0.80	2.9	3.16	0.79	145	3.43	0.86	334	3.13	0.78
4 (HT)	137	3.75	0.47	2.2	4.01	0.50	101	4.93	0.62	265	3.95	0.49

6   Experimental Evaluation

We implement our algorithms and data structures in C++ and test the performance on a real-world data set. The source code is compiled with GCC 4.6.1 setting optimization flags -O3 and mtune=native. Our test machine is an Intel Core i7-920, clocked at 2.67 GHz with four cores and 12 GiB of RAM. It runs Linux kernel version 2.6.34

Our CH variant implements the shared-memory parallel preprocessing algorithm of Vetter [16] with a hop limit of 5 and 1 000 settled nodes for witness searches and 7 hops or 2 000 settled nodes during the actual contraction of nodes. The priority function is

$$2 \cdot \text{edgeQuotient} + 4 \cdot \text{originalEdgeQuotient} + \text{nodeDepth}.$$

We experiment on the road network of Western Europe provided for the 9th DIMACS challenge on shortest paths [17] by PTV AG. The graph consists of about 18 015 449 nodes and 22 413 128 edges with 32 bit integer edge weights representing travel time. The resulting hierarchy has 39 256 327 edges. The following experiments are conducted with a transit node set of size 10 000, if not mentioned otherwise, because key results from previous work were based on the same number of transit nodes, e.g. [5].

The following design choices are used throughout the experiments. Forward and backward search spaces are merged into one set for the locality filter. Forward and backward access node sets are also merged into one set. But note that these two sets are distinct in our implementation.

As the ID of a node does not contain any particular information, node IDs can be changed to gain algorithmic advantages. This *renumbering* is done by applying a bijective permutation on the IDs, in order to ensure that each ID stays unique. We alter the labels of the nodes in V so that $\mathcal{T} = \{0, \ldots, k - 1\}$. By proceeding this way, we can easily determine if a node v is a transit node or not (during further preprocessing and during the query): $v \in \mathcal{T}$ if and only if $v < k$. $\mathcal{T}$ is renumbered with the so-called *input-level strategy*, while $V \backslash \mathcal{T}$ is ordered by the (greedy) DFS Increasing strategy. This makes the interval check a very effective "pre-filter" for the locality filter.

We test the scalability of parallel preprocessing for a varying number of cores in Table 1. The raw results of parallelizable parts (preprocessing, distance table generation and exploration) have a quite high variance of about 10%. Hence, we measured the preprocessing five times and averaged over all runs. The values reported in column *Total* are the sum of the respective averages. Column *Cores* gives the numbers of cores used. Columns *CH, Dist. Table, Exploration* measure time, speedup and efficiency of the respective parts. The bottom line reports on four CPUs with activated hyper-threading (HT).

We see that the total preprocessing time is only about a factor of two larger than plain CH preprocessing. Most additional work is due to search space exploration from each node. We see that the different parts of the algorithm scale well with an increasing number of cores. The total efficiency is slightly lower than the efficiency of the individual parts, as it includes about 23.6 seconds of non-parallelized work due to the Voronoi computation. It does not reflect the performance of real cores, but HT comes virtually for free with modern commodity processors. We choose $1\,000\,000$ million source/target pair at random. The rate of local queries is only 0.58 %. On average a non-local query takes 1.22 μs, while a local query takes 28.6 μs on average. This results in an overall average query time of 1.38 μs and the space overhead amounts to 147 Bytes per node. The high value for local queries is expected behavior.

We compare to previous approaches for our test instance. Some of these implementations were tested on an older AMD machine [7] that was still available for running the queries[1]. Table 2 shows *Reported* values as given in the respective publications denoted by *From*, while columns *Compared* give preprocessing and running times either done on or normalized to the aforementioned AMD machine using one core. Therefore, similar to the methodology in [18], a scaling factor of 1.915 is determined by measuring preprocessing and query times on both machines using a smaller graph (of Germany). Scaled numbers are indicated by a star symbol: $\star$. Values for CH were measured with our implementation. The simplest TNR implementation is GRID-TNR that splits the input graph into grid cells and computes a distance table between the cells border nodes. Note that the numbers for GRID-TNR were computed on a graph of the USA, but the characteristics should be similar to our test instance. Preprocessing is prohibitively expensive while the query is about 20 times slower than ours. The low space consumption is due to the fact that it is trivial to construct a locality filter for grid cells. For HH-TNR [7] and TNR+AF [18], preprocessing is single-threaded. The corresponding scaling factor for preprocessing is 3.551 and the fastest HH based TNR variant is still slower by about a factor of two for preprocessing and queries. Note that the HH-based methods all implement a highly tuned TNR variant with multiple levels that is much more complex than our method. While TNR+AF has faster queries by about 25%, the (scaled) preprocessing is about an order of magnitude slower and the space overhead is twice as much. Also, TNR+AF requires a sophisticated implementation with a partitioning step and the computation of arc flags.

---

[1] Note that a current off-the-shelf commodity machine is about 2–3 times faster.

Table 2. Comparison Between Various Distance Oracles

Method	From	Preprocessing			Query	
		Reported [hh:mm]	Compared [hh:mm]	Overhead [byte]	Reported [μs]	Compared [μs]
CH	-	00:03	00:05	24	103	246
Grid-TNR	[5]	≈20:00	≈20:00	21	63	63
HH-TNR-eco	[7]	00:25	00:25	120	11	11
HH-TNR-gen	[7]	01:15	01:15	247	4.30	4.30
TNR+AF	[18]	03:49	03:49	321	1.90	1.90
HL-0 local	[19]	00:03	00:35	1341	0.7	1.34 ⋆
HL-∞ global	[19]	06:12	≈120:00	1055	0.254	0.49 ⋆
HLC	[20]	00:30	00:59	100	2.989	5.74 ⋆
CH-TNR	-	00:05	00:34	147	1.38	3.27

Hub labeling (HL) allows much faster queries than CH-TNR because accessing just two label sets is very cache efficient. However, HL has much higher space consumption. Using sophisticated label compression methods, HLC [20] remedies the space consumption problem however at the cost of becoming *slower* than CH-TNR since decompression incurs a significant number of cache faults. Note that CH-TNR is not highly tuned for space consumption yet. Even the simple methods outlined in Section 7 are likely to equalize the space difference to HLC.

The preprocessing times reported in Table 2 should be interpreted with care since they are executed on different machines and sometimes with parallel preprocessing. In particular, the somewhat faster preprocessing of HL-0 compared to CH-TNR uses three times more cores, 20% faster clock speed and 50% larger L3 cache. Indeed, CH-TNR has somewhat faster sequential preprocessing time (17.4 minutes) than HL-O (17.9 minutes) despite using a slower machine.

The quality of our locality filter is compared to other TNR implementations in Table 3. These variants differ in the number of transit nodes and in the graph

Table 3. Comparison of locality filter quality

Method	From	$|\mathcal{T}|$	Local [%]	False [%]
Grid-TNR	[5]	7 426	2.6	-
Grid-TNR	[5]	24 899	0.8	-
LB-TNR	[21]	27 843	-	-
HH-TNR-eco	[7]	8 964	0.54	81.2
HH-TNR-gen	[7]	11 293	0.26	80.7
CH-TNR	-	10 000	0.58	73.6
CH-TNR	-	24 000	0.17	72.1
CH-TNR	-	28 000	0.14	72.1

used to determine them. Nevertheless, the graphs are road networks and exhibit similar characteristics. The number of transit nodes for CH-TNR is chosen to resemble data from literature. The fraction of local queries of our variant is lower than or on par with numbers from literature. Also, the rate of false positives is much lower than previous work. Most noteworthy, the recent method of LB-TNR applies sophisticated optimization techniques, but does not produce a transit node set with superior locality as the rate of local queries is virtually the same.

7   Conclusions and Future Work

We have shown that a very simple implementation of CH-TNR yields a speedup technique for route planning with an excellent trade-off between query time, preprocessing time, and space consumption. In particular, at the price of twice the (quite fast) preprocessing time of contraction hierarchies, we get two orders of magnitude faster query time. Our purely graph theoretical locality filter outperforms previously used geometric filters. To the best of our knowledge, this eliminates the last remnant of geometric techniques in competitive speedup techniques for route planning. This filter is based on intersections of CH search spaces and thus exhibits an interesting relation to the hub labelling technique.

When comparing speedup techniques, one can view this as a multi-objective optimization problem along the dimensions of query and preprocessing time, space consumption, and simplicity. Any Pareto-optimal, i.e. non-dominated, method is worthwhile considering. Good methods should have a significant advantage with respect to at least one measure without undue disadvantages for any other dimension. In this respect, CH-TNR fares very well. Only hub labelling achieves significantly better query times but at the price of much higher space consumption, in particular when comparable preprocessing times are desired. The simple variants of hub labeling have even worse space consumption and less clear advantages in query time. When looking for clearly simpler techniques than CH-TNR, plain CHs come into mind but at the price of two orders of magnitude larger query time and a surprisingly small gain in preprocessing time.

CH-TNR also has significant potential for further performance improvements. Our variant of CH-TNR focuses on maximal simplicity except for the Voronoi filter which is needed for space efficiency. But there are many further improvements that will not drastically change the position of CH-TNR in the landscape of speedup techniques. But they could yield noticeable improvements with respect to query time, preprocessing time, or space at the price of more complicated implementation. We now outline some of these possibilities:

Query Time: In Section 5 we have seen that for the special case of many-to-one queries can be accelerated by another order of magnitude being the fastest known technique for this use case. But also the general case can be further accelerated. As in [18] we expect about twice faster queries by combining CH-TNR with arc flags for an additional sense of goal direction. The additional preprocessing time could be much smaller than in [18] by using PHAST [22] for fast parallel one-to-all shortest paths. Local queries can be accelerated by introducing additional layers as in HH-TNR. Alternatively, we could use hub labelling for local queries. This is still much more space efficient than full hub labelling and very simple since we need to compute local (sub-transit-node) search spaces anyway. This variant of CH-TNR can be viewed as a generalization of hub labelling that saves space and preprocessing time at the price of larger query times.

Preprocessing Time: Besides CH construction the most time consuming part or CH-TNR preprocessing is the exploration of the sub transit node CH search spaces for finding access nodes and partition representatives. This can probably be accelerated by a top-down computation as in [19]. Note that using post-search-stalling we still get optimal sets of access nodes. Finding Voronoi regions might be parallelizable to some extent since it explores a very low diameter graph.

Space: There are a number of relatively simple low level tuning opportunities here. For example, we can more aggressively exploit overlaps between forward/backward access nodes and search space representatives. These "dual use" nodes need to be stored only in the access nodes set together with a flag indicating that they are also a region representatives. We could also encode backward distances to access nodes as differences to forward distances. As in HH-TNR we could also encode access nodes of most nodes as the union of the access nodes of their neighbors. Further details are given in the technical report [23].

Acknowledgements. We would like to thank Daniel Delling and Renato Werneck for providing additional numbers for the HL methods.

References

1. Geisberger, R., Sanders, P., Schultes, D., Delling, D.: Contraction Hierarchies: Faster and Simpler Hierarchical Routing in Road Networks. In: McGeoch, C.C. (ed.) WEA 2008. LNCS, vol. 5038, pp. 319–333. Springer, Heidelberg (2008)
2. Geisberger, R., Sanders, P., Schultes, D., Vetter, C.: Exact Routing in Large Road Networks Using Contraction Hierarchies. Transportation Science 46(3), 388–404 (2012)
3. Bast, H., Funke, S., Sanders, P., Schultes, D.: Fast Routing in Road Networks with Transit Nodes. Science 316(5824), 566 (2007)
4. Bast, H., Funke, S., Matijevic, D.: TRANSIT - Ultrafast Shortest-Path Queries with Linear-Time Preprocessing. In: Demetrescu, C., Goldberg, A.V., Johnson, D.S. (eds.) 9th DIMACS Implementation Challenge – Shortest Paths (2006)
5. Bast, H., Funke, S., Matijevic, D., Sanders, P., Schultes, D.: In Transit to Constant Shortest-Path Queries in Road Networks. In: Proceedings of the 9th Workshop on Algorithm Engineering and Experiments, ALENEX 2007, pp. 46–59. SIAM (2007)
6. Geisberger, R.: Contraction Hierarchies. Master's thesis, Universität Karlsruhe (2008), http://algo2.iti.kit.edu/1094.php
7. Schultes, D.: Route Planning in Road Networks. PhD thesis, Universität Karlsruhe (February 2008), http://algo2.iti.uka.de/schultes/hwy/schultes_diss.pdf
8. Delling, D., Sanders, P., Schultes, D., Wagner, D.: Engineering Route Planning Algorithms. In: Lerner, J., Wagner, D., Zweig, K.A. (eds.) Algorithmics. LNCS, vol. 5515, pp. 117–139. Springer, Heidelberg (2009)

9. Sanders, P., Schultes, D.: Engineering Highway Hierarchies. ACM Journal of Experimental Algorithmics 17(1), 1–40 (2012)
10. Bast, H.: Lecture slides: Efficient route planning, http://ad-wiki.informatik. uni-freiburg.de/teaching/EfficientRoutePlanningSS2012
11. Abraham, I., Delling, D., Goldberg, A.V., Werneck, R.F.: A Hub-Based Labeling Algorithm for Shortest Paths on Road Networks. In: Pardalos, P.M., Rebennack, S. (eds.) SEA 2011. LNCS, vol. 6630, pp. 230–241. Springer, Heidelberg (2011)
12. Delling, D., Goldberg, A.V., Werneck, R.F.: Faster Batched Shortest Paths in Road Networks. In: 11th Workshop on Algorithmic Approaches for Transportation Modelling, Optimization, and Systems. OpenAccess Series in Informatics (OASIcs), vol. 20, pp. 52–63 (2011)
13. Vetter, C.: Fast and Exact Mobile Navigation with OpenStreetMap Data. Master's thesis, Karlsruhe Institute of Technology (2010)
14. Knopp, S., Sanders, P., Schultes, D., Schulz, F., Wagner, D.: Computing Many-to-Many Shortest Paths Using Highway Hierarchies. In: 9th Workshop on Algorithm Engineering and Experiments, pp. 36–45. SIAM (2007)
15. Mehlhorn, K.: A faster approximation algorithm for the Steiner problem in graphs. Information Processing Letters 27(3), 125–128 (1988)
16. Vetter, C.: Parallel Time-Dependent Contraction Hierarchies. Studienarbeit, Karlsruhe Institute of Technology (2009), http://algo2.iti.kit.edu/download/vetter_sa.pdf
17. Demetrescu, C., Goldberg, A.V., Johnson, D.S. (eds.): The Shortest Path Problem: Ninth DIMACS Implementation Challenge. DIMACS Book, vol. 74. American Mathematical Society (2009)
18. Bauer, R., Delling, D., Sanders, P., Schieferdecker, D., Schultes, D., Wagner, D.: Combining Hierarchical and Goal-Directed Speed-Up Techniques for Dijkstra's Algorithm. ACM Journal of Experimental Algorithmics 15(2.3), 1–31 (2010), Special Section devoted to In: McGeoch, C.C. (ed.) WEA 2008. LNCS, vol. 5038, pp. 303–318. Springer, Heidelberg (2008)
19. Abraham, I., Delling, D., Goldberg, A.V., Werneck, R.F.: Hierarchical Hub Labelings for Shortest Paths. In: Epstein, L., Ferragina, P. (eds.) ESA 2012. LNCS, vol. 7501, pp. 24–35. Springer, Heidelberg (2012)
20. Delling, D., Goldberg, A.V., Werneck, R.F.: Hub Label Compression. In: Bonifaci, V. (ed.) SEA 2013. LNCS, vol. 7933, pp. 17–28. Springer, Heidelberg (2013)
21. Eisner, J., Funke, S.: Transit Nodes – Lower Bounds and Refined Construction. In: Proceedings of the 14th Meeting on Algorithm Engineering and Experiments, ALENEX 2012, pp. 141–149. SIAM (2012)
22. Delling, D., Goldberg, A.V., Nowatzyk, A., Werneck, R.F.: PHAST: Hardware-Accelerated Shortest Path Trees. In: 25th International Parallel and Distributed Processing Symposium, IPDPS 2011 (2011)
23. Arz, J., Luxen, D., Sanders, P.: Transit Node Routing Reconsidered. Technical report, Karlsruhe Institute of Technology (2012)

A New QEA Computing Near-Optimal Low-Discrepancy Colorings in the Hypergraph of Arithmetic Progressions
(Extended Abstract)

Lasse Kliemann[1], Ole Kliemann[1], C. Patvardhan[2],
Volkmar Sauerland[1], and Anand Srivastav[1]

[1] Christian-Albrechts-Universität zu Kiel
Institut für Informatik
Christian-Albrechts-Platz 4
24118 Kiel, Germany
lki@informatik.uni-kiel.de
[2] Dayalbagh Educational Institute
Deemed University
Agra, India
cpatvardhan@googlemail.com

Abstract. We present a new quantum-inspired evolutionary algorithm, the *attractor population QEA* (apQEA). Our benchmark problem is a classical and difficult problem from Combinatorics, namely finding low-discrepancy colorings in the hypergraph of arithmetic progressions on the first n integers, which is a massive hypergraph (e. g., with approx. 3.88×10^{11} hyperedges for $n = 250\,000$). Its optimal low-discrepancy coloring bound $\Theta(\sqrt[4]{n})$ is known and it has been a long-standing open problem to give practically and/or theoretically efficient algorithms. We show that apQEA outperforms known QEA approaches and the classical combinatorial algorithm (Sárközy 1974) by a large margin. Regarding practicability, it is also far superior to the SDP-based polynomial-time algorithm of Bansal (2010), the latter being a breakthrough work from a theoretical point of view. Thus we give the first *practical* algorithm to construct optimal colorings in this hypergraph, up to a constant factor. We hope that our work will spur further applications of Algorithm Engineering to Combinatorics.

Keywords: estimation of distribution algorithm, quantum-inspired evolutionary algorithm, hypergraph coloring, arithmetic progressions, algorithm engineering, combinatorics.

1 Introduction

Experimentation is emerging as a tool in Combinatorics. For example, experimentation is used in a Polymath project on one of the most challenging open

V. Bonifaci et al. (Eds.): SEA 2013, LNCS 7933, pp. 67–78, 2013.
© Springer-Verlag Berlin Heidelberg 2013

problems of Paul Erdős on homogeneous arithmetic progressions. In this paper we contribute to both, experimental algorithms for difficult discrepancy problems and highly-parallel evolutionary computation within the class of *estimation of distribution algorithms* (EDA).

Quantum-inspired evolutionary algorithms (QEA) belong to the class of EDAs, more precisely to the class of *univariate* EDAs. An EDA maintains a probability distribution, also called *model*, μ on the set of possible solutions, say $\{0,1\}^k$. Sampling μ yields concrete solutions, which can be used to tune μ with the intent to sample better solutions next time. In a univariate EDA, models of a simple kind are considered, namely which treat all of the k coordinates as independent random variables. Thus μ can be represented as a vector $Q = (Q_1, \ldots, Q_k) \in [0,1]^k$ with Q_i stating the probability of sampling 1 in coordinate i. Univariate EDAs have been studied since the 90ies; in 2002 [5], the term "quantum-inspired" was coined, based on the observation that the $Q_1, \ldots, Q_k$ behave similar to k qubits in a quantum computer: each is in a state between 0 and 1, and only upon observation takes on states 0 or 1 with certain probabilities. Hence what we call "sampling" is also called "observing" in the literature. We call the QEA from [5] the *standard QEA* (sQEA). It uses an *attractor*, which is the best solution found so far. The model is tuned towards the attractor in each generation. We stick to the term "quantum-inspired" since our version of univariate EDA also uses the idea of an attractor. A burden that comes with QEAs is the possibility of *premature convergence*, meaning: each Q_i moves close to one of the extremes (0 or 1), so the model Q essentially locks onto one particular solution, before a sufficiently good solution is found – *and* the algorithm does not provide a way to escape this dead end. We will show how our new QEA successfully deals with this problem.

We briefly introduce the hypergraph of arithmetic progressions and the discrepancy problem. Given $a, d, \ell \in \mathbb{N}_0 = \{0,1,2,3,\ldots\}$, the set $A_{a,d,\ell} := \{a + id; \ 0 \leqslant i < \ell\}$ is the *arithmetic progression* (AP) with starting point a, difference d, and length ℓ. It contains exactly ℓ numbers, namely $a, a + d, a + 2d, \ldots, a+(\ell-1)d$. For $n \in \mathbb{N}$ we call $A_n := \{A_{a,d,\ell} \cap \{0, \ldots, n-1\}; \ a, d, \ell \in \mathbb{N}_0\}$ the *set system* or *hypergraph of arithmetic progressions* in the first n integers. Elements of A_n are called *hyperedges* and elements of the ground set $V := \{0, \ldots, n-1\}$ are called *vertices*. The cardinality of A_n is approximately $n^2 \log(n)/2$; we will give a proof in the full version. Often, the ground set is $\{1, \ldots, n\}$ in the literature, but for our purposes starting at 0 is notationally more convenient. A *coloring* is a mapping $\chi : V \longrightarrow \{-1, +1\}$. Given a coloring χ and an AP $E \subseteq V$ we have its *discrepancy* $\mathrm{disc}_\chi(E) := \left| \sum_{v \in E} \chi(v) \right|$. The discrepancy of A_n with respect to χ is $\mathrm{disc}_\chi(A_n) := \max_{E \in A_n} \chi(E)$.

Previous and Related Work. Univariate EDAs have been studied since the 90ies, see, e. g.,, [2,13,7,5,15]. For a recent survey on general EDAs see [8] and the references therein. Particularly influential for our work have been [5] and [15], where sQEA and vQEA are presented, respectively. vQEA extends the attractor concept in a way to allow for better exploration. But for the discrepancy problem, vQEA is not well suited for reasons explained later. In recent years, variants of

QEAs have been successfully used on benchmark as well as on difficult practice problems, see, e. g.,, [1,10,11,14,6].

For $\mathcal{A}_n$, in 1964, it was shown by Roth [16] that there is no coloring with discrepancy below $\Omega(\sqrt[4]{n})$. More than 20 years later, it was shown by Matoušek and Spencer [12] that there exists a coloring with discrepancy $O(\sqrt[4]{n})$, so together with the earlier result we have $\Theta(\sqrt[4]{n})$. The proof is non-constructive, and the problem of efficiently computing such colorings remained open. For many years, Sárközy's approximation algorithm (see [4]) was the best known, provably attaining discrepancy $O(\sqrt[3]{n\log(n)})$; experiments suggest that (asymptotically) it does not perform better than this guarantee. Recently, in a pioneering work, the problem was solved by Bansal [3], using semi-definite programs (SDP). However, Bansal's algorithm requires solving a series of SDPs that grow in the number of hyperedges, making it practically problematic for $\mathcal{A}_n$. In our experiments, even for $n < 100$, it requires several hours to complete, whereas our apQEA (in a parallel implementation) only requires a couple of minutes up to $n = 100\,000$. Computing optimal low-discrepancy colorings for *general* hypergraphs is NP-hard [9].

Our Contribution. We use the problem of computing low-discrepancy colorings in $\mathcal{A}_n$ in order to show limitations of sQEA and how a new form of QEA can successfully overcome these limitations. Our new QEA uses an *attractor population* where the actual attractor is repeatedly selected from. We call it *attractor population QEA* (apQEA). The drawback of sQEA appears to be premature convergence, or in other words, a lack of exploration of the search space. We show that even by reducing the learning rate drastically in sQEA and by using local and global migration, it does not attain the speed and solution quality of apQEA. In addition to the exploration capabilities of apQEA, we show that – with an appropriate tuning of parameters – it scales well in the number of parallel processors: when doubling the number of processor cores from 96 to 192, running times reduces to roughly between 40% and 60%.

We also look at the combinatorial structure of $\mathcal{A}_n$. Based on an idea by Sárközy (see [4]) we devise a modulo coloring scheme, resulting in a search space reduction and faster fitness function evaluation. This, together with apQEA, allows us to compute low-discrepancy colorings that are optimal up to a constant factor, in the range up to $n = 250\,000$ vertices. For this n, the cardinality of $\mathcal{A}_n$ is approx. 3.88×10^{11}, which means a massive hypergraph. Precisely, we compute colorings with discrepancy not more than $3\sqrt[4]{n}$; we call $\lfloor 3\sqrt[4]{n} \rfloor$ the *target discrepancy*. We have chosen factor 3, because this appeared as an attainable goal in reasonable time in preliminary experiments. Better approximations may be possible with other parameters and more processors and/or more time. Colorings found by our algorithm can be downloaded[1] and easily verified.

Our problem sizes are a magnitude beyond that of the Polymath project. Of course, our problem is different, but related and in future work we plan to access the Erdős problem with our approach.

---

[1] http://www.informatik.uni-kiel.de/~lki/discap-results.tar.xz

---

Algorithm 1. sQEA

1   **in parallel for each** *model* $Q = (Q_1, \ldots, Q_k) \in \mathcal{M}$ **do**
2   initialize model $Q := (1/2, \ldots, 1/2)$;
3   initialize attractor $a :=$ random solution;
4   **repeat**
5       $a :=$ best attractor over all models;
6   sample Q yielding $x \in \{0,1\}^k$;
7   **if** $f(x) \leqslant f(a)$ **then**
8   **for** $i = 1, \ldots, k$ **do if** $x_i \neq a_i$ **then**
$$Q_i := \begin{cases} \max\{0, Q_i - \Delta(Q_i)\} & \text{if } a_i = 0 \\ \min\{1, Q_i + \Delta(Q_i)\} & \text{if } a_i = 1 \end{cases}$$
9   **else** $a := x$;
10   **until** *satisfied or hopeless or out of time*;

---

2   Description of Algorithms

Fitness Function and Shortcutting. As fitness function (FF), we use the negative of the discrepancy, so higher fitness is better. The sample space is of the form $\{-1, +1\}^k$, but we will often write $\{0,1\}^k$, where 0 means -1. The concrete choice of k will be explained in Sec. 3. In a QEA, given two solutions x and x^* with known fitness $f(x^*)$, it is often enough to decide whether $f(x) > f(x^*)$ and only in this case it will be required to compute $f(x)$ exactly. If we can determine that $f(x) \leqslant f(x^*)$, then we do not need the exact value of $f(x)$. Since discrepancy involves a maximum, it provides an opportunity for *shortcutting*: as soon as a hyperedge is found in which discrepancy w.r.t. x is at least as high as $\mathrm{disc}(x^*)$, evaluation can be aborted and $f(x) \leqslant f(x^*)$ can be reported. This is a big time-saver, e.g., for $n = 100\,000$ vertices, on average we require about 2 milliseconds for a shortcut FF evaluation and about 790 milliseconds for a full one – and for apQEA there are usually many more shortcut ones than full ones.

Standard QEA (sQEA). A basic version of sQEA [5] is given as Alg. 1. The set $\mathcal{M}$ typically comprises 1 to 100 models; they are distributed among the available processors. For each model, an *attractor* a is maintained. Each iteration of the repeat loop is called a *generation*. In each generation, each of the models is sampled, and if the sample x cannot beat the attractor,[2] *learning* takes place: the model is shifted slightly towards a, where x and a differ. *Linear learning* means using a fixed amount, e.g.,, $\Delta(Q_i) = \Delta = \frac{1}{100}$. In [5,15] *rotation learning* is used: the point $(\sqrt{1-Q_i}, \sqrt{Q_i})$ in the plane is rotated either clockwise (if $a_i = 0$) or counter-clockwise (if $a_i = 1$), and the new value of Q_i becomes the square root of

---

[2] The original description suggests using $f(x) < f(a)$ as the test, so the sample is not required to beat the attractor but to be at least as good as the attractor. We will comment on this later.

the new ordinate. This is inspired by quantum computing; an actual benefit could be that towards the extremes (0 and 1) shifts become smaller. We will test sQEA and vQEA with linear as well as rotation learning and stick to linear learning for apQEA. The *learning resolution* gives the number of possible values that each Q_i can assume inside $[0, 1]$. For linear learning, this is $\frac{1}{\Delta}$. For rotation learning, this is determined by the angle by which we rotate; it is typically between 0.01π and 0.001π. Since the interval is $[0, \pi/2]$, this means a learning resolution between 50 and 500. As an extension, multiple samples can be taken in line 6 and the model is only shifted if none of them beats the attractor (an arbitrary one of them is chosen for the test $x_i \neq a_i$). If one of the samples beats the attractor, the best one is used to update the attractor in line 9. We always use 10 samples.

What happens in line 5 is called *synchronization* or *migration*. Another extension, intended to prevent premature convergence, is the use of *local* and *global migration*. Models are bundled into *groups*, and the attractor of a model Q is set to the best attractor over all models in Q's group (local migration). Only every T_g generations, the best attractor over *all* models is used (global migration). We call T_g the *global migration period*.

The repeat loop stops when we are "satisfied or hopeless or out of time". We are satisfied in the discrepancy problem when the discrepancy is lesser or equal to $3\sqrt[4]{n}$. A possible criterion for hopelessness is when all the models have only very little entropy left. *Entropy* is a measure of randomness, defined as $\sum_{i=1}^{k} -\log(Q_i)$, which is at its maximum k when $Q_i = \frac{1}{2}$ for all i, and at its minimum 0 if $Q_i \in \{0, 1\}$ for all i. In all our experiments with sQEA, we will impose a simple *time limit* guided by the times needed by apQEA.

Versatile QEA (vQEA). vQEA [15] works similar to sQEA with the exception that the attractor update in line 9 is carried out *unconditionally*. The description given in [15] states that the attractor of each model in generation $t + 1$ is the best sample from generation t, over all models. This means that parallel processes have to synchronize after each generation. and also that at least one sample per generation must be fully evaluated, so only limited use of shortcutting is possible.

Attractor Population QEA (apQEA). Our new QEA, the apQEA, is given as Alg. 2. It strikes a balance between the approaches of sQEA and vQEA. In sQEA, the attractor essentially follows the best solution and only changes when better solutions are found, while in vQEA the attractor changes frequently and is also allowed to assume inferior solutions. In apQEA, the *attractor population* $\mathcal{P}$ is a set of solutions. From it, attractors are selected, e. g., using tournament selection. When a sample cannot improve the population ($f(x) \leq f_0$) the model is adjusted. Otherwise the solution is injected into the population. The number of generations that a particular attractor stays in function is called the *attractor persistence*; we fix it to 10 in all our experiments. Note that apQEA will benefit from shortcutting since $f(x)$ has only to be computed when $f(x) > f_0$. Note also that it is appropriate to treat the models *asynchronously* in apQEA, hence preventing idle time: the attractor population is there, any process may inject into it or select from it at any time (given an appropriate implementation).

Algorithm 2. apQEA

---

1 randomly initialize attractor population $\mathcal{P} \subseteq \{0,1\}^k$ of cardinality S;

2 **in parallel for each** *model* $Q = (Q_1, \ldots, Q_k) \in \mathcal{M}$ **do**

3   initialize model $Q := (1/2, \ldots, 1/2)$;

4   **repeat**

5         $a :=$ select from $\mathcal{P}$;

6   **do** *10* **times**

7             $f_0 :=$ worst fitness in $\mathcal{P}$;

8   sample Q yielding $x \in \{0,1\}^k$;

9   **if** $f(x) \leqslant f_0$ **then**

10   **for** $i = 1, \ldots, k$ **do if** $x_i \neq a_i$ **then**

$$Q_i := \begin{cases} \max\{0,\, Q_i - \Delta(Q_i)\} & \text{if } a_i = 0 \\ \min\{1,\, Q_i + \Delta(Q_i)\} & \text{if } a_i = 1 \end{cases}$$

11   **else**

12   inject x into $\mathcal{P}$;

13   trim $\mathcal{P}$ to the size of S, removing worst solutions;

14   **until** *satisfied or hopeless or out of time*;

A very important parameter is the size S of the population. We will see in experiments that larger S means better exploration abilities. For the discrepancy problem, we will have to increase S (moderately) when n increases.

Since the attractor changes often in apQEA, entropy oftentimes never reaches near zero but instead oscillates around values like 20 or 30. A more stable measure is the mean Hamming distance in the attractor population, i.e., $\frac{1}{\binom{S}{2}} \cdot \sum_{\{x,x'\} \in \binom{\mathcal{P}}{2}} |\{i;\ x_i \neq x'_i\}|$. However, it also can get stuck well above zero. To determine a hopeless situation, we instead developed the concept of a *flatline*. A flatline is a period of time in which neither the mean Hamming distance reaches a new minimum nor a better solution is found. When we encounter a flatline stretching over 25% of the total running time so far, we declare the situation hopeless. To avoid erroneously aborting in early stages, we additionally demand that the relative mean Hamming distance, which is the mean Hamming distance divided by k, falls below $1/10$. Those thresholds were found to be appropriate (for the discrepancy problem) in preliminary experiments.

3 Modulo Coloring

Let $p \leqslant n$ be an integer. Given a partial coloring $\chi' : \{0, \ldots, p-1\} \longrightarrow \{-1, +1\}$, i.e., a coloring of the first p vertices, we can construct a coloring χ by repeating χ', i.e., $\chi : V \longrightarrow \{-1, +1\}$, $v \mapsto \chi'(v \bmod p)$. We call χ' a *generating coloring*. This way of coloring, with an appropriate p, brings many benefits. Denote $E_p := A_{0,1,p} = \{0, \ldots, p-1\}$, this is an AP and also the whole set on which χ' lives.

Assume that χ' is *balanced*, i.e., $\text{disc}_{\chi'}(E_p) \leqslant 1$. Let $A_{a,d,\ell}$ be any AP and $\ell = qp + r$ with integers q, r and $r < p$. Then we have the decomposition:

$$A_{a,d,\ell} = \underbrace{\bigcup_{i=0}^{q-1} A_{a+ipd,d,p}}_{B_i :=} \uplus \underbrace{A_{a+qpd,d,r}}_{B_q :=} . \tag{1}$$

Assuming p is prime, we have $B_i \bmod p := \{v \bmod p; \; v \in B_i\} = E_p$ for each $i = 0, \ldots, q-1$, so $\text{disc}_\chi(B_i) = \text{disc}_{\chi'}(E_p) \leqslant 1$. It follows $\text{disc}_\chi(A_{a,d,\ell}) \leqslant q\,\text{disc}_{\chi'}(E_p) + \text{disc}_\chi(B_q) \leqslant q + \text{disc}_\chi(B_q)$. This is one of the essential ideas how Sárközy's $O(\sqrt[3]{n \log(n)})$ bound is proved and it gives us a hint (which was confirmed in experiments) that modulo colorings, constructed from balanced ones, might tend to have low discrepancy. It is tempting to choose p very small, but the best discrepancy we can hope for when coloring modulo p is $\lceil n/p \rceil$. Since we aim for $3\sqrt[4]{n}$, we choose p as a prime number so that $\lceil n/p \rceil$ is some way below $3\sqrt[4]{n}$, precisely we choose p prime with $n/p \approx 5/2 \cdot \sqrt[4]{n}$, i.e., $p \approx 2/5 \cdot n^{3/4}$.

Constructing balanced colorings is straightforward. Define $h := \frac{p+1}{2}$ (so $h = \Theta(n^{3/4})$) and let $x \in \{-1, +1\}^h$. Then $(x_1, \ldots, x_{h-1}, -x_{h-1}, \ldots, -x_1, x_h)$ defines a balanced coloring of E_p. We could have chosen different ways of ordering the entries of x and their negatives, but this *mirroring* construction has shown to work best so far. We additionally alternate the last entry x_h, so we use the following generating coloring of length $2p$:

$$(x_1, \ldots, x_{h-1}, -x_{h-1}, \ldots, -x_1, x_h, x_1, \ldots, x_{h-1}, -x_{h-1}, \ldots, -x_1, -x_h) . \tag{2}$$

Modulo coloring has further benefits. It reduces the search space from $\{-1, +1\}^n$ to $\{-1, +1\}^h$, where $h = \Theta(n^{3/4})$. Moreover, it allows a much faster FF evaluation: we can restrict to those $A_{a,d,\ell}$ with $a \leqslant 2p - 1$. We also make use of a decomposition similar to (1), but which is more complicated since we exploit the structure of (2); details will be given in the full version. We omit APs which are too short to bring discrepancy above the target, giving additional speedup.

4 Experiments and Results

Implementation and Setup. To fully benefit from the features of apQEA, we needed an implementation which allows *asynchronous* communication between processes. Our MPI-based implementations (version 1.2.4) exhibited unacceptable idle times when used for asynchronous communication, so we wrote our own client-server-based parallel framework. It consists of a server process that manages the attractor population. Clients can connect to it at any time via TCP/IP and do selection and injection; the server takes care of trimming the population after injection. Great care was put into making the implementation free of race conditions. Most parts of the software is written in Bigloo[3], an implementation of the Scheme programming language. The FF and a few other parts are written in C, for performance reasons and to have OpenMP available. OpenMP is used to distribute FF evaluation across multiple processor cores. So we have a

---

[3] http://www-sop.inria.fr/indes/fp/Bigloo/, version 3.9b-alpha29Nov12

two-level parallelization: on the higher level, we have multiple processes treating multiple models and communicating via the attractor population server. On the lower level, we have thread parallelization for the FF. The framework provides also means to run sQEA and vQEA. We always use 8 threads (on 8 processor cores) for the FF. If not stated otherwise, a total of 96 processor cores is used. This allows us to have 12 models fully in parallel; if we use more models then the set of models is partitioned and the models from each partition are treated sequentially. Experiments are carried out on the NEC^{TM} Linux Cluster of the Rechenzentrum at Kiel University, with SandyBridge-EP^{TM} processors.

Results for sQEA. Recall the important parameters of sQEA: number of models M, learning resolution R, global migration period T_g, and number of groups. For R, we use 50, 100 and 500, which are common settings, and also try 1000, 2000, and 3000. In [6], it is proposed to choose T_g in linear dependence on R, which in our notation and neglecting the small additive constant reads $T_g = 2R\lambda$ with $1.15 \leqslant \lambda \leqslant 1.34$. We use $\lambda = 1.25$ and $\lambda = 1.5$, so $T_g = 2.5R$ and $T_g = 3R$. In [6], the number of groups is fixed to 5 and the number of models ranges up to 100. We use 6 groups and up to 96 models. We use rotation learning, but made similar observations with linear learning.

We fix $n = 100\,000$ and do 3 runs for each set of parameters. Computation is aborted after 15 minutes, which is roughly double the time apQEA needs to reach the target discrepancy of 53. For sQEA as given in Alg. 1 best discrepancy we reach is 57. We get better results when using $f(x) < f(a)$ as the test in line 7, i. e., we also accept a sample that is as good as the attractor and not require that it is strictly better.[4] The following table gives mean discrepancies for this variant. The left number is for smaller T_g and the right for higher T_g, e. g., for $M = 12$ and $R = 50$ we have 60 for $T_g = 2.5 \cdot 50 = 125$ and 61 for $T_g = 3 \cdot 50 = 150$.

R	$M = 12$		$M = 24$		$M = 48$		$M = 96$	
50	60	61	59	59	58	57	58	58
100	59	59	57	59	59	56	56	56
500	57	56	57	57	55	56	56	57
1000	57	57	56	55	57	56	58	59
2000	57	57	57	58	59	60	63	62
3000	56	57	59	58	61	61	64	65

Target discrepancy 53 is never reached. For two runs we reach 54, namely for $(M, R, T_g) = (24, 1000, 3000)$ and $(96, 500, 1250)$. But for each of the 2 settings, only 1 of the 3 runs reached 54. There is no clear indication whether smaller or larger T_g is better. Entropy left in the end generally increases with M and R.

We pick the setting $(M, R, T_g) = (24, 1000, 3000)$, which attained 54 in 1 run and also has lowest mean value of 55, for a 5 hour run. In the 15 minutes runs with this setting, entropy in the end was 48 on average. What happens if we let the algorithm use up more of its entropy? As it turns out while entropy is brought down to 16 during the 5 hours, only discrepancy 56 is attained.

The main problem with sQEA here is that there is no clear indication which parameter to tune in order to get higher quality solutions – at least not within

---

[4] Using $f(x) < f_0$ in apQEA however has shown to be not beneficial.

reasonable time (compared to what apQEA can do). In preliminary experiments, we reached target discrepancy 53 on some occasions, but with long running times. We found no way to *reliably* reach discrepancy 55 or better with sQEA.

Finally, for $(M, R, T_g) = (24, 1000, 3000)$, we do experiments for up to $n = 200\,000$, with 3 runs for each n. The time limit is twice what apQEA requires on average, rounded to the next multiple of 5. The following table for each n gives the best result obtained over the 3 runs and the target for comparison.

$n =$	100 000	125 000	150 000	175 000	200 000
time limit in minutes	15	25	40	80	150
best sQEA result	54	60	63	68	69
target	53	56	59	61	63

Although we do multiple runs and allocate twice the time apQEA would need to attain the target, the best result for sQEA stays clearly away from the target.

Results for vQEA. Recall that vQEA in generation $t+1$ unconditionally replaces the attractor for each model with the best sample found during generation t. vQEA does not use a groups and global migration period, instead all models form a single group "to ensure convergence" [15]. Indeed, our experiments confirm that vQEA has no problem with running out of entropy. We conducted 5 runs with $R = 50$ and rotation learning for $n = 100\,000$. In all of the runs, the target of 53 was hit with about 100 of entropy left. However, the time required was almost 2 hours.[5] We also conducted 5 runs with linear learning, yielding the same solution quality at a 12% higher running time. We also did a run for $n = 125\,000$; there vQEA attained the target discrepancy after 3.5 hours.

The high running times were to be expected since vQEA can only make limited use of shortcutting. Since we take multiple samples in each generation (10 for each model), FF evaluation from the second sample on can make use of shortcutting. However, necessarily each generation takes at least the time of one full FF evaluation. We conclude that while vQEA has impressive exploration capabilities and delivers high solution quality "out of the box", i.e., without any particular parameter tuning, it is not well suited for the discrepancy problem.

Results for apQEA. Recall that the most important parameter for apQEA is the attractor population size S. We fix $R = 100$ with linear learning and $M = 12$ and vary S in steps of 10. Computation is aborted when the target of $3\lfloor \sqrt[4]{n} \rfloor$ is hit (a *success*) or a long flatline is observed (a *failure*), as explained in Sec. 2. For selecting attractors, we use tournament selection: draw two solutions randomly from $\mathcal{P}$ and use the better one with 60% probability and the inferior one with 40% probability (higher selection pressures appear to not help, this will be discussed in the full version). For each n and appropriate choices of S, we do 30 runs and record the following: whether it is a success or a failure, final discrepancy (equals target discrepancy for successes), running time (in minutes), mean final entropy.

---

[5] Even more, these 2 hours is only the time spent in FF evaluation. Total time was about 4 hours, but we suspect this to be partly due to our implementation being not particularly suited for vQEA resulting in communication overhead.

For failures, we also record the time at which the last discrepancy improvement took place (for successes, this value is equal to the running time). The following table gives results grouped into successes and failures, all numbers are mean values over the 30 runs.

$\frac{n}{1000}$	S	#	disc	successes time	entropy	#	disc	failures time	entropy	last imp.
100	20	28	$53_{\sigma=00}$	$05_{\sigma=01}$	$38_{\sigma=09}$	02	$54_{\sigma=00}$	$08_{\sigma=01}$	$22_{\sigma=01}$	$06_{\sigma=01}$
100	30	30	$53_{\sigma=00}$	$07_{\sigma=01}$	$63_{\sigma=12}$	00	na	na	na	na
125	20	17	$56_{\sigma=00}$	$09_{\sigma=02}$	$31_{\sigma=08}$	13	$57_{\sigma=00}$	$12_{\sigma=01}$	$19_{\sigma=09}$	$08_{\sigma=01}$
125	30	30	$56_{\sigma=00}$	$11_{\sigma=01}$	$48_{\sigma=11}$	00	na	na	na	na
150	30	26	$59_{\sigma=00}$	$16_{\sigma=02}$	$46_{\sigma=11}$	04	$60_{\sigma=00}$	$23_{\sigma=02}$	$32_{\sigma=09}$	$16_{\sigma=02}$
150	40	30	$59_{\sigma=00}$	$20_{\sigma=02}$	$62_{\sigma=10}$	00	na	na	na	na
175	30	16	$61_{\sigma=00}$	$24_{\sigma=04}$	$31_{\sigma=08}$	14	$62_{\sigma=00}$	$38_{\sigma=05}$	$21_{\sigma=08}$	$24_{\sigma=03}$
175	40	27	$61_{\sigma=00}$	$32_{\sigma=05}$	$42_{\sigma=09}$	03	$62_{\sigma=00}$	$47_{\sigma=02}$	$26_{\sigma=05}$	$30_{\sigma=01}$
175	50	30	$61_{\sigma=00}$	$39_{\sigma=05}$	$57_{\sigma=11}$	00	na	na	na	na
200	30	02	$63_{\sigma=00}$	$38_{\sigma=02}$	$22_{\sigma=06}$	28	$65_{\sigma=01}$	$47_{\sigma=08}$	$24_{\sigma=17}$	$30_{\sigma=05}$
200	50	28	$63_{\sigma=00}$	$53_{\sigma=08}$	$44_{\sigma=08}$	02	$64_{\sigma=00}$	$76_{\sigma=06}$	$28_{\sigma=02}$	$53_{\sigma=03}$
200	60	30	$63_{\sigma=00}$	$74_{\sigma=15}$	$53_{\sigma=12}$	00	na	na	na	na

We observe that by increasing S, we can guarantee the target to be hit. Dependence of S on n for freeness of failure appears to be approx. linear or slightly super-linear; ratios of S to $n/1000$ for no failures are 0.30, 0.24, 0.27, 0.29, and 0.30. But even if S is one step below the required size, discrepancy is only 1 away from the target (with an exception for $n = 125\,000$ and $S = 20$, where we recorded discrepancy 58 in 1 of the 30 runs). Running times for failures tend to be longer than for successes, even if the failure is for a smaller S. This is because it takes some time to detect a failure by the flatline criterion. Larger S effects larger entropy; failures tend to have lowest entropy, indicating that the problem is the models having locked onto an inferior solution. The table also shows what happens if we do *lazy S tuning*, i. e., fixing S to the first successful value $S = 30$ and then increasing n: failure rate increases and solution quality for failures deteriorates moderately. The largest difference to the target is observed for $n = 200\,000$ and $S = 30$, namely we got discrepancy 67 in 1 of the 30 runs; target is 63. For comparison, the best discrepancy we found via sQEA in 3 runs for such n was 69 and the worst was 71. We conclude that a mistuned S does not necessarily have catastrophic implications, and apQEA can still beat sQEA.

Convergence. We plot (Fig. 1) discrepancy over time for 2 runs: the first hour of the 5-hour sQEA run with $(M, R, T_g) = (24, 1000, 3000)$; and 1 for apQEA with $S = 30$, which is kept running after the target was hit (until the flatline criterion leads to termination). sQEA is shown with a dashed line and apQEA with a solid line. In the first minutes, sQEA brings discrepancy down faster, but is soon overtaken by apQEA (which reaches 51 in 10 minutes, target is 53).

Effect of Parallelization. We double number of cores from 96 to 192 and increase number of models to $M = 24$, so that they can be treated in parallel with 8 cores each. The following table shows results for 5 runs for each set of parameters. First consider $n = 175\,000$ and $200\,000$. The best failure-free settings for S are 30 and 50.

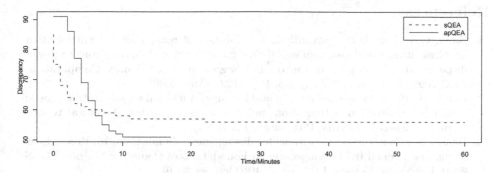

Fig. 1. Discrepancy plotted over time for sQEA and apQEA

In comparison with the best failure-free settings for 96 cores, time is reduced to $17/39 = 44\%$ and $42/74 = 57\%$, respectively. 50% or less would mean a perfect scaling. When we do not adjust S, i. e., we use 50 and 60, time is reduced to $26/39 = 67\%$ and $53/74 = 72\%$, respectively. We also compute for $n = 250\,000$.

$\frac{n}{1000}$	S	#	disc	time	entropy	#	disc	time	entropy	last imp.
				successes				failures		
175	20	02	$61_{\sigma=00}$	$14_{\sigma=02}$	$22_{\sigma=02}$	03	$62_{\sigma=00}$	$20_{\sigma=02}$	$31_{\sigma=12}$	$13_{\sigma=00}$
175	30	05	$61_{\sigma=00}$	$17_{\sigma=04}$	$43_{\sigma=07}$	00	na	na	na	na
175	50	05	$61_{\sigma=00}$	$26_{\sigma=03}$	$72_{\sigma=12}$	00	na	na	na	na
200	40	04	$63_{\sigma=00}$	$30_{\sigma=03}$	$42_{\sigma=12}$	01	$64_{\sigma=00}$	$42_{\sigma=00}$	$32_{\sigma=02}$	$28_{\sigma=00}$
200	50	05	$63_{\sigma=00}$	$42_{\sigma=10}$	$52_{\sigma=05}$	00	na	na	na	na
200	60	05	$63_{\sigma=00}$	$53_{\sigma=08}$	$60_{\sigma=11}$	00	na	na	na	na
250	50	04	$67_{\sigma=00}$	$58_{\sigma=07}$	$42_{\sigma=08}$	01	$68_{\sigma=00}$	$110_{\sigma=00}$	$29_{\sigma=02}$	$70_{\sigma=00}$
250	60	05	$67_{\sigma=00}$	$83_{\sigma=16}$	$61_{\sigma=14}$	00	na	na	na	na

5 Conclusion and Current Work

We have seen apQEA outperforming sQEA in terms of speed and solution quality by a large margin on the discrepancy problem. For this problem, apQEA is easy to tune, since a single parameter, the size S of the attractor population, has a clear and foreseeable effect: it improves solution quality (if possible) at the price of an acceptable increase in running time. vQEA has shown that it is possible to achieve the same solution quality as apQEA without parameter tuning, at the price of an enormous running time and inter-process communication overhead. It may be possible to have the best of both worlds in one algorithm, i. e., to get rid of the S parameter in apQEA. Moreover we believe that it should be attempted to mathematically analyze why vQEA and apQEA succeed where sQEA fails. We also plan new applications of apQEA in the vast area of coloring of (hyper)graphs and other combinatorial problems. Concerning arithmetic progressions, we will investigate further ways to speed up the FF by exploiting combinatorial structures.

Acknowledgements. We thank the Deutsche Forschungsgemeinschaft (DFG) for Grants Sr7/12-3 (SPP "Algorithm Engineering") and Sr7/14-1 (binational workshop). We thank Nikhil Bansal for interesting discussions.

References

1. Babu, G.S., Das, D.B., Patvardhan, C.: Solution of real-parameter optimization problems using novel quantum evolutionary algorithm with applications in power dispatch. In: Proceedings of the IEEE Congress on Evolutionary Computation, CEC 2009, Trondheim, Norway, pp. 1927–1920 (May 2009)
2. Baluja, S.: Population-based incremental learning: A method for integrating genetic search based function optimization and competitive learning. Technical report, Carnegie Mellon University, Pittsburgh, PA (1994)
3. Bansal, N.: Constructive algorithms for discrepancy minimization. In: Proceedings of the 51st Annual IEEE Symposium on Foundations of Computer Science, FOCS 2010, Las Vegas, Nevada, USA, pp. 3–10 (October 2010)
4. Erdős, P., Spencer, J.: Probabilistic Methods in Combinatorics. Akadémia Kiadó, Budapest (1974)
5. Han, K.H., Kim, J.H.: Quantum-inspired evolutionary algorithm for a class of combinatorial optimization. IEEE Transactions on Evolutionary Computation 6(6), 580–593 (2002)
6. Han, K.H., Kim, J.H.: On setting the parameters of quantum-inspired evolutionary algorithm for practical applications. In: Proceedings of the IEEE Congress on Evolutionary Computation, CEC 2003, Canberra, Australia, pp. 178–184 (December 2003)
7. Harik, G.R., Lobo, F.G., Goldberg, D.E.: The compact genetic algorithm. Technical report, Urbana, IL: University of Illinois at Urbana-Champaign, Illinois Genetic Algorithms Laboratory (1997)
8. Hauschild, M., Pelikan, M.: An introduction and survey of estimation of distribution algorithms (2011)
9. Knieper, P.: The Discrepancy of Arithmetic Progressions. PhD thesis, Institut für Informatik, Humboldt-Universität zu Berlin (1997)
10. Mani, A., Patvardhan, C.: An adaptive quantum inspired evolutionary algorithm with two populations for engineering optimization problems. In: Proceedings of the International Conference on Applied Systems Research, NSC 2009, Dayalbagh Educational Instute, Agra, India (2009)
11. Mani, A., Patvardhan, C.: A hybrid quantum evolutionary algorithm for solving engineering optimization problems. International Journal of Hybrid Intelligent Systems 7, 225–235 (2010)
12. Matoušek, J., Spencer, J.: Discrepancy in arithmetic progressions. Journal of the American Mathematical Society 9, 195–204 (1996)
13. Mühlenbein, H., Paaß, G.: From recombination of genes to the estimation of distributions I. binary parameters. In: Ebeling, W., Rechenberg, I., Voigt, H.-M., Schwefel, H.-P. (eds.) PPSN 1996. LNCS, vol. 1141, pp. 178–187. Springer, Heidelberg (1996)
14. Patvardhan, C., Prakash, P., Srivastav, A.: A novel quantum-inspired evolutionary algorithm for the quadratic knapsack problem. In: Proceedings of the International Conference on Operations Research Applications in Engineering and Management, ICOREM 2009, Tiruchirappalli, India, pp. 2061–2064 (May 2009)
15. Platel, M.D., Schliebs, S., Kasabov, N.: A versatile quantum-inspired evolutionary algorithm. In: Proceedings of the IEEE Congress on Evolutionary Computation, CEC 2007, Singapore, pp. 423–430 (September 2007)
16. Roth, K.F.: Remark concerning integer sequences. Acta Arithmetica 9, 257–260 (1964)

Computational Aspects of Ordered Integer Partition with Upper Bounds

Roland Glück[1], Dominik Köppl[1], and Günther Wirsching[2]

[1] Universität Augsburg
Universitätsstr. 6a, 86159 Augsburg, Germany
{glueck,dominik.koeppl}@informatik.uni-augsburg.de
[2] Katholische Universität Eichstätt
Ostenstrasse 26, 85072 Eichstätt, Germany
guenther.wirsching@ku-eichstaett.de

Abstract. We propose a novel algorithm for computing the number of ordered integer partitions with upper bounds. This problem's task is to compute the number of distributions of z balls into n urns with constrained capacities $i_1, \ldots, i_n$ (see [10]). Besides the fact that this elementary urn problem has no known combinatoric solution, it is interesting because of its applications in the theory of database preferences as described in [3] and [9]. The running time of our algorithm depends only on the number of urns and not on their capacities as in other previously known algorithms.

1 Introduction

There is a lot of work about (not ordered) partition with all kinds of constraints and some elementary results about ordered partitions without constraints (see e.g. [1], [2], [5] and [8]), but unfortunately, nobody has yet deeply investigated the problem of computing the number of ordered integer partition with upper bounds. Although this problem is a nearby generalisation of the concept of the composition of a natural number (see e.g. [8]) the problem with the additional constraint of upper bounds seems not to be treated on a broader way till now. A special case of the problem is treated in [10], but not from an algorithmic point of view.

Our attention to this problem was attracted by the work in [3] and [9]. There and in subsequent ongoing work it appears in the context of Preference SQL, an extension of SQL allowing soft-conditions for filtering a selection in order to avoid empty results. Therefore, the select-statement in Preference SQL is enhanced by classic hard conditions in a post- and prefiltering phase, while providing preference evaluation as an interim stage. Preferences are distinguished as base preferences evaluating on a single attribute, and complex preferences that are a composition of multiple preferences. Complex preferences conjugate by ranking, prioritisation or Skyline-composition. For the latter one, there are a lot of propositions how Skyline-computation can be done in an efficient manner. Preference SQL embraces the modelling of Pareto-Skylines by using a so called

V. Bonifaci et al. (Eds.): SEA 2013, LNCS 7933, pp. 79–90, 2013.
© Springer-Verlag Berlin Heidelberg 2013

better-than-graph, the Hasse diagram of an order on a (temporary) database relation induced by the Pareto-preference. Having solved our tackled problem, it is easy for us to determine the width of this graph. Beside this application, it is an interesting expansion of the well-known urn problem and thus a solution of everyday's life problems.

The paper is organised as follows: Section 2 introduces the problem and fixes some writing conventions. In Section 3 the basic idea of our algorithm is sketched and illustrated by an example to which we will refer also in the following sections. A formal proof of the correctness is given in Section 4, whereas Section 5 is dedicated to the implementation and analysis of its running time. The predictions of the theoretical analysis are reviewed in Section 6 which is dedicated to our experimental results. Finally, we summarise our results and give an outlook to future work in Section 7.

2   Basic Definitions

Definition 2.1. *Given a sequence $I = (i_1, i_2, \ldots, i_n)$ of n positive integers and an integer number z we call a sequence $J = (j_1, j_2, \ldots, j_n)$ of n nonnegative integers an* ordered partition *of z wrt. I if the following two requirements are fulfilled:*

1. $\forall l \in \{1 \ldots n\} : j_l \leq i_l$
2. $\sum\limits_{k=1}^{n} j_k = z$

We call I the upper bounds *and z the* target value.

Our goal is to compute the number of distinct ordered partitions of z wrt. I. We denote this number by $\#(I, z)$.

Since we consider in this paper only ordered partitions we will from now on use the term partition instead of ordered partition. With upper case letters I and J we refer to sequences of nonnegative (or positive, depending on the context) integers. By $|I|$ we denote the length of such a sequence.

Clearly, for all sequences I we have $\#(I, z) = 0$ if $z < 0$ and $\#(I, 0) = 1$. Moreover, for a sequence $I = (i_1, i_2, \ldots, i_n)$ the equalities $\#(I, \sum\limits_{k=1}^{n} i_k) = 1$ and $\#(I, z) = 0$ for arbitrary $z > \sum\limits_{k=1}^{n} i_k$ hold trivially. In the case $I = (i_1)$ we obtain $\#((i_1), z) = 1$ if $0 \leq z \leq i_1$ and $\#((i_1), z) = 0$ otherwise.

3   The Idea of the Algorithm

[9] and [3] give the following recursion formula for the number of partitions:

Lemma 3.1. *The equality* $\#((i_1, i_2, \ldots, i_n), z) = \sum\limits_{k=0}^{z} \#((i_1, i_2, \ldots, i_{n-m}), k) \cdot$ $\#((i_{n-m+1}, i_{n-m+2}, \ldots, i_n), z - k)$ *holds for arbitrary m with $1 \leq m \leq n - 1$.*

Using this formula one can reduce in a divide-and-conquer manner the computation to smaller problems of the form $\#(I, z)$ with $|I| \in \{1, 2\}$ (for these cases there are shortcuts running in constant time). Although easy to prove (see again [3] and [9]) and intuitively clear, this formula is unsatisfying from a computational point of view. Its running time obviously depends on the i'_ks and the value of z, even if there are heuristics how to split up the sequence I advantageously. In contrast, we will develop an algorithm with a running time depending only on the length of the sequence of upper bounds under the assumption that elementary arithmetic operation can be executed in constant time.

By setting $m = n - 1$ in Lemma 3.1 we obtain the following formula:

$$\#((i_1, i_2, \ldots, i_n), z) = \sum_{k=0}^{z} \#((i_1, i_2, \ldots, i_{n-1}), k) \cdot \#((i_n), z - k) \qquad (1)$$

Since (i_n) is a single valued sequence the factor $\#(i_n, z - k)$ becomes either zero or one. Our aim is to show that $\#(I, z)$ for a fixed sequence I can be written as piecewise defined polynomials in z of degree at most $|I| - 1$.

z	$\#((30), z)$
$-\infty < z \leq -1$	0
$0 \leq z \leq 30$	1
$31 \leq z < \infty$	0

z	$\#((30, 50), z)$
$-\infty < z \leq -1$	0
$0 \leq z \leq 30$	$z + 1$
$31 \leq z \leq 49$	31
$50 \leq z \leq 80$	$81 - z$
$81 \leq z \leq \infty$	0

z	$\#((30, 50, 10), z)$
$-\infty < z \leq -1$	0
$0 \leq z \leq 9$	$\frac{z^2}{2} + \frac{3z}{2} + 1$
$10 \leq z \leq 30$	$11z - 44$
$31 \leq z \leq 40$	$-\frac{z^2}{2} + \frac{81z}{2} - 479$
$41 \leq z \leq 49$	341
$50 \leq z \leq 59$	$-\frac{z^2}{2} + \frac{99z}{2} - 884$
$60 \leq z \leq 80$	$-11z + 946$
$81 \leq z \leq 90$	$\frac{z^2}{2} - \frac{183z}{2} + 4186$
$91 \leq z < \infty$	0

Fig. 1. The Stages of the Algorithm

Before proving this formally we will illustrate our approach on an example. Consider the sequence $I_e = (30, 50, 10)$. We begin the recursion from Formula 1 with the sequence $I_e^1 = (30)$. Clearly we have $\#((30), z) = 0$ for $z < 0$, $\#((30), z) = 1$ for $0 \leq z \leq 30$ and $\#((30), z) = 0$ for $z \geq 31$. This is illustrated in the upper left table of Figure 1. To evaluate the next recurrence step, namely $\#((30,50),z) = \sum_{k=0}^{z} \#((30),k) \cdot \#((50),z-k)$, we investigate the relative behaviour of both the terms $\#((30),k)$ and $\#((50),z-k)$ occurring in the sum. For $z < 0$ the sum is empty and therefore evaluates to zero, so we have $\#((30,50),z) = 0$ for $z < 0$. The next stage is the case $z \in [0,30]$. Then the term $\#((50),z-k)$ evaluates to one, and we have $\#((30,50),z) = \sum_{k=0}^{z} \#((30),z) = z+1$. For $z \in [31,49]$ the term $\#((30),k)$ becomes zero for $k \in [31,49]$, whereas $\#((50),z-k)$ equals always one. So we have $\#((30,50),z) = 31$. Next we consider $z \in [50,80]$. Then $\#((30),k)$ equals one for $k \in [0,30]$ and $\#((50),z-k)$ equals one for $k \in [0,z-50]$, so both factors become simultaneously one iff $k \in [z-50,30]$, so here we have $\#((30,50),z) = 81 - z$. For $z > 81$ there are no values for k where both the first and the second factor evaluate to a value different from zero, so we have $\#((30,50),z) = 0$ for $z > 81$. Note that we did compute the values of $\#((30,50),z)$ via piecewise defined affine linear functions which determine the value of $\#((30,50),z)$. The situation is depicted in the upper right table of Figure 1.

Till now we have piecewise defined polynomials in z of degree 1 for the value of $\#((30,50),z)$. For the next step we have to evaluate the term $\sum_{k=0}^{z} \#((30,50),k) \cdot \#((10),z-k)$. Since $\#((10),z-k)$ equals 1 iff $0 \leq z - k \leq 10$ and equals 0 otherwise the previous sum can be written as $\sum_{k=z-10}^{z} \#((30,50),k)$. This means, we have to sum up eleven consecutive values of $\#((30,50),k)$, which can be done using Gauß's summation formula for the first n natural numbers, i.e. $\sum_{i=0}^{n} = \frac{n(n+1)}{2}$. To execute this purpose we have to examine the relative positions of the interval $[z-10,z]$ and the validity intervals of the polynomials of $\#((30,50),z)$, i.e. the intervals $]-\infty,-1]$, $[0,30]$, $[31,49]$, $[50,80]$ and $[81,\infty[$. We will illustrate this on a few examples; the rest is left to the reader.

First, for $z < 0$ we clearly have $\#((30,50,10),z) = 0$.

The next case $(z - 10 \in]-\infty,-1] \wedge z \in [0,30])$ is the range $z \in [0,9]$. Here we have $\sum_{k=z-10}^{z} \#((30,50),k) = \sum_{k=0}^{z} \#((30,50),k) = \sum_{k=0}^{z} (k+1) = \frac{z^2}{2} + \frac{3z}{2} + 1$.

Now we consider the interval $[10,30]$, where $[z-10,z] \subseteq [0,30]$. Here we get $\sum_{k=z-10}^{z} \#((30,50),k) = \sum_{k=z-10}^{z} (k+1)$, and simple arithmetic and Gauß' formula lead to the result $11z - 44$.

Next we consider the case $z - 10 \in [0,30] \wedge z \in [31,49]$. This happens if $z \in [31,40]$ and gives the formula $\sum_{k=z-10}^{z} \#((30,50),k) = \sum_{k=z-10}^{30} (k+1) + \sum_{k=31}^{z} 31$, which evaluates to $-\frac{z^2}{2} + \frac{81z}{2} - 479$.

The next cases are omitted here, but the reader can find the final result in the lower table of Figure 1 and may compare it with his own results.

4   Formal Proof

We will now formalise the ideas from the previous section. At the beginning we need some definitions.

Definition 4.1. *An* interval partition *of the integers is a finite sequence* $\mathbb{I} = (\mathcal{I}_0, \mathcal{I}_1, \ldots, \mathcal{I}_n)$ *of nonempty intervals of the integers such that* $\max(\mathcal{I}_k) = \min(\mathcal{I}_{k+1}) - 1$ *for all* $k \in \{0, \ldots, n-1\}$ *holds and* $\bigcup_{i=0}^{n} \mathcal{I}_n = \mathbb{Z}$.

Obviously, for an interval partition $(\mathcal{I}_0, \mathcal{I}_1, \ldots, \mathcal{I}_n)$ we have $\mathcal{I}_0 =]-\infty, z_1]$ and $\mathcal{I}_n = [z_2, \infty[$ for some integers z_1 and z_2. An interval partition $(\mathcal{I}_0, \mathcal{I}_1, \ldots, \mathcal{I}_n)$ is called *normalised* if $\max(\mathcal{I}_0) = -1$ holds. The number of intervals of an interval partition $\mathbb{I}$ is also called its *length* and denoted by $l(\mathbb{I})$.

During the example execution in Section 3 we got the interval partition $\mathbb{I}_e = (]-\infty, -1], [0, 30], [31, 50], [51, 80], [81, \infty[)$ with $l(\mathbb{I}) = 5$, which is even normalised.

Next we will formalise the splitting of an interval partition:

Definition 4.2. *Given an interval partition* $\mathbb{I} = (\mathcal{I}_0, \mathcal{I}_1, \ldots, \mathcal{I}_n)$ *and a positive integer* z *we define a relation* $\sim_z \subseteq \mathbb{Z} \times \mathbb{Z}$ *by* $x \sim_z y =_{df} \exists i, j : x - z \in \mathcal{I}_i \land y - z \in \mathcal{I}_i \land x \in \mathcal{I}_j \land y \in \mathcal{I}_j$.

Clearly, $\sim_z$ is an equivalence relation. Because $x \sim_z y$ implies $x \sim_z y'$ for all y' with $x \leq y' \leq y$ (which is easy to verify) the equivalence classes of $\sim_z$ are intervals of $\mathbb{Z}$. Moreover, $\mathcal{I}_0$ and $[\min(\mathcal{I}_n) + z, \infty[$ are equivalence classes of $\sim_z$, so ordering the equivalence classes of $\sim_z$ according to their minima or maxima yields again an interval partition, called the *interval partition induced by* $\mathbb{I}$ *and* z and denoted it by $\mathbb{P}(\mathbb{I}, z)$. It is straightforward to see that the interval partition induced by a normalised interval partition and an arbitrary positive integer is also a normalised interval partition. For an interval partition $\mathbb{I} = (\mathcal{I}_0, \mathcal{I}_1, \ldots, \mathcal{I}_n)$, a $z \in \mathbb{N}^+$ and an interval $\mathcal{J}$ of $\mathbb{P}(\mathbb{I}, z)$ we call the unique pair $(\mathcal{I}_i, \mathcal{I}_j)$, given by $\forall x \in \mathcal{J} : x - z \in \mathcal{I}_i \land x \in \mathcal{I}_j$, the *witness intervals* of $\mathcal{J}$ and denote it by $\iota(\mathcal{J})$ Note that ι is an injective mapping due to the definition of $\sim_z$.

In our example the interval partition induced by $\mathbb{I}_e$ and 10 is the sequence $(]-\infty, -1], [0, 10], [11, 30], [31, 40], [41, 50], [51, 60], [61, 80], [81, 90], [91, \infty[)$. The first five members of this sequence were also considered in Section 3. We have e.g. $\iota([0, 10]) = (]-\infty, -1], [0, 30])$, $\iota([11, 30]) = ([0, 30], [0, 30])$

There is an upper bound for the length of an induced interval partition:

Lemma 4.3. *For an interval partition* $\mathbb{I} = (\mathcal{I}_0, \mathcal{I}_1, \ldots, \mathcal{I}_n)$ *and a positive integer* z *we have the inequality* $l(\mathbb{P}(\mathbb{I}, z)) \leq 2l(\mathbb{I}) - 1$.

Proof. We introduce a linear order on the set of witness intervals by $(\mathcal{I}_{i_1}, \mathcal{I}_{j_1}) <_{\iota} (\mathcal{I}_{i_2}, \mathcal{I}_{j_2}) \Leftrightarrow i_1 < i_2 \lor (i_1 = i_2 \land j_1 < j_2)$. Clearly, $(\mathcal{I}_0, \mathcal{I}_0)$ is the least element wrt. this order, and $(\mathcal{I}_n, \mathcal{I}_n)$ is its greatest element (note that both $(\mathcal{I}_0, \mathcal{I}_0)$ and $(\mathcal{I}_n, \mathcal{I}_n)$ are indeed interval witnesses). Now we order the set of all interval witnesses according to $<_\iota$ and obtain a chain $(\mathcal{I}_{i_0}, \mathcal{I}_{j_0}) <_\iota (\mathcal{I}_{i_1}, \mathcal{I}_{j_1}) <_\iota \ldots <_\iota$ $(\mathcal{I}_{i_m}, \mathcal{I}_{j_m})$ with $i_0 = j_0 = 0$ and $i_m = j_m = n$. For two consecutive pairs $(\mathcal{I}_{i_k}, \mathcal{I}_{j_k})$ and $(\mathcal{I}_{i_{k+1}}, \mathcal{I}_{j_{k+1}})$ of this sequence we have $(i_{k+1} + j_{k+1}) - (i_k + j_k) \in \{1, 2\}$ due to the underlying definition of $\sim_z$. Since $i_0 + j_0 = 0$ and $i_m + j_m = 2n$ the claim follows.   ∎

After these definitions and lemmas concerning the partition of $\mathbb{Z}$ into intervals we will now turn our attention back to integer partitions.

Lemma 4.4. *Let $I = (i_1, i_2, \ldots, i_n)$ with $n \geq 2$ be a sequence of n positive integers and $z \in \mathbb{Z}$. Then $\#(I, z) = \sum\limits_{k=z-i_n}^{z} \#((i_1, i_2, \ldots, i_{n-1}), k)$ holds.*

Proof. Let $J = (j_1, j_2, \ldots, j_n)$ be a partition of z wrt. I. Then clearly $\sum\limits_{k=1}^{n-1} j_k = z - j_n$ holds. So for a fixed j_n there are exactly $\#((i_1, i_2, \ldots, i_{n-1}), z - j_n)$ partitions. Since j_n has to be drawn from the interval $[0, i_n]$ we have the equality $\#(I, z) = \sum\limits_{k=0}^{i_n} \#((i_1, i_2, \ldots, i_{n-1}), z - k)$, which leads to the claim after an elementary index shift.   ∎

A basic fact we rely on is the so called Faulhaber's formula (see e.g. [2] or [6]). It states the equality $\sum\limits_{i=1}^{n} i^p = \sum\limits_{j=0}^{p} \binom{p}{j} \frac{B_{p-j}}{j+1} n^{j+1}$ for $p \in \mathbb{N}$. Here B_l denotes the l-th Bernoulli number (see also [2] or [6]).

An easy consequence of this formula is the following lemma:

Lemma 4.5. *Consider two arbitrary natural numbers n and γ. Then the function $\sigma_\gamma(z) : [\gamma, \infty[\to \mathbb{Z}$, defined by $\sigma_\gamma(z) = \sum\limits_{k=0}^{z-\gamma} k^n$, is a polynomial in z of degree $n + 1$.*

Proof. We show not only the claim but also how to compute the coefficients. First there are coefficients $a_0, a_1, \ldots, a_{n+1}$ with $\sum\limits_{k=0}^{z-\gamma} k^n = \sum\limits_{l=0}^{n+1} a_l \cdot (z - \gamma)^l$ according to Faulhaber's formula, as described in [6]. Due to the binomial theorem this sum equals the sum $\sum\limits_{l=0}^{n+1} (a_l \cdot \sum\limits_{m=0}^{l} \binom{l}{m} \cdot z^m \cdot (-\gamma)^{l-m})$. We define $\beta_{lm} = \binom{l}{m}(-\gamma)^{l-m}$ and obtain the term $\sum\limits_{l=0}^{n+1} (a_l \cdot \sum\limits_{m=0}^{l} \beta_{lm} \cdot z^m)$. After defining $\delta_m = \sum\limits_{l=m}^{n+1} a_l \cdot \beta_{lm}$ this leads to the desired result $\sum\limits_{m=0}^{n+1} \delta_m \cdot z^m$.   ∎

The following corollary follows from the previous lemma by simple arithmetic operations:

Corollary 4.6. *Let* $p(x) = \sum_{i=0}^{n} a_i \cdot x^i$ *be a polynomial of degree* n *and fix an arbitrary natural number* y. *Then the function* $\sigma_{\geq y} : [y, \infty[\to \mathbb{R}$, *defined by* $\sigma_{\geq y}(z) = \sum_{i=y}^{z} p(i)$, *is a polynomial of degree at most* $n + 1$. *An analogous claim holds for the function* $\sigma_{\leq y} : [0, y] \to \mathbb{R}$, *defined by* $\sigma_{\leq y}(z) = \sum_{i=z}^{y} p(i)$. *Moreover, the function* $\sigma_{-k} : [k, \infty[\to \mathbb{R}$, *defined by* $\sigma_{-k}(z) = \sum_{i=z-k}^{z} p(i)$, *is polynomial of degree at most* n.

Note that the coefficients of $\sigma_{\geq y}$, $\sigma_{\leq y}$ and σ_{-k} can be easily determined using the coefficients from the proof of Lemma 4.5.

Now we are ready to prove the main result of this section:

Theorem 4.7. *Let* $I = (i_1, i_2, \ldots, i_n)$ *be a sequence of* n *positive integers and* z *a natural number. Then there are a normalised interval partition* $\mathcal{I}_0$, $\mathcal{I}_1$, $\ldots$, $\mathcal{I}_r$ *and polynomials* $p_0(x)$, $p_1(x)$, $\ldots$, $p_r(x)$ *of degree at most* $n - 1$ *such that* $\#(I, z) = p_x(z)$ *where* $\mathcal{I}_x$ *is the (unique) interval containing* z. *For* r *the inequality* $r \leq 2^n + 1$ *holds.*

Proof. The proof is done via induction over $|I|$.

Induction Base: In the case $|I| = 1$ we chose the intervals $\mathcal{I}_0 =]-\infty, -1]$, $\mathcal{I}_1 = [0, z]$ and $\mathcal{I}_2 = [z + 1, \infty[$. For the polynomials we choose $p_0(x) = 0$, $p_1(x) = 1$ and $p_2(x) = 0$, and we are done.

Induction Step: Consider a sequence $I^{n+1} = (i_1, i_2, \ldots, i_{n+1})$ of $n+1$ positive integers. Let $\mathbb{I}^n = (\mathcal{I}_0^n, \mathcal{I}_1^n, \ldots, \mathcal{I}_r^n)$ and $p_0^n(x), p_1^n(x), \ldots, p_r^n(x)$ be a normalised interval partition and polynomials with the properties from Theorem 4.7 wrt. to the sequence $(i_1, i_2, \ldots, i_n)$. Denote by $\mathbb{P}(\mathbb{I}^n, i_{n+1}) = (\mathcal{I}_0^{n+1}, \mathcal{I}_1^{n+1}, \ldots, \mathcal{I}_s^{n+1})$ the partition induced by $\mathbb{I}^n$ and i_{n+1}.

The inequality $s \leq 2^n + 1$ follows immediately from the induction hypothesis and Lemma 4.3.

For the rest of the claim we consider an arbitrary interval $\mathcal{I}_m^{n+1} = [z_m^1, z_m^2]$ from $\mathbb{P}(\mathbb{I}^n, z)$, and denote its witness intervals by $\iota(\mathcal{I}_m^{n+1}) = (\mathcal{I}_l^n, \mathcal{I}_u^n) = ([z_l^1, z_l^2], [z_u^1, z_u^2])$. Fix now an arbitrary $z \in \mathcal{I}_m^{n+1}$. According to Lemma 4.4 we have the equality $\#(I^{n+1}, z) = \sum_{k=z-i_n}^{z} \#((i_1, i_2, \ldots, i_n), k)$. Moreover, by definition we have $z - i_n \in \mathcal{I}_l^n$. Assume first that $\mathcal{I}_l^n \neq \mathcal{I}_u^n$ holds. Then the sum can be rewritten as $\sum_{k=z-i_n}^{z_l^2} \#((i_1, i_2, \ldots, i_n), k) + \sum_{k \in \mathcal{I}_j^n, l < j < u} \#((i_1, i_2, \ldots, i_n), k) +$

$\sum_{k=z_u^1}^{z} \#((i_1, i_2, \ldots, i_n), k)$. The second sum is a constant, and the first and third sum can be written as polynomials in z of degree at most n according to the induction hypothesis and Corollary 4.6. In the case $\mathcal{I}_l^n = \mathcal{I}_u^n$ we have according to the induction hypothesis $\sum_{k=z-i_n}^{z} \#((i_1, i_2, \ldots, i_n), k) = \sum_{k=z-i_n}^{z} p(k)$ for some polynomial p with degree at most $n - 1$. But this sum is even a polynomial in z of degree at most $n - 1$ due to Corollary 4.6.   ∎

5   Implementation and Running Time

Now we show how the proof of Theorem 4.7 leads to an algorithm for determining $\#(I, z)$. For the running time analysis we assume that the basic arithmetic operations addition, subtraction, multiplication and division can be carried out in constant time.

The considerations of the previous section can be used to develop a data structure which computes efficiently the value of $\#(I, z)$ for a fixed sequence $I = (i_1, i_2, \ldots, i_n)$ and arbitrary $z \in \mathbb{Z}$. The idea is to proceed iteratively along the lines of the proof of Theorem 4.7. So in the $k + 1$-th iteration we compute the normalised interval partition $\mathbb{P}(\mathbb{I}^k, i_{k+1}) = (\mathcal{I}_0^{k+1}, \mathcal{I}_1^{k+1}, \ldots, \mathcal{I}_s^{k+1})$ and the associated polynomials as described in the proof of Theorem 4.7 using the interval partition and polynomials after the k-th step. After the n-th step we know the validity intervals and the associated polynomials for the sequence I. So facing an integer z we will search the interval which contains z and evaluate the associated polynomial at z to obtain $\#(I, z)$.

We will now describe this in more detail.

In a precomputation step we compute the coefficients of Faulhaber's formula for all exponents between 1 and $n - 1$. Moreover, we compute all binomial coefficients $\binom{m}{l}$ with $0 \leq l \leq m \leq n+1$ (they can be stored during the computation of the Faulhaber coefficients). This can be done in polynomial time in n, which will not influence the asymptotic running time as we will see later.

Since the interval partitions during the execution are always normalised we can identify such a normalised interval partition $\mathbb{I} =] - \infty, -1], [0, z_1], [z_1 + 1, z_2], \ldots, [z_{j-1} + 1, z_j], [z_j + 1, \infty[$ after the k-th iteration with the sequence $(z_1, z_2, \ldots, z_j)$. In the $k + 1$-th step we have to compute the interval partition induced by $\mathbb{I}$ and i_{k+1} which can be done in $\mathcal{O}(j)$ time. To see this we first observe that the new sequence arises from ordering the set $\{z_l \mid 1 \leq l \leq j\} \cup \{i_{k+1} - 1\} \cup \{z_l + i_{k+1} \mid 1 \leq l \leq j\}$. So we compute the (ordered!) sequence $(i_{k+1} - 1, z_1 + i_{k+1}, z_2 + i_{k+1}, \ldots, z_j + i_{k+1})$ and merge it with the ordered sequence $(z_1, z_2, \ldots, z_j)$ into the new (ordered) sequence $(y_1, y_2, \ldots, y_l)$ while removing duplicates.

During this merging process we can for each interval $\mathcal{I}_m^{k+1}$ of $\mathbb{P}(\mathbb{I}^k, i_{k+1})$ determine its witness intervals $(\mathcal{I}_{m_l}^k, \mathcal{I}_{m_r}^k) = [(l_{m_l}, r_{m_l}), (l_{m_r}, r_{m_r})]$ in constant time by simple case distinction.

The most demanding part is the computation of the coefficients for each interval $\mathcal{I}_m^{k+1}$ of $\mathbb{P}(\mathbb{I}^k, i_{k+1})$. If $(\mathcal{I}_{m_l}^k \neq \mathcal{I}_{m_r}^k)$ we have analogously to the proof of

Theorem 4.7 to determine the coefficients of the term $\sum_{k=z-i_n}^{r_{m_l}} \#((i_1, i_2, \ldots, i_n), k)$

$+ \sum_{k \in \mathcal{I}_j, m_l < j < m_r} \#((i_1, i_2, \ldots, i_n), k) + \sum_{k=l_{m_r}}^{z} \#((i_1, i_2, \ldots, i_n), k)$. The computation of the coefficients of the first and last sum can be carried out along the lines of the proof of Lemma 4.5 and Corollary 4.6 in $\mathcal{O}(k^3)$ time (for a monomial of the form $c_j z^j$ the coefficients can be computed in $\mathcal{O}(j^2)$ time, and we do so for monomials $c_0 z^0$, $c_1 z^1$, ..., $c_k z^k$). The middle sum need not to be computed newly in every step but can be handled using stepping technique. Therefore we process the intervals given by $(y_1, y_2, \ldots, y_l)$ in the order defined as in the proof of Lemma 4.3. Then the value of this sum between two consecutive steps either remains zero, is incremented or decremented by the value $\sum_{k \in \mathcal{I}_j^k} \#((i_1, i_2, \ldots, i_n), k)$ for an interval $\mathcal{I}_j^k$ of $\mathbb{I}^k$, or is incremented and decremented by such a term. The value of this terms can be computed analogously as sketched above in $\mathcal{O}(k^3)$ time. In the case $\mathcal{I}_{m_l}^k = \mathcal{I}_{m_r}^k$ we can compute the coefficients analogously to above also in $\mathcal{O}(k^3)$ time. So the overall running time for the k-th iteration is in $\mathcal{O}(2^k k^3)$ (remember that we have at most $2^k - 1$ validity intervals). Since we have n iterations the total running time is in $\mathcal{O}(\sum_{k=1}^{n} 2^k k^3) = \mathcal{O}(2^n n^3)$.

Assume now that we have executed this construction. Then for a given z the search for the validity interval containing z can be done by means of binary search in $\mathcal{O}(log(2^n + 1)) = \mathcal{O}(n)$ time (remember that at the end we have at most $2^n + 1$ validity intervals). The evaluation of a polynomial of degree at most $n - 1$ can be done in time $\mathcal{O}(n - 1)$ using the Horner scheme, so the computation of $\#(I, z)$ can now be carried out in time $\mathcal{O}(n)$. This considerations lead to the following theorem:

Theorem 5.1. *Given a fixed sequence $I = (i_1, i_2, \ldots, i_n)$ of n positive integer numbers there is a data structure which can be constructed in $\mathcal{O}(2^n n^3)$ time and determines the value $\#(I, z)$ for every $z \in \mathbb{Z}$ in $\mathcal{O}(n)$ time.*

6 Experimental Results

We compared our Faulhaber based algorithm with a naïve algorithm based on Lemma 3.1. This algorithm splits in every step a sequence of upper bounds $(i_1, i_2, \ldots, i_n)$ into the two sequences $(i_1, i_2, \ldots, i_{\lfloor \frac{n}{2} \rfloor})$ and $(i_{\lfloor \frac{n}{2} \rfloor+1}, i_{\lfloor \frac{n}{2} \rfloor+2}, \ldots, i_n)$ which are processed recursively. It terminates if it reaches a sequence of length one or two and uses shortcuts for these cases which run in constant time (as described in [3]). The running time of this algorithm is hard to specify since it depends both on the upper bounds, the target value and hence of the final result.

The implementations were written in C++11 using the GNU Compiler Collection without any optimisation (-O0 -ggdb). For the arithmetic operations we used the GMP library (see http://gmplib.org). For further improvement we stored the coefficients of already considered polynomials in order to avoid multiple computations. The calculation run kvm-virtualised on linux (Arch Linux gcc version 4.7.1 20120721 (prerelease)) with one core of Intel(R) Xeon(R) CPU E5540 clocked at 2.53GHz. Under these circumstances we obtained the results from Figure 2. The running times of the Faulhaber based algorithm include also the precomputation stage. We stopped a test run after one hour.

Nr.	I, z	Faulhaber based algorithm	naïve algorithm
1	$(3, 4, 5), 1$	50 ms	4 ms
2	$(3, 4, 5), 15$	49 ms	4 ms
3	$(3000, 4000, 5000), 1$	49 ms	4 ms
4	$(3000, 4000, 5000), 6000$	48 ms	149 ms
5	$(10^4, 10^4, 10^4, 10^4, 10^4), 1$	54 ms	4 ms
6	$(10^4, 10^4, 10^4, 10^4, 10^4), 300$	53 ms	302 ms
7	$(10^4, 10^4, 10^4, 10^4, 10^4), 2.5 \cdot 10^4$	57 ms	11 min(!)
8	$(10000, 10005, 10010, 10015, 10020),$ 25015	64 ms	11 min(!)
9	$(10993, 10520, 10856, 10346, 10039),$ 1	70 ms	4 ms
10	$(10993, 10520, 10856, 10346, 10039),$ 26377	68 ms	13 min(!)
11	$(33, 29, 42, 34, 59, 76, 54, 33), 180$	345 ms	78 ms
12	$(10000, 10005, 10010, 10015,$ $10021, 10027, 10039, 10063),$ 40090	458 ms	> 1 h
13	$(10000, 10000, 10000, 10000, 10000,$ $10000, 10000, 10000, 10000, 10000),$ 5000	74 ms	> 1 h
14	$(10993, 10520, 10856, 10346, 10039,$ $10644, 10005, 10941, 10718, 10305),$ 52683	4,5 s(!)	> 1 h
15	$(12184, 12324, 14685, 11098, 13357, 13863,$ $10796, 10914, 10989, 11115, 10937),$ 66131	17 s(!)	> 1 h
16	$(12184, 12324, 14685, 11098, 13357, 13863,$ $10796, 10914, 10989, 11115, 10937, 13634),$ 1	88 s(!)	4 ms
17	$(12184, 12324, 14685, 11098, 13357, 13863,$ $10796, 10914, 10989, 11115, 10937, 13634),$ 72948	86 s(!)	> 1 h
18	$(3696, 3894, 4137, 7588, 7816), 2856$	57 ms	61 s(!)
19	$(5641, 9314, 969, 8643, 6291,$ $6241, 8747, 7041), 26433$	371 ms	> 1 h

Fig. 2. Experimental Results

At a first glance one may wonder about the high running times even for small instances as Nr.1 and Nr. 2. This is due to the use of data types of the GMP library. The use of elementary data types decreases the running time to values under one microsecond for both algorithms. However, for big input instances the result exceeds the range of `int` and `float` so the use of the GMP library is justified. Moreover, it helps to avoid rounding errors. We applied it also to small instances to obtain consistent comparison results.

In order to test and demonstrate some properties we constructed the examples Nr.1-17 by hand and used randomly generated instances to test the average behaviour (Nr. 18 and Nr. 19). The experimental data shows some expected and some interesting results:

- For small instances the naïve algorithm achieves the better running times because it has no expensive precomputation part like the Faulhaber based algorithm (cf. Nr. 1, 2 and 3).
- The naïve algorithm runs fast on instances with a small target value, even if the upper bounds are great (cf. Nr. 5, 9, 11 and 16; Nr. 6 is an intermediate result). In this case the sum from Lemma 3.1 has only a small number of summands and can be evaluated almost immediately.
- The running time of the Faulhaber based algorithm depends on the number of the upper bounds (cf. Nr. 1-4, 5-10 and Nr. 11/12; in the last pair we have also a growth due to larger numbers).
- The Faulhaber based algorithm has a running time depending also on the number of validity intervals in the constructed data structure. For a sequence $(i_1, i_2, \ldots, i_n)$ with $i_1 = i_2 = \ldots = i_n$ there are exactly $2n + 1$ validity intervals in the final result. The sequences from Nr. 9-12 are constructed such that they have maximal number of validity intervals in the final data structure. This explains the difference between Nr. 6 and Nr. 10. A drastic example for this phenomenon is given by Nr. 13 and Nr. 14.
- The examples Nr. 4, 10, 12, 14, 15 and 17 (which have the maximal number of validity intervals in the final data structure) show the hyperexponential growth of the runtime (recall the precomputation time in $\mathcal{O}(2^n n^3)$) of the Faulhaber based algorithm However, it performs for big input instances and great target values better than the naïve algorithm.
- Seemingly surprisingly, the Faulhaber based algorithm performs for a fixed sequence better for values of z near to $\sum_{j=1}^{n} i_j/2$ than for values near 1 (see Nr. 1/2, 3/4, 5/6, 9/10 and 16/17). There are two reasons for this behaviour: first, the binary search starts in the middle. Second, around $\sum_{j=1}^{n} i_j/2$, the polynomials often have a simpler structure than near 1 (cf. the final result from Figure 1) and can hence be evaluated faster.
- The test examples Nr. 18 and Nr. 19 are randomly generated examples with five and eight upper bounds, resp. In these cases the Faulhaber based algorithm performs much better. The running times for similar randomly generated instances behaved in a comparable way.

In future practical applications based on [3] and [9], especially parallel algorithms for the evaluation of Preference SQL expressions, the number of upper bounds will likely exceeding five, and their size can be arbitrarily up to around 10^4 (big data). In this domain, the Faulhaber based algorithm offers for the very first time the possibility for practical computation.

7   Conclusion and Outlook

Our novel algorithm performed well on big instances (which are also of practical interest) whereas for simple small instances the naïve algorithm is preferable.

The main drawback of our algorithm is the possibly exponential number of validity intervals. This can not be avoided by permutation of the upper bounds (this will lead to roughly the same validity intervals; the value of $\#(I, z)$ remains the same if the upper bounds of I are permuted), so a substantial improvement of an approach based on validity intervals is hard to expect.

One idea of improvement is to exploit the property $\#((i_1, i_2, \ldots, i_n), z) = \#((i_1, i_2, \ldots, i_n), \sum_{j=1}^{n} i_j - z)$ which means that $\#((i_1, i_2, \ldots, i_n), z)$ is symmetric with respect to $z_0 = \sum_{j=1}^{n} \frac{i_j}{2}$ (see again [3] and [9]). However, this should lead to a speed up of at most a factor two, so the asymptotic running time remains the same. Also there could be improvements for the computing of the coefficients as already mentioned in Section 5.

A totally different approach consists in exploiting results for discrete convolutions, as the equation from Lemma 3.1 is a discrete convolution of the functions $\#((i_1, i_2, \ldots, i_{n-m}), k)$ and $\#((i_{n-m+1}, i_{n-m+2}, \ldots, i_n), z - k)$. So it seems to be reasonable to use known algorithms for fast discrete convolution (see e.g. [4] and [7]). Here further research is needed to investigate this approach.

Acknowledgements. We are grateful to Markus Endres and the anonymous referees for valuable discussions and remarks.

References

1. Andrews, G.E.: The Theory of Partitions, 1st edn. Cambridge University Press (1998)
2. Conway, H., Guy, R.: The Book of Numbers, 1st edn. Copernicus (1995)
3. Endres, M.: Semi-Skylines and Skyline Snippets - Theory and Applications. Books on Demand (2011)
4. von zur Gathen, J., Gerhard, J.: Modern Computer Algebra, 1st edn. Cambridge University Press (2003)
5. Hardy, G., Wright, E.: An Introduction to the Theory of Numbers, 3rd edn. Oxford University Press (1954)
6. Knuth, D.E.: Johann Faulhaber and Sums of Powers. Math. Comp. 61(203), 277–294 (1993)
7. Knuth, D.E.: Seminumerical Algorithms, 3rd edn. Addison-Wesley (1997)
8. Matoušek, J., Nešetřil, J.: Invitation to Discrete Mathematics. Springer (2002)
9. Preisinger, T.: Graph-based Algorithms for Pareto Preference Query. Books on Demand (2009)
10. Wirsching, G.: Balls in constrained urns and cantor-like sets. Zeitschrift für Analysis und ihre Anwendungen 17, 979–996 (1998)

Hypergraph Transversal Computation
with Binary Decision Diagrams

Takahisa Toda

ERATO MINATO Discrete Structure Manipulation System Project, Japan Science
and Technology Agency, at Hokkaido University, Sapporo 060-0814, Japan
toda@erato.ist.hokudai.ac.jp, toda.takahisa@gmail.com

Abstract. We study a hypergraph transversal computation: given a
hypergraph, the problem is to generate all minimal transversals. This
problem is related to many applications in computer science and various algorithms have been proposed. We present a new efficient algorithm
using the compressed data structures BDDs and ZDDs, and we analyze
the time complexity for it. By conducting computational experiments, we
show that our algorithm is highly competitive with existing algorithms.

Keywords: hitting set, BDD, ZDD, transversal hypergraph, Boolean
function, data mining, logic, artificial intelligence, monotone dualization.

1 Introduction

A *hypergraph* is a pair $\mathcal{H} = (V, \mathcal{E})$ of a set V and a family $\mathcal{E}$ of subsets of V,
where the sets in $\mathcal{E}$ are called *hyperedges*. A *hitting set* (or *transversal*) for $\mathcal{E}$
is a set $T \subseteq V$ such that T "hits" every hyperedge in $\mathcal{E}$, that is, $T \cap U \neq \emptyset$
for all $U \in \mathcal{E}$. A hitting set is *minimal* if no proper subsets are hitting sets.
The *transversal hypergraph* of $\mathcal{H}$ is a hypergraph whose ground set is V and
whose hyperedges are all minimal hitting sets for $\mathcal{E}$. The *hypergraph transversal
computation* is, given a hypergraph, to compute the transversal hypergraph by
generating all minimal hitting sets.

The hypergraph transversal computation has attracted the attention of many
researchers in computer science, since it is related to a fundamental aspect of
set families and hence there are many important applications in a wide variety
of areas in computer science, especially in data mining, logic, and artificial intelligence. On detailed description of applications and known results, the reader
is referred to the survey papers [1,2], as well as to the references therein.

Many efforts have been made to clarify the exact complexity. The breakthrough result of Fredman and Khachiyan [3] shows that the problem of decid-
ing, given two hypergraphs $\mathcal{G}$ and $\mathcal{H}$, if $\mathcal{G}$ is the transversal hypergraph of $\mathcal{H}$
can be solved in quasi-polynomial time $N^{o(\log N)}$, where N is the combined size
of the input $\mathcal{G}$ and $\mathcal{H}$. Furthermore, it is known ([4,5]) that the hypergraph
transversal computation can be solved in *quasi-polynomial total time*, i.e. quasi-
polynomial time in the combined size of input and output hypergraphs, which is

V. Bonifaci et al. (Eds.): SEA 2013, LNCS 7933, pp. 91–102, 2013.

the theoretically best known result so far. Note that the complexity is measured by quasi-polynomial total time, since the size of an output hypergraph can be exponentially larger than the size of an input hypergraph. It remains still open whether there is a polynomial total time algorithm.

On the other hand, because of the wide applicability, research activities with an emphasis on practical efficiency have been pursued and many algorithms have been proposed. Only recently Murakami and Uno [6] have developed two algorithms based on reverse search (RS) and depth-first search (DFS), and experiments showed that their algorithms outperform the existing algorithms [7–10] in almost all datasets they used. Yet another algorithm was given by Knuth as an exercise of his famous book [11, pp.669–670]. This algorithm is based on different paradigm from the algorithms ever proposed.

In this paper we present a new algorithm for the hypergraph transversal computation. Our algorithm makes use of the two special data structures for Boolean functions and set families: $BDDs$ and $ZDDs$, respectively. They are known as efficient compression techniques: Boolean functions and set families can be respectively compressed into BDDs and ZDDs; various operations can be efficiently performed on these data structures without decompression. Our algorithm can be considered as a variant of Knuth algorithm, since Knuth algorithm uses only ZDDs, while our algorithm in addition uses BDDs. As far as we know, the ZDD-based approach was initially invented by Knuth. Unfortunately this approach seems to be not well-known (indeed, it is buried in a large number of exercises) and there is almost no knowledge of performance. For this, we give an explanation not only of our algorithm but also of necessary notions and results of BDDs and ZDDs. We furthermore conduct experiments with many datasets, including comparisons with Knuth, RS and DFS algorithms.

This paper is organized as follows. In Section 2 we introduce the two data structures BDDs and ZDDs. In Section 3 we present our algorithm based on BDDs and ZDDs together with theoretical analysis. Section 4 provides experimental results. We conclude in the final section.

2   Data Structures for Set Families

2.1   Introduction to BDDs

A *binary decision diagram* (*BDD*) is a graph representation of Boolean functions, which was introduced by Bryant [12] for an application to VLSI logic design and verification (see also [11, pp.257–258]). The advantages of BDDs are: commonly encountered functions are represented as BDDs of reasonable sizes and various Boolean operations can be efficiently performed. BDDs can also be viewed as a data structure for set families, since a Boolean function $f(x_1, \ldots, x_n)$ corresponds to the set of solutions $\{v \in \{0,1\}^n : f(v) = 1\}$ and each solution $(v_1, \ldots, v_n)$ corresponds to the set of variable indices $\{i : v_i = 1\}$.

Figure 1 shows an example of BDD. The node at the top is called the *root*. Each internal node has the three fields V, LO, and HI. The V holds the index of a variable. The fields LO and HI point to other nodes, which are called LO and HI *children*, respectively. The arc to a LO child is called a LO *arc* and illustrated by a dashed arrow. Similarly, the arc to a HI child is called a HI *arc* and illustrated by a solid arrow. There are only two terminal nodes, denoted by $\top$ and $\bot$. In order to distinguish between the terminal nodes of a BDD and those of a ZDD, we denote the former ones by $\bot_{BDD}$, $\top_{BDD}$ and the latter ones by $\bot_{ZDD}$, $\top_{ZDD}$.

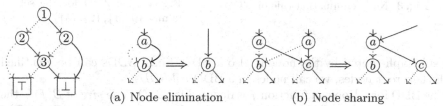

Fig. 1. The BDD for the set family $\{\emptyset, \{1\}, \{2\}, \{3\}\}$

(a) Node elimination   (b) Node sharing

Fig. 2. Reduction rules on BDDs

BDDs satisfy the following two conditions. They must be *ordered*: If an internal node u points to an internal node v, then $V(u) < V(v)$. They must be *reduced*: the following two reduction operations can not be applied.

1. For each internal node u whose two arcs point to the same node v, redirect all the incoming arcs of u to v, and then eliminate u (see Fig. 2(a)).
2. For any nodes u and v, if the subgraphs rooted by u and v are equivalent, then share the two subgraphs (see Fig. 2(b)).

We can understand BDDs as follows. Each path from the root to a terminal node represents a $(0, 1)$-assignment for arguments and the value of a Boolean function. For example, in Fig. 1 the path ① $\rightarrow$ ② $\dashrightarrow$ ③ $\dashrightarrow$ $\top$ means $f(1, 0, 0) = 1$ and ① $\rightarrow$ ② $\rightarrow$ $\bot$ means $f(1, 1, 0) = f(1, 1, 1) = 0$. Note that a ③ node is eliminated from the latter path thus the value of x_3 does not influence the value of f. When the BDD is considered as a representation of a set family on $\{1, 2, 3\}$, the paths ① $\rightarrow$ ② $\dashrightarrow$ ③ $\dashrightarrow$ $\top$ and ① $\rightarrow$ ② $\rightarrow$ $\bot$ mean that the BDD has $\{1\}$ but neither $\{1, 2\}$ nor $\{1, 2, 3\}$.

It is known (see for example [11, 12]) that every Boolean function has one and only one representation as a BDD and that if the number of variables is fixed, then every BDD represents a unique Boolean function. BDD nodes are uniquely represented by using a hash table, called *uniquetable*. The function BDD_UNIQUE manipulates the uniquetable as follows. Given the triple of an index k and nodes l, h, the function BDD_UNIQUE returns a node associated with the key (k, l, h) if exists; otherwise, create a new node p such that $V(p) = k$, $LO(p) = l$, and $HI(p) = h$; register p to the uniquetable and return p. The uniquetable guarantees that two nodes are different if and only if the subgraphs rooted by them represent different Boolean functions. Thus, for example, equivalence checking of BDDs can be done in constant time. For any node in a BDD,

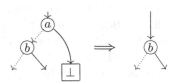

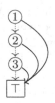

Fig. 3. Node elimination rule on ZDDs

Fig. 4. The ZDD for the set family $\{\emptyset, \{1\}, \{2\}, \{3\}\}$

the subgraph rooted by the node is also a BDD. Since BDDs can be identified with their root nodes, we call nodes in a BDD *subBDDs*.

The BDD for a Boolean function f is denoted by $B(f)$. The *size* of $B(f)$ is the number of nodes in $B(f)$, including terminal nodes, and denoted by $|B(f)|$. The operation AND $(B(f), B(g)) := B(f \wedge g)$ can be computed in time proportional to $|B(f)| \cdot |B(g)|$ (see [13]).

2.2   Introduction to ZDDs

When a family of sparse sets[1] is represented as a BDD, it is likely that there are many nodes whose HI arcs point to $\perp$. Minato [14] introduced a variety of BDDs specialized for such set families, called *zero-suppressed binary decision diagrams* (*ZDDs*). Specifically, ZDDs are ordered BDDs with the following reduction rules.

1. For each internal node u whose HI arc points to $\perp$, redirect all the incoming arcs of u to the LO child, and then eliminate u (see Fig. 3).
2. For any nodes u and v, if the subgraphs rooted by u and v are equivalent, then share the two subgraphs (see Fig. 2(b)).

Note that ZDDs need not satisfy the node elimination rule of the original BDDs.

Each path in a ZDD exactly corresponds to a single set. The ZDD in Fig. 4 represents the same set family to the BDD in Fig. 1. The paths ① → ⊤ and ① --→ ② --→ ③ --→ ⊤ correspond to $\{1\}$ and $\emptyset$, respectively. Note that the two arcs of the ③ node both point to ⊤, but it must not be eliminated.

As in BDDs, similar results are known (see for example [11,14]). Given a set V, every hypergraph on V has a unique form as a ZDD if the order of the vertices is fixed. The ZDD for a set family $\mathcal{E}$ is denoted by $Z(\mathcal{E})$. The two terminal nodes $\perp_{ZDD}$ and $\top_{ZDD}$ correspond to $\emptyset$ and $\{\emptyset\}$, respectively. Note that $\perp_{BDD}$ and $\top_{BDD}$ correspond to $\emptyset$ and 2^V. As in BDDs, ZDD nodes are maintained by their uniquetable, and the function ZDD_UNIQUE (k, l, h) returns a unique node associated with the key (k, l, h) in constant time. For any node v in a ZDD, the subgraph rooted by v is also a ZDD; thus we identify v with the subgraph rooted by v if no danger of confusion. We call nodes in a ZDD *subZDDs*. The operations UNION $(Z(\mathcal{U}), Z(\mathcal{V})) := Z(\mathcal{U} \cup \mathcal{V})$ and DIFF $(Z(\mathcal{U}), Z(\mathcal{V})) := Z(\mathcal{U} \setminus \mathcal{V})$ can be computed in time proportional to $|Z(\mathcal{U})| \cdot |Z(\mathcal{V})|$ (see [14]).

---

[1] The size of a set tends to be much smaller than the size of a ground set.

3   Algorithm

Our entire algorithm consists of the following 4 parts. Note that $S(x)$ denotes the set family for a BDD (or ZDD) x. For simplicity, we assume that the ground set V of an input set family $\mathcal{E}$ is $\{1, \ldots, n\}$ and each hyperedge in $\mathcal{E}$ is sorted.

1. Compress an input set family $\mathcal{E}$ into a ZDD p.
2. Compute the BDD q for all hitting sets for $S(p)$.
3. Compute the ZDD r for all minimal sets in $S(q)$.
4. Decompress r to a set family $\mathcal{E}^*$ and output $\mathcal{E}^*$.

This approach, in particular the use of BDDs as intermediate representations and Algorithm 2 computing the 2nd part, is a new result. In practice, it would be better to output the ZDD obtained in the 3rd part, since otherwise additional time and space are required, and what is worse, a huge number of sets can be dumped. Nevertheless, in the experimental comparison in a later section, the last part is included, because we want to compare algorithms under the same input and output conditions. We analyze the time complexity for the 1st part, since there seems no literature which explicitly mentions it. The 3rd part is computed by Algorithm 3. Although this algorithm is implicitly mentioned in [11, pp.255–256], for the completeness we include and outline it.

Compression of a set family $\mathcal{E}$ is given in Algorithm 1. Let $U \in \mathcal{E}$. Let i_k denote the k-th number in U in decreasing order. The following recursion holds.

$$Z(\{\{i_1, \ldots, i_k\}\}) = \begin{cases} \text{ZDD\_UNIQUE}\,(i_1, \bot_{\text{ZDD}}, \top_{\text{ZDD}}) & (\text{if } k = 1) \\ \text{ZDD\_UNIQUE}\,(i_k, \bot_{\text{ZDD}}, Z(\{\{i_1, \ldots, i_{k-1}\}\})) & (\text{if } 1 < k) \end{cases}$$

Thus $Z(\{U\})$ can be constructed in a bottom up fashion in $O(|U|)$ time. Since ZDDs must be ordered, it is essential to select numbers in U in decreasing order. Suppose that we have constructed the ZDD $p := Z(\{U_1, \ldots, U_{m-1}\})$. For a new set $U_m \in \mathcal{E}$, the UNION $(p, Z(\{U_m\}))$ produces $Z(\{U_1, \ldots, U_{m-1}, U_m\})$. In general, this function requires time proportional to the product of the sizes of two input ZDDs. However, in this case it can be done in time proportional to the size of the ground set. We show this while referring to the algorithm for UNION described in [14]. Suppose that UNION (P, Q) is called in computing UNION $(p, Z(\{U_m\}))$. There are three cases. If $V(P) = V(Q)$, then since Q is a subZDD of $Z(\{U_m\})$ and its LO arc points to $\bot$, the UNION $(\text{LO}\,(P), \text{LO}\,(Q))$ immediately returns LO (P), thus UNION $(\text{HI}\,(P), \text{HI}\,(Q))$ is then called. If $V(P) < V(Q)$, then for the same reason UNION $(P, \text{LO}\,(Q))$ immediately returns P and no further call is required. If $V(P) > V(Q)$, then UNION $(\text{LO}\,(P), Q)$ is called. One can observe that UNION (P, Q) essentially calls at most one function UNION (P', Q') with $V(P) < V(P')$. Since $V(P)$ is bounded above by the size of the ground set V, the time required to compute UNION $(p, Z(\{U_m\}))$ is $O(|V|)$. Therefore, Algorithm 1 requires $O(|V| \cdot |\mathcal{E}|)$ time.

Conversely, decompression of a ZDD can be done as follows. Since paths from the root to $\top_{\text{ZDD}}$ correspond in a one-to-one way to sets stored in the ZDD, it suffices, for each such path, to compute the corresponding set. Since the length

---

Algorithm 1. Compress a set family $\mathcal{E}$ into a ZDD p

---

function COMP($\mathcal{E}$)

    $p \leftarrow \perp_{\text{ZDD}}$;

    for each set $U \in \mathcal{E}$ **do**

        $t \leftarrow \top_{\text{ZDD}}$;

        for each number $i \in U$ in decreasing order **do**

            $t \leftarrow \text{ZDD\_UNIQUE}\,(i, \perp_{\text{ZDD}}, t)$;

        end for

        $p \leftarrow \text{UNION}\,(p, t)$;

    end for

    return p;

end function

---

of a path is at most the size of a ground set V, the time required is $O(|V| \cdot |\mathcal{E}^*|)$, where $\mathcal{E}^*$ denotes the output set family.

For the 2nd part of our algorithm, let HIT denote the function given in Algorithm 2. We show by structural induction on an input ZDD p that the HIT correctly returns the BDD for all hitting sets for $\mathcal{S}(p)$. The case that p is a terminal node is immediate. For the other case, let pl and ph denote the LO and HI children of p, respectively. Observe that $\mathcal{S}(p)$ is the disjoint union of $\mathcal{S}(pl)$ and $\{\{V\,(p)\} \cup U : U \in \mathcal{S}(ph)\}$. Thus a necessary and sufficient condition for a set T to be a hitting set for $\mathcal{S}(p)$ is that (1) T is a hitting set for $\mathcal{S}(pl)$ and (2) $V\,(p) \in T$ or T is a hitting set for $\mathcal{S}(ph)$. By induction hypothesis, all sets T satisfying the condition (1) are enumerated by HIT (pl), while those satisfying the condition (2) are enumerated by BDD_UNIQUE $(V\,(p), \text{HIT}\,(ph), \top_{\text{BDD}})$. Thus, all sets with the both conditions are enumerated by the following BDD

$$\text{AND}\,(\text{HIT}\,(pl), \text{BDD\_UNIQUE}\,(V\,(p), \text{HIT}\,(ph), \top_{\text{BDD}}))\ .$$

Therefore, the output HIT (p) is correct.

Algorithm 2. Given a ZDD p, compute the BDD for all hitting sets for $\mathcal{S}(p)$

---

function HIT(p)

    if $p = \top_{\text{ZDD}}$ **then**

        return $\perp_{\text{BDD}}$;

    end if

    if $p = \perp_{\text{ZDD}}$ **then**

        return $\top_{\text{BDD}}$;

    end if

    $hl \leftarrow \text{HIT}\,(\text{LO}\,(p))$; $hh \leftarrow \text{HIT}\,(\text{HI}\,(p))$;

    $t \leftarrow \text{BDD\_UNIQUE}\,(V\,(p), hh, \top_{\text{BDD}})$;

    $q \leftarrow \text{AND}\,(hl, t)$;

    return q;

end function

Theorem 1. *Given a ZDD p, Algorithm 2 can be implemented to run in time proportional to $|p| \cdot N(p)^2$, where $N(p) = \max\{|\text{HIT}(p')| : p'$ is a subZDD of $p\}$.*

Proof. Use a hash table to memorize the output BDDs HIT (p') for subZDDs p' of p. For each subZDD p' of p, the computation of HIT (p') is executed exactly once. The BDD_UNIQUE can be computed in constant time. Furthermore, AND (hl, t) can be computed in time proportional to $|hl| \cdot |hh|$ (since $|t| \leq |hh| + 2$), bounded above by $N(p)^2$. Thus, the time necessary to compute HIT (p) is $O(|p| \cdot N(p)^2)$. □

Let us proceed to the 3rd part. This part is to extract minimal sets from an output BDD q of Algorithm 2. For this, we first consider a corresponding notion to such minimal sets in terms of Boolean functions and then consider the ZDD representing them. Let f_q denote the Boolean function for the BDD q above. Observe that $\mathcal{S}(q)$ is *upward closed*, that is, for any set $U \in \mathcal{S}(q)$, if $U \subseteq U'$, then $U' \in \mathcal{S}(q)$. From this, it follows that f_q is monotone, that is, for all $u, v \in \{0,1\}^n$, if $u \leq v$, then $f_q(u) \leq f_q(v)$. It is known (see for example [11, pp.54–55]) that minimal solutions S of a monotone Boolean function h exactly correspond to prime implicants g of h in such a way that $g = \bigwedge_{i \in S} x_i$, where recall that a set S of variable indices is a solution of h if $h(v) = 1$ for $v \in \{0,1\}^n$ such that for all $1 \leq j \leq n$, the j-th component of v equals 1 if and only if $j \in S$. Therefore, minimal sets in $\mathcal{S}(q)$ can be considered as prime implicants of f_q. From the recursion of the ZDD for prime implicants of f_q, thus also for minimal sets in $\mathcal{S}(q)$, described in [11, pp.256], we immediately obtain Algorithm 3.

Algorithm 3. Given a BDD q such that $\mathcal{S}(q)$ is upward closed, compute the ZDD for all minimal sets in $\mathcal{S}(q)$

```
function MIN(q)
    if q = ⊥BDD then
        return ⊥ZDD;
    end if
    if q = ⊤BDD then
        return ⊤ZDD;
    end if
    mh ← MIN (HI (q)); ml ← MIN (LO (q));
    t ← DIFF (mh, ml);
    r ← ZDD_UNIQUE (V (q), ml, t);
    return r;
end function
```

Let MIN denote the function given in Algorithm 3. The following theorem can be proved in a similar way to Theorem 1, and thus we omit the proof.

Theorem 2. *Given a BDD q, Algorithm 3 can be implemented to run in time proportional to $|q| \cdot L(q)^2$, where $L(q) = \max\{|\text{MIN}(q')| : q'$ is a subBDD of $q\}$.*

4 Experiments

Implementation and Environment. We implemented our algorithm and Knuth algorithm [11, pp.669–670] in C. This program is released in [15]. We used the BDD Package SAPPORO-Edition-1.0 developed by Minato, in which not only BDDs but also ZDDs are available and basic operations for BDDs and ZDDs are provided. The implementation of Murakami-Uno algorithms SHD version 3.1 was obtained from the Hypergraph Dualization Repository [16]. All experiments were performed on a 2.67GHz Xeon®E7-8837 with 1.5TB RAM, running SUSE Linux Enterprise Server 11. We compiled our code with version 4.3.4 of the gcc compiler. Note that our implementation does not make use of multi-cores.

Problem Instances. We used total 90 instances, which are classified into the 10 types listed below. These instances have been commonly used in previous studies [5, 6, 9] and can be obtained from [16]. For detailed information, the reader is referred to [6, 16].

1. Matching graph (M(n)): a hypergraph with n vertices (n is even) and $n/2$ edges forming a matching. The parameter n runs over every other number from 20 to 46 except for 22 and 26.
2. Dual Matching graph (DM(n)): the transversal hypergraph of M(n), where n runs over every other number from 20 to 46 except for 22 and 26.
3. Threshold graph (TH(n)): a hypergraph with n vertices (n is even) and the edge set $\{\{i, j\} : 1 \leq i < j \leq n,\ j$ is even$\}$. The parameter n runs over every 20th number from 40 to 200.
4. Self-Dual Threshold graph (SDTH(n)): a hypergraph whose hyperedges are obtained from a TH($n - 2$) and its transversal hypergraph. The parameter n runs over every 20th number from 42 to 202 and every 40th number from 242 to 402.
5. Self-Dual Fano-Plane graph (SDFP(n)): A hypergraph with n vertices and $(k_n - 2)^2/4 + k_n/2 + 1$ hyperedges, where $k_n := (n - 2)/7$. The parameter n runs over every 7th number from 9 to 51 (see [9] for details).
6. accidents (ac(n)): the complements of the sets of maximal frequent itemsets with support threshold $n \cdot 10^3$ from a dataset "accident", where $n \in \{30, 50, 70, 90, 130, 150, 200\}$.
7. BMS-WebView-2 (bms(n)): this is constructed in the same way as ac(n) from a dataset "BMS-WebView-2", where $n \in \{10, 20, 30, 50, 100, 200, 400, 800\}$.
8. Connect-4 win (win(n)): a hypergraph with n hyperedges corresponding to minimal winning stages of the first player in a board game "connect-4". The parameter n runs over $\{100, 200, 400, 800, 1600, 3200, 6400, 12800, 25600\}$.
9. Connect-4 lose (lose(n)): a hypergraph with n hyperedges corresponding to minimal losing stages of the first player in a board game "connect-4". The parameter n runs over $\{100, 200, 400, 800, 1600, 3200, 6400, 12800\}$.
10. Uniform random (rand(n)): a hypergraph such that each vertex is included in a hyperedge with probability $n/10$ ($n \in \{6, 7, 8, 9\}$) and the number of hyperedges is 1000.

Dominating Factor in Algorithm 3. According to Theorem 2, Algorithm 3 depends on the size of an intermediate ZDD, i.e. $|\mathrm{MIN}(q')|$ for a subBDD q' of an input BDD q. It is natural to conjecture $|\mathrm{MIN}(q')| \leq |q'|$. Although we obtained a counterexample to this conjecture by the computational experiment exhaustively conducted on all recursive calls of MIN for all 90 instances, it simultaneously turned out that in all cases the size of $\mathrm{MIN}(q')$ was not more than double the size of q' (the largest ratio was about 1.8). This suggests that $|\mathrm{MIN}(q')|$ is likely to be bounded above by a constant factor of $|q'|$. A similar observation is done in [11, pp.674]. Recall that $\mathrm{DIFF}(mh, ml)$ can be computed in $O(|mh| \cdot |ml|)$ time. Since the size of any subBDD of q is at most the size of q, the experimental observation above implies that $|mh| \cdot |ml|$ is bounded above by a constant factor of $|q|^2$, thus the time required for MIN is $O(|q|^3)$.

Dominating Factor in Algorithm 2. The same experiment was conducted on all recursive calls of the function HIT. We observed that the largest ratio of $|\mathrm{HIT}(p')|/|p'|$ depends on instances: the ratio was always 1.0 in the instances $\mathrm{TH}(n)$, $\mathrm{DM}(n)$, $\mathrm{M}(n)$ with all possible parameters, while it drastically changed in instances lose(n) and rand(n). In particular, the largest ratio was about 1378, achieved by rand(6). Thus $|\mathrm{HIT}(p')|$ is not likely to be bounded by a constant factor of $|p'|$. On the other hand, the smallest ratio among all cases was about 0.5. Thus, for the present instances, $|p|$ is bounded above by a constant factor of $|\mathrm{HIT}(p)|$, thus Algorithm 2 is dominated only by the maximum size of an intermediate BDD $N(p)$. However, in general this can not be applied to every case. An extreme example is a ZDD p with $\emptyset \in \mathcal{S}(p)$. Since no set can hit $\emptyset$, the family of hitting sets for $\mathcal{S}(p)$ is empty and the corresponding ZDD is $\bot$. On the other hand, clearly $|p|$ can not be bounded above by a constant. Therefore, we conclude that Algorithm 2 is dominated by both $|p|$ and $N(p)$.

Running Time for Algorithm 2 and 3. For convenience, we introduce the following terminology: for a ZDD p, a BDD x is called a *dominating* BDD if there

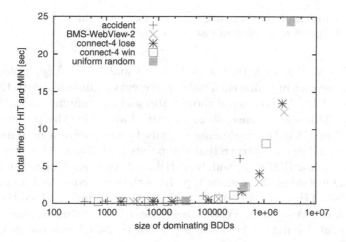

Fig. 5. The size of dominating BDDs and the total time required for Algorithm 2 and 3

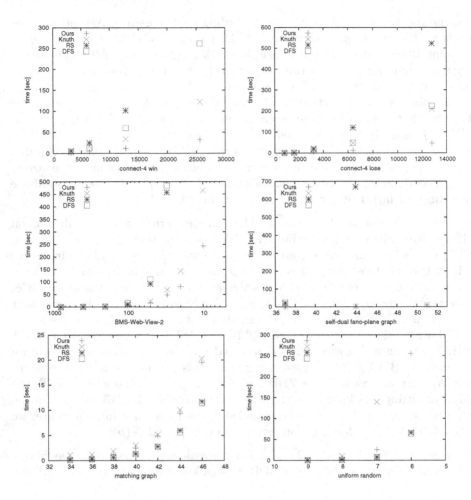

Fig. 6. Comparisons of running time with existing algorithms, where the horizontal coordinate of a point represents an instance parameter

is a subZDD p' of p such that $x = \mathrm{HIT}\,(p')$ and $|x| = N(p)$ hold. Figure 5 shows a scatter plot in which each point represents an instance with the size of a dominating BDD (the horizontal coordinate) and the running time (the vertical coordinate). This makes sense, since as argued above for the present instances Algorithm 2 and 3 both are dominated only by the size of a dominating BDD in Algorithm 2. We can observe that the points form a smooth curve. Since the sizes of dominating BDDs are widely distributed, we used a logarithmic horizontal axis. The instances of the same type have the same color and all instances in $M(n)$, $DM(n)$, $TH(n)$, $SDTH(n)$, or $SDFP(n)$ are excluded, since they finished within at most 0.3 seconds. Furthermore, rand(6) is excluded, since the corresponding point, having the coordinate $(25640938, 245)$, has a too larger vertical coordinate than the other points have, although this point is roughly placed along the extension of that curve.

Running Time for Entire Algorithm. We compared running time between our algorithm, Knuth algorithm and Murakami-Uno algorithms (RS and DFS). We selected some characteristic results and the corresponding figures are given in Fig. 6 (see [15] for detailed results). When a computation did not terminate in 1000 seconds, we forcefully closed it and no point for the instance was plotted.

- For the instances in win(n), lose(n), bms(n), and SDFP(n), our algorithm was significantly faster than the other algorithms.
- For ac(30), which was the most time-consuming instance in ac(n), our algorithm was about 2 times faster than Murakami-Uno algorithms and about 7 times faster than Knuth algorithm.
- For the instances in M(n), most of the running time in our algorithm was spent by decompression of an output ZDD.
- For rand(6), in which the worst size of a dominating BDD was achieved, our algorithm was about 4 times slower than Murakami-Uno algorithms.

We remark that our approach needs much memory due to uniquetable. See [15] for comparisons of maximum memory usages.

5 Conclusion

We presented a new algorithm for the hypergraph transversal computation. Experiments for total 90 instances suggested that this algorithm is highly competitive with the algorithm of Knuth [11, pp.669–670] and the two recently developed algorithms of Murakami and Uno [6], where it is shown that Murakami-Uno algorithms outperform the existing algorithms [7–10] in almost all datasets they used. Furthermore, we experimentally observed that the main part of our algorithm is dominated by both the size of an input ZDD of the function HIT and the largest size of an intermediate BDD generated by HIT.

A future work will be to theoretically prove the experimental observation above. Parallel algorithms for the hypergraph transversal computation have been studied (see [2] [17]). It would be interesting to parallelize our algorithm and compare with existing algorithms. It would be also interesting to apply our algorithm to problems described in [1]. For this, there is a big advantage of BDD and ZDD-based approach: as demonstrated in [18], BDDs and ZDDs provide useful operations to solve combinatorial problems, and compression technique allows us to solve large-scale instances that cannot be handled otherwise.

Acknowledgments. The author would like to thank Professor Shin-ichi Minato and Dr. Takeru Inoue for inspiring discussions and comments, and anonymous reviewers for their helpful suggestions.

References

1. Eiter, T., Gottlob, G.: Hypergraph transversal computation and related problems in logic and AI. In: Flesca, S., Greco, S., Leone, N., Ianni, G. (eds.) JELIA 2002. LNCS (LNAI), vol. 2424, pp. 549–564. Springer, Heidelberg (2002)
2. Eiter, T., Makino, K., Gottlob, G.: Computational aspects of monotone dualization: A brief survey. Discrete Applied Mathematics 156, 2035–2049 (2008)
3. Fredman, M., Khachiyan, L.: On the complexity of dualization of monotone disjunctive normal forms. Journal of Algorithms 21, 618–628 (1996)
4. Gurvich, V., Khachiyan, L.: On generating the irredundant conjunctive and disjunctive normal forms of monotone Boolean functions. Discrete Applied Mathematics 96-97, 363–373 (1999)
5. Khachiyan, L., Boros, E., Elbassioni, K., Gurvich, V.: An efficient implementation of a quasi-polynomial algorithm for generating hypergraph transversals and its application in joint generation. Discrete Applied Mathematics 154, 2350–2372 (2006)
6. Murakami, K., Uno, T.: Efficient algorithms for dualizing large-scale hypergraphs. In: Proc. of the Meeting on Algorithm Engineering & Experiments, ALENEX, New Orleans, Louisiana, USA, pp. 1–13 (January 2013)
7. Dong, G., Li, J.: Mining border descriptions of emerging patterns from dataset pairs. Knowledge and Information Systems 8, 178–202 (2005)
8. Bailey, J., Manoukian, T., Ramamohanarao, K.: A fast algorithm for computing hypergraph transversals and its application in mining emerging patterns. In: Proc. of the 3rd IEEE International Conference on Data Mining, pp. 485–488. IEEE Computer Society (November 2003)
9. Kavvadias, D., Stavropoulos, E.: An efficient algorithm for the transversal hypergraph generation. Journal of Graph Algorithms and Applications 9(2), 239–264 (2005)
10. Hérbert, C., Bretto, A., Crémilleux, B.: A data mining formalization to improve hypergraph minimal transversal computation. Fundamental Informaticae 80, 415–433 (2007)
11. Knuth, D.: The Art of Computer Programming, vol. 4A. Addison-Wesley Professional, New Jersey (2011)
12. Bryant, R.: Graph-based algorithms for Boolean function manipulation. IEEE Transaction on Computers 35, 677–691 (1986)
13. Yoshinaka, R., Kawahara, J., Denzumi, S., Arimura, H., Minato, S.: Counterexamples to the long-standing conjecture on the complexity of BDD binary operations. Information Processing Letters 112, 636–640 (2012)
14. Minato, S.: Zero-suppressed BDDs for set manipulation in combinatorial problems. In: Proc. of 30th ACM/IEEE Design Automation Conference, DAC 1993, Dallas, Texas, USA, pp. 272–277 (June 1993)
15. Toda, T.: HTC-BDD: Hypergraph Transversal Computation with Binary Decision Diagrams (2013), http://kuma-san.net/htcbdd.html (accessed on March 28)
16. Murakami, K., Uno, T.: Hypergraph Dualization Repository (2013), http://research.nii.ac.jp/~uno/dualization.html (accessed on January 19)
17. Khachiyan, L., Boros, E., Elbassioni, K., Gurvich, V.: A global parallel algorithm for the hypergraph transversal problem. Information Processing Letters 101, 148–155 (2007)
18. Coudert, O.: Solving graph optimization problems with ZBDDs. In: Proc. of the 1997 European Conference on Design and Test, Paris, France, pp. 224–228 (March 1997)

Efficient Counting of Maximal Independent Sets in Sparse Graphs

Fredrik Manne and Sadia Sharmin

Dep. of Informatics, Univ. of Bergen, Norway
{fredrikm,Sadia.Sharmin}@ii.uib.no

Abstract. There are a number of problems that require the counting or the enumeration of all occurrences of a certain structure within a given data set. We consider one such problem, namely that of counting the number of maximal independent sets (MISs) in a graph. Along with its complement problem of counting all maximal cliques, this is a well studied problem with applications in several research areas.

We present a new efficient algorithm for counting all MISs suitable for sparse graphs. Similar to previous algorithms for this problem, our algorithm is based on branching and exhaustively considering vertices to be either in or out of the current MIS. What is new is that we consider the vertices in a predefined order so that it is likely that the graph will decompose into multiple connected components. When this happens, we show that it is sufficient to solve the problem for each connected component, thus considerably speeding up the algorithm. We have performed extensive experiments comparing our algorithm with the previous best algorithms for this problem using both real world as well as synthetic input graphs. The results from this show that our algorithm outperforms the other algorithms and that it enables the solution of graphs where other approaches are clearly infeasible.

As there is a one-to-one correspondence between the MISs of a graph and the maximal cliques of its complement graph, it follows that our algorithm also solves the problem of counting the number of maximal cliques in a dense graph. To our knowledge, this is the first algorithm that can handle this problem.

1 Introduction

Enumerating all configurations that conforms with a given specification is a well studied problem in combinatorics. Graph theory deals with many interesting problem of this type. Enumerating all maximal independent sets (MISs) of a graph is one of these problems that has attracted considerable attention in the past [5,10,12,14,20]. This problem is also equivalent to enumerating all the maximal cliques of a graph as there is a one-to-one correspondence between the MISs of a graph and the maximal cliques of its complement graph. For a recent overview of applications of this problem, see [6] and the references therein.

In the classical MIS enumeration problem, the number of configurations to be generated is potentially exponential in the size of the input. Moon and Moser showed that a graph on n vertices can have at most $3^{\frac{n}{3}}$ MISs and that this bound is tight [16]. Thus for

V. Bonifaci et al. (Eds.): SEA 2013, LNCS 7933, pp. 103–114, 2013.

graphs coming close to this bound one can only expect to be able to list or enumerate all MISs of graphs of fairly limited size.

In this paper we study algorithms for *counting* the number of MISs, a problem one might expect can be solved more efficiently than the enumeration problem. However, it is known that the counting problem is ♯P-complete even when restricted to chordal graphs [17], and therefore no polynomial time algorithm exists unless P=NP [12].

The currently fastest algorithm for solving the MIS counting problem on general graphs is by Gaspers et al. who presented a branch and bound algorithm that runs in $O(1.3642^n)$ time [8]. As $3^{\frac{n}{3}} \approx 1.44^n$ this shows that it is possible to count MISs faster than by generating each one. For the case of sub-cubic graphs Junosza-Szaniawski and Tuczyński recently gave an algorithm with running time $O(1.2570^n)$ [11]. For trees, Wilf presented a simple linear time dynamic programming algorithm [21]. He also showed that the number of MISs in a tree is at most $2^{n/2-1} + 1$ and that there are graphs that meet this bound.

The counting problem can obviously be solved by enumeration, a problem which has seen a variety of approaches by a number of authors, see [5] for an overview. The standard algorithm for this problem is the Bron–Kerbosch algorithm [1] which is a recursive backtracking algorithm that searches for all maximal cliques in a given graph G, (which in the complement graph corresponds to the MISs). This algorithm was later improved by Tomita et al. [18] using a pivoting heuristic that reduces the number of recursive calls. We also note that Eppstein and Strash gave a variation of the Tomita algorithm by initially reordering the vertices using a degeneracy ordering [7], something that is advantageous for very sparse graphs.

Experimental work on these (and other) algorithms for enumerating cliques has mainly focused on sparse graphs [2,3,7,18]. This means that they are applicable on dense graphs for enumerating (or counting) MISs. Enumerating MISs on sparse graphs (or enumerating cliques in dense graphs) is a substantially harder problem as one would expect the number of MISs to decrease as the graph becomes denser. We are not aware of any experimental studies of algorithms for this problem.

Our Results: We present the first algorithm specifically suited for counting MISs in sparse graphs. The algorithm combines a branching approach with a divide and conquer strategy. This is achieved by making the branching follow vertex separators in the graph. In this way the remaining graph will become disconnected and one can solve the problem for each connected component separately. To find the separators we initially use graph partitioning software to compute a nested dissection ordering on the graph.

We apply the algorithm to both real world as well as synthetic sparse graphs and show that it outperforms other suggested algorithms designed for counting or enumerating MISs. Although the individual aspects of our algorithm are not new, this is, as far as we know, the first time they have been combined together to create an efficient code for counting MISs in sparse graphs.

2   Notation

We consider an undirected finite graph $G = (V, E)$ without loops, where V is the set of vertices of G and E is the set of edges. We denote the neighborhood of a vertex v in the graph G by $N_G(v)$, that is the set of vertices u such that the edge $(u, v) \in E$. The closed neighborhood of a vertex v is denoted by $N_G[v]$, which is $N_G(v) \cup \{v\}$. The degree of a vertex $d_G(v) = |N_G(v)|$. If I is a set of vertices in G then $N_G(I) = (\cup_{v \in I} N_G(v)) \setminus I$. Also, the induced subgraph of I in G is the graph $G[I] = (I, E_I)$ where $(v, w) \in E_I$ if and only if $v, w \in I$ and $(v, w) \in E$.

A set of vertices S is an independent set if no two vertices in S are adjacent. A vertex $v \notin S$ which is adjacent to a vertex $w \in S$ is said to be *dominated* by w, or just dominated. A vertex not in S which is not dominated is undominated. A maximal independent set is an independent set that is not a subset of any other independent set.

3   Previous Algorithms for Counting and Enumerating MISs

In the following we present previously suggested algorithms for counting or enumerating all MISs of a graph G. For the enumeration problem we present algorithms that have been used in experimental studies and that are fairly straight forward to implement. We also outline the counting algorithm by Gaspers et al. Since our main interest is to count the number of MISs, we describe all algorithms as applied to this problem.

The Bron-Kerbosch algorithm in its basic form uses recursive backtracking to list all maximal cliques in a given graph [1]. In the following we present the dual of this algorithm, so that instead of cliques the algorithm counts all MISs in G.

Given three vertex sets R, P, and X, Algorithm 1: BKMIS(R, P, X) finds all MISs that include all vertices in R, any possible legal subset of the vertices from P, and none of the vertices in X. The recursion is initiated by setting both R and X to $\emptyset$ and $P = V$. Within each recursive call, the algorithm considers in turn every vertex in P for inclusion in R. Thus for each $v \in P$ the algorithm makes a recursive call in which v is moved from P to R and any neighbor of v is removed from P and X. In any subsequent call where both P and X are empty, R is counted as a MIS. This will find all maximal independent set extensions of R that contain v. When the recursive call returns, v is moved from P to X before the algorithm continues with the next vertex in P.

Intuitively, one can think of the algorithm as having already found the MISs that contain any vertex from X. Thus any set that does not dominate every vertex in X cannot be a new MIS.

Algorithm 1. BKMIS(R, P, X)

Input: Three vertex sets R, P, and X.
Output: Number of MISs containing all vertices in R, some from P and none from X.
if $P \cup X = \emptyset$ **then**
    Count R as a MIS
for each vertex $v \in P$ **do**
    BKMIS($R \cup \{v\}, P \setminus N_G[v], X \setminus N_G(v)$)
    $P \leftarrow P \setminus \{v\}$
    $X \leftarrow X \cup \{v\}$

The Bron–Kerbosch algorithm is not output-sensitive meaning that it does not run in polynomial time per generated set. The worst-case running time of the Bron–Kerbosch algorithm is $O(3^{\frac{n}{3}})$, matching the Moon and Moser bound [18].

Tomita et al. presented an improved variant of the Bron-Kerbosch algorithm by using a pivoting heuristic [18]. Here we present its dual for computing MISs. In Algorithm 1, $|P|$ recursive calls are made, one for each vertex in P. The pivoting strategy seeks to reduce this number. Consider a vertex $u \in P \cup X$. It follows that no vertex in $N_G[u]$ has been added to R so far. But for the current R to be expanded to a MIS at least one vertex of $P \cap N_G[u]$ must be included in R, otherwise R will not be maximal. Thus once the *pivot* u has been selected, it is sufficient to iterate over the vertices in $P \cap N_G[u]$ for inclusion in R. The idea in Algorithm 2: TOMITAMIS(R, P, X) is then to choose u such that this number is as small as possible. Computing both the pivot and the vertex

---

Algorithm 2. TOMITAMIS(R, P, X)

Input: Three vertex sets R, P and X.
Output: Number of MISs containing all vertices in R, some vertices from P and no vertex from X.
if $P \cup X = \emptyset$ **then**
    Count R as a MIS
Choose a pivot $u \in P \cup X$ that minimizes $|P \cap N_G(u)|$
for each vertex $v \in P \cap N_G[u]$ **do**
    TOMITAMIS$(R \cup \{v\}, P \setminus N_G[v], X \setminus N_G(v))$
    $P \leftarrow P \setminus \{v\}$
    $X \leftarrow X \cup \{v\}$

---

sets for the recursive calls can be done in time $O(|P|(|P| + |X|))$ within each call to the algorithm using an adjacency matrix, giving an overall running time of $O(3^{\frac{n}{3}})$. Experimental comparisons have shown that the maximal clique algorithm by Tomita et al. is faster by orders of magnitude compared to other algorithms [18]. However, both the theoretical analysis and implementation rely on the use of an adjacency matrix representation of the input graph. For this reason, the algorithm has limited applicability for large graphs, whose adjacency matrix may not fit into working memory [7].

Eppstein et al. [6] also proposed a variant of the Bron-Kerbosch algorithm. On the top level this algorithm is similar to the Bron-Kerbosch algorithm, although the vertices are processed according to a degeneracy ordering. Such an ordering can be found by repeatedly selecting and removing a minimum degree vertex. The algorithm then makes $|V|$ calls to the algorithm by Tomita et al., each time with R initially set to the next vertex in the ordering and with P and X updated accordingly. With this setup the algorithm can be implemented to list all maximal cliques of an n-vertex graph in time $O(dn3^{\frac{d}{3}})$, where a graph has degeneracy d if every subgraph has a vertex of degree at most d. In a recent study Eppstein and Strash [7] show that the algorithm is highly competitive with the algorithm by Tomita et al. This is particularly true for large sparse graphs where it in many cases outperform the algorithm by Tomita et al. by orders of magnitude.

Gaspers et al. gave a fast exponential time algorithm of complexity $O(1.3642^n)$ for counting the number of MISs in a graph [8]. This running time is lower than the Moon and Moser bound, something that is possible since the algorithm, unlike the previous mentioned ones, does not enumerate the MISs but only counts their number.

The structure of the algorithm is similar to TOMITAMIS in that a vertex $u \in P \cup X$ is selected as a pivot according to a degree based criterion before branching on the vertices in $P \cap N[u]$. But unlike the previous algorithms, it will in each call first try if any of seven reduction rules can be applied to achieve a smaller but equivalent instance. If this is possible then the instance is reduced accordingly before calling the recursive function again. We note that all rules but one, will return the value given by the following recursive call. The only exception being a rule which checks if there exist two vertices u and v such that their current neighborhoods are identical. In this case v will be removed from the graph and the value of the recursive call will be returned plus the number of MISs discovered in this call that contained u. Another difference is that the algorithm tests if there is a vertex in X having no neighbor in P indicating that the current configuration cannot be expanded to a MIS. If this is the case then the algorithm returns immediately. In the paper it is also noted that if the graph at some stage should become disconnected then the algorithm is called (recursively) for each of its connected components, and the product of the returned values then gives the number of MISs. As far as we know there has been no study of how practical the algorithm is. We refer the interested reader to [8] for the details of the algorithm.

4   A New Algorithm

In the following we present a simple recursive branching algorithm for counting the number of MISs in a graph. Our algorithm is based on locating and exploiting vertex separators of the graph, and is similar in spirit to the algorithm by Lipton and Tarjan for computing a maximum independent set in a planar graph [13].

The Lipton and Tarjan algorithm initially finds a vertex separator $S \subset V$ such that $|S| = O(\sqrt{n})$ and such that no component of $G \setminus S$ contains more than $\frac{2}{3}|V|$ vertices. This is possible since G is assumed to be planar. Then for every independent set I_S of S the algorithm recursively finds a maximum independent set for each connected component of $G \setminus (S \cup N_G(I_S))$. The solution giving the combined largest solution is then the maximum independent set of G. The running time of the algorithm is $2^{O(\sqrt{n})}$.

We modify the Lipton and Tarjan algorithm to compute the number of MISs by using ideas from BKMIS and TOMITAMIS. Note however first that it is not possible to use the algorithm of Lipton and Tarjan to count MISs. The reason for this is that if we pick a particular independent set I_S from a separator S in G and (recursively) calculate the number of MISs in each component of $G[V \setminus (S \cup N_G(I_S))]$, then it is not given that I_S together with every combination of MISs from each of the components will form a MIS in G as some combinations might leave undominated vertices in $S \setminus I_S$.

The new algorithm, Algorithm 3: CCMIS, is recursive and uses two vertex sets P and X to count the number of MISs in $G[P \cup X]$ containing any combination of vertices from P while using none of the vertices in X. Thus if $P \cup X = \emptyset$ this will be counted as one MIS. Also, similar to the algorithm by Gaspers et al. if there exist a vertex in X that is not adjacent to any vertex in P then the algorithm will return 0, as this indicates that the current solution cannot be expanded into a complete MIS. The algorithm also tests at each level of recursion if $G[P \cup X]$ is connected. If this is not the case then the recursive procedure will be called once for each connected component and the product of the number of MISs in each component will be returned. Checking for connectedness and listing the components is done using a linear depth first search through $G[P \cup X]$.

In the case that none of the mentioned conditions apply, the algorithm picks one remaining vertex v from P and then performs two recursive calls, first to compute the number of MISs containing v and then to compute the number of MISs excluding v. Finally, the sum of these two numbers is returned. When counting the number of MISs containing v, any vertex in $N[v]$ is first removed from P and X as these will be dominated by v. Similarly, when counting the number of MISs not containing v, the vertex v is moved from P to X as it must then be dominated by some other vertex in P in a MIS. Note that it is only following a recursive call where v is set to be in the current MIS that the structure of $G[P \cup X]$ will change so that there is any possibility of getting a disconnected graph. The recursion is initiated by setting $X = \emptyset$ and $P = V$.

Algorithm 3. CCMIS(P, X)

Input: Two vertex sets P and X.
Output: Number of MISs in $G[P \cup X]$ containing only vertices from P.
if $P \cup X = \emptyset$ **then**
    return 1
if $\exists w \in X$ with no neighbor in P **then**
    return 0
if $G[P \cup X]$ is not connected **then**
    $count \leftarrow 1$
    for each connected component $CC(V_{CC}, E_{CC})$ of $G[P \cup X]$ **do**
        $count \leftarrow count * $ CCMIS$(V_{cc} \cap P, V_{cc} \cap X)$
    return $count$
Select a vertex $v \in P$ to branch on
$count \leftarrow$ CCMIS$(P \setminus N_G[v], X \setminus N_G(v))$
$count \leftarrow count + $ CCMIS$(P \setminus \{v\}, X \cup \{v\})$
return $count$

As we explain in the following CCMIS differs substantially from the previous algorithms in which order the vertices are selected from P to branch on. It is clear from the description of CCMIS that one can select any vertex $v \in P$ to branch on. Thus one could similar to the previous algorithms use degree based information when selecting the branching vertex v. Picking a maximum degree vertex could be advantageous for the the first recursive call as it would give a maximum reduction in the size of $G[P \cup X]$, thus making it more likely that the remaining graph is disconnected. Picking a minimum degree vertex could be advantageous for the second recursive call as there would be fewer remaining vertices in P that could dominate v. However, as our main interest is in computing the number of MISs for sparse graphs we use a different selection criterion that exploits this. Algorithm 3: CCMIS has a considerable advantage over the Bron-Kerbosh type enumeration algorithms whenever the remaining graph becomes disconnected. This follows since the CCMIS algorithm does not have to generate every MIS but only needs to find the number of MISs in each connected component and then to multiply these numbers together. Although the algorithm by Gaspers et al. also exploit connected components in this way, their algorithm is bound to using a degree based criterion when selecting a pivot. Thus this might limit how often the remaining graph becomes disconnected. Since we have no restrictions in CCMIS when selecting the branching vertex $v \in P$ we do so with the sole objective that the remaining graph should become disconnected.

Prior to running the algorithm we compute a *nested dissection* ordering $\alpha = \{v_1, v_2, \ldots, v_{|V|}\}$ on the vertices of G [9]. Such an ordering strives to number vertices that make up a (preferably small) separator S of G first, with the added constraint that the remaining components of $G \setminus S$ should be of roughly equal size. This is then repeated recursively for each connected component. One can also view a nested dissection ordering as an *elimination tree* [4]. This tree displays the separators in α, with vertices in a separator S making up a path hanging of the preceding separator S' on the component containing S. Within each separator, a vertex $v_j \in S$ will be a child of the highest numbered vertex $v_k \in S$ where $k < j$. If $v_j, j \neq 1$, is the first ordered vertex in S then v_j will be a child of the last ordered vertex of S', where S' is as defined above. It follows that a low elimination tree height is an indication that it was possible to (recursively) partition the graph using small separators.

As an example of a nested dissection ordering, consider the graph in Figure 1a. Then a possible α could be $\{d, c, a, b, f, e, g, h\}$. The d vertex is the first separator and c and f the two remaining ones. Note that the relative ordering between the vertices of the two components of $G \setminus \{d\}$ can be changed as long as c is ordered before a and b, and f is ordered before $e, g,$ and h. Also, when a separator consists of multiple vertices their relative order is not necessarily important.

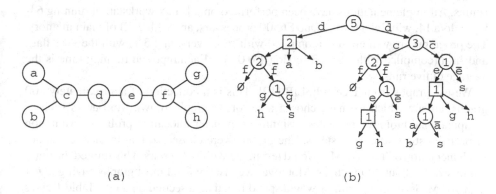

(a)   (b)

Fig. 1. A possible execution of Algorithm 3

The strategy we employ is now to choose the first vertex $w \in \alpha$ that is also in $P \cup X$. We have two cases for selecting the vertex v to branch on. If $w \in P$ then we set $v = w$ and if $w \in X$ then we select v to be a vertex in $P \cap N_G(w)$. Such a vertex must exist since the algorithm would already have returned if $w \in X$ had no neighbor in P. The effect of following α in this way is that we will only expand solutions where each vertex in S has either been included in the current MIS or is being dominated. Thus we are ensured that the remaining graph will be disconnected. Note that the strategy of picking a vertex in $N_G(w)$ to branch on whenever $w \in X$ is similar to the pivoting strategy in

TOMITAMIS. Comparing with the algorithm by Lipton and Tarjan the difference is that even though we follow the separator structure, for a particular separator S we allow for vertices in $N_G(S)$ to be assigned values before deciding exactly which vertices from S should be in the MIS.

The tree in Figure 1b shows the recursion tree of the algorithm when applied to the graph in Figure 1a. Each time the algorithm branches on a particular vertex is denoted by a round node, where the left branch denotes that the branching vertex is in the current MIS and the right branch that it is not. Whenever the remaining graph consist of just one vertex in P we only show the name of the vertex as it must be in any MIS. When the remaining graph is empty we write $\emptyset$, and if a particular branch cannot be extended to a MIS we write s. We use a square node to indicate when the graph has become disconnected and then draw one branch for each connected component. The number inside each node is the number of MISs returned by a particular branch.

With the current description of the algorithm there is still some freedom as to the order in which the branching vertices are selected. As already pointed out, we can re-order the vertices within a separator in α. Also, once a vertex $v \in S$ has been chosen to branch on then in the configuration where v is considered to be out of the current MIS, we are free to decide the order in which we pick vertices from $N_G(v) \cap P$ to dominate v. We will expand further on these issues in Section 5.

5   Experiments

In the following we describe experiments performed to evaluate the presented algorithms. All implementations have been performed on a Linux workstation running 64-bit Fedora 14, with Intel Core 2 Duo E6500 processors, and with 2GB of main memory. The programs are written in C (compiled with `gcc` (version 4.5.1) with the $-$O3 flag) and Java (compiled with `javac` version 1.6.0_30). Each reported running time is the average of five runs.

We use graphs from TreewidthLIB [19]. This is a collection of approximately 700 graphs, among which we have chosen a set of 22 graphs drawn from areas such as computational biology, frequency assignment, register allocation problem, evaluation of probabilistic inference systems. The graphs were chosen so that in most cases our implementation of TOMITAMIS would terminate within 24 hours. This limited the maximum size to about 200 vertices. Moreover we also avoided most graphs having fewer than 10^6 MISs as all algorithms would spend less than a second on these. Table 1 gives the statistics for the chosen 22 graphs. Here p gives the edge density, eth gives the elimination tree height, while $MISs$ gives the number of maximal independent sets. In addition to these graphs we have performed experiments using rectangular grids.

Our first set of experiments concerns a comparison between the algorithm by Gaspers et al. and TOMITAMIS. In addition to the regular algorithm by Gaspers et al. we also implemented variants of it where we only apply the reduction rules at regular intervals, the most extreme case being when the reduction rules are not used at all. Since the algorithm by Gaspers et al. is by far the most complex of the considered algorithms, we have performed these comparisons using Java as this offers better support for more complex data structures such as sets. The results of the comparisons on nine representative graphs can be seen in the left plot of Figure 2. Here the first seven graphs are the ones marked with a * in Table 1, while the 8th graph is a path on 40 vertices, and the 9th and 10th graphs are grids of size 7×7 and 8×8, respectively. The numbers are reported relative to the performance of the regular algorithm by Gaspers et al. (G100). G50 denotes the algorithm where the reduction rules are only applied in 50% of the recursive calls and G0 where they are not used at all.

Table 1. Description for benchmark real world graphs from TreewidthLIB [19]

Graph No.	Graph name	V	E	p	eth	MISs
1*	risk	42	83	0.01	13	66498
2*	pigs-pp	48	137	0.12	17	131402
3*	1sem	57	570	0.35	41	12405
4*	BN_100	58	273	0.16	31	134201
5*	1r69	63	692	0.35	46	22993
6*	1ail	69	631	0.26	44	160312
7	macaque71	71	444	0.18	30	182044
8	jean	80	508	0.16	22	1251960
9*	1aba	85	886	0.25	54	1067404
10	david	87	406	0.11	22	4.41×10^7
11	celar02	100	311	0.06	29	2.87×10^{10}
12	celar06	100	350	0.07	22	2.72×10^{10}
13	1lkk	103	1162	0.22	62	1.44×10^7
14	1fs1	114	1351	0.21	73	5.10×10^7
15	1a62-pp	120	1507	0.21	73	7.56×10^7
16	miles250	128	387	0.05	36	1.75×10^{13}
17	anna	138	493	0.05	23	2.75×10^{10}
18	mulsol1.i.5	186	3973	0.23	47	3.33×10^9
19	celar05	200	681	0.03	36	7.86×10^{20}
20	zeroin.i.3	206	3540	0.17	43	1.29×10^7
21	zeroin.i.2	211	3541	0.16	43	1.81×10^7
22	BN_93	422	1705	0.02	38	4.55×10^{11}

As can be observed there is no advantage in using the reduction rules, and when they are not used at all the performance is very similar to that of TOMITAMIS. Based on these results we did not pursue the algorithm by Gaspers et al. any further. For the remaining experiments all algorithms have been implemented in C as this gave considerable faster code compared to using Java.

We then compared BKMIS, TOMITAMIS, and the algorithm by Eppstein et al. These experiments showed that, as expected, TOMITAMIS outperformed BKMIS, while there was little difference between TOMITAMIS and the algorithm by Eppstein et al. We note that this last observation does not contradict the results in [7] as these were concerned with enumerating cliques in sparse graphs which is equivalent to enumerating MISs in dense graphs, while we are enumerating MISs in sparse graphs. Due to space constraints we omit these results.

Our next set of experiments concerns different variants of CCMIS where we use Metis [15] to precompute a nested dissection ordering. The time spent on this was insignificant compared to the algorithm itself and is not included in the timings.

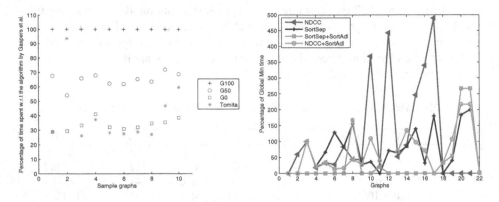

Fig. 2. Relative performance of TOMITAMIS compared to the algorithm by Gaspers et al. (left), and relative performance of different CCMIS algorithms (right).

The versions we tried include the basic algorithm (NDCC) where the vertices are processed for branching according to the ordering given by Metis and versions where we reorder the vertices within each separator and also the relative order of the neighbor lists. Similar in spirit with TOMITAMIS we tried a version where one branches on a vertex v in the current separator such that $|P \cap N_G[v]|$ is minimized. This slowed down the algorithm compared to NDCC and we therefore switched to presorting each separator based on their degree in G. We label this algorithm SortSep. Next we considered the order in which the neighbor lists are ordered. This is of importance when trying to dominate a vertex v currently in X. Consider a vertex w with several undominated neighbors in the current separator S. In the configuration where w is in the current MIS all neighbors of w will be dominated, thus reducing the number of undominated vertices in S. In the configuration where w is in X each undominated neighbor of w will have one vertex less that must be tried to dominate it. Based on these observations we implemented a version (SortAdl) where the adjacency list of every vertex v was presorted according to the number of neighbors each vertex has in the same separator as v belonged to. We also tried to compute this ordering on the fly using the number of remaining undominated vertices in the current separator but this only increased the running time. In the right plot of Figure 2 we display the relative running time for all four combinations of these approaches. For each graph we report the relative performance compared to the best algorithm for that graph. In all of these implementations we only check if the graph is disconnected if the previous call to CCMIS moved a vertex into the current MIS.

The average distances from the best algorithm was for SortSep + SortAdl 36%, for NDCC 185%, for SortSep 172%, for NDCC + SortAdl 167%. Thus it is clear that sorting both the the separators and the neighbor lists is crucial for performance.

Finally we tried two versions of CCMIS where the selection criterion for which vertex to branch on was strictly based on the degree of the remaining vertices, one where we always selected the vertex of minimum degree and one where we selected the vertex of maximum degree (MaxDegCC). Both of these were considerably slower than any of the other CCMIS variations. The absolute running times for MaxDegCC,

Table 2. CPU time(sec) for benchmark real world graphs from TreewidthLIB [19]

Graph	1	2	3	4	5	6	7	8	9	10	11
TomitaMIS	0.07	0.22	0.02	0.25	0.03	0.14	0.31	0.69	1.09	29.05	20095.5
NDCC	0.01	0.08	0.02	0.26	0.04	0.09	0.11	0.13	1.04	1.03	0.06
MaxDegCC	0.02	0.23	0.02	0.46	0.05	0.14	0.15	0.09	1.59	0.31	4.56
SortSep+SortAdl	0.01	0.05	0.01	0.22	0.03	0.07	0.06	0.24	0.73	0.22	0.06
Graph	12	13	14	15	16	17	18	19	20	21	22
TomitaMIS	10648.1	16.24	61.5	87.85	-	32716.1	2722.0	-	23.81	32.66	135407.1
NDCC	0.38	8.7	16.4	84.38	3.56	1.18	0.03	187.63	0.06	0.06	1303.0
MaxDegCC	8.67	15.3	40.9	82.06	7.17	0.69	0.18	-	0.84	0.86	1658.85
SortSep+SortAdl	0.07	5.7	8.7	24.37	0.81	0.2	0.04	290.57	0.22	0.22	76.12

TomitaMIS, NDCC, and SortSEp+SortAdl are given in Table 2. We note that the average distance from the best algorithm for each graph was for MaxDegCC 1371% and for TomitaMIS $1.6 \times 10^6\%$.

As can be seen the running time of TomitaMIS is by far the highest, for some graphs the algorithm did not finish. Also, following a nested dissection ordering is advantageous in most cases, and as already noted presorting the separators and neighbor lists further emphasizes this effect.

We have also experimented with how often one should check if the graph is disconnected in CCMIS. We tried version where we only checked for a certain percentage of the calls, where we only checked once a separator had been dominated, and checking when the remaining graph is at least of some predefined size. From these tests we conclude that when the remaining graph has at least 10 vertices, then checking every time after some vertex has been added be in the current MIS was the best option.

6   Conclusion

We have shown the first practical algorithm for counting MISs in moderately sized sparse graphs. Comparisons with other algorithms showed that our algorithm is highly competitive for this problem. One can get an indication of how good the algorithm is likely to be by looking at the height of the elimination tree. These results also extend to counting cliques in dense graphs. We note that searching for a (small) separator in a graph is equivalent to searching for a (large) complete r-partite graph, $r \geq 2$, in its complement graph. For $r = 2$ this is equivalent to searching for a (not necessarily induced) bi-clique.

We are currently working on implementing the algorithm by Gaspers et al. in C to be able to perform a more complete comparison of it with the other algorithm, although we do not expect that this will change any of our conclusions. We would also like to experiment further with what impact the partitioning strategy has on the running time.

Finally, we note that the presented ideas could be used to compute a maximum independent set in a graph in a similar fashion as the algorithm by Lipton and Tarjan. We are not aware of any practical studies of how to solve this problem on sparse graphs, although the complement problem of finding the maximum size clique has been studied extensively on sparse graphs.

References

1. Bron, C., Kerbosch, J.: Finding all cliques of an undirected graph. Com. ACM 16, 575–577 (1973)
2. Cheng, J., Ke, Y., Fuu, W., Xu Yu, J.: Finding maximal cliques in massive networks. ACM Trans. Database Syst. 36(4), 21:1 – 21:34 (2011)
3. Cheng, J., Zhu, L., Ke, Y., Chu, S.: Fast algorithms for maximal clique enumeration with limited memory. In: Proc. of the 18th ACM SIGKDD International Conference on Knowledge Discovery and Data Mining, KDD 2012, pp. 1240–1248. ACM, New York (2012)
4. Eisenstat, S.C., Liu, J.W.H.: The theory of elimination trees for sparse unsymmetric matrices. SIAM J. Mat. Anal. App. 26(3), 686–705 (2005)
5. Eppstein, D.: All maximal independent sets and dynamic dominance for sparse graphs. ACM Trans. on Alg. 5 (2009)
6. Eppstein, D., Löffler, M., Strash, D.: Listing all maximal cliques in sparse graphs in near-optimal time. In: Cheong, O., Chwa, K.-Y., Park, K. (eds.) ISAAC 2010, Part I. LNCS, vol. 6506, pp. 403–414. Springer, Heidelberg (2010)
7. Eppstein, D., Strash, D.: Listing all maximal cliques in large sparse real-world graphs. In: Pardalos, P.M., Rebennack, S. (eds.) SEA 2011. LNCS, vol. 6630, pp. 364–375. Springer, Heidelberg (2011)
8. Gaspers, S., Kratsch, D., Liedloff, M.: On independent sets and bicliques in graphs. Journal of Graph Theoritic Concepts in Computer Science, 171–182 (2008)
9. George, A.: Nested dissection of a regular finite element mesh. SIAM J. on Num. Anal. 10(2), 345–363 (1973)
10. Johnson, D.S., Yannakakis, M., Papadimitriou, C.H.: On generating all maximal independent sets. Information Processing Letters 27, 119–123 (1988)
11. Junosza-Szaniawski, K., Tuczyński, M.: Counting maximal independent sets in subcubic graphs. In: Bieliková, M., Friedrich, G., Gottlob, G., Katzenbeisser, S., Turán, G. (eds.) SOFSEM 2012. LNCS, vol. 7147, pp. 325–336. Springer, Heidelberg (2012)
12. Lawler, E.L., Lenstra, J.K., Rinnooy Kan, A.H.G.: Generating all maximal independent sets: NP-hardness and polynomial time algorithms. SIAM J. Comp. 9, 558–565 (1980)
13. Lipton, R.J., Tarjan, R.E.: Applications of a planar separator theorem. SIAM J. Comp. 9(3), 615–627 (1980)
14. Loukakis, E., Tsouros, C.: A depth first search algorithm to generate the family of maximal independent sets of a graph lexicographically. Computing 4, 349–366 (1981)
15. Metis - serial graph partitioning and fill-reducing matrix ordering, http://glaros.dtc.umn.edu/gkhome/views/metis/
16. Moon, J.W., Moser, L.: On cliques in graphs. Israel J. of Math., 23–28 (1965)
17. Okamoto, Y., Uno, T., Uehara, R.: Linear-time counting algorithms for independent sets in chordal graphs. In: Kratsch, D. (ed.) WG 2005. LNCS, vol. 3787, pp. 433–444. Springer, Heidelberg (2005)
18. Tomita, E., Tanaka, A., Takahashi, H.: The worst-case time complexity for generating all maximal cliques and computational experiments. Theor. Comput. Sci. 363, 28–42 (2006)
19. Treewidthlib (2004-.), http://www.cs.uu.nl/people/hansb/treewidthlib
20. Tsukiyama, S., Ide, M., Ariyoshi, H., Shirakawa, I.: A new algorithm for generating all maximal independent sets. SIAM J. Comp. 6, 505–517 (1977)
21. Wilf, H.S.: The number of maximal independent sets in a tree. SIAM J. Alg. Disc. Meth. 7(1), 125–130 (1986)

An Edge Quadtree for External Memory

Herman Haverkort[1], Mark McGranaghan[2,*], and Laura Toma[2,**]

[1] Eindhoven University of Technology, The Netherlands
[2] Bowdoin College, USA

Abstract. We consider the problem of building a quadtree subdivision for a set $\mathcal{E}$ of n non-intersecting edges in the plane. Our approach is to first build a quadtree on the vertices corresponding to the endpoints of the edges, and then compute the intersections between $\mathcal{E}$ and the cells in the subdivision. For any $k \geq 1$, we call a K-quadtree a linear compressed quadtree that has $O(n/k)$ cells with $O(k)$ vertices each, where each cell stores the edges intersecting the cell. We show how to build a K-quadtree in $O(sort(n + l))$ I/O's, where $l = O(n^2/k)$ is the number of such intersections. The value of k can be chosen to trade off between the number of cells and the size of a cell in the quadtree. We give an empirical evaluation in external memory on triangulated terrains and USA TIGER data. As an application, we consider the problem of map overlay, or finding the pairwise intersections between two sets of edges. Our findings confirm that the K-quadtree is viable for these types of data and its construction is scalable to hundreds of millions of edges.

1 Introduction

The word *quadtree* describes a class of data structures that partition the space hierarchically and are defined by a stopping criterion that decides when a region is not subdivided further. In 2D, the quadtree recursively divides a square containing the data into four equal regions (quadrants or cells), until each region satisfies the stopping condition (usually, when a cell is "small" enough). The set of cells that are not split further define the leaves of the tree and represent a subdivision of the input region. Quadtrees have been used for many types of data (points, line segments, polygons, rectangles, curves) and many types of applications. For an ample survey we refer to [12].

In this paper we are interested in quadtrees for data sets that are so large that they do not fit in the internal memory of the computer, so that at any time, most of the data has to reside in external memory. To analyze the efficiency of the construction and query algorithms in this case, we use the standard I/O-model by Aggarwal and Vitter [2]. In this model, a computer has an internal memory of size M and an arbitrarily large disk. The data is stored on disk in blocks of size B, and, whenever the algorithms needs to access data not present in memory, it loads the block(s) containing the data from disk. The I/O-complexity of an algorithm is the number of I/O's it performs, that is, the number of blocks

* Supported by Bowdoin Freedman Fellowship and NSF award no. 0728780.
** Supported by NSF award no. 0728780.

V. Bonifaci et al. (Eds.): SEA 2013, LNCS 7933, pp. 115–126, 2013.
© Springer-Verlag Berlin Heidelberg 2013

transferred (read or written) between main memory and disk. Sorting takes $sort(n) = \Theta(\frac{n}{B} \log_{M/B} \frac{n}{B})$ I/O's [2]; scanning takes $scan(n) = \Theta(n/B)$ I/O's.

Quadtrees can be viewed as trees representing the hierarchical space decomposition, or as the set of leaf cells ordered along a space-filling curve. The latter variant of quadtree, called the *linear quadtree*, was introduced by Gargantini [5]. The linear quadtree is particularly useful when dealing with disk-based structures, because its space requirements are smaller. Quadtrees are known to perform well empirically in many different applications, but their worst-case behaviour is not ideal, except in the simplest cases. Given a set of n points in the plane, a quadtree that splits a region until it contains at most one point can have unbounded size. However, it is known how to construct a *compressed* quadtree of $O(n)$ cells which each have at most one point. In a compressed quadtree, paths consisting of nodes with only one non-empty child are replaced by a single node, with all empty children merged into one. Throughout this paper, the concept of quadtrees will encompass both compressed and uncompressed quadtrees.

Building a quadtree on a set of n non-intersecting *edges* in the plane, rather than points, is harder. We refer to a quadtree for a set of edges as an *edge quadtree*, and we denote by l the number of intersections between the edges and the cells in the quadtree subdivision. One way to build an edge quadtree is to first build a compressed quadtree on the endpoints of the edges, and then compute the intersections between the edges and the cells in the subdivision. In the worst case, each edge can intersect almost all cells, giving a quadtree of quadratic size. Another type of edge quadtree may split a region until it intersects a single edge. Since the distance between two edges can be arbitrarily small, the resulting quadtree has unbounded size. Other edge quadtrees can be defined by formulating specific stopping criteria. Such structures were described by Samet *et al.* [14,13,10]. The *PM quadtree* [14] allows a region to contain more than one edge if the edges meet at a vertex inside the region. Variants of PM quadtrees differ in how to handle regions that contain no vertices. The *segment quadtree* [13] is a linear quadtree in which a leaf cell is either empty, contains one edge and no vertices, or contains precisely one vertex and its incident edges. The *PMR quadtree* [10] is a linear quadtree where each region may have a variable number of segments and regions are split if they contain more than a predetermined threshold. Hoel and Samet [9] compared the PMR quadtree with some variants of R-trees on TIGER data, in terms of storage requirements, construction time (disk I/O's), and a number of queries. They find that the PMR quadtree performs well compared to the R-tree for map overlay. Subsequently, improved algorithms for the construction of the PMR quadtree have been proposed [7,6,8]. The algorithms perform well in practice, but there are several disadvantages: First, the stopping rule of the PMR quadtree means that the size of a leaf depends on both the splitting threshold and the depth of the leaf, and the quadtree depends on the insertion order. Second, the complexity is analysed in terms of various parameters that depend on the data, in a way that is not well understood. Finally, the algorithms are fairly complex, and the performance is not worst-case optimal.

Quadtrees in the I/O-model were described by Agarwal *et al.* [1] and De Berg *et al.* [4]. Agarwal *et al.* describe an algorithm for constructing a quadtree on a set of n vertices in the plane such that each cell contains $O(k)$ vertices (for any $k \geq 1$), that runs in $O(\frac{n}{B} \frac{h}{\log M/B})$ I/O's, where h is the height of the quadtree. This is $O(sort(n))$ I/O's when $h = O(\log n)$ i.e. the vertices are nicely distributed. Their algorithm was implemented and tested in practice as part of an application to interpolate LIDAR datasets into grids. De Berg *et al.* described the star-quadtree for triangulations, and the guard-quadtree for sets of edges in the plane, which contain at most one vertex per cell and can be constructed in $O(sort(n + l))$ I/O's, where $l = O(n^2)$ is the number of edge-cell intersections. The star- and guard-quadtrees are designed to exploit fatness and density: for fat triangulations[1] and sets of edges of low density[2], respectively, the star-quadtree and guard-quadtree have the property that each cell intersects $O(1)$ edges, thus $l = O(n)$. An experimental evaluation of these structures has not been reported.

Our Contribution.   We consider building an edge quadtree for a set $\mathcal{E}$ of n non-intersecting edges. Let $k \geq 1$ be a user defined parameter. Our algorithm has two steps: First it builds, in $O(sort(n))$ I/O's, a compressed linear quadtree on the endpoints of $\mathcal{E}$ with $O(n/k)$ cells in total and such that each cell has $O(k)$ vertices. Second, it computes the intersections between the edges and the quadtree subdivision in $O(sort(n + l))$ I/O's (where $l = O(n^2/k)$ is the total number of intersections). We refer to the resulting quadtree as a K-quadtree.

The first step, constructing the quadtree subdivision, is a generalization of the algorithm for building guard-quadtrees in [4]. Compared to the algorithm by Agarwal *et al.* [1], our algorithm has better complexity, is much simpler, and gives an upper bound on the number of cells in the subdivision. The second step, which we refer to as edge distribution, is based on an idea communicated to us by Doron Nussbaum. For $k = 1$ the algorithm has the same complexity as in [4], but it is simpler and faster.

In Section 4 we give an empirical evaluation of K-quadtrees on triangulated terrains (in GIS: TINs) and USA TIGER data. We examine the size of the quadtree, the size of a cell and the construction time for different values of k. We use test datasets up to 427 million edges, two orders of magnitude larger than in related work [7,6,8]. On TINs and TIGER data the K-quadtrees have linear size, which matches the results of [9,7,6,8]. In terms of construction time (or bulk loading), a comparison with previous work is difficult. The running times in [9] are given in terms of disk block accesses, not the total execution time. The tests in [7,6,8] are performed on three TIGER data sets, the largest one having approx. 200,000 edges, on a machine with 64MB RAM. Our largest TIGER bundle has 427 million edges (6.8 GB), and we use machines with 512MB RAM. Furthermore a precise comparison is not possible without knowing all the tuning parameters used in [8].

---

[1] A triangulation such that every angle is larger than some fixed positive constant δ.
[2] Any disk D is intersected by at most λ edges whose length is at least the diameter of D, for some fixed constant λ.

As an application of quadtrees we consider one of the basic operations in GIS and spatial data structures, map overlay: computing the pairwise segment intersections (overlay) between two sets of edges. Given two sets of edges, each pre-processed as a quadtree, their intersections can be computed in a very simple manner by scanning the two quadtrees as in [4]. We implemented map overlay and report on the running time using various values of k.

Overall, our experimental results confirm that the K-quadtree is viable for very large TIN and TIGER data. These represent relatively simple classes of inputs; however they arise frequently in practice and have been used extensively as tests beds for spatial index structures. Further experiments are necessary for other types of data, and we leave this as a topic for future work.

2   Preliminaries

For simplicity, we assume that the edges $\mathcal{E}$ lie in the unit square. For quadtree background and notation see e.g. [11,4]. A square that is obtained by recursively dividing the input square into quadrants is called a *canonical square*. To order the quadrants, we use the z-order space-filling curve that visits the 4 quadrants, recursively, in order SW, NW, SE, NE. z-order gives a well-defined ordering between the cells in the quadtree subdivision, as well as between any two points. For a point $p = (p_x, p_y)$ in the unit square, define its z-index $Z(p)$ to be the value in the range $[0, 1)$ obtained by interleaving the bits in the fractional parts of p_x and p_y. The value $Z(p)$ is sometimes called the *Morton block index* of p. The z-order of two points is the order of their z-indices. The z-indices of all points in a canonical square σ form an interval $[z_1, z_2)$ of $[0, 1)$, where z_1 is the z-index of the bottom left corner of σ. A compressed quadtree subdivision has two types of cells: canonical squares, and *donut* cells, corresponding to empty nodes that were merged together. A donut cell is the difference between two canonical squares $[z_1, z_2] - [z_3, z_4]$ and is represented as the union of two intervals $[z_1, z_3] \bigcup [z_4, z_2]$.

With this notation, a (compressed) quadtree subdivision corresponds to a subdivision $\mathcal{Q}$ of the z-order curve, and it can be viewed as a set of consecutive, adjacent, non-overlapping intervals, covering $[0, 1)$, in z-order: $\mathcal{Q} = \{[z_1 = 0, z_2), [z_2, z_3), [z_3, z_4), ...\}$; Each interval corresponds to a cell σ_i, which is either a canonical square or a part of a donut. We represent a K-quadtree as a subdivision of the z-order curve where each intersection of an edge e with a cell σ corresponding to the interval $[z_1, z_2)$ is represented by storing edge e with key z_1. A K-quadtree is thus a list of pairs $\{(z_1, e)\}$, stored in order of z_1.

In the rest of the paper we denote by l the number of intersections between $\mathcal{E}$ and the cells in the quadtree subdivision, and we use the terms quadtree, quadtree subdivision and subdivision interchangeably.

3   Constructing a K-Quadtree

In this section we describe our algorithm for building a K-quadtree. Let $k \geq 1$ be a user defined parameter. Our algorithm has two steps: In the first step it ignores the edges and builds, in $O(sort(n))$ I/O's, a linear quadtree on the endpoints of the edges. The quadtree has $O(n/k)$ cells in total, each containing $O(k)$ vertices. Second, it computes the intersections between the edges and the quadtree subdivision in $O(sort(n+l))$ I/O's. We describe the two steps below.

Constructing the subdivision. Let $\mathcal{P} = \{p_0, p_1, p_2, ...\}$ be the vertices of $\mathcal{E}$. A straightforward idea to build a quadtree with $O(k)$ vertices per cell would be to start with one of the standard algorithms for building a quadtree with at most one vertex per cell, and then traverse the subdivision and merge cells into cells of size $O(k)$. However, we would like to avoid generating first a larger subdivision and then merging its cells to get a smaller subdivision. Another approach might be to build the quadtree top-down: stop if the cell contains $O(k)$ vertices, otherwise split the cell and distribute the points among the four children, and continue on the children recursively; however, this may take $\Theta(n^2)$ time as the quadtree may have height $\Theta(n)$.

Our idea to generate a quadtree subdivision with $O(n/k)$ cells and $O(k)$ vertices in each cell directly, is a simple and elegant generalization of an algorithm in [4]. Assume that $\mathcal{P}$ has been sorted in z-order, and denote $\mathcal{P}_k$ the set of every k^{th} point in $\mathcal{P}$: $\mathcal{P}_k = \{p_0, p_k, p_{2k}, ...\} \subset \mathcal{P}$. The idea is to build the quadtree subdivision induced by $\mathcal{P}_k$: for every pair of consecutive points in $\mathcal{P}_k$, we find their smallest enclosing canonical square, and output the five z-indices corresponding to the four z-intervals of the quadrants of this square. We claim that:

Lemma 1. *The resulting list of z-indices represents a compressed quadtree subdivision with $O(n/k)$ cells and $O(k)$ vertices per cell.*
Proof. Every pair of consecutive points of $\mathcal{P}_k$ causes a split, and generates 4 cells, therefore $O(n/k)$ cells; each cell contains at most one point of $\mathcal{P}_k$ inside (or otherwise it would have been split), therefore $O(k)$ points of $\mathcal{P}$.   □

Assuming that the operations involving z-indices take $O(1)$ time, this step runs in $O(n)$ time and $O(scan(n))$ I/O's. With the help of the stack described in the appendix of [4], we can actually output the z-indices in increasing order without additional I/O. Thus we get a compressed quadtree subdivision represented by a list of z-intervals, in z-order of their first endpoint: $\mathcal{Q} = \{[z_1 = 0, z_2], [z_2, z_3], ...\}$ $= \{I_1, I_2, ...\}$. We note that in practice we represent the second endpoint of the intervals implicitly.

An algorithm for edge distribution when $k = 1$. Let $\mathcal{Q} = \{I_1, I_2, ...\}$ be a subdivision of the endpoints of $\mathcal{E}$ obtained by the algorithm described above, and assume $\mathcal{Q}$ is given in z-order. We will now first consider the case $k = 1$, i.e. every cell contains at most one vertex, and the total number of cells is $O(n)$. We describe how to find the intersections between $\mathcal{Q}$ and $\mathcal{E}$ in $O(sort(n+l))$ I/O's. Later we will show how to generalize this process to a subdivision with $O(k)$ vertices in a cell, where $k > 1$.

We assume edges are oriented from left to right, vertical segments are oriented upwards, and let $\mathcal{E}_+$ and $\mathcal{E}_-$ denote the edges of positive and negative slope, respectively. The crux of the algorithm is to process the edges of positive and negative slope separately. We describe below the two steps.

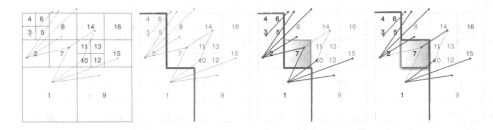

Fig. 1. (a) $\mathcal{E}_+$ (b) B_7 (c) X_7 and processing cell σ_7 (d) X_8

Distributing the edges of positive slope. The idea is to scan $\mathcal{Q}$, one interval at a time, and find all the edges in $\mathcal{E}_+$ intersecting the cell corresponding to the current interval. Let $I_j = [z_j, z_{j+1}]$ be the next interval we read from $\mathcal{Q}$, and let σ_j be the corresponding cell in the subdivision. There are two types of intersections between σ_j and $\mathcal{E}_+$, see Fig. 1:

- First, there may be edges that intersect σ_j and originate in σ_j;
- Second, there may be edges that intersect σ_j and originate outside σ_j.

Intersections of the first type can be detected by scanning $\mathcal{Q}$ and $\mathcal{E}_+$ in sync, as follows. Let $\mathcal{E}_+$ be sorted in z-order of the first endpoints of the edges. Let I_j be the current interval in $\mathcal{Q}$, and let $e = (p, q)$ be the next edge in $\mathcal{E}_+$. To check whether e originates in σ_j means checking if $z(p) \in I_j$. This leads to the following algorithm: For each interval $I_j \in \mathcal{Q}$, we read from $\mathcal{E}_+$ all the edges that originate in σ_j, and stop when encountering the first edge $e' = (p', q')$ with $z(p') > z_{j+1}$. Then we continue with the next interval from $\mathcal{Q}$, in the same fashion. Because the edges in $\mathcal{E}_+$ are stored in z-order of their first endpoint, we know that once we encounter an edge with $z(p') > z_{j+1}$, then all subsequent edges have the same property and none of them can originate in σ_j. This runs in $O(scan(|\mathcal{Q}| + |\mathcal{E}_+|)) = O(scan(n))$ I/O's.

The harder problem is finding the intersections of σ_j with the edges that originate outside σ_j. It is here that we exploit that $\mathcal{E}_+$ and $\mathcal{E}_-$ are processed separately. The key observation is that any edge of positive slope that intersects σ_j originates in a cell that comes *before* σ_j, in z-order. In general we have:

Lemma 2. *An edge of positive slope intersects the cells in $\mathcal{Q}$ in z-order.*

Consider the current interval I_j in $\mathcal{Q}$. By Lemma 2 it follows that all the edges that intersect σ_j and do not start in σ_j must originate in a cell σ_i *before* σ_j, that is $i < j$. Let B_j denote the boundary between the cells explored before σ_j, $\bigcup_{i<j} \sigma_i$ and the rest of the cells $\bigcup_{i \geq j} \sigma_i$, for any $j \geq 1$. The edges in $\mathcal{E}_+$ that originate (but do not end) in a cell before σ_j will intersect the boundary B_j; let X_j be the set of these edges. See Fig. 1. More precisely, let XL_j be the edges of X_j that intersect B_j between the left edge of the unit square and the lower left corner of σ_j, and let XB_j be the edges of X_j that intersect B_j between the lower left corner of σ_j and the bottom edge of the unit square. Here, if σ_j is the

second part of a donut, we define its lower left corner as the upper right corner of the hole, which is, in fact, the upper right corner of σ_{j-1}.

Lemma 3. B_j is a monotone staircase and the intersection of σ_j and B_j covers a connected part of B_j.

The algorithm will maintain X_j on two stacks SL and SB, keeping the following invariant: before processing an interval I_j from Q, the stack SL contains, from bottom to top, the edges of XL_j in the order of their intersections with B_j from the left edge of the unit square to the lower left corner of σ_j; the stack SB contains, from bottom to top, the edges of XB_j in the order of their intersections with B_j from the bottom edge of the unit square to the lower left corner of σ_j. Initially, for $j = 1$, the boundary B_1 is empty and both stacks are empty.

The algorithm now scans Q and $\mathcal{E}_+$. When I_j is the next interval in Q, the algorithm reads all edges that originate in σ_j from $\mathcal{E}_+$, and pops all edges that intersect σ_j "from before" from SL and SB. Out of these edges, we push those that leave σ_j between the upper left and upper right corner onto SL, and those that leave σ_j between the lower right and the upper right corner onto SB, in order of their intersections with the boundary of σ_j towards the upper right corner (if σ_j is part of a donut surrounding σ_{j+1}, we take the lower left corner of σ_{j+1} as the upper right corner of σ_j). Finally we establish the invariant for the next interval: if the lower left corner of σ_{j+1} lies above the lower left corner of σ_j, we do this by popping edges from SL and pushing them onto SB one by one until SL is empty or the top of SL intersects B_{j+1} between the left edge of the unit square and the lower left corner of σ_{j+1}; otherwise we establish the invariant in a symmetric way by moving edges from SB to SL.

From Lemma 3 and the invariant it follows that before processing cell σ_j, all edges of X_j that intersect σ_j are on top of SL or SB, and thus the algorithm correctly finds all edges intersecting σ_j and correctly restores the invariant after every step. It remains to analyse the efficiency of the algorithm. We claim that:

Lemma 4. Each edge of $\mathcal{E}_+$ is pushed onto SB at most once and pushed onto SL at most once for each intersection with Q.

For a brief justification, consider a square and its four quadrants. Let h be the left half of the horizontal midline of the square, and let H be the edges intersecting h. These edges leave the lower left quadrant across its top edge and are therefore pushed onto SL; they are moved to SB just before processing the first cell in the upper left quadrant. From there, the edges of H will never move back to SL while still representing the intersection with h, as this would only happen if the lower left corner of the next cell σ_{j+1} is to the right of the intersections of H with h. However, by the monotonicity of B_{j+1}, this can only happen after all cells that touch h from above and to the left of σ_{j+1} have already been processed, at which time the edges of H must have been removed from the stack. Similarly, the edges crossing the vertical midline of the square leave the quadrants on the left across their right edges and are therefore pushed onto SB; they are moved to SL just before processing the first cell in the lower right quadrant. From there they are removed as we traverse the leftmost cells within the quadrants on the right from bottom to top.

Let l^+ be the number of intersections between $\mathcal{E}_+$ and $\mathcal{Q}$. Putting everything together it follows that the intersections of $\mathcal{E}_+$ and $\mathcal{Q}$ can be found in $O(scan(n + l^+))$ I/O's once $\mathcal{E}_+$ and $\mathcal{Q}$ are sorted.

Distributing the edges of negative slope. To distribute the edges of negative slope, we observe that Lemma 2 holds for edges of negative slope if we consider a different z-order: Z'= NW, NE, SW, SE. We convert $\mathcal{Q}$ to a subdivision $\mathcal{Q}'$ onto the Z'-order curve, find the intersections with $\mathcal{E}_-$ using the same algorithm as above, and map the intersections back to the cells in $\mathcal{Q}$. All these steps run in $O(sort(n + l^-))$ I/O's, where l^- stands for the number of intersections between $\mathcal{E}_-$ and $\mathcal{Q}$. Overall, the intersections between $\mathcal{Q}$, $\mathcal{E}_+$ and $\mathcal{E}_-$ can be found in $O(sort(n + l))$ I/O's, where $l = l^+ + l^-$ is the total number of intersections.

Distributing edges in a K-quadtree. Above we described how to find the intersections between $\mathcal{E}$ and a quadtree subdivision where each cell contains at most one vertex ($k = 1$). We now describe briefly how to extend the algorithm to $k > 1$.

Recall that the algorithm for $k = 1$ reads intervals in order from $\mathcal{Q}$ while maintaining the stacks SL and SB. For each interval I_j it: (a) finds the edges that originate in σ_j; (b) finds the edges that intersect σ_j and originate outside σ_j; (c) merges these two groups of edges in order onto the stacks. The only step that is different when $k > 1$ is (c). In this case the edges that originate in σ_j need to be carefully interleaved with the edges of X_j. Note that we read the edges originating in σ_j from $\mathcal{E}_+$ in z-order of their start point, which is not necessarily the order in which they will appear in X_{j+1}. For each edge we find the intersection with σ_j, and then sort all edges intersecting σ_j (the edges found on the stacks and the edges originating in σ_j) by the point where they leave σ_j. Since the boundary of σ_j is a monotone staircase, sorting the edges by these exit points gives them in the order in which they appear on B_{j+1}. Overall the algorithm runs in $O(sort(n + l))$ I/O's.

4   Experimental Results

In this section we present an empirical evaluation of K-quadtrees on two types of data commonly used in GIS applications, triangulated terrains (TINs) and TIGER data. We implemented the construction algorithm described in Section 3 and experimented with various values of k. The current implementation assumes that $k = O(M)$ and the number of edges that intersect a cell fit in memory; they are sorted using system qsort. We compare the resulting subdivisions in terms of total number of edge intersections, average number of edge intersections per cell, maximum number of intersections per cell, and construction time. For comparison we also implemented the construction algorithm in [4], denote QDT-1-OLD. As an application we consider the time to compute the pairwise segment intersections (overlay) between two sets of edges, which is one of the standard operations in GIS and spatial databases. Given two sets of edges, each pre-processed as a K-quadtree, their intersections can be computed in a very simple and efficient manner, while scanning the two quadtrees, see e.g. [4].

Let e denote the number of edges in the input dataset, c the number of cells in the quadtree subdivision, and l the number of edge-cell intersections in the quadtree subdivision. For each quadtree we measured the following average quantities: (i) the average number of cells per input edge, c/e; (ii) the average number of edge-cell intersections per edge, l/e (indicates the total size of the quadtree, relative to the input size); (iii) the average number of edges intersecting a cell, l/c (indicates the average size of a cell in the quadtree).

Datasets. In the first set of experiments we built quadtrees on triangulated terrains, for which we ignored the elevation, with size up to $53.9 \cdot 10^6$ edges, or 411GB (with 8B per edge). The datasets represent Delaunay triangulations of elevation samples of real terrains. They have not been filtered to eliminate narrow triangles. For all our test datasets, the minimum angle is on the order of $0.001°$ and the maximum angle close to $180°$; 5% of the angles are below $18°$ and 5% above $108°$; the average minimum angle is around $33°$; and the median angle $57°$. The maximum number of edges incident on a vertex varies widely across all datasets, ranging between 31 and 356; the average incidence across all datasets is approx. 6.

In the second set of experiments we used USA TIGER2006SE data. This consists of 50 datasets, one for each state, containing the roads, railways, boundaries and hydrography in the state. The size of a dataset ranges from 115,626 edges (DE), to 40.4 million edges (TX). We assembled 4 (larger) datasets: *New England* (25.8 million edges, or 197MB), *East Coast* ($113.0 \cdot 10^6$ edges or 862MB), *Eastern Half* ($208.3 \cdot 10^6$ edges or 1.5GB) and *All US* ($427.7 \cdot 10^6$ edges or 3.2GB).

Platform. The algorithms are implemented in C and compiled with g++ 4.1.2 with optimization level -O3. All experiments were run on HP 220 blade servers, with an Intel 2.83 GHz processor, 512MB of RAM and a 5400 rpm SATA hard drive. The hard disk is standard speed for laptop hard-drives. As I/O-library we used IOStreams [15], an I/O-kernel derived from TPIE [3]. The only components used were scanning and sorting, so other I/O-libraries can be plugged in.

Results on triangulations. In the first set of experiments we computed K-quadtrees on TIN data for various values of $k \geq 1$, denoted QDT-K. Some results are shown in Fig 2. Our construction algorithm is more than 5 times faster than QDT-1-OLD (210 minutes vs. 1071 minutes on a TIN with $e = 54 \cdot 10^6$). As expected, when k increases, the construction time decreases (Fig 2(a)); the number of cells in the quadtree decreases (Fig 2(b)) and the overall size of the quadtree decreases (since fewer cells lead to fewer edge-cell intersections, Fig 2(c)). On the other hand the average number of edge intersections per cell, l/c, increases (Fig 2(d)). For example, on a TIN with $e = 54 \cdot 10^6$, QDT-1 is built in 210 minutes, has $c = .6e, l = 2.9e$ and $l/c = 4.8$; QDT-100 is built in 57 minutes, and has $c = .004e, l = 1.2e$ and $l/c = 257$. Note that l/c represents an average quantity over the entire TIN and

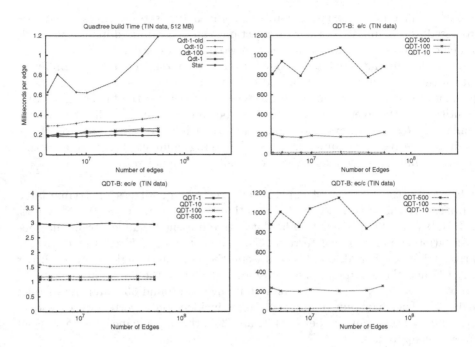

Fig. 2. Quadtree build times and sizes on TIN data (512MB RAM)

the maximum number of edges per cell can be much higher. In summary, for increasing k, QDT-k is built faster, has smaller overall size and larger cell size. The total size of the quadtree stays consistently small across all TINs, and appears to grow linearly with the number of edges. More detailed results will appear in the long version of this paper.

Results on TIGER data. In the second set of experiments we computed K-quadtrees for TIGER data. The results are shown in Fig. 3. Same as for TINs, the build time gets faster up to $k = 100$, and then levels. E.g., on EastHalf ($e = 208 \cdot 10^6$), it takes 24.7 hours to build QDT-1, 9.0h to build QDT-10, 4.8h to build QDT-100, and 4.5h to build QDT-500; on AllUSA ($e = 428 \cdot 10^6$), QDT-100 can be built in 9.7h. The algorithms run at 70% CPU utilization. Similar to [7,6,8], we found that the bottleneck in quadtree construction is edge distribution; in our case it accounts for more than 90% of the total running time, and runs at more than 70% CPU. On TIGER data, our new algorithm is up to 2.9 times faster than QDT-1-OLD; For example, on NEWENGLAND ($e = 25.8 \cdot 10^6$), QDT-1-OLD runs in 545 min, while QDT-1 runs in 186 min. Even with our new algorithm, building a QDT-1 is practically infeasible on moderately large data, taking more than 20h.

The average quadtrees sizes are relatively consistent across all datasets, which is somewhat surprising. QDT-1 has one edge per cell ($l/c = 1$) on average and an overall size $l = 3e$. We also computed the maximum cell size (Fig. 3(c)), which

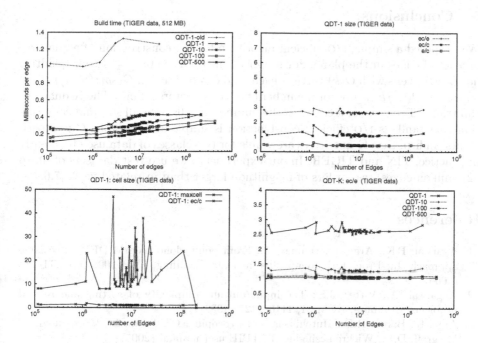

Fig. 3. Quadtree build times and sizes on TIGER data (512MB RAM)

varies widely from state to state; for example, the largest cell in the `EastHalf` bundle intersects 58 edges, while for states like ME and VT, the largest cell intersects 8 edges. For increasing values of k, QDT-K has a larger average cell, but has fewer cells and an overall smaller size. Empirically, for $k = 10, 100, 500, 1000$, QDT-K has $l = 1.5e, 1.1e, 1.04e$ and $1.03e$, respectively.

Segment intersection using quadtrees. To test the efficiency of segment intersection using quadtrees, we ran a set of experiments using a TIN ($e = 53.9 \cdot 10^6$) stored as QDT-1, and the TIGER datasets stored as QDT-K, for various values of k. To force all datasets to cover the same area, we scaled them to the unit square; the resulting intersections are artificial, and we only use them for run time analysis.

Overall, computing the intersecting segments is fast and scalable. For example, computing the overlay of 262 million TIN edges can be done in under 1.8 hours if the quadtrees are given. For the larger TIGER sets, we see two competing effects in the running time. On one hand, as k increases, the size of a cell increases, and the time to compute the intersections between two cells increases. On the other hand, the overall number of cells and edge-cell intersections decrease, resulting in fewer cell-to-cell comparisons. The two effects, combined, cause the total time to first decrease as k increases from 1 to 100, and again increase for $k = 500$. The optimal K-quadtree for segment intersection against QDT-1 is not the one with $k = 1$, as one might have expected, but seems to be one with $k \in [100, 500]$.

5   Conclusions

We proposed a simple, I/O-efficient algorithm for the construction of a quadtree of a set of edges in the plane. For a user defined parameter $k \geq 1$, our quadtree has $O(n/k)$ cells with $O(k)$ vertices each, and can be built in $O(sort(n+l))$ I/O's, where $l = O(n^2/k)$ is the total number of edge-cell intersections. The K-quadtree can trade off the size of a cell with the number of cells, overall size and construction time, and its I/O-efficient construction is simple and scalable. Our experiments confirm that K-quadtrees are viable for two classes of data used frequently in practice, TIN and TIGER. In our experiments we use test datasets of up to 427 million edges, two orders of magnitude larger than in related work [7,6,8].

References

1. Agarwal, P.K., Arge, L., Danner, A.: From point cloud to grid DEM: a scalable approach. In: Proc. 12th Symp. Spatial Data Handling, SDH 2006, pp. 771–788 (2006)
2. Aggarwal, A., Vitter, J.S.: The input/output complexity of sorting and related problems. Commun. ACM 31, 1116–1127 (1988)
3. Arge, L., Barve, R.D., Hutchinson, D., Procopiuc, O., Toma, L., Vahrenhold, J., Vengroff, D.E., Wickremesinghe, R.: TPIE user manual (2005)
4. de Berg, M., Haverkort, H., Thite, S., Toma, L.: Star-quadtrees and guard-quadtrees: I/O-efficient indexes for fat triangulations and low-density planar subdivisions. Computational Geometry 43(5), 493–513 (2010)
5. Gargantini, I.: An effective way to represent quadtrees. Commun. ACM 25(12), 905–910 (1982)
6. Hjaltason, G., Samet, H.: Improved bulk-loading algorithms for quadtrees. In: Proc. ACM International Symposium on Advances in GIS, pp. 110–115 (1999)
7. Hjaltason, G., Samet, H., Sussmann, Y.: Speeding up bulk-loading of quadtrees. In: Proc. ACM International Symposium on Advances in GIS (1997)
8. Hjaltason, G.R., Samet, H.: Speeding up construction of PMR quadtree-based spatial indexes. VLDB Journal 11, 109–137 (2002)
9. Hoel, E., Samet, H.: A qualitative comparison study of data structures for large segment databases. In: Proc. SIGMOD, pp. 205–213 (1992)
10. Nelson, R., Samet, H.: A population analysis for hierarchical data structures. In: Proc. SIGMOD, pp. 270–277 (1987)
11. Samet, H.: Spatial Data Structures: Quadtrees, Octrees, and Other Hierarchical Methods. Addison-Wesley, Reading (1989)
12. Samet, H.: Foundations of Multidimensional and Metric Data Structures. Morgan-Kaufmann (2006)
13. Samet, H., Shaffer, C., Webber, R.: The segment quadtree: a linear quadtree-based representation for linear features. Data Structures for Raster Graphics, 91–123 (1986)
14. Samet, H., Webber, R.: Storing a collection of polygons using quadtrees. ACM Transactions on Graphics 4(3), 182–222 (1985)
15. Toma, L.: External Memory Graph Algorithms and Applications to Geographic Information Systems. PhD thesis, Duke University (2003)

Branchless Search Programs

Amr Elmasry[1] and Jyrki Katajainen[2]

[1] Department of Computer Engineering and Systems, Alexandria University
Alexandria 21544, Egypt
[2] Department of Computer Science, University of Copenhagen
Universitetsparken 5, 2100 Copenhagen East, Denmark

Abstract. It was reported in a study by Brodal and Moruz that, due to branch mispredictions, skewed search trees may perform better than perfectly balanced search trees. In this paper we take the search procedures under microscopic examination, and show that perfectly balanced search trees—when programmed carefully—are better than skewed search trees. As in the previous study, we only focus on the static case. We demonstrate that, by decoupling element comparisons from conditional branches and by writing branchless code in general, harmful effects caused by branch mispredictions can be avoided. Being able to store perfectly balanced search trees implicitly, such trees get a further advantage over skewed search trees following an improved cache behaviour.

1 Introduction

In traditional algorithm analysis each instruction is assumed to take a constant amount of time. In real computers pipelining and caching are omnipresent, so the unit-cost assumption may not always be valid. In this paper we study the impact of hardware effects on the efficiency of search programs for search trees. The outcome of the element comparisons performed in searches is often hard to predict. Also, sequential access is to be avoided in order to support searches in sublinear time. Both of these aspects make search algorithms interesting subjects of study.

We focus on the following hardware phenomena (for more details, see [15]):

Branch misprediction: In a pipelined processor, instructions are executed in parallel in a pipelined fashion. Conditional branches are problematic since the next instruction may not be known when its execution should be started. By maintaining a table of the previous choices, a prediction can be made. But, if this prediction is wrong, the partially processed instructions are discarded and correct instructions are fed into the pipeline.

Cache miss: In a hierarchical memory, the data is transferred in blocks between different memory levels. We are specifically interested in what happens between the last-level cache and main memory. Block reads and block writes are called by a common name: *I/Os*. When an I/O is necessary, the processor should wait until the block transfer is completed.

V. Bonifaci et al. (Eds.): SEA 2013, LNCS 7933, pp. 127–138, 2013.

The motivation for the present study came from a paper by Brodal and Moruz [4], where they showed that skewed search trees can perform better than perfectly balanced search trees. This anomaly is mainly caused by branch mispredictions. Earlier, Kaligosi and Sanders [11] had observed a similar anomaly in quicksort; namely, when the pivot is given for free, a skewed pivot-selection strategy can perform better than the exact-median pivot-selection strategy.

In a companion paper [6] we proved that, when a simple static branch predictor is in use, any program can be transformed into an equivalent program that only contains $O(1)$ conditional branches and induces at most $O(1)$ branch mispredictions. That is, in most cases branch-misprediction anomalies can be avoided, but this program transformation may increase the number of instructions executed. In spite of the existence of this general transformation, current compilers can do branch optimization only in some special cases, and it is difficult for the programmer to force the compiler to do it.

As in the study of Brodal and Moruz [4], our goal is to find the best data representation for a static collection of N integers so that random membership searches can be supported as efficiently as possible. We restrict this study to classical comparison-based methods, so we will not utilize the universe size in any way. Our hypothesis is that, in this setup, balanced search trees are better than skewed search trees. We provide both theoretical and experimental evidence for this proposition. To summarize, the following facts support its validity.

1. The search procedure for a perfectly balanced search tree can be programmed such that it performs at most $\lg N + O(1)$ (two-way) element comparisons (Section 2). When element comparisons are expensive, their cost will dominate the overall costs.
2. The search procedure can be modified, without a significant slowdown, such that it induces $O(1)$ branch mispredictions (Section 4) and performs $O(1)$ conditional branches (Section 7). Hereafter most problems related to branch mispredictions are avoided and skewing does not give any advantage.
3. In several earlier studies (see, e.g. [2,5,14,16,17,18]) it has been pointed out that the layout of a search tree can be improved such that it incurs $O(\log_B N)$ cache misses per search (Section 5), where B is the size of the cache lines measured in elements. By making the layout implicit, the representation can be made even more compact and more cache-friendly (Section 6).

In addition to providing branchless implementations of search algorithms, we investigated their practical performance. The experiments were carried out on a laptop (Intel® Core™ i5-2520M CPU @ 2.50GHz × 4) running Ubuntu 12.04 (Linux kernel 3.2.0-36-generic) using g++ compiler (gcc version 4.6.3) with optimization -O3. The size of 12-way-associative L3 cache was 3 MB, the size of cache lines 64 B, and the size of the main memory 3.8 GB. Micro-benchmarks showed that in this computer unpredictable conditional branches were slow, whereas predictable conditional branches resulted in faster code than conditional-move primitives available at the assembly-language level. As expected, random access was much slower than sequential access, but the last-level cache (L3) was relatively large so cache effects were first visible for large problem instances.

In all experiments the elements manipulated were 4-byte integers; it was ensured that the input elements were distinct. All execution times were measured using the function gettimeofday accessible in the C standard library. All branch-misprediction and cache-miss measurements were carried out using the simulators available in valgrind (version 3.7.0). Each experiment was repeated 10^6 times and the mean over all test runs was reported.

Our main purpose was to use the experiments as a sanity check, not to provide a thorough experimental evaluation of the search procedures. We ported the programs to a couple of other computers and the behaviour was similar to that on our test computer. On the other hand, a few tests were enough to reveal that the observed running times are highly dependent on the environment—like the computer architecture, compiler, and operating system. Readers interested in more detailed comparison of the programs are advised to verify the results on their own platforms. The search programs discussed in this paper are in the public domain [12].

2   Search Procedure

We assume that a node in a binary search tree stores an element and three pointers pointing to the left child, right child, and parent of that node, respectively. For a node x, we let (*x).element(), (*x).left(), (*x).right(), and (*x).parent() denote the values of these four fields.

In a textbook description, search algorithms often rely on three-way comparisons having three possible outcomes: less, greater, or equal. We, however, assume that only two-way comparisons are possible. The simulation of a three-way comparison involves two two-way comparisons. Hence, a naive implementation of the search procedure essentially performs two element comparisons per visited node. The search procedure using two-way branching was discussed and experimentally evaluated by Andersson [1]. With this procedure only one element comparison per level is performed, even though more nodes may be visited. According to Knuth [13, Section 6.2.1], the idea was described in 1962. This optimization is often discussed in textbooks in the context of binary search, but it works for every search tree.

Assume that we search for element v. The idea is to test whether to go to the left at the current node x (when v < (*x).element()) or not. If not, we may have equality; but, if there is a right child, we go to the right anyway and continue the search until x refers to null. The only modification is that we remember the last node y on the search path for which the test failed and we went to the right. Upon the end of the search, we do one more element comparison to see whether (*y).element() < v (we already know that (*y).element() ≤ v). It is not hard to see that, if v = (*y).element(), the last node on the search path is either y itself (if (*y).right() is null) or its successor (all subsequent tests fail and we follow the left spine from (*y).right()).

Let N denote the type of the nodes and V the type of the elements, and let less be the comparison function used in element comparisons. Using C++, the implementation of the search procedure requires a dozen lines of code.

```
1  bool is_member(V const & v) {
2    N* y = nullptr; // candidate node
3    N* x = root; // current node
4    while (x != nullptr) {
5      if (less(v, (*x).element())) {
6        x = (*x).left();
7      }
8      else {
9        y = x;
10       x = (*x).right();
11     }
12   }
13   if (y == nullptr || less((*y).element(), v)) {
14     return false;
15   }
16   return true;
17 }
```

3 Models

We wanted to study the branch-prediction and cache behaviour separately. Therefore, instead of testing the search procedure in a real-world scenario, we decided to use models that capture its branch-prediction behaviour (see Fig. 1). The *three-way model* emulates the search procedure for a random search when three-way branching is used; the *two-way model* does the same when two-way branching is used; and the *branchless model* mixes Boolean and integer arithmetic to avoid the conditional branch altogether. In principle, all the models work in the same way: They reduce the search range from N to ℓ or to $N - 1 - \ell$ depending on the outcome of the branch executed; here ℓ is the border between the left and right portions. In the actual implementation the random numbers were generated beforehand, stored in an array, and retrieved from there.

We measured the execution time and the number of branch mispredictions induced by these models for different values of N. We report the results of

```
1  k = random() * N;
2  while (N > 1) {
3    ℓ = α * N;
4    if (k < ℓ) {
5      N = ℓ;
6    }
7    else if (k > ℓ) {
8      k = k - 1 - ℓ;
9      N = N - 1 - ℓ;
10   }
11   else {
12     N = 1;
13   }
14 }
```

Three-way model

```
1  k = random() * N;
2  while (N > 1) {
3    ℓ = α * N;
4    if (k < ℓ) {
5      N = ℓ;
6    }
7    else {
8      k = k - ℓ;
9      N = N - 1 - ℓ;
10   }
11 }
```

Two-way model

```
1  k = random() * N;
2  while (N > 1) {
3    ℓ = α * N;
4    Δ = (k < ℓ);
5    k = k - ℓ + Δ * ℓ;
6    N = Δ * ℓ + (1 - Δ) * (N - 1 - ℓ);
7  }
```

Branchless model

Fig. 1. The models considered

Table 1. Runtime performance of the models; execution time per $\lg N$ [in nanoseconds]

N	Three-way $\alpha = \frac{1}{2}$	$\frac{1}{3}$	$\frac{1}{4}$	Two-way $\alpha = \frac{1}{2}$	$\frac{1}{3}$	$\frac{1}{4}$	Branchless $\alpha = \frac{1}{2}$	$\frac{1}{3}$	$\frac{1}{4}$
2^{16}	7.6	7.4	7.8	7.7	7.6	8.2	7.0	8.4	9.8
2^{24}	7.7	7.5	8.0	7.8	7.6	8.1	7.2	8.4	9.7
2^{32}	8.0	7.9	8.3	8.1	7.8	8.4	7.6	8.7	10.0

Table 2. Branch behaviour of the models; number of conditional branches executed ($\oslash$) and branch mispredictions induced per search, both divided by $\lg N$

N	Three-way $\alpha = \frac{1}{3}$ $\oslash$	Mispred.	Two-way $\alpha = \frac{1}{3}$ $\oslash$	Mispred.	Branchless $\alpha = \frac{1}{2}$ $\oslash$	Mispred.
2^{16}	2.67	0.44	2.15	0.45	1.00	0.06
2^{24}	2.67	0.44	2.16	0.44	1.00	0.04
2^{32}	2.79	0.44	2.17	0.44	1.00	0.03

these experiments in Tables 1 and 2. There are three things to note. First, the differences in the running times are not that big, but the branchless version is the fastest. Second, the model relying on three-way branching executes more conditional branches than the model relying on two-way branching. Third, since the outcome of at least one of the conditional branches inside the loop is difficult to predict, the first two models may suffer from branch mispredictions.

These experiments seem to confirm the validity of the experimental results reported by Brodal and Moruz [4]. For example, for the three-way model, for $N = 2^{32}$, the number of element comparisons increased from $2.4 \lg N$ ($\alpha = \frac{1}{2}$, not shown in Table 2) to $2.79 \lg N$ ($\alpha = \frac{1}{3}$), but the branch-misprediction rate went down from 0.51 ($\alpha = \frac{1}{2}$) to 0.44 ($\alpha = \frac{1}{3}$). This was enough to obtain an improvement in the running time. For larger values of α the running times got again higher because of the larger amount of work done. The behaviour of the two-way model was very similar to that of the three-way model. On the other hand, the situation was different for the branchless model; the value $\alpha = \frac{1}{2}$ always gave the best results and the branch-misprediction rate was 0.06 or lower.

4   Skewed Search Trees

A *skewed binary search tree* [4] is a binary search tree in which the left subtree of a node is always lighter than the right subtree. Moreover, this bias is exact so that, if $weight(x)$ denotes the number of nodes stored in the subtree rooted at node x, $weight((*x).\texttt{left}()) = \lfloor \alpha \cdot weight(x) \rfloor$ for a fixed constant α, $0 < \alpha \leq \frac{1}{2}$. A *perfectly balanced search tree* is a special case where $\alpha = \frac{1}{2}$. We implemented skewed search trees and profiled their performance for different values of α.

To start with, we consider a memory layout where each node is allocated randomly from a contiguous pool of nodes. In addition to the search procedure relying on a two-way element comparison described in Section 2, we implemented a variant where the `if` statement inside the `while` loop was eliminated. The resulting program is interestingly simple.

```
 1  N* choose(bool c, N* x, N* y) {
 2    return (N*)((char*) y + c * ((char*) x - (char*) y));
 3  }
 4
 5  bool is_member(V const & v) {
 6    N* y = nullptr; // candidate node
 7    N* x = root; // current node
 8    while (x != nullptr) {
 9      bool smaller = less(v, (*x).element());
10      y = choose(smaller, y, x);
11      x = choose(smaller, (*x).left(), (*x).right());
12    }
13    if (y == nullptr || less((*y).element(), v)) {
14      return false;
15    }
16    return true;
17  }
```

The statement `z = choose(c, x, y)` has the same effect as the C statement `z = (c) ? x : y`, i.e. it executes a conditional assignment. Here it would have been natural to use the conditional-move primitive provided by the hardware, but we wanted to avoid it for two reasons. First, it would have been necessary to implement this in assembly language, because there was no guarantee that the primitive would be used by the compiler. Second, our micro-benchmarks showed that the conditional-move primitive was slow in our test computer.

Assuming that the underlying branch predictor is static and that a `while` loop is translated by ending the loop conditionally, the outcome of the conditional branch at the end of the `while` loop is easy to predict. If we assume that backward branches are taken, this prediction is only incorrect when we step out of the loop. Including the branch mispredictions induced by the last `if` statement, the whole procedure may only induce $O(1)$ branch mispredictions per search.

The performance of the search procedures is summed up in Tables 3 and 4. Compared to the models discussed in the previous section, the running times are higher because of memory accesses. The results obtained are consistent with our earlier experiments [6,7], where branch removal turned out to be beneficial for small problem instances. For the problem sizes we considered, when the

Table 3. Runtime performance of the search procedures for skewed search trees (random layout); execution time per $\lg N$ [in nanoseconds]

N	Skewed random Two-way			Skewed random Branchless		
	$\alpha = \frac{1}{2}$	$\frac{1}{3}$	$\frac{1}{4}$	$\alpha = \frac{1}{2}$	$\frac{1}{3}$	$\frac{1}{4}$
2^{15}	10.3	10.5	10.7	7.6	11.0	12.3
2^{20}	38.5	40.6	44.5	36.1	42.6	48.8
2^{25}	75.8	82.6	91.9	76.8	86.3	97.8

Table 4. Branch behaviour of the search procedures for skewed search trees (random layout); number of conditional branches executed ($\lessdot$) and branch mispredictions induced per search, both divided by $\lg N$

N	Balanced random Two-way $\alpha = \frac{1}{2}$		Skewed random Two-way $\alpha = \frac{1}{3}$		Skewed random Two-way $\alpha = \frac{1}{4}$		Balanced random Branchless $\alpha = \frac{1}{2}$	
	$\lessdot$	Mispred.	$\lessdot$	Mispred.	$\lessdot$	Mispred.	$\lessdot$	Mispred.
2^{15}	2.13	0.57	2.28	0.52	2.53	0.46	1.13	0.07
2^{20}	2.10	0.55	2.26	0.50	2.52	0.44	1.10	0.05
2^{25}	2.08	0.54	2.24	0.49	2.51	0.42	1.08	0.04

memory layout was random, we could not repeat the runtime results of Brodal and Moruz [4], although the branch-misprediction rate went down. For the branchless procedure there is a perfect match between theory and practice: It executes $\lg N + O(1)$ conditional branches but only induces $O(1)$ branch mispredictions.

5   Local Search Trees

By increasing the locality of memory references, improvements in performance can be seen at two levels:

- At the cache level, when more elements are in the same cache line, fewer cache misses will be incurred.
- At the memory level, when more elements are in the same page, fewer TLB (translation-lookaside buffer) misses will be incurred. At each memory access, the virtual memory address has to be translated into a physical memory address. The purpose of the TLB is to store this mapping for the most recently used memory addresses to avoid an access to the page table.

By running a space-utilization benchmark [3, Appendix 3], we observed that a binary-search-tree node storing one 4-byte integer and three 8-byte pointers required 48 bytes of memory (even though the raw data is only 28 bytes). Since in our test computer the size of the cache lines is 64 bytes, we did not expect much improvement at the cache level. However, as elucidated in an early study by Oksanen and Malmi [14], improvements at the memory level can be noticeable.

In the literature many schemes have been proposed to improve memory-access patterns for data structures supporting searching. The cache-sensitive schemes can be classified into three categories: one-level layouts [2,14], two-level layouts [5,17], and more complex multi-level layouts [2,16,18]. When deciding which one to choose, our primary criterion was the simplicity of programming. We were not interested in schemes solely providing good big-Oh estimates for the critical performance measures; good practical performance was imperative. Furthermore, cache-obliviousness [16] was not an important issue for us since we were willing to perform some simple benchmarks to find the best values for the tuning parameters in our test environment.

In a *local search tree* the goal is to lay out the nodes such that, when a path from the root to a leaf is traversed, as few pages are visited as possible. An interesting variant is obtained by seeing the tree as an F-ary tree, where each so-called *fat node* stores a complete binary tree. Since the subtrees inside the fat nodes are complete, F must be of the form 2^h for an integer h, $h > 0$, and each fat node of size (up to) $2^h - 1$. The F-ary tree can be laid out in memory as an F-ary heap [10] and the trees inside the fat nodes as a binary heap [19].

Jensen et al. [9] gave formulas how to get from a node to its children and parent in this layout. Using these formulas the pointers at each node can be set in a single loop. After this, it is easy to populate the tree provided that the elements are given in sorted order. Searching can be done as before. Each search will visit at most $\lceil \lceil \lg(1 + N) \rceil / \lg F \rceil$ fat nodes. The fat nodes may not be perfectly aligned with the actual memory pages, but this is not a big problem since F is expected to be relatively large.

In our experiments we first determined the optimal value of F; in our test computer this turned out to be 16, but all values between 8 and 64 worked well. Then we compared our implementation of local search trees to the C++ standard-library implementation of red-black trees [8]. The results of this comparison are given in Tables 5 and 6. As seen from these and the previous results, red-black trees perform almost equally badly as perfectly balanced search trees when the memory layout is random. By placing the nodes more locally, almost a factor of two speed-up in the running time was experienced in our test environment.

Table 5. Runtime performance of the search procedures for two search trees; running time per search divided by $\lg N$ [in nanoseconds].

N	Local		Red-black
	Two-way	Branchless	Two-way
2^{15}	7.6	5.9	10.3
2^{20}	21.9	20.9	36.6
2^{25}	33.5	36.1	64.7

Table 6. Branch behaviour of the search procedures for two search trees; number of conditional branches executed ($\lessgtr$) and branch mispredictions induced per search, both divided by $\lg N$

N	Local Two-way		Local Branchless		Red-black Two-way	
	$\lessgtr$	Mispred.	$\lessgtr$	Mispred.	$\lessgtr$	Mispred.
2^{15}	2.16	0.57	1.12	0.07	2.12	0.58
2^{20}	2.05	0.55	1.05	0.05	2.09	0.56
2^{25}	2.13	0.54	1.09	0.04	2.08	0.59

6   Implicit Search Trees

If the data set is static, an observant reader may wonder why to use a search tree when a sorted array will do. A sorted array is an implicit binary search tree where arithmetic operations are used to move from one node to another; no explicit pointers are needed. Local search trees can also be made implicit using the formulas given in [9].

It was easy to implement implicit local search trees starting from the code for local search trees. Each time the left child (or the right child or the parent) of a node x was accessed, instead of writing (*x).left(), we replaced it with the corresponding formula. As for local search trees, we did not align the fat nodes perfectly with the cache lines. This would have been possible by adding some padding between the elements, but we wanted to keep the data structure compact and the formulas leading to the neighbouring nodes unchanged.

The results of our tests for implicit search trees are given in Tables 7 (running time), 8 (branch mispredictions), and 9 (cache misses). As a competitor to our search procedures, we considered two implementations of binary search (std::binary_search and our branchless modification of it) applied for a sorted array. Also here, before the final tests, we determined the best value for the parameter F; in our environment it was 16.

As to the running time (Table 7), for large problem instances implicit local search trees were the fastest of all structures considered in this study, but for small problem instances the overhead caused by the address calculations was clearly visible and implicit local search trees were slow.

Table 7. Runtime performance of the search procedures for implicit search trees; running time per search divided by lg N [in nanoseconds]

N	Implicit local		Sorted array	
	Two-way	Branchless	Two-way	Branchless
2^{15}	15.0	14.2	6.9	7.2
2^{20}	16.3	15.6	14.7	20.7
2^{25}	20.1	22.8	32.1	45.8

Table 8. Branch behaviour of the search procedures for implicit search trees; number of conditional branches executed ($\oslash$) and branch mispredictions induced per search, both divided by lg N

N	Implicit local Two-way		Implicit local Branchless		Sorted array Two-way		Sorted array Branchless	
	$\oslash$	Mispred.	$\oslash$	Mispred.	$\oslash$	Mispred.	$\oslash$	Mispred.
2^{15}	3.21	0.86	1.12	0.07	2.07	0.57	1.13	0.10
2^{20}	3.05	0.84	1.05	0.05	2.05	0.55	1.10	0.05
2^{25}	3.18	0.82	1.09	0.04	2.04	0.54	1.08	0.04

Table 9. Cache behaviour of three search trees; number of memory references performed, cache I/Os performed, and cache misses incurred per search, all divided by $\log_B N$, where B is the number of elements that fit in a cache line (16 in our test)

N	Local Two-way			Implicit local Two-way			Sorted array Two-way		
	Refs.	I/Os	Misses	Refs.	I/Os	Misses	Refs.	I/Os	Misses
2^{15}	9.19	2.43	0.00	9.46	0.40	0.00	6.93	2.00	0.00
2^{20}	8.60	3.27	1.32	8.80	0.81	0.12	6.70	3.20	0.73
2^{25}	8.84	3.77	2.27	9.00	1.01	0.51	6.56	3.37	3.01

One interesting fact of the formulas used for computing the indices of the neighbouring nodes is that they all contain an `if` statement. This is because a separate handling is necessary depending on whether we are inside a fat node or whether we move from one fat node to another. By inspecting the assembly-language code generated by the compiler, we observed that for the branchless version the compiler used conditional moves to eliminate these `if` statements, whereas for the non-optimized version conditional branches were used. This explains the discrepancies in the numbers in Table 8 (approximatively 3 vs. 1 conditional branches per iteration).

Finally, we compared the cache behaviour of local, implicit local, and implicit search trees (Table 9). We measured the number of memory references, cache I/Os, and cache misses. For small problem instances there were no cache misses since the data structures could be kept inside the cache. For implicit local search trees both the number of cache I/Os and the number of cache misses were smaller than the corresponding numbers for other structures. For $N = 2^{25}$, the $2.25 \lg N$ memory accesses generated $1.01 \log_B N$ cache I/Os, which is basically optimal when the size of the cache blocks is B; only half of the I/Os ended up to be a miss. One can explain the low number of cache misses by observing that the cache of our test computer is large enough so that, with an ideal cache replacement, the top portion of the search tree will be kept inside the cache at all times. A sorted array was another extreme; it made more than three times as many cache I/Os and almost every cache I/O incurred a cache miss.

7   Unrolling the Loop

It is well-known that loop unrolling can be used to improve the performance of programs in many respects (see, e.g. [3, Appendix 4]). Bentley [3, Column 4] gave a nice description of an ancient idea how to unroll binary search. The same technique applies for the search procedure of perfectly balanced search trees. In this section we describe how to do this unrolling so that the search procedure only contains a few branches and has no loops. An immediate corollary is that such a straight-line program cannot induce more than a constant number of branch mispredictions per search.

When the search tree is perfectly balanced such that the lengths of root-to-leaf paths only vary by one, the following kind of procedure will do the job.

```
 1  bool is_member(V const & v) {
 2    N* y = nullptr; // candidate node
 3    N* x = root; // current node
 4    bool smaller;
 5    switch (height) {
 6    case 31:
 7      smaller = less(v, (*x).element());
 8      y = choose(smaller, y, x);
 9      x = choose(smaller, (*x).left(), (*x).right());
      ⋮
125   case 1:
126     smaller = less(v, (*x).element());
127     y = choose(smaller, y, x);
128     x = choose(smaller, (*x).left(), (*x).right());
129   default:
130     smaller = (x == nullptr) || less(v, (*x).element());
131     y = choose(smaller, y, x);
132   }
133   if ((y == nullptr) || less((*y).element(), v)) {
134     return false;
135   }
136   return true;
137 }
```

To manage skewed trees the bottom-most levels must be handled as in the normal search procedure, because we cannot be sure when x refers to null. For skewed trees the procedure should also be able to tolerate larger heights.

A theoretician may oppose this solution because the length of the program is not a constant. A practitioner may be worried about the portability since we assumed that the maximum height of the tree is 31. Both of these objections are reasonable, but neither is critical. The maximum height could be made larger and the program could even be generated on the fly after the user has specified the height of the search tree.

As a curiosity we tested the efficiency of the unrolled search procedure; we only considered perfectly balanced trees with random memory layout. The runtime performance did not improve at all. The main reason for this seems to be that in our test computer the cost of easy-to-predict branches is so low. On the other hand, both the branch-count rate and the branch-misprediction rate went to zero when N increased.

8   Conclusion

We were mainly interested in understanding the impact of branch mispredictions on the performance of the search procedures. We hope that we could make our case clear: Branch prediction is not a good enough reason to switch from balanced search trees to skewed search trees. In theory, any program can be transformed into a form that induces $O(1)$ branch mispredictions [6]. As shown in the present paper, in the context of searching, simple transformations can be used to eliminate branches and thereby avoid branch mispredictions. In most cases, these transformations work well in practice.

References

1. Andersson, A.: A note on searching in a binary search tree. Software Pract. Exper. 21(10), 1125–1128 (1991)
2. Bender, M.A., Demaine, E.D., Farach-Colton, M.: Efficient tree layout in a multilevel memory hierarchy. In: Möhring, R.H., Raman, R. (eds.) ESA 2002. LNCS, vol. 2461, pp. 165–173. Springer, Heidelberg (2002)
3. Bentley, J.: Programming Pearls, 2nd edn. Addison-Wesley, Reading (2000)
4. Brodal, G.S., Moruz, G.: Skewed binary search trees. In: Azar, Y., Erlebach, T. (eds.) ESA 2006. LNCS, vol. 4168, pp. 708–719. Springer, Heidelberg (2006)
5. Chen, S., Gibbons, P.B., Mowry, T.C., Valentin, G.: Fractal prefetching B^+-trees: Optimizing both cache and disk performance. In: SIGMOD 2002, pp. 157–168. ACM, New York (2002)
6. Elmasry, A., Katajainen, J.: Lean programs, branch mispredictions, and sorting. In: Kranakis, E., Krizanc, D., Luccio, F. (eds.) FUN 2012. LNCS, vol. 7288, pp. 119–130. Springer, Heidelberg (2012)
7. Elmasry, A., Katajainen, J., Stenmark, M.: Branch mispredictions don't affect mergesort. In: Klasing, R. (ed.) SEA 2012. LNCS, vol. 7276, pp. 160–171. Springer, Heidelberg (2012)
8. Guibas, L.J., Sedgewick, R.: A dichromatic framework for balanced trees. In: FOCS 1978, pp. 8–21. IEEE Computer Society, Los Alamitos (1978)
9. Jensen, C., Katajainen, J., Vitale, F.: Experimental evaluation of local heaps. CPH STL Report 2006-1, Department of Computer Science, University of Copenhagen, Copenhagen (2006)
10. Johnson, D.B.: Priority queues with update and finding minimum spanning trees. Inform. Process. Lett. 4(3), 53–57 (1975)
11. Kaligosi, K., Sanders, P.: How branch mispredictions affect quicksort. In: Azar, Y., Erlebach, T. (eds.) ESA 2006. LNCS, vol. 4168, pp. 780–791. Springer, Heidelberg (2006)
12. Katajainen, J.: Branchless search programs: Electronic appendix. CPH STL Report 2012-1, Department of Computer Science, University of Copenhagen, Copenhagen (2012)
13. Knuth, D.E.: Sorting and Searching, The Art of Computer Programming, 2nd edn., vol. 3. Addison-Wesley, Reading (1998)
14. Oksanen, K., Malmi, L.: Memory reference locality and periodic relocation in main memory search trees. In: 5th Hellenic Conference on Informatics, pp. 679–687. Greek Computer Society, Athens (1995)
15. Patterson, D.A., Hennessy, J.L.: Computer Organization and Design: The Hardware/Software Interface, revised 4th edn. Morgan Kaufmann, Waltham (2012)
16. Prokop, H.: Cache-oblivious algorithms. Master's Thesis, Department of Electrical Engineering and Computer Science, Massachusetts Institute of Technology, Cambridge (1999)
17. Rahman, N., Cole, R., Raman, R.: Optimised predecessor data structures for internal memory. In: Brodal, G.S., Frigioni, D., Marchetti-Spaccamela, A. (eds.) WAE 2001. LNCS, vol. 2141, pp. 67–78. Springer, Heidelberg (2001)
18. Saikkonen, R., Soisalon-Soininen, E.: Cache-sensitive memory layout for binary trees. In: Ausiello, G., Karhumäki, J., Mauri, G., Ong, L. (eds.) IFIP TCS 2008. IFIP, vol. 273, pp. 241–255. Springer, Boston (2008)
19. Williams, J.W.J.: Algorithm 232: Heapsort. Commun. ACM 7(6), 347–348 (1964)

Lightweight Lempel-Ziv Parsing*

Juha Kärkkäinen, Dominik Kempa, and Simon J. Puglisi

Department of Computer Science,
University of Helsinki
Helsinki, Finland
firstname.lastname@cs.helsinki.fi

Abstract. We introduce a new approach to LZ77 factorization that uses $O(n/d)$ words of working space and $O(dn)$ time for any $d \geq 1$ (for polylogarithmic alphabet sizes). We also describe carefully engineered implementations of alternative approaches to lightweight LZ77 factorization. Extensive experiments show that the new algorithm is superior, and particularly so at the lowest memory levels and for highly repetitive data. As a part of the algorithm, we describe new methods for computing matching statistics which may be of independent interest.

1 Introduction

The Lempel-Ziv factorization [28], also known as the LZ77 factorization, or LZ77 parsing, is a fundamental tool for compressing data and string processing, and has recently become the basis for several compressed full-text pattern matching indexes [17,11]. These indexes are designed to efficiently store and search massive, highly-repetitive data sets — such as web crawls, genome collections, and versioned code repositories — which are increasingly common [22].

In traditional compression settings (for example the popular `gzip` tool) LZ77 factorization is kept timely by factorizing relative to only a small, recent window of data, or by breaking the data up into blocks and factorizing each block separately. This approach fails to capture widely spaced repetitions in the input, and anyway, in many applications, including construction of the above mentioned LZ77-based text indexes, whole-string LZ77 factorizations are required.

The fastest LZ77 algorithms (see [12] for the latest comparison) use a lot of space, at least $6n$ bytes for an input of n symbols and often more. This prevents them from scaling to really large inputs. Space-efficient algorithms are desirable even on smaller inputs, as they place less burden on the underlying system.

One approach to more space efficient LZ factorization is to use compressed suffix arrays and succinct data structures [21]. Two proposals in this direction are due to Kreft and Navarro [16] and Ohlebusch and Gog [23]. In this paper, we describe carefully engineered implementations of these algorithms. We also propose a new, space-efficient variant of the recent ISA family of algorithms [15].

* This research is partially supported by Academy of Finland grants 118653 (ALGO-DAN) and 250345 (CoECGR).

V. Bonifaci et al. (Eds.): SEA 2013, LNCS 7933, pp. 139–150, 2013.

Compressed indexes are usually built from the uncompressed suffix array (SA) which requires $4n$ bytes. Our implementations are instead based on the Burrows-Wheeler transform (BWT), constructed directly in about 2–$2.5n$ bytes using the algorithm of Okanohara and Sadakane [26]. There also exists two online algorithms based on compressed indexes [25,27] but they are not competitive in practice in the offline context.

The main contribution of this paper is a new algorithm to compute the LZ77 factorization without ever constructing SA or BWT for the whole input. At a high-level, the algorithm divides the input up into blocks, and processes each block in turn, by first computing a pattern matching index for the block, then scanning the prefix of the input prior to the block through the index to compute longest-matches, which are then massaged into LZ77 factors. For a string of length n and σ distinct symbols, the algorithm uses $n \log \sigma + O(n \log n / d)$ bits of space, and $O(dnt_{rank})$ time, where d is the number of blocks, and t_{rank} is the time complexity of the rank operation over sequences with alphabet size σ (see e.g. [2]). The $n \log \sigma$ bits in the space bound is for the input string itself which is treated as read-only.

Our implementation of the new algorithm does not, for the most part, use compressed or succinct data structures. The goal is to optimize speed rather than space in the data structures, because we can use the parameter d to control the tradeoff. Our experiments demonstrate that this approach is superior to algorithms using compressed indexes.

As a part of the new algorithm, we describe new techniques for computing matching statistics [5] that may be of independent interest. In particular, we show how to invert matching statistics, i.e., to compute the matching statistics of a string B w.r.t. a string A from the matching statistics of A w.r.t. B, which saves a lot of space when A is much longer than B.

All our implementations operate in main memory only and thus need at least n bytes just to hold the input. Reducing the memory consumption further requires some use of external memory, a direction largely unexplored in the literature so far. We speculate that the scanning, block-oriented nature of the new algorithm will allow efficient secondary memory implementations, but that study is left for the future.

2   Basic Notation and Algorithmic Machinery

Strings. Throughout we consider a string $X = X[1..n] = X[1]X[2]\ldots X[n]$ of $|X| = n$ symbols drawn from the alphabet $[0..\sigma - 1]$. We assume $X[n]$ is a special "end of string" symbol, $\$$, smaller than all other symbols in the alphabet. The reverse of X is denoted $\hat{X}$. For $i = 1, \ldots, n$ we write $X[i..n]$ to denote the *suffix* of X of length $n - i + 1$, that is $X[i..n] = X[i]X[i + 1]\ldots X[n]$. We will often refer to suffix $X[i..n]$ simply as "suffix i". Similarly, we write $X[1..i]$ to denote the *prefix* of X of length i. $X[i..j]$ is the *substring* $X[i]X[i + 1]\ldots X[j]$ of X that starts at position i and ends at position j. By $X[i..j)$ we denote $X[i..j - 1]$. If $j < i$ we define $X[i..j]$ to be the empty string, also denoted by ε.

Suffix Arrays. The suffix array [19] $\mathsf{SA_X}$ (we drop subscripts when they are clear from the context) of a string X is an array $\mathsf{SA}[1..n]$ which contains a permutation of the integers $[1..n]$ such that $\mathsf{X}[\mathsf{SA}[1]..n] < \mathsf{X}[\mathsf{SA}[2]..n] < \cdots < \mathsf{X}[\mathsf{SA}[n]..n]$. In other words, $\mathsf{SA}[j] = i$ iff $\mathsf{X}[i..n]$ is the jth suffix of X in ascending lexicographical order. The inverse suffix array ISA is the inverse permutation of SA, that is $\mathsf{ISA}[i] = j$ iff $\mathsf{SA}[j] = i$.

Let $\mathsf{lcp}(i, j)$ denote the length of the longest-common-prefix of suffix i and suffix j. For example, in the string $\mathsf{X} = zzzzzapzap$, $\mathsf{lcp}(1, 4) = 2 = |zz|$, and $\mathsf{lcp}(5, 8) = 3 = |zap|$. The longest-common-prefix (LCP) array [14,13], $\mathsf{LCP_X} = \mathsf{LCP}[1..n]$, is defined such that $\mathsf{LCP}[1] = 0$, and $\mathsf{LCP}[i] = \mathsf{lcp}(\mathsf{SA}[i], \mathsf{SA}[i-1])$ for $i \in [2..n]$.

For a string Y, the Y-interval in the suffix array $\mathsf{SA_X}$ is the interval $\mathsf{SA}[s..e]$ that contains all suffixes having Y as a prefix. The Y-interval is a representation of the occurrences of Y in X. For a character c and a string Y, the computation of the $c\mathsf{Y}$-interval from the Y-interval is called a *left extension* and the computation of the Y-interval from the $\mathsf{Y}c$-interval is called a *right contraction*. *Left contraction* and *right extension* are defined symmetrically.

BWT and backward search. The Burrows-Wheeler Transform [3] $\mathsf{BWT}[1..n]$ is a permutation of X such that $\mathsf{BWT}[i] = \mathsf{X}[\mathsf{SA}[i] - 1]$ if $\mathsf{SA}[i] > 1$ and $\$$ otherwise. We also define $\mathsf{LF}[i] = j$ iff $\mathsf{SA}[j] = \mathsf{SA}[i] - 1$, except when $\mathsf{SA}[i] = 1$, in which case $\mathsf{LF}[i] = \mathsf{ISA}[n]$. Let $C[c]$, for symbol c, be the number of symbols in X lexicographically smaller than c. The function $\mathsf{rank}(\mathsf{X}, c, i)$, for string X, symbol c, and integer i, returns the number of occurrences of c in $\mathsf{X}[1..i]$. It is well known that $\mathsf{LF}[i] = C[\mathsf{BWT}[i]] + \mathsf{rank}(\mathsf{BWT}, \mathsf{BWT}[i], i)$. Furthermore, we can compute the left extension using C and rank. If $\mathsf{SA}[s..e]$ is the Y-interval, then $\mathsf{SA}[C[c] + \mathsf{rank}(\mathsf{BWT}, c, s), C[c] + \mathsf{rank}(\mathsf{BWT}, c, e)]$ is the $c\mathsf{Y}$-interval. This is called *backward search* [8].

NSV/PSV and RMQ. For an array A, the *next and previous smaller value* (NSV/PSV) operations are defined as $\mathsf{NSV}(i) = \min\{j \in [i+1..n] \mid \mathsf{A}[j] < \mathsf{A}[i]\}$ and $\mathsf{PSV}(i) = \max\{j \in [1..i-1] \mid \mathsf{A}[j] < \mathsf{A}[i]\}$. A related operation on A is *range minimum query*: $\mathsf{RMQ}(\mathsf{A}, i, j)$ is $k \in [i..j]$ such that $\mathsf{A}[k]$ is the minimum value in $\mathsf{A}[i..j]$. Both NSV/PSV operations and RMQ operations over the LCP array can be used for implementing right contraction (see Section 4).

LZ77. Before defining the LZ77 factorization, we introduce the concept of a *longest previous factor* (LPF). The LPF at position i in string X is a pair $\mathsf{LPF_X}[i] = (p_i, \ell_i)$ such that, $p_i < i$, $\mathsf{X}[p_i..p_i + \ell_i) = \mathsf{X}[i..i + \ell_i)$, and ℓ_i is maximized. In other words, $\mathsf{X}[i..i + \ell_i)$ is the longest prefix of $\mathsf{X}[i..n]$ which also occurs at some position $p_i < i$ in X. Note also that there may be more than one potential source (that is, p_i value), and we do not care which one is used.

The LZ77 factorization (or LZ77 parsing) of a string X is then just a greedy, left-to-right parsing of X into longest previous factors. More precisely, if the jth LZ factor (or *phrase*) in the parsing is to start at position i, then we output (p_i, ℓ_i) (to represent the jth phrase), and then the $(j + 1)$th phrase starts at position $i + \ell_i$. The exception is the case $\ell_i = 0$, which happens iff $X[i]$ is the leftmost occurrence of a symbol in X. In this case we output $(X[i], 0)$ (to represent $X[i..i]$) and the next phrase starts at position $i + 1$. When $\ell_i > 0$, the substring $X[p_i..p_i + \ell_i)$ is called the *source* of phrase $X[i..i + \ell_i)$. We denote the number of phrases in the LZ77 parsing of X by z.

Matching Statistics. Given two strings Y and Z, the matching statistics of Y w.r.t. Z, denoted $MS_{Y|Z}$, is an array of $|Y|$ pairs, $(p_1, \ell_1), (p_2, \ell_2), ..., (p_{|Y|}, \ell_{|Y|})$, such that for all $i \in [1..|Y|]$, $Y[i..i + \ell_i) = Z[p_i..p_i + \ell_i)$ is the longest substring starting at position i in Y that is also a substring of Z. The observant reader will note the resemblance to the LPF array. Indeed, if we replace LPF_Y with $MS_{Y|Z}$ in the computation of the LZ factorization of Y, the result is the relative LZ factorization of Y w.r.t. Z [18].

3   Lightweight, Scan-Based LZ77 Parsing

In this section we present a new algorithm for LZ77 factorization called LZscan.

Basic Algorithm. Conceptually LZscan divides X up into $d = \lceil n/b \rceil$ fixed size blocks of length b: $X[1..b], X[b + 1..2b], ...$. The last block could be smaller than b, but this does not change the operation of the algorithm. In the description that follows we will refer to the block currently under consideration as B, and to the prefix of X that ends just before B as A. Thus, if $B = X[kb + 1..(k + 1)b]$, then $A = X[1..kb]$.

To begin, we will assume no LZ factor or its source crosses a boundary of the block B. Later we will show how to remove these assumptions.

The outline of the algorithm for processing a block B is shown below.

1. Compute $MS_{A|B}$
2. Compute $MS_{B|A}$ from $MS_{A|B}$, SA_B and LCP_B
3. Compute $LPF_{AB}[kb + 1..(k + 1)b]$ from $MS_{B|A}$ and LPF_B
4. Factorize B using $LPF_{AB}[kb + 1..(k + 1)b]$

Step 1 is the computational bottleneck of the algorithm in theory and practice. Theoretically, the time complexity of Step 1 is $O((|A| + |B|)t_{rank})$, where t_{rank} is the time complexity of the rank operation on BWT_B (see, e.g., [2]). Thus the total time complexity of LZscan is $O(dnt_{rank})$ using $O(b)$ words of space in addition to input and output. The practical implementation of Step 1 is described in Section 4. In the rest of this section, we describe the details of the other steps.

Step 2: Inverting Matching Statistics. We want to compute $MS_{B|A}$ but we cannot afford the space of the large data structures on A required by standard methods [1,23]. Instead, we compute first $MS_{A|B}$ involving large data structures on B, which we can afford, and only a scan of A (see Section 4 for details). We then *invert* $MS_{A|B}$ to obtain $MS_{B|A}$. The inversion algorithm is given in Fig. 1.

Algorithm MS-Invert
1: **for** $i \leftarrow 1$ **to** $|B|$ **do** $MS_{B|A}[i] \leftarrow (0,0)$
2: **for** $i \leftarrow 1$ **to** $|A|$ **do**   // transfer information from $MS_{A|B}$ to $MS_{B|A}$
3:      $(p_A, \ell_A) \leftarrow MS_{A|B}[i]$
4:      $(p_B, \ell_B) \leftarrow MS_{B|A}[p_A]$
5:      **if** $\ell_A > \ell_B$ **then** $MS_{B|A}[p_A] \leftarrow (i, \ell_A)$
6: $(p, \ell) \leftarrow MS_{B|A}[SA_B[1]]$ // spread information in $MS_{B|A}$
7: **for** $i \leftarrow 2$ **to** $|B|$ **do**   // in lexicograhically ascending direction
8:      $\ell \leftarrow \min(\ell, LCP_B[i])$
9:      $(p_B, \ell_B) \leftarrow MS_{B|A}[SA_B[i]]$
10:     **if** $\ell > \ell_B$ **then** $MS_{B|A}[SA_B[i]] \leftarrow (p, \ell)$
11:     **else** $(p, \ell) \leftarrow (p_B, \ell_B)$
12: $(p, \ell) \leftarrow MS_{B|A}[SA_B[|B|]]$ // spread information in $MS_{B|A}$
13: **for** $i \leftarrow |B| - 1$ **downto** 1 **do**   // in lexicograhically descending direction
14:     $\ell \leftarrow \min(\ell, LCP_B[i + 1])$
15:     $(p_B, \ell_B) \leftarrow MS_{B|A}[SA_B[i]]$
16:     **if** $\ell > \ell_B$ **then** $MS_{B|A}[SA_B[i]] \leftarrow (p, \ell)$
17:     **else** $(p, \ell) \leftarrow (p_B, \ell_B)$

Fig. 1. Inverting matching statistics

Note that the algorithm accesses each entry of $MS_{A|B}$ only once and the order of these accesses does not matter. Thus we can execute the code on lines 3–5 immediately after computing $MS_{A|B}[i]$ in Step 1 and then discard that value. This way we can avoid storing $MS_{A|B}$.

Step 3: Computing LPF. Consider the pair $(p, \ell) = LPF_{AB}[i]$ for $i \in [kb + 1..(k + 1)b]$ that we want to compute and assume $\ell > 0$ (otherwise i is the position of the leftmost occurrence of $X[i]$ in X, which we can easily detect). Clearly, either $p \leq kb$ and $LPF_{AB}[i] = MS_{B|A}[i]$, or $kb < p < i$ and $LPF_{AB}[i] = (kb + p_B, \ell_B)$, where $(p_B, \ell_B) = LPF_B[i - kb]$. Thus computing LPF_{AB} from $MS_{B|A}[i]$ and LPF_B is easy.

The above is true if the sources do not cross the block boundary, but the case where $p \leq kb$ but $p + \ell > kb + 1$ is not handled correctly. An easy correction is to replace $MS_{A|B}$ with $MS_{AB|B}[1..kb]$ in all of the steps.

Step 4: Parsing. We use the standard LZ77 parsing to factorize B except LPF_B is replaced with $LPF_{AB}[kb + 1..(k + 1)b]$.

So far we have assumed that every block starts with a new phrase, or, put another way, that a phrase ends at the end of every block. Let $X[i..(k + 1)b]$ be the last factor in B, after we have factorized B as described above. This may not be a true LZ factor when considering the whole X but may continue beyond the end of B. To find the true end point, we treat $X[i..n]$ as a pattern, and apply the constant extra space pattern matching algorithm of Crochemore [7], looking for the longest prefix of $X[i..n]$ starting in $X[1..i - 1]$. We must modify the algorithm of Crochemore so that it finds the longest matching prefix of the pattern rather than a full match, but this is possible without increasing its time or space complexity.

4   Computation of Matching Statistics

In this section, we describe how to compute the matching statistics $\mathsf{MS}_{\mathsf{A}|\mathsf{B}}$. As mentioned in Section 3, what we actually want is $\mathsf{MS}_{\mathsf{AB}|\mathsf{B}}[1..kb]$. However, the only difference is that the starting point of the computation is the B-interval in SA_B instead of the ε-interval.

Similarly to most algorithms for computing the matching statistics, we first construct some data structures on B and then scan A. During the whole LZ factorization, most of the time is spend on the scanning and the time for constructing the data structures is insignificant in practice. Thus we omit the construction details here. The space requirement of the data structures is more important but not critical as we can compensate for increased space by reducing the block size b. Using more space (per character of B) is worth doing if it increases scanning speed more than it increases space. Consequently, we mostly use plain, uncompressed arrays.

Standard approach. The standard approach of computing the matching statistics using the suffix array is to compute for each position i the longest prefix $\mathsf{P}_i = \mathsf{A}[i..i + \ell_i)$ of the suffix $\mathsf{A}[i..|\mathsf{A}|]$ such that the P_i-interval in SA_B is non-empty. Then $\mathsf{MS}_{\mathsf{A}|\mathsf{B}}[i] = (p_i, \ell_i)$, where p_i is any suffix in the P_i-interval. This can be done either with a forward scan of A, computing each P_i-interval from P_{i-1}-interval using the extend right and contract left operations [1], or with a backward scan computing each P_i-interval from P_{i+1}-interval using the extend left and contract right operations [24]. We use the latter alternative but with bigger and faster data structures.

The extend left operation is implemented by backward search. We need the array C of size σ and an implementation of the rank function on BWT. For the latter, we use the fast rank data structure of Ferragina et al. [9], which uses $4b$ bytes.

The contract right operation is implemented using the NSV and PSV operations on LCP_B similarly to Ohlebusch and Gog [24], but instead of a compressed representation, we store the NSV/PSV values as plain arrays. As a nod towards reducing space, we store the NSV/PSV values as offsets using 2 bytes each. If the offset is too large (which is very rare), we obtain the value using the NSV/PSV data structure of Cánovas and Navarro [4], which needs less than $0.1b$ bytes. Here the space saving was worth it as it had essentially no effect on speed.

The peak memory use of the resulting algorithm is $n + (24.1)b + O(\sigma)$ bytes.

New approach. Our second approach is similar to the first, but instead of maintaining both end points of the P_i-interval, we keep just one, arbitrary position s_i within the interval. In principle, we perform left extension by backward search, i.e., $s_i = C[X[i]] + \text{rank}(BWT, X[i], s_{i+1})$. However, checking whether the resulting interval is empty and performing right contractions if it is, is more involved. To compute s_i and ℓ_i from s_{i+1} and ℓ_{i+1}, we execute the following steps:

1. Let $c = X[i]$. If $BWT[s_{i+1}] = c$, set $s_i = C[c] + \text{rank}(BWT, c, s_{i+1})$ and $\ell_i = \ell_{i+1} + 1$.
2. Otherwise, let $BWT[u]$ be the nearest occurrence of c in BWT before the position s_{i+1}. Compute the rank of that occurrence $r = \text{rank}(BWT, c, u)$ and $\ell_u = LCP[RMQ(LCP, u + 1, s_{i+1})]$. If $\ell_u \geq \ell_{i+1}$, set $s_i = C[c] + r$ and $\ell_i = \ell_{i+1} + 1$.
3. Otherwise, let $BWT[v]$ be the nearest occurrence of c in BWT after the position s_{i+1} and compute $\ell_v = LCP[RMQ(LCP, s_{i+1} + 1, v)]$. If $\ell_v \leq \ell_u$, set $s_i = C[c] + r$ and $\ell_i = \ell_u + 1$.
4. Otherwise, set $s_i = C[c] + r + 1$ and $\ell_i = \min(\ell_{i+1}, \ell_v) + 1$.

The implementation of the above algorithm is based on the arrays BWT, LCP and $R[1..b]$, where $R[i] = \text{rank}(BWT, BWT[i], i)$. All the above operations can be performed by scanning BWT and LCP starting from the position s_{i+1} and accessing one value in R. To avoid long scans, we divide BWT and LCP into blocks of size 2σ, and store for each block and each symbol c that occurs in B, the values r, ℓ_u and ℓ_v that would get computed if scans starting inside the block continued beyond the block boundaries.

The peak memory use is $n + 27b + O(\sigma)$ bytes. This is more than in the first approach, but this is more than compensated by increased scanning speed.

Skipping repetitions. During the preceding stages of the LZ factorization, we have built up knowledge of repetition present in A, which can be exploited to skip (sometimes large) parts of A during the matching-statistics scan. Consider an LZ factor $A[i..i+\ell]$. Because, by definition, $A[i..i+\ell]$ occurs earlier in A too, any source of an LZ factor of B that is completely inside $A[i..i+\ell]$ could be replaced with an equivalent source in that earlier occurrence. Thus such factors can be skipped during the computation of $MS_{A|B}$ without an effect on the factorization.

More precisely, if during the scan we compute $MS_{A|B}[j] = (p, k)$ and find that $i \leq j < j+k \leq i+\ell$ for an LZ factor $A[i..i+\ell)$, we will compute $MS_{A|B}[i-1]$ and continue the scanning from $i - 1$. However, we will do this only for long phrases with $\ell \geq 40$. To compute $MS_{A|B}[i - 1]$ from scratch, we use right extension operations implemented by a binary search on SA.

To implement this "skipping trick" we use a bitvector of n bits to mark LZ77 phrase boundaries adding $0.125n$ bytes to the peak memory.

5   Algorithms Based on Compressed Indexes

We went to some effort to ensure the baseline system used to evaluate LZscan in our experiments was not a "straw man". This required careful study and improvement of some existing approaches, which we now describe.

FM-Index. The main data structure in all the algorithms below is an implementation of the FM-index (FMI) [8]. It consists of two main components:

- BWT$_X$ *with support for the rank operation.* This enables backward search and the LF operation as described in Section 2. We have tried several rank data structures and found the one by Navarro [20, Sect. 7.1] to be the best in practice.
- A *sampling of* SA$_X$. This together with the LF operation enables arbitrary SA access since $SA[i] = SA[LF^k[i]] + k$ for any $k < SA[i]$. The sampling rate is a major space–time tradeoff parameter.

In many implementations of FMI, the construction starts with computing the uncompressed suffix array but we cannot afford the space. Instead, we construct BWT directly using the algorithm of Okanohara and Sadakane [26]. The method uses roughly 2–2.5n bytes of space but destroys the text, which is required later during LZ parsing. Thus, once we have BWT, we build a rank structure over it and use it to invert the BWT. During the inversion process we recover and store the text and gather the SA sample values.

CPS2 simulation. The CPS2 algorithm [6] is an LZ parsing algorithm based on SA$_X$. To compute the LZ factor starting at i, it computes the $X[i..i + \ell)$-interval for $\ell = 1, 2, 3, \ldots$ as long as the $X[i..i+\ell)$-interval contains a value $p < i$, indicating an occurrence of $X[i..i + \ell)$ starting at p.

The key operations in CPS2 are right extension and checking whether an SA interval contains a value smaller than i. Kreft and Navarro [16] as well as Ohlebusch and Gog [23] are using FMI for $\hat{X}$, the reverse of X, which allows simulating right extension on SA$_X$ by left extension on SA$_{\hat{X}}$. The two algorithms differ in the way they implement the interval checks:

- Kreft and Navarro use the RMQ operation. They use the RMQ data structure by Fischer and Heun [10] but we use the one by Cánovas and Navarro [4]. The latter is easy and fast to construct during BWT inversion but queries are slow without an explicit SA. We speed up queries by replacing a general RMQ with the check whether the interval contains a value smaller than i. This implementation is called LZ-FMI-RMQ.
- Ohlebusch and Gog use NSV/PSV queries. The position s of i in SA must be in the $X[i..i+\ell)$-interval. Thus we just need to check whether either NSV(s) or PSV(s) is in the interval too. They as well as we implement NSV/PSV using a balanced parentheses representation (BPR). This representation is initialized by accessing the values of SA left-to-right, which makes the construction slow using FMI. However, NSV/PSV queries with this data structure are fast, as they do not require accessing SA. This implementation is called LZ-FMI-BPR.

ISA variant. Among the most space efficient prior LZ factorization algorithms are those of the ISA family [15] that use a sampled ISA, a full SA and a rank/LF implementation that relies on the presence of the full SA. We reduce the space further by replacing SA and the rank/LF data structure with the FM-index described above to obtain an algorithm called LZ-FMI-ISA.

Table 1. Data set used in the experiments. The files are from the Pizza & Chili standard corpus[1] (S) and the Pizza & Chili repetitive corpus[2] (R). The value of n/z (average length of an LZ77 phrase) is included as a measure of repetitiveness. We use 100MiB prefixes of original files in order to reduce the time required to run the experiments with several algorithms and a large number of parameter combinations.

Name	σ	n/z	$n/2^{20}$	Source	Description
dna	16	14.2	100	S	Human genome
english	215	14.1	100	S	Gutenberg Project
sources	227	16.8	100	S	Linux and GCC sources
cere	5	84	100	R	yeast genome
einstein	121	2947	100	R	Wikipedia articles
kernel	160	156	100	R	Linux Kernel sources

6   Experiments

We performed experiments with the files listed in Table 1. All tests were conducted on a 2.53GHz Intel Xeon Duo CPU with 32GiB main memory and 8192K L2 Cache. The machine had no other significant CPU tasks running. The operating system was Linux (Ubuntu 10.04) running kernel 3.0.0-26. The compiler was g++ (gcc version 4.4.3) executed with the -O3 -static -DNDEBUG options. Times were recorded with the C `clock` function. All algorithms operate strictly in-memory. The implementations are available at http://www.cs.helsinki. fi/group/pads/.

LZscan vs. other algorithms. We compared the LZscan implementation using our new approach for matching statistics boosted with the "skipping trick" (Section 4) to algorithms based on compressed indexes (Section 5). The experiments measured the LZ factorization time and the memory usage with varying parameter settings for each algorithm. The results are shown in Fig. 2. In all cases LZscan outperforms other algorithm across the whole tradeoff spectrum. Moreover, it can operate with very small memory (close to n bytes) unlike other algorithms, which all require at least $2n$ bytes to compute BWT. It achieves a superior performance for highly repetitive data even at very low memory levels.

Variants of LZscan. We made a separate comparison of LZscan with the different variants of the matching statistics computation (see Section 4). As can be seen from Fig. 3, our new algorithm for matching statistics computation is a significant improvement over the standard approach. Adding the skipping trick usually improves further (english) but can also slightly deteriorate the speed (dna). On the other hand, for highly repetitive data, the skipping trick alone gives a dramatic time reduction (einstein).

---

[1] http://pizzachili.dcc.uchile.cl/texts.html
[2] http://pizzachili.dcc.uchile.cl/repcorpus.html

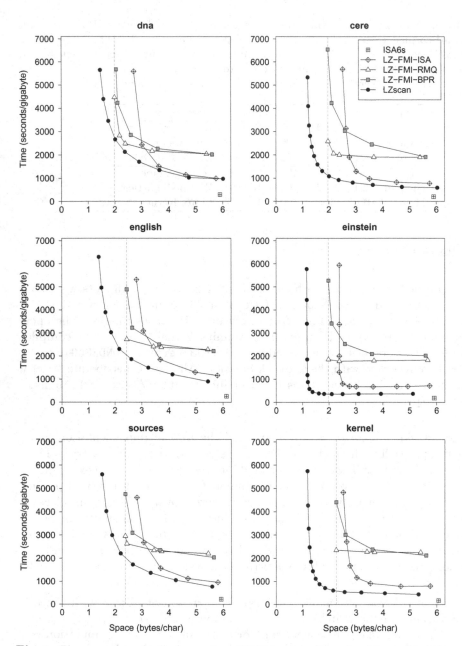

Fig. 2. Time-space tradeoffs for various LZ77 factorization algorithms. The times do not include reading from or writing to disk. For algorithms with multiple parameters controlling time/space we show only the optimal points, that is, points forming the lower convex hull of the points "cloud". The vertical line is the peak memory usage of the BWT construction algorithm [26], which is a space lower bound for all algorithms except LZscan. For comparison, we show the runtimes of ISA6s [15], currently the fastest LZ77 factorization algorithm using $6n$ bytes.

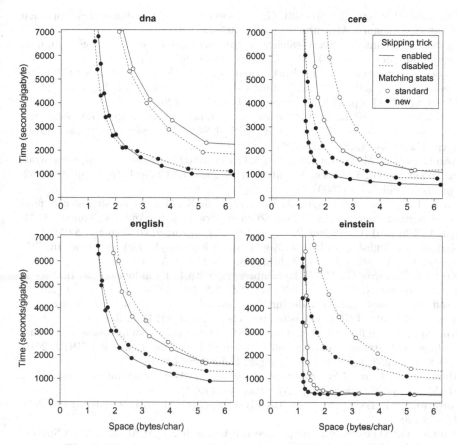

Fig. 3. Time-space tradeoffs for variants of LZscan (see Section 4)

References

1. Abouelhoda, M.I., Kurtz, S., Ohlebusch, E.: Replacing suffix trees with enhanced suffix arrays. J. Discrete Algorithms 2(1), 53–86 (2004)
2. Barbay, J., Gagie, T., Navarro, G., Nekrich, Y.: Alphabet partitioning for compressed rank/select and applications. In: Cheong, O., Chwa, K.-Y., Park, K. (eds.) ISAAC 2010, Part II. LNCS, vol. 6507, pp. 315–326. Springer, Heidelberg (2010)
3. Burrows, M., Wheeler, D.: A block sorting lossless data compression algorithm. Tech. Rep. 124, Digital Equipment Corporation, Palo Alto, California (1994)
4. Cánovas, R., Navarro, G.: Practical compressed suffix trees. In: Festa, P. (ed.) SEA 2010. LNCS, vol. 6049, pp. 94–105. Springer, Heidelberg (2010)
5. Chang, W.I., Lawler, E.L.: Sublinear approximate string matching and biological applications. Algorithmica 12(4-5), 327–344 (1994)
6. Chen, G., Puglisi, S.J., Smyth, W.F.: Lempel-Ziv factorization using less time and space. Mathematics in Computer Science 1(4), 605–623 (2008)
7. Crochemore, M.: String-matching on ordered alphabets. Theoretical Computer Science 92, 33–47 (1992)
8. Ferragina, P., Manzini, G.: Indexing compressed text. J. ACM 52(4), 552–581 (2005)

9. Ferragina, P., Gagie, T., Manzini, G.: Lightweight data indexing and compression in external memory. Algorithmica 63(3), 707–730 (2012)
10. Fischer, J., Heun, V.: Space-efficient preprocessing schemes for range minimum queries on static arrays. SIAM J. Comput. 40(2), 465–492 (2011)
11. Gagie, T., Gawrychowski, P., Kärkkäinen, J., Nekrich, Y., Puglisi, S.J.: A faster grammar-based self-index. In: Dediu, A.-H., Martín-Vide, C. (eds.) LATA 2012. LNCS, vol. 7183, pp. 240–251. Springer, Heidelberg (2012)
12. Kärkkäinen, J., Kempa, D., Puglisi, S.J.: Linear time Lempel–Ziv factorization: Simple, fast, small. In: CPM 2013. LNCS. Springer (to appear, 2013), http://arxiv.org/abs/1212.2952
13. Kärkkäinen, J., Manzini, G., Puglisi, S.J.: Permuted longest-common-prefix array. In: Kucherov, G., Ukkonen, E. (eds.) CPM 2009. LNCS, vol. 5577, pp. 181–192. Springer, Heidelberg (2009)
14. Kasai, T., Lee, G., Arimura, H., Arikawa, S., Park, K.: Linear-time longest-common-prefix computation in suffix arrays and its applications. In: Amir, A., Landau, G.M. (eds.) CPM 2001. LNCS, vol. 2089, pp. 181–192. Springer, Heidelberg (2001)
15. Kempa, D., Puglisi, S.J.: Lempel-Ziv factorization: simple, fast, practical. In: Zeh, N., Sanders, P. (eds.) ALENEX 2013, pp. 103–112. SIAM (2013)
16. Kreft, S., Navarro, G.: LZ77-like compression with fast random access. In: Storer, J.A., Marcellin, M.W. (eds.) DCC, pp. 239–248. IEEE Computer Society (2010)
17. Kreft, S., Navarro, G.: Self-indexing based on LZ77. In: Giancarlo, R., Manzini, G. (eds.) CPM 2011. LNCS, vol. 6661, pp. 41–54. Springer, Heidelberg (2011)
18. Kuruppu, S., Puglisi, S.J., Zobel, J.: Relative Lempel-Ziv compression of genomes for large-scale storage and retrieval. In: Chavez, E., Lonardi, S. (eds.) SPIRE 2010. LNCS, vol. 6393, pp. 201–206. Springer, Heidelberg (2010)
19. Manber, U., Myers, G.W.: Suffix arrays: a new method for on-line string searches. SIAM Journal on Computing 22(5), 935–948 (1993)
20. Navarro, G.: Indexing text using the Ziv-Lempel trie. J. Discrete Algorithms 2(1), 87–114 (2004)
21. Navarro, G., Mäkinen, V.: Compressed full-text indexes. ACM Computing Surveys 39(1), article 2 (2007)
22. Navarro, G.: Indexing highly repetitive collections. In: Arumugam, S., Smyth, B. (eds.) IWOCA 2012. LNCS, vol. 7643, pp. 274–279. Springer, Heidelberg (2012)
23. Ohlebusch, E., Gog, S.: Lempel-Ziv factorization revisited. In: Giancarlo, R., Manzini, G. (eds.) CPM 2011. LNCS, vol. 6661, pp. 15–26. Springer, Heidelberg (2011)
24. Ohlebusch, E., Gog, S., Kügel, A.: Computing matching statistics and maximal exact matches on compressed full-text indexes. In: Chavez, E., Lonardi, S. (eds.) SPIRE 2010. LNCS, vol. 6393, pp. 347–358. Springer, Heidelberg (2010)
25. Okanohara, D., Sadakane, K.: An online algorithm for finding the longest previous factors. In: Halperin, D., Mehlhorn, K. (eds.) ESA 2008. LNCS, vol. 5193, pp. 696–707. Springer, Heidelberg (2008)
26. Okanohara, D., Sadakane, K.: A linear-time Burrows-Wheeler transform using induced sorting. In: Karlgren, J., Tarhio, J., Hyyrö, H. (eds.) SPIRE 2009. LNCS, vol. 5721, pp. 90–101. Springer, Heidelberg (2009)
27. Starikovskaya, T.: Computing Lempel-Ziv factorization online. In: Rovan, B., Sassone, V., Widmayer, P. (eds.) MFCS 2012. LNCS, vol. 7464, pp. 789–799. Springer, Heidelberg (2012)
28. Ziv, J., Lempel, A.: A universal algorithm for sequential data compression. IEEE Transactions on Information Theory 23(3), 337–343 (1977)

Space-Efficient, High-Performance Rank and Select Structures on Uncompressed Bit Sequences

Dong Zhou[1], David G. Andersen[1], and Michael Kaminsky[2]

[1] Carnegie Mellon University
[2] Intel Labs

Abstract. Rank & select data structures are one of the fundamental building blocks for many modern *succinct data structures*. With the continued growth of massive-scale information services, the space efficiency of succinct data structures is becoming increasingly attractive in practice. In this paper, we re-examine the design of rank & select data structures from the bottom up, applying an architectural perspective to optimize their operation. We present our results in the form of a recipe for constructing space and time efficient rank & select data structures for a given hardware architecture. By adopting a *cache-centric* design approach, our rank & select structures impose space overhead as low as the most space-efficient, but slower, prior designs—only 3.2% and 0.39% extra space respectively—while offering performance competitive with the highest-performance prior designs.

1 Introduction

Rank & select data structures [6] are one of the fundamental building blocks for many modern *succinct data structures*. Asympototically, these data structures use only the minimum amount of space indicated by information theory. With the continued growth of massive-scale information services, taking advantage of the space efficiency of succinct data structures is becoming increasingly attractive in practice. Examples of succinct structures that commonly use rank & select include storing monotone sequences of integers [2,3] and binary or n-ary trees [6,1]. These structures in turn form the basis for applications such as compressed text or genome searching, and more.

For a zero-based bit array B of length n, the two operations under consideration are:

1. Rank(x) - Count the number of 1s up to position x;
2. Select(y) - Find the position of the y-th 1.

More formally, let B_i be the i-th bit of B, then

$$Rank(x) = \sum_{0 \leq i < x} B_i, \quad 1 \leq x \leq n$$

and

$$Select(y) = \min\{ k \mid Rank(k) = y \}, \quad 1 \leq y \leq Rank(n)$$

For example, in the bit array $0,1,0,1,0$, using zero-based indexing, Rank(2)=1, and Select(1)=2.

V. Bonifaci et al. (Eds.): SEA 2013, LNCS 7933, pp. 151–163, 2013.

In this paper, we consider the design of rank & select data structures for large in-memory bit arrays—those occupying more space than can fit in the CPU caches in modern processors, where n ranges from a few million up to a few tens of billions. We present our results in the form of a recipe for constructing space and time efficient rank & select data structures for a given hardware architecture. Our design, like several practical implementations that precede it [10,5,9], is not strictly optimal in an asymptotic sense, but uses little space in practice on 64-bit architectures.

The core techniques behind our improved rank & select structures arise from an aggressive focus on cache-centric design: It begins with an extremely small (and thus, cache-resident) first-layer index with 64-bit entries. This index permits the second-layer index to use only 32-bit entries, but maintains high performance by not incurring additional cache misses. This first-layer index is followed by an interleaved second and third layer index that is carefully sized so that accessing both of these indices requires only one memory fetch. The result of this design is a structure that simultaneously matches the performance of the fastest available rank & select structure, while using as little space as the (different) most space-efficient approach, adding only 3.2% and 0.39% space overhead for rank and select, respectively.

2 Design Overview and Related Work

Before we dive into the detailed design of our rank & select data structures, we first provide an overview of previous approaches, identify common design frameworks shared among them, and examine their merits and drawbacks. Because rank & select are often implemented in different ways, we discuss them separately.

2.1 Rank

For rank, almost all previous approaches embrace the following design framework:

1. Determine the size of the *basic block*, along with an efficient way to count the number of bits inside a basic block. Because the basic block is the lowest level in the rank structure, we should be able to do counting directly upon the original bit array.
2. Design an *index*, with one or multiple layers, that provides the number of 1s preceding the basic block in which x is located. Each index entry maintains aggregation information for a group of consecutive basic blocks, or *superblocks*.

Figure 1 illustrates a typical two-layer rank structure. In this example, basic blocks have a size of 8 bits. Entries in the first layer index are absolute counts, while entries in the second layer index count relative to the superblock start, rather than the very beginning of the bit array. Whenever a query for $\texttt{rank}(x)$ comes in,

1. First, look in the first layer index to find p, the number of 1s preceding the superblock into which x falls.
2. Second, look in the second layer index to find q, the number of 1s within that superblock that are to the left of the basic block into which x falls.
3. Finally, count the number of 1s to the left of x within that basic block, r.

The answer to $\texttt{rank}(x)$ is then $p + q + r$.

To demonstrate the generality of this design framework, we summarize several representative approaches, along with our rank structure, in Table 1. RG 37 is a variant

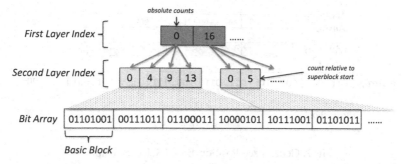

Fig. 1. Generalized Rank Structure

Table 1. Previous & Our Rank Structures

Approach	Basic Block Size	In-block Counting Method	Index Design	Space Overhead	Max Supported Size
Classical solution [5]	$\lfloor \log(n)/2 \rfloor$ bits	Precomputed table	Two-layer index	66.85%	2^{32}
RG 37 [5]	256 bits	Precomputed table	Two-layer index	37.5%	2^{32}
rank9 [10]	64 bits	Broadword programming	Two-layer index	25%	2^{64}
combined sampling [9]	1024 bits	Precomputed Table	One-layer index	3.125%	2^{32}
Ours (poppy)	512 bits	popcnt instruction	Three-layer index	3.125%	2^{64}

of the classical constant-time solution proposed by González et al. [5], and adds 37.5% extra space above the raw bit array. rank9 [10] employs broadword programming [7] to efficiently count the number of one bits inside a 64-bit word[1], and stores the first and second layer of index in an *interleaved form*—each first layer index entry is followed by its second layer entries, which reduces cache misses. combined sampling [9] explores the fact that the space overhead is inversely proportional to the size of the basic block, and achieves low space overhead ($\sim$ 3%) by using 1024-bit basic blocks. However, this space efficiency comes at the expense of performance. It is roughly 50%–80% slower than rank9. Therefore, our goal is to match the performance of rank9 and the space overhead of combined sampling.

Notice that except for rank9, all of the previous rank structures can only support bit arrays that have up to 2^{32} bits. However, as the author of rank9 observes, efficient rank & select structures are particularly useful for extremely large datasets: a bit array of size 2^{32} substantially limits the utility of building compressed data structures based on rank & select. Unfortunately, naively extending existing structures to support larger bit arrays by replacing 32 bit counters with 64 bit counters causes their space overhead to nearly double.

From the above overview, we identify three important features for a rank structure:

1. **Support bit arrays with up to 2^{64} bits.**
2. **Add no more than 3.125% extra space.**
3. **Offer performance competitive to the state-of-art.**

Section 3.1 explores the design of our new rank structure, called poppy, which fulfills all three requirements.

---

[1] Broadword programming, also termed as "SWAR" (SIMD Within A Register), can count the number of one bits in $O(\log d)$ instructions, where d is the number of bits in the word. The latest version of rank9 replaces broadword programming with popcnt instruction.

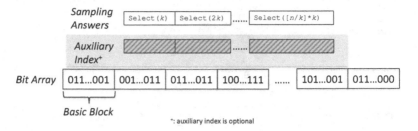

Fig. 2. Generalized Position-Based Select Structure

2.2 Select

Two general approaches are used to implement select. One is *rank-based selection*, and the other is *position-based selection*. For both methods, the first step is the same:

1. Determine the size of the *basic block*, and an efficient way to select within a basic block. This size need not be the same as the basic block size for rank, but making them the same is a common design choice.

For *rank-based selection*, the second step is:

2a. Design an *index* with one or multiple layers that identifies the location of the basic block in which the x-th one bit is located. This index is similar to the index for rank, but the demand imposed on it is different. In rank, we know exactly which entry is needed. For example, given a rank structure with 8-bit basic blocks, bit 15 is always in the second basic block. However, the 15-th *one bit* might be located in the 10000-th basic block! Therefore, in rank-based select, we must find the correct entry by *searching* (most commonly, by binary searching). These two distinct access patterns give us an intuitive understanding of why select is more difficult than rank. Although the index for select is similar to that for rank, they are not necessarily the same.

And for *position-based selection*, the second step is:

2b. Store a sampling of select answers and possibly an auxiliary index. Using these two structures, we can reach a position that is very close to the basic block in which the x-th one bit is located. Then, scan sequentially to find the correct basic block.

Figure 2 presents a typical position-based select structure, which stores select results for every k ones. To answer select(y), we first find the largest j such that $jk \leq y$. Because select(jk) is stored in a precomputed table, we can obtain the position of the jk-th one bit by a lookup in that table. Then, we locate the basic block containing the y-th one bit with or without the help of an auxiliary index. After finding the target basic block, we perform an in-block select to find the correct bit position.

Table 2 lists several previous approaches for select. The space overhead, *excludes* the space occupied by the rank structure, though several select structures rely on their corresponding rank structure to answer queries. Similar to rank, three features are desirable for a select index:

Table 2. Previous Select Structures

Approach	Type	Basic Block Size	In-block Selection Method	Space Overhead	Max Supported Size
Clark's structure [1]	Position-based	$\lceil \log n \rceil$ bits	Precomputed Table	60%	2^{32}
Hinted bsearch [10]	Rank-based	64 bits	Broadword programming	$\sim 37.38\%$	2^{64}
select9 [10]	Position-based	64 bits	Broadword programming	$\sim 50\%$	2^{64}
simple select [10]	Position-based	64 bits	Broadword programming[2]	9.01%-45.94%	2^{64}
combined sampling [9]	Position-based	1024 bits	Byte-wise table lookup + bit-wise scan	$\sim 0.39\%$	2^{32}
Ours (cs-poppy)	Position-based	512 bits	popcnt + broadword programming	$\sim 0.39\%$	2^{64}

1. **Support bit arrays with up to 2^{64} bits.**
2. **Add no more than 0.39% extra space.**
3. **Offer performance competitive to the state-of-art.**

Section 3.2 explores the design of our new select structure, called combined sampling with poppy or cs-poppy for short, which fulfills all three requirements.

3 Design

In light of the above observations, we now present our design recipe for a rank & select data structure. Like most previous solutions, we use a hierarchical approach to rank & select. Our recipe stems from three underlying insights from computer architecture:

For large bit arrays, the overall performance is strongly determined by cache misses. In a bit array occupying hundreds of megabytes of space, it is necessary to fetch at least one block from memory into the cache. Thus, optimizing the computation to be much faster than this fetch time does not provide additional benefit. A fetch from memory requires approximately 100ns, enough time to allow the execution of hundreds of arithmetic operations.

Parallel operations are cheap. Executing a few operations in parallel often takes only modestly longer than executing only one. This observation applies to both arithmetic operations (fast CPUs execute up to 4 instructions at a time) and memory operations (modern CPUs can have 8 or more memory requests in flight at a time).

Optimize for cache misses, then branches, then arithmetic/logical operations. There is over an order of magnitude difference in the cost of these items: 100ns, 5ns, and $< \frac{1}{4}$ns, respectively. A related consequence of this rule is that it is worth engineering the rank/select structures to be cache-aligned (else a retrieval may fetch two cachelines), and also to be 64-bit aligned (else a retrieval may cost more operations).

In the rest of this section, we describe our design as optimized for recent 64-bit x86 CPUs. When useful, we use as a running example the machine from our evaluation (2.3 GHz Intel Core i7 "Sandy Bridge" processor, 8 MB shared L3 cache, 8 GB of DRAM).

3.1 Rank

Basic Block for Rank The basic block is the lowest level of aggregation. Within a basic block, both Rank and Select work by counting the bits set up to a particular position

---

[2] As rank9, the latest version of simple select uses popcnt + broadword programming.

Table 3. Performance for different methods of popcounting a 64M bit array 300 times

Method	Time (ms)
Precomputed table (byte-wise)	729.0
popcnt instruction	191.7
SSE2	336.0
SSSE3	237.7
Broadword programming	798.9

(often referred as *population count* or *popcount*), without using any summary information. Both theoretical analysis as well as previous approaches demonstrate that the space overhead of rank & select is inversely proportional to the size of the basic block. Larger basic blocks of bits mean that fewer superblocks are needed in the index. Meanwhile, excessively enlarging the size of the basic block degrades performance, because operating on larger blocks requires more computation and more memory accesses, which are extremely expensive. Specifically, the number of memory accesses grows *linearly* as the size of the basic block increases. Therefore, algorithm implementers should focus most of their effort on finding techniques to *efficiently* increase the number of bits that can be processed at the lowest level with no auxilary information.

Previous work showed that we can set this size to 32 bits and perform popcount using lookups in a precomputed table [5,9], or set the size to 64 bits and use the broadword programming bit-hacking trick to implement popcount in $O(\log d)$ instructions where d is the number of bits in the word [10]. Other choices include using the vector SSE instructions (SSE2), the PSHUFB instruction (SSSE3) which looks up 4 bits at a time in a table in parallel, or as proposed recently by Ladra et al. [8], the popcnt instruction which is available in newer Intel processors (Nehalem and later architectures).

We ran microbenchmarks and measured the performance of each method. The microbenchmark creates a bit array of 64M bits and measures the performance of each method by popcounting the entire bit array 300 times. Because we count the number of one bits over the entire array, multiple *popcounts* can be in flight at the same time. The results (Table 3) show that the popcnt instruction is substantially faster than other approaches.

Next we must choose the basic block size. One straightforward design is to use 64-bit basic blocks, as in the design by Vigna [10]. However, as we noted above, larger basic blocks reduce the space overhead of the index; furthermore, executing several operations in parallel often takes only modestly longer than executing a single instruction. We therefore want to find the largest *effective* size. We call a size effective if moving up to that size yields performance benefit from parallel operations. If, instead, when we double the size of the basic block, the amount of time to popcount also doubles, this is a strong indicator that it is time to stop increasing the block size. Table 4 shows the performance of popcounting different basic block sizes using 10^8 random positions over a bit array with 2^{32} bits.

Before 512 bits, each doubling of the basic block slows execution by less than 2x, which implies that 512 is the right answer to the question. This also matches our expectation from a computer architecture perspective: the overwhelming factor in performance is cache misses. The size of a cache line is 512 bits. Hence, for well-aligned

Table 4. Performance for popcounting 10^8 randomly chosen blocks of increasing sizes

Size (bits)	Time (seconds)	# of cache misses
64	0.13	1
128	0.19	1
256	0.30	1
512	0.50	1
1024	0.99	2
2048	2.01	4

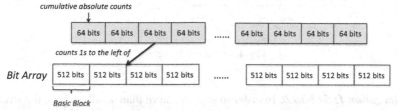

Fig. 3. Strawman Design of Rank Structure

bit arrays, popcounting 512 bits leads to exactly one cache miss. In short, not only can we popcount 512 bits extremely quickly, but doing so does not steal memory bandwidth from other operations. This choice contributes greatly to the speed of our space-efficient design for rank & select data structures.

Ladra et al. [8] also observed that varying the basic block size of the auxiliary data structure for rank and select offers a space/time tradeoff, which they can leverage to improve their space overhead. Here, we provide additional insight about how to best use their observation: by incorporating knowledge about the underlying memory hierarchy, our proposed guideline can help algorithm implementors understand how to make this space/time tradeoff.

Layered Index. With popcount efficiently supporting blocks of 512 (2^9) bits, we have considerable flexbility in designing the index without sacrificing space. For example, an index that supports up to 4 billion bits (2^{32}) could simply directly index each 512-bit basic block, adding only 6.25% extra space. However, efficient rank & select data structures are particularly useful for extremely large datasets, and thus we would like to support a larger bit array.

Strawman Design. The strawman design (Figure 3) is to directly index each 512-bit basic block using a 64-bit counter to store the number of one bits to the left of that basic block. This solution offers good performance (roughly two cache misses per query: one for looking up in the rank structure, the other for examing bit vector itself), and adds 12.5% extra space.

To reduce the space overhead, we adopt two optimizations, each of which halves the index space, as illustrated in Figure 4.

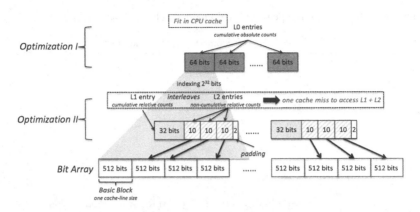

Fig. 4. Our Rank Structure

Optimization I: 64 bit L0. In order to support more than 4 billion bits, the strawman design used a 64-bit counter for each basic block. Supporting up to 2^{64} bits is important, but in practice, bit arrays are not *too* large. Therefore, for each 2^{32} bits (an *upper block*), we store a 64-bit counter to store the number of one bits to the left of that upper block. These 64-bit counters create a new index layer, called the first layer (L0) index. When answering a query for rank(x), we examine this index to find the number of one bits preceding the upper block in which x is located, and look up the underlying structure to find out the number of one bits preceding x within that upper block.

Accessing this additional index does not significantly affect performance for two reasons: First, the L0 index is small enough to fit in fast cache memory. It contains only 64 bits for each 2^{32} bits in the original array. For a bit array of 16 billion bits (2GB), it requires only 128 bytes. Second, the lookup in this index is independent of the lookup in the second-layer index, so these operations can be issued in parallel. This additional layer of index confers an important space advantage: The underlying indexes now need only support 2^{32} bits, so we can represent each 512-bit basic block using only a 32-bit counter. This design results in a rank structure with about 6.25% extra space.

Optimization II: Interleaved L1/L2. We can further improve the space overhead by adding an additional layer to the index. Recall our architectural insight that the overall performance is strongly determined by the number of cache misses, which implies that if no more cache misses are introduced, *slightly* more computation will have minimal performance impact. According to this idea, we designed a two-layer index to support rank queries for 2^{32} bit ranges. For each *four* consecutive basic blocks (a *lower block*, containing 2048 bits), we use a 32-bit counter to store the number of one bits preceding that lower block. These counters make up the second layer (L1) index. Underneath, for each lower block, we use three 10-bit counters, each storing the popcount value for one of the first three basic blocks within that lower block. These 10-bit counters make up the third layer (L2) index. To look up a block, it is necessary to *sum* the appropriate third-layer index entries, but because there are only three such entries, the cost of this linear operation is low.

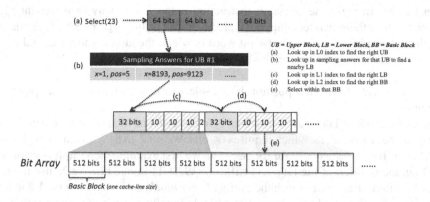

Fig. 5. Process of Answering a Select Query

To avoid causing extra cache misses, we leverage the technique of Vigna [10]: storing the L1 and L2 index entries in an *interleaved form*. Each L1 index entry is followed by its L2 index entries. Since the total size of an L1 index entry and its L2 index entries is 62 bits, which fits in one cache line, this design guarantees that by paying exactly one cache miss, we are able to fetch all the necessary data to answer a rank query. (We pad the structure by two bits to ensure that it is both cache and word aligned.) Even though several additional comparisons and arithmetic operations must be performed, the overall performance is only slightly reduced. Because each 2048 bits of the bit array require 64 bits, the space overhead is 3.125%.

Of note is that each layer of the index uses a different type of count: The first layer uses 64-bit cumulative, absolute counts. The second layer uses cumulative counts relative to the beginning of the upper block, and so fits in 32 bits. The third layer uses non-cumulative, relative counts in order to fit all three of them into less than 32 bits, a design constraint required to ensure that the L1/L2 index entries could always be cache-line aligned. The combination of these three types of counts makes our high-performance, space-efficient rank structure possible.

3.2 Select

`combined sampling` [9] is the highest-performing of the space-efficient variants of select, which uses position-based selection. We therefore focus on it as a target for applying our cache-centric optimization and improvements from rank. Our goal, as with `combined sampling`, is to enable maximal re-use of index space already devoted to rank. In contrast, many prior approaches [1,10] create an entirely separate index to use for position-based selection, which requires considerable extra space. As we show, our rank structure, `poppy`, is a natural match for the combined design, and enables support for larger (up to 2^{64} bits) bit arrays while offering competitive or even better performance.

Basic Block for Select. Similar to rank, we first microbenchmark the best in-block select method. The result shows that broadword programming [10] is the best method to select within 64 bits. Because the popcnt instruction is the fastest way to popcount 512 bits, we combine these two techniques to select within 512 bits—popcnt sequentially through the basic block to find the 64-bit word in which the target is located, and then use broadword select to find the individual bit within the word.

Sampling Answers. Like other position-based select structures, we store a sampling of select answers.

Strawman Design. The strawman design is for every L one bits, we store the position of the first one among them, which requires 64 bits. We set L to 8192, as in combined sampling. To answer a query for select(y), we first examine the sampling answers to find out the position of the $(\lfloor (y-1)/8192 \rfloor \cdot 8192 + 1)$-th one bit. We re-use the L1 index of the rank structure to reach the correct *lower block*, and look up in the L2 index of that lower block to find the correct basic block. Finally, we use the combination of popcnt and broadword programming to select within that basic block. In the worst case, such a structure adds about 0.78% extra space.

Optimization: 64 bit L0. The same idea from our rank optimization can be used for select, splitting the index into a 64-bit upper part and 32-bit lower part. We binary search the L0 index of the rank structure to find out the upper block in which the y-th one bit is located. For each upper block, we store a sampling of answers similar to the strawman design, but this time only 32 bits are required to store a position. Then the process of answering a select query is similar to that of strawman design, except that it requires one more look up, as shown in Figure 5. This re-use of the L0 index halves the space overhead, from 0.78% to 0.39%. Because we re-use our poppy structure as a building block, we call this select structure cs-poppy.

This design is similar to combined sampling, with two important differences. First, cs-poppy can support up to 2^{64} bits. Second, the rank index allows cs-poppy to locate the correct position to within 512 bits, instead of combined sampling's 1024-bit basic block, requiring (potentially) one less cache miss when performing select directly within a basic block. cs-poppy thus outperforms combined sampling and is performance competitive with the much less space-efficient simple select.

Micro-optimization: Jumping to offset. The sampling index indicates the L1 block containing the sampled bit. Our select performs a small optimization to potentially skip several L1 blocks: Each L1 block can only contain 2048 one bits. Therefore, for select(y), it is safe to skip forward by $\frac{y\%8192}{2048}$ L1 entries. This optimization improves select performance by 12.6%, 2.4%, and 0.5% for 2^{30}-entry bit arrays consisting of 90%, 50%, and 10% ones, respectively.

4   Evaluation

To evaluate our rank & select structures, we performed several experiments on the Intel Core i7-based machine mentioned above. The source code was compiled using gcc 4.7.1 with options -O9, -march=native and -mpopcnt. We measure elapsed time using the function gettimeofday.

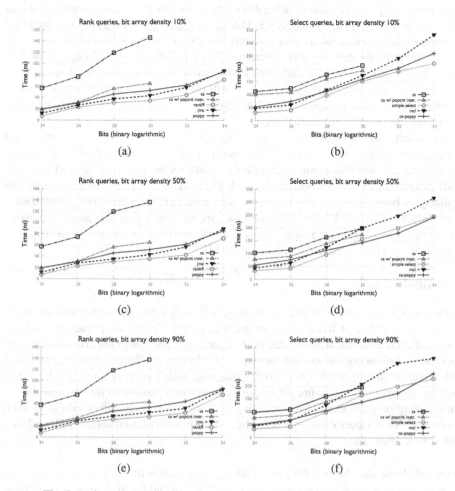

Fig. 6. Performance of rank & select operations in bit arrays of increasing size

We pre-generate random bit arrays and 1,000,000 test queries before measurement begins. Each test is repeated 10 times. Because the deviation among these runs is small, we report the mean performance. We execute rank & select queries over bit arrays of densities 10%, 50%, and 90%, similar to the experiments by Navarro et al. [9].

For rank, because our goal is to provide the performance of rank9 while matching the space overhead of combined sampling, we compare poppy with these two. We also compare with SDSL [4]'s rank_support_jmc, which implements the classical solution [6]. Figure 6 shows the results. For small bit arrays, our rank structure, poppy performs slower than rank9 and rank_support_jmc. The performance gap shrinks as the size of bit array increases. When the size is increased to 2^{34}, poppy's performance is competitive or even better. For small arrays, the index structure fits in cache, and the relative cost of poppy's extra computation adds measureable overhead, but as the index grows to exceed cache, this extra arithmetic is overshadowed by the cost that all schemes must pay to access DRAM. In fact, poppy can out-peform the other schemes for index sizes where poppy's smaller index fits in cache and the others do not. On the other hand, poppy is substantially faster than the original implementation of combind sampling To understand why, we modified the original implementation to use the popcnt instruction. poppy still outperforms this modified implementation of combined sampling by 20%-30%, which we believe is mainly from the new cache-aware index structure design.

For select, we compare cs-poppy against simple select, combined sampling, combined sampling using popcnt, and SDSL's select_support_mcl, which is an implementation of Clark's structure [1] enhanced by broardword programming. As shown in Figure 6 (b), (d), and (f), cs-poppy performs similarly or better than simple select and select_support_mcl, and always outperforms combined sampling and its variant. This result matches our analysis that combined sampling may require one cache miss more than cs-poppy, because its basic block occupies two cache lines (1024 bits).

5   Conclusion

In this paper, we overview several representative rank & select data structures and summarize common design frameworks for such structures. Then, we present our design recipe for each component, motivated both algorithmic and computer architecture considerations. Following our design recipe, we build space-efficient, high-performance rank & select structures on a commodity machine which support up to 2^{64} bits. The resulting poppy rank structure offers performance competitive to the state of the art while adding only 3% extra space; building upon it, cs-poppy offers similar or even better select performance than the best alternative position-based select, while adding only 0.39% extra space.

Acknowledgments. We gratefully acknowledge: the authors of combined sampling for providing their source code for comparison; Bin Fan and the SEA reviewers for their feedback and suggestions; and Google, the Intel Science and Technology Center for Cloud Computing, and the National Science Foundation under award CCF-0964474 for their financial support of this research.

References

1. Clark, D.R.: Compact pat trees. PhD thesis, Waterloo, Ont., Canada, Canada (1998)
2. Elias, P.: Efficient Storage and Retrieval by Content and Address of Static Files. J. ACM 21(2), 246–260 (1974)
3. Fano, R.M.: On the number of bits required to implement an associative memory. Memorandum 61, Computer Structures Group, Project MAC (1971)
4. Gog, S.: https://github.com/simongog/sdsl
5. González, R., Grabowski, S., Mäkinen, V., Navarro, G.: Practical implementation of rank and select queries. In: Poster Proceedings Volume of 4th Workshop on Efficient and Experimental Algorithms (WEA 2005), pp. 27–38 (2005)

6. Jacobson, G.: Space-efficient static trees and graphs. In: Proc. Symposium on Foundations of Computer Science, SFCS 1989, pp. 549–554. IEEE Computer Society, Washington, DC (1989)
7. Knuth, D.E.: The Art of Computer Programming. Fascicle: Bitwise Tricks & Techniques; Binary Decision Diagrams, vol. 4. Addison-Wesley Professional (2009)
8. Ladra, S., Pedreira, O., Duato, J., Brisaboa, N.R.: Exploiting SIMD instructions in current processors to improve classical string algorithms. In: Morzy, T., Härder, T., Wrembel, R. (eds.) ADBIS 2012. LNCS, vol. 7503, pp. 254–267. Springer, Heidelberg (2012)
9. Navarro, G., Providel, E.: Fast, Small, Simple Rank/Select on Bitmaps. In: Klasing, R. (ed.) SEA 2012. LNCS, vol. 7276, pp. 295–306. Springer, Heidelberg (2012)
10. Vigna, S.: Broadword implementation of rank/select queries. In: McGeoch, C.C. (ed.) WEA 2008. LNCS, vol. 5038, pp. 154–168. Springer, Heidelberg (2008)

Think Locally, Act Globally: Highly Balanced Graph Partitioning*

Peter Sanders and Christian Schulz

Karlsruhe Institute of Technology, Karlsruhe, Germany
{sanders,christian.schulz}@kit.edu

Abstract. We present a novel local improvement scheme for graph partitions that allows to enforce strict balance constraints. Using negative cycle detection algorithms this scheme combines local searches that individually violate the balance constraint into a more global feasible improvement. We combine this technique with an algorithm to balance unbalanced solutions and integrate it into a parallel multi-level evolutionary algorithm, KaFFPaE, to tackle the problem. Overall, we obtain a system that is fast on the one hand and on the other hand is able to improve or reproduce many of the best known *perfectly* balanced partitioning results reported in the Walshaw benchmark.

1 Introduction

In computer science, engineering, and related fields *graph partitioning* is a common technique. For example, in parallel computing good partitionings of unstructured graphs are very valuable. In this area, graph partitioning is mostly used to partition the underlying graph model of computation and communication. Generally speaking, nodes in this graph represent computation units and edges denote communication. This graph needs to be partitioned such that there are few edges between the blocks (pieces). In particular, if we want to use k processors we want to partition the graph into k blocks of about equal size. Here we focus on the case when the bounds on the size are very strict, including the case of *perfect balance* when the maximal block size has to equal the average block size.

The problem is NP-hard and hard to approximate on general graphs so that mostly heuristics are used in practice. A successful heuristic for partitioning large graphs is the *multi-level* approach. Here, the graph is recursively *contracted* to achieve a smaller graph with the same basic structure. After applying an *initial partitioning* algorithm to the smallest graph in the hierarchy, the contraction is undone and, at each level, a *local refinement* method is used to improve the partitioning induced by the coarser level.

During the last years we started to put all aspects of the multi-level graph partitioning (MGP) scheme on trial since we had the impression that certain aspects of the method are not well understood. Our main focus is partition quality rather than partitioning speed. In our sequential MGP framework KaFFPa (Karlsruhe Fast Flow Partitioner) [12], we presented novel local search as well as global search algorithms.

---

* This paper is a short version of the TR [14].

V. Bonifaci et al. (Eds.): SEA 2013, LNCS 7933, pp. 164–175, 2013.

In the Walshaw benchmark [15], KaFFPa was beaten mostly for small graphs that combine multi-level partitioning with an evolutionary algorithm. We therefore developed an improved evolutionary algorithm, KaFFPaE (KaFFPa Evolutionary) [13], that also employs coarse grained parallelism. Both of these algorithms are able to compute partitions of very high quality in a reasonable amount of time when some imbalance $\epsilon > 0$ is allowed. However, they are not yet very good for small values of ϵ, in particular for the perfectly balanced case $\epsilon = 0$.

State-of-the-art local search algorithms exchange nodes between blocks of the partition trying to decrease the cut size while also maintaining balance. This highly restricts the set of possible improvements. We introduce new techniques that relax the balance constraint for node movements but globally maintain balance by combining multiple local searches. We reduce the combination problem to finding negative cycles in a graph, exploiting the existence of very efficient algorithms for this problem. We also provide balancing variants of these techniques that are able to make infeasible partitions feasible. This makes our partitioner the only current system which is able to guarantee any balance constraint. From a meta heuristic point of view our techniques are an interesting example for a local improvement technique that vastly increases the size of the neighborhood by efficiently combining many highly localized infeasible improvements into a feasible one.

The paper is organized as follows. We begin in Section 2 by introducing basic concepts. After presenting some related work in Section 3 we describe novel improvement and balancing algorithms in Section 4. Here we start by explaining the very basic idea that allows us to find combinations of simple node movements. We then explain directed local searches and extend the basic idea to a complex model containing more node movements. This is followed by a description on how these techniques are integrated into KaFFPaE. A summary of extensive experiments done to evaluate the performance of our algorithms is presented in Section 5.

2 Preliminaries

Consider an undirected graph $G = (V, E, \omega)$ with edge weights $\omega : E \to \mathbb{R}_{>0}$, $n = |V|$, and $m = |E|$. We extend ω to sets, i.e., $\omega(E') := \sum_{e \in E'} \omega(e)$. $\Gamma(v) := \{u : \{v, u\} \in E\}$ denotes the neighbors of v. We are looking for *blocks* of nodes $V_1,\ldots,V_k$ that partition V, i.e., $V_1 \cup \cdots \cup V_k = V$ and $V_i \cap V_j = \emptyset$ for $i \neq j$. A *balancing constraint* demands that $\forall i \in \{1..k\} : |V_i| \leq L_{\max} := (1 + \epsilon)\lceil |V|/k \rceil$. In the perfectly balanced case the imbalance parameter ϵ is set to zero. The objective is to minimize the total *cut* $\sum_{i<j} w(E_{ij})$ where $E_{ij} := \{\{u, v\} \in E : u \in V_i, v \in V_j\}$. A block V_i is called underloaded if $|V_i| < L_{\max}$ and overloaded if $|V_i| > L_{\max}$. A node $v \in V_i$ that has a neighbor $w \in V_j, i \neq j$, is a boundary node. An abstract view of the partitioned graph is the so called *quotient graph*, where nodes represent blocks and edges are induced by connectivity between blocks. Given a partition, the gain of a node v in block A with respect to a block B is defined as $g_{(A,B)} = \omega(\{(v, w) \mid w \in \Gamma(v) \cap B\}) - \omega(\{(v, w) \mid w \in \Gamma(v) \cap A\})$, i.e. the reduction in the cut when v is moved from block A to block B. By default, our initial inputs will have unit node weights. However, the proposed algorithms can be easily extended to handle weighted nodes.

3 Related Work

There has been a huge amount of research on graph partitioning so that we refer the reader to [3]. Well known software packages based on this multi-level approach include, Jostle [17], Metis [9], and Scotch [11]. However, for various reasons they are not able guarantee strict balance constraints. KaFFPaE [13] is a distributed parallel evolutionary algorithm that uses our multi-level graph partitioning framework KaFFPa [12] to create individuals and modifies the coarsening phase to provide new effective combine operations. It currently holds the best results for many graphs in Walshaw's Benchmark Archive [15] when some imbalance is allowed. Benlic et al. [2] provided multi-level memetic algorithms for perfectly balanced graph partitioning. Their approach is able to compute many entries in Walshaw's Benchmark Archive [15] for the case $\epsilon = 0$. However, they are not able to guarantee that the computed partition is perfectly balanced especially for larger values of k.

4 Globalized Local Search by Negative Cycle Detection

In this section we describe our local search and balancing algorithms for strictly balanced graph partitioning. Roughly speaking, all of our algorithms consist of two components. The *first component* are local searches on pairs of blocks that share a non-empty boundary, i.e. all edges in the quotient graph. These local searches are not restricted to the balance constraint of the graph partitioning problem and are undone after they have been performed. The *second component* uses the information gathered in the first component. That means we build a model using the node movements performed in the first step enabling us to find combinations of those node movements that *maintain balance*.

We begin by describing the very basic algorithm and go on by presenting an advanced model which enables us to combine complex local searches. This is followed by a description on how local search and balancing algorithms are put together. At the end of this section we show how we integrate these algorithms into our evolutionary framework KaFFPaE.

Basic Idea – Using a Negative Cycle Detection Algorithm. We start with a very simple case where the first component only moves single nodes. A node in the graph G can have two states *marked* and *unmarked*. By default a node is unmarked. It is called *eligible* if it is not adjacent to a previously marked node. We now build the model of the underlying partition of the graph G, $\mathcal{Q} = (\{1, \cdots, k\}, \mathcal{E})$ where $(A, B) \in \mathcal{E}$ if there is an edge in G that runs between the blocks A and B. We define edge weights $\omega_{\mathcal{Q}} : \mathcal{E} \to \mathbb{R}$ in the following way: for each *directed* edge $e = (A, B) \in \mathcal{E}$ in a random order, find a *eligible* boundary node v in block A having maximum gain $g_{\max}(A, B)$, i.e. a node v that maximizes the reduction in cut size when moving it from block A to block B. If there is more than one such node, we break ties randomly. Node v is then marked. The weight of e is then $\omega_{\mathcal{Q}}(e) := -g_{\max}(A, B)$, i.e., the negative gain value associated with moving v from A to B. Note that, in general, $\omega_{\mathcal{Q}}((A, B)) \neq \omega_{\mathcal{Q}}((B, A))$. An example for this basic model is shown in Figure 1. Observe that the basic model is a

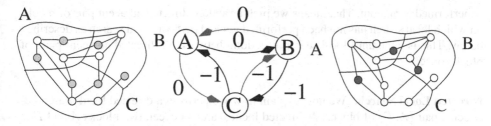

Fig. 1. Left: example graph partitioned into three parts (A, B and C). Possible candidates are highlighted. Middle: corresponding model and one negative cycle is highlighted. Right: updated partition after associated node movements of cycle are performed. Moved nodes are highlighted.

directed and weighted version of the quotient graph and that the selected nodes form an independent set.

Note that each cycle in this model defines a set of node movements and furthermore when the associated nodes of a cycle are moved, then each block contains the same number of nodes as before. Also the weight of a cycle in the model is equal to the reduction in the cut when the associated node movements are performed. However, the most important aspect is that a *negative cycle* in the model corresponds to a set of node movements that will decrease the overall cut and maintain the balance of the partition. To detect a negative cycle in this model we introduce a node s and connect it to all nodes in Q. The weight of the inserted edges is set to zero. We can apply a standard shortest path algorithm [4] that can handle negative edge weights to detect a negative cycle. If the model contains a negative cycle we can perform a set of node movements that will not alter the balance of the blocks since each block obtains and emits a node.

We can find additional useful augmentations by connecting blocks which can take at least one node without becoming overloaded to s by a zero weight edge. Now, negative cycles containing s change some block weights but will not violate any additional balance constraints. Indeed, when the node following s is overloaded initially, this overload will be reduced.

If there is no negative cycle in the model, we apply a diversification strategy based on cycles of weight zero. This strategy is explained in the TR [14]. Moreover, we apply a balancing algorithm which is explained in the following sections. An interesting observation is that the algorithm can be seen as an extension of the classical FM algorithm [6] which swaps nodes between two adjacent blocks (two at a time) which is basically a negative cycle of length two in our model if the gain of the two node movements is positive.

Advanced Model. We now integrate advanced local search algorithms. Each edge in the advanced model stands for a *set* of node movements found by a local search. Hence, a negative cycle corresponds to a combination of local searches with positive overall gain that maintain balance or that can improve balance. Before we build the advanced model we perform *directed local search* on each pair of blocks that share a non-empty boundary, i.e. each pair of blocks that is adjacent in the quotient graph. A local search on a directed pair of blocks (A, B) is only allowed to move nodes from block A to block B. The order in which the directed local search between a directed pair of blocks is performed is random. That means we pick a random directed adjacent pair of blocks on which local search has not been performed yet and perform local search as described below. This is done until local search was done between all directed adjacent pairs of blocks once.

Directed Local Search. We now explain how we perform a directed local search between a pair (A, B) of blocks. A directed local search between two blocks A and B is very localized akin to the multi-try method used in KaFFPa [12]. However, a directed local search between A and B is restricted to move nodes from block A to block B. It is similar to the FM-algorithm: We start with a *single* random eligible boundary node of block A having maximum gain $g_{max}(A, B)$ and put this node into a priority queue. The priority queue contains nodes of the block A that are valid to move. The priority is based on the *gain*, i.e. the decrease in edge cut when the node is moved from block A to block B. We always move the node that has the highest priority to block B. After a node is moved its eligible neighbors that are in block A are inserted into the priority queue. We perform at most τ steps per directed local search, where τ is a parameter. Note that during a directed local search we only move nodes that are not incident to a node moved during a previous directed local search. This restriction is necessary to keep the model described below accurate. Thus we *mark all nodes* touched during a directed local search *after* it was performed which also implies that each node is moved at most once. In addition all moved nodes are *moved back* to their origin, since these movements would make the partition imbalanced. We stress that all nodes incident to nodes that have been moved during a directed local search are not *eligible* for any later local search during the construction since this would make the gain values computed imprecise.

The Model Graph. The advanced model allows us to find combinations of directed local searches such that the balance of the given partition is at least maintained. The challenge here is that, in contrast to movements of single nodes, we cannot combine arbitrary local searches since they do not all move the same number of nodes. Hence, we specify a more sophisticated graph with the property that a negative cycle maintains feasibility.

The local search process described above yields for each pair of blocks $e = (A, B)$ in the quotient graph a sequence of node movements S_e and a sequence of gain values g_e. The d'th value in g_e corresponds to the reduction in the cut between the pair of blocks (A, B) when the first d nodes in S_e are moved from their source block A to their target block B. By construction, a node $v \in V$ can occur in at most one of the sequences created and in its sequence only once.

Generally speaking, the *advanced model* consists of τ layers. Essentially each layer is a copy of the quotient graph. An edge starting and ending in layer d of this model corresponds to the movement of exactly d nodes. The weight of an edge $e = (A, B)$ in layer d of the model is set to the negative value of the d'th entry in g_e. In other words it encodes the negative value of the gain, when the first d nodes in S_e are moved from block A to block B. Hence, a negative cycle whose nodes are all in layer d will move exactly d nodes between each of the respective block pairs contained in the cycle and results in a overall decrease in the edge cut. We add additional edges to the model such that it contains *more possibilities* in presence of underloaded blocks. To be more precise, in these cases we want to get rid of the restriction that each block sends and emits the same number of nodes. To do so we insert *forward* edges between all consecutive layers, i.e. block k in layer d is connected by an edge of weight zero to block k in layer $d + 1$. These edges are not associated with node movements. Furthermore, we add *backward* edges as follows: for an edge (A, B) in layer d we add an edge with the same weight between block A in layer d and block B in layer $d - \ell$ if block B can take ℓ nodes without becoming overloaded. The newly inserted edge is associated with the same node movements as the initial edge (A, B) within layer d. This way we encode movements in the model where a block can emit more nodes then it gets and vice versa without violating the balance constraint. Additionally we connect each node in layer d back to s if the associated block can take at least d nodes without becoming overloaded. Again this means that the model might contain cycles through s which stand for paths in the quotient graph being associated with node movements that decrease the overall cut. Moreover, these moves never increase the imbalance of the input partition. An example for the advanced model can be found in the TR [14]. We can apply the same zero weight cycle diversification as in the basic model. The advanced model can contain *conflicting* cycles that cannot be used. Due to space constraints, we explain when conflicts occur and how we handle them in the TR [14].

Multiple Directed Local Searches. The algorithm can be further improved by performing multiple directed local searches (MDLS) between each pair of blocks that share a non-empty boundary. More precisely, after we have computed node movements on *each* pair of blocks $e = (A, B)$, we start again using the nodes that are still eligible. This is done μ times. The model is then slightly modified in the following way: For the creation of edges in the model that correspond to the movement of d nodes from block A to block B we use the directed local search on $e = (A, B)$ from the process above with the best gain when moving d nodes from block A to block B (and use this gain value for the computation of the weight of corresponding edges).

Balancing. As we will see, to create ϵ-balanced partitions we start our algorithm with partitions where larger imbalance is allowed. Hence, to satisfy the balance constraint, we have to think about balancing strategies. A balancing step will only be applied if the model does not contain a negative cycle (see next section for more details). Hence, we can modify the advanced model such that we can find a set of node movements that will decrease the total number of overloaded nodes by at least one and minimizes the increase in the number of edges cut. Specifically, we introduce a second node t. Now instead of connecting s to all vertices, we connect it only to nodes representing overloaded blocks, i.e. $|V_i| > \lceil |V|/k \rceil$. Additionally, we connect a node in layer ℓ to t if the associated block can take at least ℓ nodes without becoming overloaded. Since the underlying model does not contain negative cycles, we can apply a shortest path algorithm to find a shortest path from s to t. We use a variant of the algorithm of Bellman and Ford since edge weights might still be negative (for more details see Section 5).

It is now easy to see that a shortest path in this model yields a set of node movements with the smallest increase in the number of cut edges and that the total number of overloaded nodes decreases by at least one. If τ is set to one we call this algorithm basic balancing otherwise advanced balancing.

However, we have to make sure that there is at least one s-t path in the model. Let us assume for now that the graph is connected. If the graph is connected then the directed version of the quotient graph is strongly connected. Hence an s-t path exists in the model if we are able to perform local search between *all* pairs of blocks that share a non-empty boundary. Because a directed local search can only start from an eligible node we might not be able to perform directed local search between all adjacent pairs of blocks, e.g. if there is no eligible node between a pair of blocks left. We try to ensure that there is at least one s-t path in the model by doing the following. Roughly speaking we try to integrate a s-t path in the model by changing the order in which directed local searches are performed. First we perform a breadth first search (BFS) in the quotient graph which is initialized with all nodes that correspond to overloaded blocks in a random order. We then pick a random node in the quotient graph that corresponds to a block A that can take nodes without becoming overloaded. Using the BFS-forest we find a path $\mathcal{P} = B \to \cdots \to A$ from an overloaded block B to A.

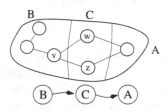

Fig. 2. Top: a graph partitioned into three parts. Bottom: BFS-tree in the quotient graph starting in overloaded block B. This path cannot be integrated into the model. After a directed local search on pair (B, C), v is marked and there is no eligible node left for the local search on pair (C, A). A similar argument holds if local search is done on the pair (C, A) first.

We now first perform directed local search on all consecutive pairs of blocks in $\mathcal{P}$. Here we use $\tau = 1$ for the number of node movements to minimize the number of non-eligible nodes. If this was successful, i.e. we have been able to move one node between all directed pairs of blocks in that path, we perform directed local searches as before on *all* pairs of blocks that share a non-empty boundary. Otherwise we undo the searches done (every node is eligible again) and start with the next random block that can take a node without becoming overloaded. In some rare cases the algorithm fails to find such a path, i.e. each time we look at a path we have one directed pair of blocks where no eligible node is left. An example is shown in Figure 2. In this case we apply a fallback balance routine that guarantees to reduce the total number of overloaded nodes by one if the input graph is connected. Given the BFS-forest of the quotient graph from above, we look at all paths in it from an overloaded block to a block that can take a node without becoming overloaded. At this point there are at most $\mathcal{O}(k)$ such paths in our BFS-forest. Specifically for a path $\mathcal{P} = Z \to Y \to X \to \cdots \to A$ we select a node having maximum gain $g_{Z,Y}$ in Z and move it to Y. We then look at Y and do the same with respect to X and so on until we move a node to block A. Note that this time we can ensure to find nodes because after a node has been moved it is not blocked for later movements. After the operations have been performed they are undone and we continue with the next path. In the end we use the movements of the path that resulted in the smallest number of edges cut. If the graph contains more than one connected component then the algorithms described above may not work. If this is the case we use a fall back algorithm which is described in the TR [14].

Putting Things Together. In practice we start our algorithms with an unbalanced input partition. We define two algorithms, basic and advanced, depending on the models used. Both the basic and the advanced algorithm operate in rounds. In each round we iterate the negative cycle based local search algorithm until there are no negative cycles in the corresponding model (basic or advanced). After each negative cycle local search step we try to find zero weight cycles in the model to introduce some diversification. Since we have random tie breaking at multiple places, we iterate this part of the algorithm. If we do not succeed to find an improved cut using these two operations for λ iterations, we perform a single balancing step if the partition is still unbalanced; otherwise we stop. The parameter λ basically controls how fast the unbalanced input partition is transformed into a partition that satisfies the balance constraint. After the balancing operation, the total number of overloaded nodes is reduced by at least one depending on the balancing model. In the basic algorithm we use the basic balancing model ($\tau = 1$) and in the advanced algorithm we use the advanced balancing model. Since the balance operation can introduce new negative cycles in the model we start the next round. The refinement techniques introduced within this paper are called Karlsruhe Balanced Refinement (KaBaR).

Integration into KaFFPaE. We now describe how we integrate our new algorithms into our distributed evolutionary algorithm KaFFPaE [13]. An evolutionary algorithm starts with a population of individuals (in our case partitions of the graph) and evolves the population into different populations over several rounds. In each round, the evolutionary algorithm uses a selection rule based on the fitness of the individuals (in our case the edge cut) of the population to select good individuals and combine them to obtain improved offspring. Roughly speaking, KaFFPaE uses KaFFPa to create individuals and modifies the coarsening phase to provide new effective combine operations.

We adopt the idea of allowing larger imbalance since this is useful to create good partitions [16]. To do so, we modify the create and combine operations as follows: each time we perform such an operation, we randomly choose an imbalance parameter $\epsilon' \in [0.005, \hat{\epsilon}]$ where $\hat{\epsilon}$ is an upper bound for the allowed imbalance (a tuning parameter). This imbalance is then used to perform the operation, i.e. after the operation is performed, the offspring/partition has blocks with size at most $(1 + \epsilon')\lceil |V|/k \rceil$. After the respective operation is performed, we apply our advanced algorithms to obtain a partition of the graph that fulfils the required balance constraint. This individual is the final offspring of the operation. We insert it into the population using the techniques of KaFFPaE [13]. Note that *at all times* each individual in the population of the evolutionary algorithm fulfils the balance constraint. Also note that allowing larger imbalance enables us to use previously developed techniques that otherwise would not be applicable, e.g. max-flow min-cut based local search methods from [12]. We call the overall algorithm Karlsruhe Balanced Partitioner Evolutionary (KaBaPE). When we use KaBaPE to create ϵ-balanced partitions we choose $\epsilon' \in [\epsilon + 0.005, \epsilon + \hat{\epsilon}]$ for the combine and create operations and transform the offspring into an ϵ-balanced partition afterwards.

5   Experiments

Implementation. We have implemented the algorithm described above using C++. We implemented negative cycle detection with subtree disassembly and distance updates as described in [4]. Overall, our program (including KaFFPa(E)) consists of about 23 000 lines of code. The implementation of the presented local search algorithms has about 3 400 lines of code.

System. Experiments have been done on two machines. Machine A has four Quad-core Opteron 8350 (2.0GHz), 64GB RAM, running Ubuntu 10.04. Machine B is a cluster where each node has two Quad-core Intel Xeon processors (X5355, 2.667 GHz) and 16 GB RAM, 2x4 MB of L2 cache and runs Suse Linux Enterprise 11 SP 1. All programs were compiled using GCC Version 4.7 and optimization level 3 using OpenMPI 1.5.5.

Parameters. After an extensive evaluation of the parameters we fixed the number of multiple directed local searches to $\mu = 20$ (larger values of μ, e.g. iterating until no boundary node is eligible did not yield further improvements). The maximum number of node movements per directed local search is set to $\tau = 15$ for $k \leq 8$ and to $\tau = 7$ for $k > 8$. The number of unsuccessful iterations until we perform a balancing step λ is set to three. Each time we perform a create or combine operation we pick a random number of node movements per directed local search $\tau \in [1, 30]$, a random number of multiple directed local searches $\mu \in [1, 20]$ and $\lambda \in [1, 10]$ and use these parameters for the balancing and negative cycle detection strategies.

5.1   Walshaw Benchmark

In this section we apply our techniques to all graphs in Chris Walshaw's benchmark archive [15]. This archive is a collection of real-world instances for the graph partitioning problem. The rules used there imply that the running time is not an issue, but one wants to achieve minimal cut values for $k \in \{2, 4, 8, 16, 32, 64\}$ and balance parameters $\epsilon \in \{0, 0.01, 0.03, 0.05\}$. It is the most used graph partitioning benchmark in the literature. Most of the graphs of the benchmark come from finite-element applications, VLSI design. A road network is also included.

Improving Existing Partitions. When we started to look at perfectly balanced partitioning we counted the number of perfectly balanced partitions in the benchmark archive that contain nodes having positive gain, i.e. nodes that could reduce the cut when being moved to a different block. Astonishingly, we found that 55% of the perfectly balanced partitions in the archive contain nodes with positive gain (some of them have up to 1400 of such nodes). These nodes usually cannot be moved by simple local search due to the balance constraint. Therefore, we now use the existing perfectly balanced partitions in the benchmark archive and use them as input to our local search algorithms KaBaR. This experiment has been performed on machine A and for all configurations of the algorithm we used $\lambda = 20$ for the number of unsuccessful tries. Table 1 shows the relative number of partitions that have been improved by different algorithm configurations and k (in total there are 34 graphs per number of blocks k).

Table 1. Rel. no. of improved instances in the Walshaw Benchmark. Configurations: Basic (Basic Neg. Cycle Impr.), +ZG (Basic + Cycle Diversification), Adv. (Adv. Model + Cycle Div.), +MDLS. (Adv. + MDLS)

k	Basic	+ZG	Adv.	+MDLS
2	0%	0%	0%	**0%**
4	18%	24%	41%	**44%**
8	38%	50%	64%	**74%**
16	64%	68%	71%	**79%**
32	76%	76%	88%	**91%**
64	82%	82%	79%	**88%**
sum	47%	50%	57%	**63%**

It is somewhat surprising that already the most basic variant of the algorithm, i.e. negative cycle detection without the zero weight cycle diversification mechanism, can improve 47% of the existing entries. All of the algorithms have a tendency to improve more partitions when the number of blocks k increases. Less surprisingly, more advanced local searches and models increase this percentage further. When applying the advanced algorithm with multiple directed local searches (the most expensive configuration of the algorithm), we are able to improve 128 partitions, i.e. 63% of the entries. Note that it took *overall* roughly two hours to compute these entries using one core of machine A. This is *very affordable* considering the fact that some of the previous approaches, such as Soper et. al. [15], have taken many days to compute *one* entry to the benchmark tables. Of course in practice we want to find high quality partitions without using input partitions generated by other algorithms. We therefore compute partitions from scratch in the next section.

Computing Partitions from Scratch. We now compute perfectly balanced partitions from scratch. We use machine B and run KaBaPE with a time limit $t_k = 225 \cdot k$ seconds using 32 cores (four nodes of the cluster) per graph and $k > 2$. On the eight largest graphs of the archive we gave KaBaPE a time limit of $\hat{t}_k = 4 \cdot t_k$ per graph and $k > 2$. For $k = 2$ we gave KaBaPE one hour of time. $\hat{\epsilon}$ was set to 4% for the small graphs and to 3% for the eight largest graph in the archive. We summarize the results in Table 2 and report the complete list of results obtained in the TR [14]. Currently we are able to improve or reproduce 86% of the entries reported in this benchmark[1]. In the bipartition case we mostly reproduce the entries reported in the benchmark (instead of improving). This is not surprising since the models presented in this paper can contain only trivial cycles of length two in this case. Also recently it has been shown by Delling et. al [5] that some of the balanced bipartitions reported there are optimal. We also applied our algorithm for larger imbalances, i.e. 1%, 3% and 5%, in the Walshaw Benchmark. For the case $\epsilon = 1\%$ we run our algorithm KaBaPE on all instances using

Table 2. Number of improvements computed from scratch for the perfectly balanced case

k	2	4	8	16	32	64	$\sum$
<	4	19	24	25	30	29	64%
≤	29	31	27	27	31	30	86%

the same parameters $\hat{\epsilon}$ and t_k as above. Here we are able to improve or reproduce the cut in 160 out of 204 cases. A table reporting detailed results can be found in the TR [14]. Afterwards we performed additional partitioning trials on all instances where our systems (including [8]. [10]. [12], [13]) currently *not* have been able to reproduce or improve the entry reported there using different parameters and different machines.

---

[1] 1. Oct. 2012.

We now improved or reproduced 97%, 99%, 99%, 99% of the entries reported there for the cases $\epsilon = 0, 1\%, 3\%, 5\%$ respectively. These numbers include the entries where we used the current record as an input to our algorithms and actually improved the input partition. They contribute roughly 4%, 7%, 11%, 9% for the cases $\epsilon = 0, 1\%, 3\%, 5\%$ respectively.

Costs for Perfect Balance. It is hard to perform a meaningful comparison to other partitioners since publicly available tools such as Scotch [11], Jostle [17] and Metis [9] are either not able to take the desired balance as an input parameter or are not able to guarantee perfect balance. This is a major problem for the comparison with these tools since allowing larger imbalances, i.e. $\epsilon = 3\%$, decreases the number of edges cut significantly. However, we have shown in [12] that KaFFPa produces better partitions compared to Scotch and Metis. Hence, we have a look at the number of edges cut by our algorithm when perfect balance is enforced, i.e. the increase in the number of edges cut when we seek a perfectly balanced partition. We use machine B and KaFFPaStrong to create partitions having an imbalance of $\epsilon = 1\%$ and

Table 3. Cost for Perfect Balance, Rel. to KaFFPa with $\epsilon = 1\%$ imbalance. Rel. EC average increase in cut after 1% partitions are balanced and Rel. t is average time used by KaBaR rel. time of KaFFPa.

k	2	4	8	16	32	64
Rel. EC [%]	9	7	5	6	4	3
Rel. t [%]	12	56	99	107	134	163

then create perfectly balanced partitions using our advanced negative cycle model and advanced balancing. KaFFPaStrong is designed to achieve very good partition quality. For each instance (graph, k) we repeat the experiment ten times using different random seeds. We compare the final cuts of the perfectly balanced partitions to the number of edges cut before the balancing and negative cycle search started. We further measure the runtime consumed by the algorithm and report it relative to the runtime of KaFFPa. The instances used for this experiment are the same as in KaFFPa [12] and are available for download at [1]. The main properties of these graphs can be found in the TR [14]. Table 3 summarizes the results, detailed results are reported in the TR [14].

6   Conclusion and Future Work

In this paper we have presented novel algorithms to tackle the balanced graph partitioning problem, including the case of *perfect balance* when the maximal block size has to equal the average block size. These algorithms combine local searches by a model in which a cycle corresponds to a set of node movements in the original partitioned graph that roughly speaking do not alter the balance of the partition. Experiments indicate that previous algorithms have not been able to find such rather complex movements. In contrast to previous algorithms such as Scotch [11], Jostle [17] and Metis [9], our algorithms are able to *guarantee* that the output partition is feasible.

An open question is whether it is possible to define a *conflict-free* model that encodes the same kind of node movements as our advanced model. In future work, it could be interesting to see if one can integrate other types of local searches from KaFFPa [12] into our models. The MDLS algorithm can be improved such that it finds the best combination of the computed local searches. It will be interesting to see whether our techniques are useful for other problems where local search is restricted by constraints, e.g. multi-constraint or hypergraph partitioning.

Shortly *after* we submitted our results to the benchmark archive we lost entries to an implementation of [7] by Frank Schneider (the original work does not provide perfectly balanced partitions). However, we are still able to improve more than half of these entries when using those as input to KaBaR. Furthermore, we integrated the techniques of [7] and again have been able to improve many entries. We conclude that the algorithms presented in this paper are still very useful.

Acknowledgements. Financial support by the Deutsche Forschungsgemeinschaft (DFG) is gratefully acknowledged (DFG grant SA 933/10-1).

References

1. Bader, D., Meyerhenke, H., Sanders, P., Wagner, D.: 10th DIMACS Implementation Challenge - Graph Partitioning and Graph Clustering
2. Benlic, U., Hao, J.-K.: An effective multilevel tabu search approach for balanced graph partitioning. Computers & OR 38(7), 1066–1075 (2011)
3. Bichot, C., Siarry, P. (eds.): Graph Partitioning. Wiley (2011)
4. Cherkassky, B.V., Goldberg, A.V.: Negative-cycle detection algorithms. In: Díaz, J. (ed.) ESA 1996. LNCS, vol. 1136, pp. 349–363. Springer, Heidelberg (1996)
5. Delling, D., Werneck, R.F.: Better bounds for graph bisection. In: Epstein, L., Ferragina, P. (eds.) ESA 2012. LNCS, vol. 7501, pp. 407–418. Springer, Heidelberg (2012)
6. Fiduccia, C.M., Mattheyses, R.M.: A Linear-Time Heuristic for Improving Network Partitions. In: 19th Conference on Design Automation, pp. 175–181 (1982)
7. Galinier, P., Boujbel, Z., Coutinho Fernandes, M.: An efficient memetic algorithm for the graph partitioning problem. Annals of Operations Research, 1–22 (2011)
8. Holtgrewe, M., Sanders, P., Schulz, C.: Engineering a Scalable High Quality Graph Partitioner. In: 24th IEEE IPDPS, pp. 1–12 (2010)
9. Karypis, G., Kumar, V.: Parallel multilevel k-way partitioning scheme for irregular graphs. SIAM Review 41(2), 278–300 (1999)
10. Osipov, V., Sanders, P.: n-level graph partitioning. In: de Berg, M., Meyer, U. (eds.) ESA 2010, Part I. LNCS, vol. 6346, pp. 278–289. Springer, Heidelberg (2010)
11. Pellegrini, F.: http://www.labri.fr/perso/pelegrin/scotch/
12. Sanders, P., Schulz, C.: Engineering multilevel graph partitioning algorithms. In: Demetrescu, C., Halldórsson, M.M. (eds.) ESA 2011. LNCS, vol. 6942, pp. 469–480. Springer, Heidelberg (2011)
13. Sanders, P., Schulz, C.: Distributed evolutionary graph partitioning. In: ALENEX, pp. 16–29. SIAM/Omnipress (2012)
14. Sanders, P., Schulz, C.: Think Locally, Act Globally: Perfectly Balanced Graph Partitioning. Technical Report. arXiv:1210.0477 (2012)
15. Soper, A.J., Walshaw, C., Cross, M.: A combined evolutionary search and multilevel optimisation approach to graph-partitioning. J. of Global Optimization 29(2), 225–241 (2004)
16. Walshaw, C., Cross, M.: Mesh Partitioning: A Multilevel Balancing and Refinement Algorithm. SIAM Journal on Scientific Computing 22(1), 63–80 (2000)
17. Walshaw, C., Cross, M.: JOSTLE: Parallel Multilevel Graph-Partitioning Software – An Overview. In: Mesh Partitioning Techniques and Domain Decomposition Techniques, pp. 27–58. Civil-Comp Ltd. (2007)

Evaluation of ILP-Based Approaches for Partitioning into Colorful Components

Sharon Bruckner[1,*], Falk Hüffner[2,**],
Christian Komusiewicz[2], and Rolf Niedermeier[2]

[1] Institut für Mathematik, Freie Universität Berlin, Germany
`sharonb@mi.fu-berlin.de`
[2] Institut für Softwaretechnik und Theoretische Informatik, TU Berlin, Germany
`{falk.hueffner,christian.komusiewicz,rolf.niedermeier}@tu-berlin.de`

Abstract. The NP-hard COLORFUL COMPONENTS problem is a graph partitioning problem on vertex-colored graphs. We identify a new application of COLORFUL COMPONENTS in the correction of Wikipedia interlanguage links, and describe and compare three exact and two heuristic approaches. In particular, we devise two ILP formulations, one based on HITTING SET and one based on CLIQUE PARTITION. Furthermore, we use the recently proposed implicit hitting set framework [Karp, JCSS 2011; Chandrasekaran et al., SODA 2011] to solve COLORFUL COMPONENTS. Finally, we study a move-based and a merge-based heuristic for COLORFUL COMPONENTS. We can optimally solve COLORFUL COMPONENTS for Wikipedia link correction data; while the CLIQUE PARTITION-based ILP outperforms the other two exact approaches, the implicit hitting set is a simple and competitive alternative. The merge-based heuristic is very accurate and outperforms the move-based one. The above results for Wikipedia data are confirmed by experiments with synthetic instances.

1 Introduction

Each entry in Wikipedia has links to the same entry in other languages. Sometimes, these links are wrong or missing, since they are added and updated manually or by naïve bots. These errors can be detected by a graph model [3, 13, 14]: Each entry in a language corresponds to a vertex, and an interlanguage link corresponds to an edge. Then, ideally, a connected component in this graph would be a clique that corresponds to a single Wikipedia term in multiple languages, and, under the plausible assumption that for every language there is at most one Wikipedia entry on a particular term, each language should occur at most once in a connected component. However, due to errors this is not the case. Our goal is to recover the correct terms by removing a minimum number of incorrect links and completing the resulting components. This can be done using a partitioning problem on a vertex-colored graph where vertices correspond to entries and colors correspond to languages.

---

* Supported by project NANOPOLY (PITN-GA-2009-238700).
** Supported by DFG project PABI (NI 369/7-2).

V. Bonifaci et al. (Eds.): SEA 2013, LNCS 7933, pp. 176–187, 2013.

COLORFUL COMPONENTS
Instance: An undirected graph $G = (V, E)$ and a coloring of the vertices $\chi : V \to \{1, \ldots, c\}$.
Task: Find a minimum-size edge set $E' \subseteq E$ such that in $G' = (V, E \backslash E')$, all connected components are *colorful*, that is, they do not contain two vertices of the same color.

We remark that the plain model naturally generalizes to an edge-weighted version and our solution strategies also apply to this. To solve COLORFUL COMPONENTS, we need to separate vertices of the same color. A *bad path* is a simple (i. e., cycle-free) path between two vertices of the same color.

Related work. Implicitly, COLORFUL COMPONENTS has first been considered in a biological context as part of a multiple sequence alignment process, where it is solved by a simple min-cut heuristic [8]. Previously, we showed that it is NP-hard even in three-colored graphs with maximum degree six [4], and proposed an exact branching algorithm with running time $O((c - 1)^k \cdot |E|)$ where k is the number of deleted edges. We also developed a merge-based heuristic which outperformed that of Corel et al. [8] on multiple sequence alignment data.

Avidor and Langberg [2] introduced WEIGHTED MULTI-MULTIWAY CUT and provided results on its polynomial-time approximability (with non-constant approximation factors).

WEIGHTED MULTI-MULTIWAY CUT
Instance: An undirected graph $G = (V, E)$ with edge weights $w : E \to \{x \in \mathbb{Q} : x \geq 1\}$ and vertex sets $S_1, \ldots, S_c \subseteq V$.
Task: Find a minimum-weight edge set $E' \subseteq E$ such that in $G' = (V, E \backslash E')$ no connected component contains two vertices from the same S_i.

COLORFUL COMPONENTS is the special case of WEIGHTED MULTI-MULTIWAY CUT when the vertex sets form a partition.

A previous formalization of the Wikipedia link correction problem leads to a harder problem: it uses several separation criteria (instead of using only language data) and also allows to "ignore" the separation criterion for some vertices [13, 14]. The resulting optimization problem is a generalization of WEIGHTED MULTI-MULTIWAY CUT. Since solving to optimality turned out to be too time-costly and non-scalable, a linear programming approach was followed [12, 13].

2 Solution Methods

We examine three approaches to finding optimal solutions for COLORFUL COMPONENTS: One based on the implicit hitting set model by Moreno-Centeno and Karp [11, 15], and two based on integer linear programming (ILP) with row generation. We then present cutting planes to enhance the performance of all three approaches. Finally, we describe a previous and a new heuristic for COLORFUL COMPONENTS.

2.1 Implicit Hitting Set

Many NP-hard problems are naturally related to the well-known NP-hard HIT-
TING SET problem, which is defined as follows:

HITTING SET
Instance: A ground set U and a set of *circuits* $S_1, \ldots, S_\ell$ with $S_i \subseteq U$
for $1 \leq i \leq \ell$.
Task: Find a minimum-size *hitting set*, that is, a set $H \subseteq U$ with $H \cap$
$S_i \neq \emptyset$ for all $1 \leq i \leq \ell$.

We can easily reduce COLORFUL COMPONENTS to HITTING SET: The ground
set U is the set of edges, and the circuits to be hit are all bad paths. Unfortunately, this can produce an exponentially-sized instance, and thus this approach
is not feasible. However, we can model COLORFUL COMPONENTS as an *implicit
hitting set* problem [1, 6, 11, 15]: the circuits have an implicit description, and
a polynomial-time oracle is available that, given a putative hitting set H, either
confirms that H is a hitting set or produces a circuit that is not hit by H. In
our case, the implicit description is simply the colored graph, and the oracle
either returns a bad path that is not hit by H or confirms that H is a solution
to COLORFUL COMPONENTS.

Implicit hitting set models are useful for finding approximation algorithms [1,
6], but also for implementing exact solving strategies [15]. In the latter case, the
approach is as follows. We maintain a list of circuits which have to be hit, initially
empty. Then, we compute an optimal hitting set H for these circuits. If H is a
feasible solution to the implicit hitting set instance, then it is also an optimal
solution to COLORFUL COMPONENTS. Otherwise, the oracle yields a bad path
that is not destroyed by H. This bad path is added to the list of circuits, and we
compute again a hitting set for this new list of circuits. This process is repeated
until an optimal solution is found. The hitting set instances can be solved by
using any HITTING SET solver as a black box; Moreno-Centeno and Karp [15]
suggest an ILP solver, using a standard set-cover-constraint formulation.

As suggested by Moreno-Centeno and Karp [15], we use the following two
speed-up tricks. First, we initially solve each hitting set problem using a heuristic,
and only use the ILP solver in case the oracle confirms that the heuristic returns a
valid (but possibly non-optimal) solution for COLORFUL COMPONENTS. Second,
instead of adding only one new circuit in each iteration, we greedily compute a
set of disjoint shortest bad paths that are added to the circuit set.

2.2 Hitting Set ILP Formulation

Moreno-Centeno and Karp [15] mention that their approach is related to column (variable) generation schemes for ILP solvers. Possibly even more straightforward, we can solve any implicit hitting set problem with an ILP solver by a
row (constraint) generation scheme (also called "lazy constraints" in the well-
known CPLEX solver). For this, we declare binary variables $h_1, \ldots, h_{|U|}$, where
the value of h_i is to indicate whether the ith element of U (under some arbitrary

order) is in H. The objective is to minimize $\sum_i h_i$. We then start the branch-and-bound process with an empty constraint set and, in a callback, query the oracle for further constraints once an integer feasible solution is obtained. If new constraints are generated, they are added to the problem, cutting off some parts of the search tree. Otherwise, we have found a valid solution. Note that adding lazy constraints is different from adding cutting planes, since cutting planes are only allowed to cut off fractional solutions that are not integer feasible, whereas lazy constraints can also cut off integer feasible solutions.

More concretely, for COLORFUL COMPONENTS, we have a variable $d_{uv}, u < v$, for each $\{u, v\} \in E$, where $d_{uv} = 1$ indicates that edge $\{u, v\}$ gets deleted. We then want to minimize $\sum_{e \in E} d_e$. The oracle deletes all edges $\{u, v\}$ with $d_{uv} = 1$, and then looks for a bad path $u_1, \ldots, u_l$. If it finds one, it yields the *path inequality*

$$\sum_{i=1}^{l-1} d_{u_i u_{i+1}} \geq 1. \tag{1}$$

We could hope that this process is more effective than the general implicit hitting set approach which uses an ILP solver as a black-box solver, since constraints are generated early on without the need for the solver to optimally solve subproblems that yield solutions that are not globally feasible.

The main disadvantage of this approach, compared to the implicit hitting set formulation, is that it requires a solver-specific implementation; further, some ILP solvers such as Coin CBC 2.7 or Gurobi 4.6 do not support adding lazy constraints without starting the solving process from scratch (the recently released Gurobi 5.0 adds this feature).

2.3   Clique Partitioning ILP Formulation

It is known for a long time (e.g. [7]) that multicut problems can be reduced to CLIQUE PARTITIONING. In this problem, vertex pairs are annotated as being similar or as being dissimilar, and the goal is to find a partition of the vertices that maximizes consistency with these annotations. We model the partition of the vertices as a *cluster graph*, that is, a graph where every connected component is a clique. The formal problem definition is then as follows:

CLIQUE PARTITIONING
Instance: A vertex set V with a weight function $\delta : \binom{V}{2} \to \mathbb{Q}$.
Task: Find a cluster graph (V, E) that minimizes $\sum_{\{u,v\} \in E} \delta(u, v)$.

Herein, $\delta(u, v)$ denotes the *dissimilarity* between u and v. To obtain a CLIQUE PARTITIONING instance from a COLORFUL COMPONENTS instance, we set

$$\delta(u, v) = \begin{cases} \infty & \text{if } \chi(u) = \chi(v), \\ -1 & \text{if } \{u, v\} \in E, \\ 0 & \text{otherwise.} \end{cases} \tag{2}$$

A component in a feasible solution for this CLIQUE PARTITIONING instance cannot contain more than one vertex of a color, since the component is a clique and the two vertices would be connected, incurring a cost of ∞. Thus, the solution also is a feasible solution for COLORFUL COMPONENTS, and the cost is the number of edges between components, and therefore equals the number of edges that need to be deleted for COLORFUL COMPONENTS.

There is a well-known ILP formulation of CLIQUE PARTITIONING [9, 17], which we can adapt for COLORFUL COMPONENTS. It has been successfully implemented and augmented with cutting planes [5, 9, 16]. It uses binary variables e_{uv} for $u, v \in V, u < v$, where $e_{uv} = 1$ iff the edge $\{u, v\}$ is part of the solution cluster graph. Cluster graphs are exactly those graphs that do not contain a P_3 as induced subgraph, that is, three distinct vertices u, v, w with $\{u, v\} \in E$ and $\{v, w\} \in E$ but $\{u, w\} \notin E$. Thus, we can ensure that the graph is a cluster graph by avoiding a P_3 for each possible triple of vertices $u < v < w \in V$:

$$e_{uv} + e_{vw} - e_{uw} \leq 1 \qquad (3)$$

$$e_{uv} - e_{vw} + e_{uw} \leq 1 \qquad (4)$$

$$-e_{uv} + e_{vw} + e_{uw} \leq 1 \qquad (5)$$

We can shrink the ILP by substituting $e_{uv} = 0$ for $u \neq v \in V, \chi(u) = \chi(v)$. Finally, the objective is to minimize $\sum_{\{u,v\} \in E} \delta(u, v) e_{uv}$.

Compared to the formulation from Section 2.2, an advantage of this formulation is that it has only polynomially many constraints, as opposed to exponentially many, and therefore it can often be stated explicitly. However, the number of constraints is $3\binom{n}{3} = \Theta(n^3)$ which can get easily too large for memory. Therefore, we also implement here a row generation scheme. We find violated inequalities by a simple brute-force search. When finding a violated inequality involving vertices u, v, w, we add all three inequalities (3)–(5), since we found this to be more efficient in our experiments.

2.4   Cutting Planes

As mentioned, the effectiveness of ILP solvers comes from the power of the relaxation. We enhance this by adding *cutting planes*, which are valid constraints that cut off fractional solutions. This scheme is called *branch-and-cut*. As demonstrated in Section 3, this addition to generic CLIQUE PARTITIONING or HITTING SET approaches is necessary to obtain competitive performance.

First, since for both the HITTING SET and the CLIQUE PARTITIONING formulation we are using row generation, we can check if already a fractional solution violates a problem-defining constraint. This allows to improve the relaxation and to cut off infeasible solutions earlier in the search tree. For the HITTING SET formulation, violated constraints can be found by a modified breadth-first search from each vertex that considers the current variable values, and for the clique partitioning model, violated constraints can be found by brute force.

Chopra and Rao [7] suggest several cutting planes for MULTIWAY CUT, the special case of MULTI-MULTIWAY CUT with $c = 1$. Each color of a COLORFUL

COMPONENTS instance induces a MULTIWAY CUT polytope. The COLORFUL COMPONENTS polytope is thus an intersection of several multiway cut polytopes. Therefore, these cutting planes are valid for COLORFUL COMPONENTS, too. We present them here for the CLIQUE PARTITIONING formulation.

Let $T = (V_T, E_T)$ be a subgraph of G that is a tree such that all leaves L of the tree have color c, but no inner vertex has. Then the inequality

$$\sum_{uv \in E_T} e_{uv} \leq (|E_T| - |L|) + 1 \tag{6}$$

is called a *tree inequality*. Note that for $|L| = 2$, we get a path inequality (1). There are exponentially many tree inequalities. Therefore, we consider only tree inequalities with one (called *star inequalities*) or two internal vertices. We can also apply these cuts for the HITTING SET formulation: we need to substitute $e_{uv} = 1 - d_{uv}$ and restrict the sums to edges present in the graph. In our implementation of the HITTING SET row generation scheme, we add all initially violated star inequalities at once. This already covers all length-2 bad paths.

2.5   Heuristics

One advantage of having an algorithm that is able to solve large-scale instances optimally is that we can evaluate heuristics more precisely. We can thus further examine our previous heuristic [4], which outperformed the one proposed by Corel et al. [8] on multiple sequence alignment data. For completeness, we briefly recall this greedy heuristic [4]. The idea is to repeatedly merge the two vertices "most likely" to be in the same component. During the process, we immediately delete edges connecting vertices with identical colors. Thus, we can determine the *merge cost* of two vertices u and v as the weight of the edges that would need to be deleted in this way when merging u and v. The *cut cost* is an approximation of the minimum cut between u and v obtained by looking only at their common neighbors. We then repeatedly merge the endpoints of the edge that maximizes cut cost minus merge cost.

For density-based partitioning it was shown that *greedy vertex moving* outperforms merge-based heuristics [10]. Hence, it is interesting to consider greedy vertex moving also for COLORFUL COMPONENTS. We follow the approach by Görke et al. [10]. Here we start with singleton clusters, that is, every cluster contains exactly one vertex. Then, we consider all possible ways of moving one vertex from a cluster to another. Of these possibilities, we greedily perform the one that decreases the number of inter-cluster edges the most without violating the colorfulness condition. Once no further improvement is possible, clusters are merged into vertices and the procedure is applied recursively.

3   Experiments

We performed experiments both to evaluate the COLORFUL COMPONENTS model for Wikipedia interlanguage link correction and to compare the five solution approaches. The ILP approaches were implemented in C++ using the CPLEX 12.4 ILP solver. The experimental data and the code are available at `http://fpt.akt.tu-berlin.de/colcom/`. The test machine is a 3.6 GHz Intel Xeon E5-1620 with 10 MB L3 cache and 64 GB main memory, running under Debian GNU/Linux 7.0. Only a single thread was used.

Each connected component is solved separately. Before starting the solver, we use data reduction as described before [4]. This actually yields an instance of the more general weighted MULTI-MULTIWAY CUT problem. Adapting the ILP formulations above to this problem is straightforward. We further use the result of the merge-based heuristic (Section 2.5) as *MIP start* (that is, we pass this solution to the ILP solver such that it can start with a good upper bound).

3.1   Medium-Sized Wikipedia Graphs

To construct the Wikipedia interlanguage graph, we downloaded the freely-available Wikipedia interlanguage links and page data dumps from January 9th, 2012. We chose a set of seven languages: Chinese, English, French, German, Hebrew, Russian, and Spanish. We then created the graph as described in the introduction, with vertices as pages and a link between two pages if one has an interlanguage link to the other. As suggested by de Melo and Weikum [13], we weigh the edges as follows: If two pages link to each other, the edge receives a weight of 2. Otherwise, the weight is 1. The graph contains 4,090,160 vertices and 9,666,439 edges in 1,332,253 connected components. The largest connected component has size 409. Of these components, 1,252,627 are already colorful.

We then found the colorful components using the CLIQUE PARTITIONING algorithm from Section 2.3, in 54 seconds. The merge-based heuristic obtained a solution that was 0.22% off the optimum, taking 38 seconds, and the greedy vertex moving was 0.90 % off, taking 37 seconds. Note that our implementations of the heuristics have not been optimized for speed.

The cost of the optimal solution is 188,843, deleting 184,759 edges. After removing the edges of the solution we had 1,432,822 colorful components. We obtained 1,355,641 colorful components of size > 1, each corresponding to an entry in several languages. Almost half of the colorful components comprise two vertices, 20% comprise three vertices, 11%, 6%, 3%, 2% comprise four, five, six and seven vertices, respectively. The remaining components are singletons. In each colorful component that is not already a clique, two vertices that are not already connected by an edge represent two pages in different languages that should have a new inter-language link between them. Overall, we found 52,058 such new links. To get an idea of the correctness of these links we looked at the Hebrew and English pages and manually checked the new links between them. For example, we identified missing links between "data compression" and its Hebrew counterpart, and "scientific literature" and the equivalent Hebrew entry, which was previously linking to the less fitting "academic publishing".

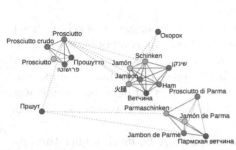 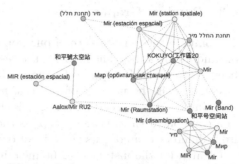

(a) A single connected component in the Wikipedia graph, disambiguating different types of pork. The English page "Prosciutto" is connected both to the correct "Prosciutto" cluster and to the "Ham" cluster. The COLORFUL COMPONENTS algorithm separates the two correctly.

(b) A single connected component in the Wikipedia graph, regarding the term "MIR". The algorithm successfully separates the cluster of entries corresponding to the disambiguation of the term from the cluster centering on the MIR space station. The outlier for the band MIR is now also disconnected.

Fig. 1. Two connected components in the Wikipedia graph. Green edges have been inserted and dotted red edges have been deleted by the algorithm.

Figures 1(a) and 1(b) demonstrate results of the algorithm. In both cases the algorithm successfully separates clusters representing related, but not identical terms and identifies outliers.

3.2 Large-Scale Wikipedia Graphs

To test our fastest ILP formulation (CLIQUE PARTITIONING) and the heuristics on even larger inputs, we downloaded Wikipedia interlanguage link data for the largest 30 languages[1] on 7 June 2012. To decrease noise, we excluded user pages and other special pages. The resulting instance has 11,977,500 vertices and 46,695,719 edges. Of the 2,698,241 connected components, 225,760 are not colorful, the largest of which has 1,828 vertices and 14,403 edges. The instance can be solved optimally in about 80 minutes (we cannot give a more precise figure since, because of memory constraints, we had to run our implementation on a different machine that was also loaded with other tasks). In the solution, 618,660 edges are deleted, and the insertion of 434,849 can be inferred. The merge-based heuristic has an error of 0.81 %. Solving the largest component takes 182 seconds; 10.2 % of the edges are deleted. The merge-based heuristic takes 13.4 seconds, with 1.15 % error. The languages in the component are similarly distributed as in the overall graph. It contains mostly terms related to companies, in particular different legal forms of these, and family relationships. We noted that many inconsistencies have been introduced by bots that aim to fill in "missing" translations; for example, the Hungarian word "Részvény" (stock) is wrongly linked to the term for "free float" in many languages.

---

[1] http://meta.wikimedia.org/wiki/List_of_Wikipedias

3.3   Random Graphs

To compare the performance of our approaches, we generated a benchmark set of random instances. The model is the recovery of colorful components that have been perturbed. More precisely, the model has five parameters: c is the number of colors; n is the number of vertices; p_v is the probability that a component contains a vertex of a certain color; p_e is the probability that between two vertices in a component there is an edge; p_x is the probability that between two vertices from different components there is an edge.

Clearly, for the instances to be meaningful, p_e must be much higher than p_x. We first generate a benchmark set of 243 instances with parameters similar to those of the largest connected components in the 7-language Wikipedia instance. Note that since each instance models a connected component, they are much smaller than a typical real-world instance.

Based on the parameters corresponding to the largest connected components in the Wikipedia instance, we choose the parameters as follows: $c \in \{3, 5, 8\}$, $n \in \{60, 100, 170\}$, $p_v \in \{0.4, 0.6, 0.9\}$, $p_e \in \{0.4, 0.6, 0.9\}$, and $p_x \in \{0.01, 0.02, 0.04\}$. In Fig. 2(a), we compare the running times for the three approaches and additionally the branching algorithm from [4], with a time limit of 15 minutes. The branching algorithm is clearly not competitive. Among the ILP-based approaches, the CLIQUE PARTITIONING formulation eventually comes out as a winner. All instances with $n = 60$ are solved, and only 4 of the $n = 100$ instances remain unsolved, all of which have $p_x = 0.04$. The performance of the row generation scheme is somewhat disappointing, solving less instances than the implicit hitting set formulation. One possible reason is that for the implicit hitting set formulation, the solver is able to employ its presolve functions to simplify the instance. Further tuning and application of cutting planes might give the row generation scheme an advantage, though.

We now compare the effect of varying a single parameter, starting with the base parameters $n = 70$, $c = 6$, $p_v = 0.6$, $p_e = 0.7$, and $p_x = 0.03$. We set a timeout of 5 minutes. In Fig. 2(b), the exponential growth of the running time when increasing the instance size is clearly visible. This is as expected for an exact approach to an NP-hard problem. In Fig. 2(c), we vary the number c of colors. The running time grows with more colors, but remains manageable. The parameter p_v does not seem to have a large effect on running times (Fig. 2(d)); the approach copes well even with components with many missing vertices. For the parameter p_e (Fig. 2(e)), we note lower running times for high values. This matches intuition, since dense clusters should be easier to identify. The running time is also lower for small values; this can probably be explained by the fact that such instances have overall very few edges. Finally, we see that the parameter p_x, which models the number of "errors" in the instance, has a large influence on running time (Fig. 2(f)); in fact, the running time also seems to grow exponentially with this parameter.

Finally, we examine the performance of the heuristics (Section 2.5) on the benchmark set for those 213 instances where we know the optimal solution. The maximum running time for an instance is 0.4 s for both heuristics. The merge-based

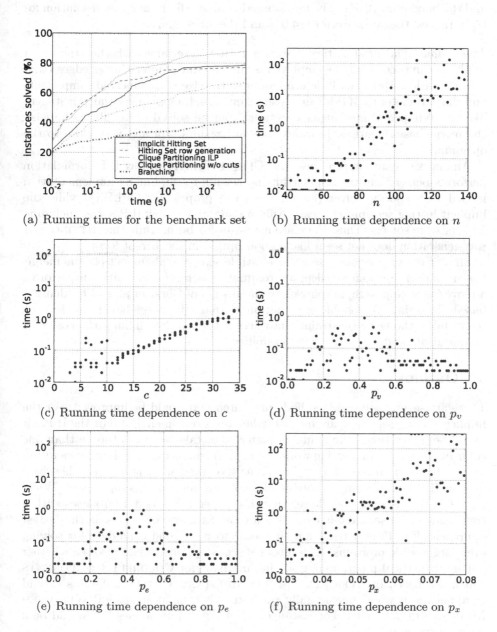

(a) Running times for the benchmark set   (b) Running time dependence on n

(c) Running time dependence on c (d) Running time dependence on p_v

(e) Running time dependence on p_e (f) Running time dependence on p_x

Fig. 2. Running times for synthetic COLORFUL COMPONENTS instances: (a) method comparison; (b)–(f) running time dependencies on parameters for the CLIQUE PARTITIONING ILP

heuristic finds an optimal solution for 124 instances; the average error is 0.86 % and the maximum 12.5 %. The move-based heuristic finds an optimal solution for 55 instances; the average error is 4.9 % and the maximum 38.7 %.

Discussion. The most critical parameter that determines whether the exact methods can successfully be employed is the amount of inter-cluster edges (that is, the solution size), since it determines both the size n of connected components and the parameter p_x. If this value is small enough, then even very large instances like the Wikipedia interlanguage network can be solved optimally. Otherwise, the merge-based heuristic provides excellent results typically very close to the optimum.

Among the exact approaches, the CLIQUE PARTITIONING ILP formulation performs better than the implicit hitting set approach, but its implementation is tied to a specific solver (in our case, the proprietary CPLEX), while the implicit hitting set approach can easily be adapted to any ILP solver including free software solvers. Thus, there are use cases for both, while the HITTING SET row generation does not seem like a good option in its current form.

Similar to our previous results for multiple sequence alignment [4], the merge-based heuristic gives an excellent approximation here. It also clearly outperforms a move-based approach, in contrast to the results of Görke et al. [10] for density-based clustering. A possible explanation is that the merge-based heuristic already takes the color constraints into account when determining the cost of a modification, and not only for its feasibility.

4 Outlook

There are several ways the methods presented here could be improved. For the implicit hitting set, there are many further ways to tune it [15]. For the ILPs, it would be interesting to find cutting planes that take vertices of more than one color into account. In ongoing work, we experimented with a column generation approach based on the CLIQUE PARTITIONING model, using a greedy heuristic and an ILP formulation for solving the column generation subproblem. While on the synthetic data it is slower than the other ILP-based approaches with a time limit of 10 seconds, it can solve almost as many instances as the fastest approach after 15 minutes. Thus, it seems to be a good candidate for solving even larger-scale problems. As a next natural step, one should also see whether our mathematical programming solving methods for COLORFUL COMPONENTS which are based on WEIGHTED MULTI-MULTIWAY CUT formulations extend to applications where one actually *needs* to solve the more general WEIGHTED MULTI-MULTIWAY CUT. For instance, cleansing of taxonomies [12] would be a natural candidate.

Concerning applications and modeling, there are several ways to expand our results. First, we currently only demand a cluster to be a connected subgraph; further restrictions on its density might be useful. Also, for some applications the conditions on colors in a component might be relaxed, for example by allowing a constant number of duplicates per component. Finally, finding further applications would be interesting. We briefly sketch one candidate application here. Consider a graph where each vertex corresponds to a user profile in a social network, two profiles are adjacent when they are similar, and the color of a vertex is the network (Twitter etc.). Then, COLORFUL COMPONENTS could be used to identify groups of profiles that correspond to the same natural person, assuming that every person has at most one profile in each network.

References

[1] Ashley, M.V., Berger-Wolf, T.Y., Chaovalitwongse, W., DasGupta, B., Khokhar, A., Sheikh, S.: An implicit cover problem in wild population study. Discrete Mathematics, Algorithms and Applications 2(1), 21–31 (2010)
[2] Avidor, A., Langberg, M.: The multi-multiway cut problem. Theoretical Computer Science 377(1-3), 35–42 (2007)
[3] Bolikowski, Ł.: Scale-free topology of the interlanguage links in Wikipedia. Technical Report arXiv:0904.0564v2, arXiv (2009)
[4] Bruckner, S., Hüffner, F., Komusiewicz, C., Niedermeier, R., Thiel, S., Uhlmann, J.: Partitioning into colorful components by minimum edge deletions. In: Kärkkäinen, J., Stoye, J. (eds.) CPM 2012. LNCS, vol. 7354, pp. 56–69. Springer, Heidelberg (2012)
[5] Böcker, S., Briesemeister, S., Klau, G.W.: Exact algorithms for cluster editing: Evaluation and experiments. Algorithmica 60(2), 316–334 (2011)
[6] Chandrasekaran, K., Karp, R.M., Moreno-Centeno, E., Vempala, S.: Algorithms for implicit hitting set problems. In: Proc. 22nd SODA, pp. 614–629. SIAM (2011)
[7] Chopra, S., Rao, M.R.: On the multiway cut polyhedron. Networks 21(1), 51–89 (1991)
[8] Corel, E., Pitschi, F., Morgenstern, B.: A min-cut algorithm for the consistency problem in multiple sequence alignment. Bioinformatics 26(8), 1015–1021 (2010)
[9] Grötschel, M., Wakabayashi, Y.: A cutting plane algorithm for a clustering problem. Mathematical Programming 45(1-3), 59–96 (1989)
[10] Görke, R., Schumm, A., Wagner, D.: Experiments on density-constrained graph clustering. In: Proc. 2012 ALENEX, pp. 1–15. SIAM (2012)
[11] Karp, R.M.: Heuristic algorithms in computational molecular biology. Journal of Computer and System Sciences 77(1), 122–128 (2011)
[12] Lee, T., Wang, Z., Wang, H., Hwang, S.: Web scale taxonomy cleansing. In: Proceedings of the VLDB Endowment, vol. 4, pp. 1295–1306 (2011)
[13] de Melo, G., Weikum, G.: Untangling the cross-lingual link structure of Wikipedia. In: Proc. 48th ACL, pp. 844–853. ACM (2010)
[14] de Melo, G., Weikum, G.: MENTA: inducing multilingual taxonomies from Wikipedia. In: Proc. 19th CIKM, pp. 1099–1108. ACM (2010)
[15] Moreno-Centeno, E., Karp, R.M.: The implicit hitting set approach to solve combinatorial optimization problems with an application to multigenome alignment. Operations Research (to appear, 2013)
[16] Oosten, M., Rutten, J.H.G.C., Spieksma, F.C.R.: The clique partitioning problem: Facets and patching facets. Networks 38(4), 209–226 (2001)
[17] Régnier, S.: Sur quelques aspects mathématiques des problèmes de classification automatique. I.C.C. Bulletin 4, 175–191 (1965)

Finding Modules in Networks
with Non-modular Regions

Sharon Bruckner*, Bastian Kayser, and Tim O.F. Conrad

Institut für Mathematik, Freie Universität Berlin, Germany
{sharonb,conrad}@math.fu-berlin.de, bastiankayser84@googlemail.com

Abstract. Most network clustering methods share the assumption that the network can be completely decomposed into modules, that is, every node belongs to (usually exactly one) module. Forcing this constraint can lead to misidentification of modules where none exist, while the true modules are drowned out in the noise, as has been observed e. g. for protein interaction networks. We thus propose a clustering model where networks contain both a *modular region* consisting of nodes that can be partitioned into modules, and a *transition region* containing nodes that lie between or outside modules. We propose two scores based on spectral properties to determine how well a network fits this model. We then evaluate three (partially adapted) clustering algorithms from the literature on random networks that fit our model, based on the scores and comparison to the ground truth. This allows to pinpoint the types of networks for which the different algorithms perform well.

1 Introduction

A common way of analyzing networks is to partition them into *clusters* (or *modules, communities*) where similar or interacting nodes are grouped together. This is known as *Graph Clustering*. Such a grouping can help identifying the underlying structure of the network and extract insights from it. For example, modules in a protein–protein interaction (PPI) network can correspond to protein complexes (see, e.g. [9, 27]). Accordingly, many clustering methods have been developed, varying in their definition of the optimal clustering and in the approach taken to compute it [17]. However, in most of these methods, the partition must be a *full partition*, meaning every node must belong to exactly one module. This constraint both limits the classes of networks that can be clustered, and the insights that can be gained from them.

In this work, we propose a more flexible model, where networks have two parts: a *modular region*, which can be fully partitioned into individual modules, and a *transition region*, containing nodes that cannot be assigned to any module. We call these networks "not completely clusterable" (NCC networks). Consider for example the PPI network mentioned above. It is well-known that there are proteins that are involved in more than one protein complex [10]. Additionally,

---

* Supported by project NANOPOLY (PITN-GA-2009–238700).

V. Bonifaci et al. (Eds.): SEA 2013, LNCS 7933, pp. 188–199, 2013.

not every protein takes part in a complex, meaning that many proteins should not be assigned to *any* cluster [24]. By forcing each node into a cluster, we could fail to single out these proteins and regions and could introduce errors and meaningless clusters.

Previous work. In recent years, there has been interest in methods that challenge one part of the "full partition" assumption: Methods to find overlapping clusters, or perform *fuzzy clustering* on the network, allow each node in the network to belong to more than one module. Overlapping Clusters approaches, such as those based on clique percolation [14], variants of modularity [13, 28] or other concepts [22], identify modules that can share nodes. In fuzzy graph clustering, each node receives a probability of being assigned to each module and the modules can then be determined by thresholding the probabilities [19, 25]. Fewer methods exist, to our knowledge, that further weaken the assumptions above: methods like SCAN [26] and the one described by Feng et al. [6] define three types of nodes: nodes that belong to a single module, nodes that can belong to more than one module (*hubs*), and nodes that belong to no module at all (*outliers*). This is close to the framework that we present in this paper; however, we assume here that the nodes that do not belong in modules (either hubs or outliers, as termed by the other methods) need not necessarily comprise a small and negligible part of the network.

The MSM (Markov State Model) algorithm was proposed by some of us [20] to directly address the problem of identifying modules in NCC networks. It uses the concept of *metastability* and tries to identify metastable sets, which are then equated with modules.

Our Contributions. We discuss NCC networks exhibiting the properties defined above: (1) presence of a transition region, (2) presence of modules. In Section 2, we propose a simple characterization of such networks, *partial modularity*. Intuitively, the structures of the modular region of our networks are dense, while the transition region is relatively sparse. We propose and discuss two scores to formalize these notions, and demonstrate their behavior on simple networks.

The next natural question is, given that a network has a high partial modularity, how can we identify its modules? In Section 3, we first present a score for the quality of a clustering resulting from an algorithm by comparing it with the clustering dictated by some ground truth. We use this scoring function to compare the performance of state-of-the-art module finding methods on benchmark networks. Two of these algorithms are adaptations of existing popular algorithms, and the third, MSM, is designed for NCC networks.

2   Scoring Partial Modularity

To formalize the notion of an NCC network, we want a quantitative score for a network that evaluates the degree to which it contains modules. The natural candidate for this is Newman's modularity [12], designed to measure the strength of a full partition of a network into modules: dense connections within modules, sparse connections between the modules. However, this score has a quality that makes it not suitable for NCC networks: Networks that are tree and tree-like have a high modularity [2]. This means that sparse networks without any dense subgraphs have a high modularity, despite having no dense modules.

Since Newman's modularity is not adequate, we search in a different direction. It has been known (see, e.g. [11, 23]) that the number of eigenvalues of the transition matrix of a network that are close to 1 are linked to the number of modules in the network, and that the size of the eigenvalue gap could then indicate the difference between modules and the rest of the network. We take this idea further by looking at the gap of a different transition matrix: that of an embedded Markov chain of the continuous Markov process defined by Sarich et al. [20], where the continuous random walk is generated by a custom generator L such that the process stays for extended periods of time in dense regions of the network. The advantage is a better correspondence between eigenvalues close to 1 and dense modules, and a clearer gap. We therefore consider as a network score the size of the largest gap of this matrix $P = \exp(\alpha L)$, where α is the lag time defined in [20], which acts as a granularity parameter. Let λ_u be the eigenvalue of L that lies above the gap, and λ_l the eigenvalue below the gap. Then we define the gap score as

$$Q_\gamma := \exp(\alpha \lambda_u) - \exp(\alpha \lambda_l). \tag{1}$$

We first note that sparse networks do not have a high gap score. For example, we tested several road networks.[1] These networks have a very low average clustering coefficient (~ 0.01) and density (~ 0.0002). While their Newman modularity [12] is > 0.95, the gap score is < 0.002, considerably lower and better reflecting the absence of modules.

The gap score has several drawbacks. First, there is not one "true" gap in the spectrum, just as there is not one "true" clustering of the network. Different gaps induce different network partitions, and the choice of largest gap can be arbitrary. Second, there can be networks that contain modules, but do not have a clear gap. This can occur, e.g. when the density of the modules is close to that of the transition region. We will demonstrate this in Section 3.

To try and overcome these problems, we use the concept of *metastability* to define "good" NCC networks. In [20] we introduced the following definition of a metastable partition (see also [3]):

$$R := \max_{y \notin \mathcal{M}} \mathbb{E}_y(\tau(\mathcal{M})) \ll \min_{i=1,\dots,m} \mathbb{E}_i(\tau(\mathcal{M}_i)) =: W, \tag{2}$$

Here, $\mathbb{E}_y(\tau(\mathcal{M}))$ is the expected entry time of the process into an arbitrary module, if started in some node $y \in T$ in the transition region $T = V \setminus \mathcal{M}$. Likewise $\mathbb{E}_i(\tau(\mathcal{M}_i))$ denotes the expected entry time into a module M_j with $j \neq i$ if started from M_i. In other words, the *return time* R the random walk

---

[1] Downloaded from http://www.cs.fsu.edu/~lifeifei/SpatialDataset.htm

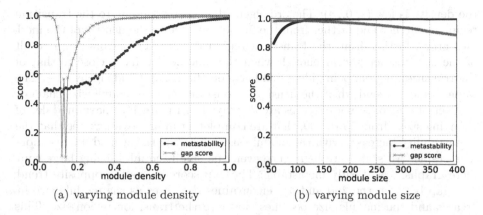

(a) varying module density   (b) varying module size

Fig. 1. Metastability and gap scores for random networks with transition region 1000 nodes and 2 modules

needs to enter one of the modules, if in the transition region, is small compared to its typical *waiting time* W between transitions from one module to another. Based on this, we define the metastability score as

$$Q_m := 1 - R/W. \tag{3}$$

This score, unlike the gap score above, explicitly takes the transition region into account, therefore might be better suited for NCC networks. The main drawback is that this is not a global score, but rather a score for single partition. We would have liked to continue, analogously to Newman modularity [12], by then finding the partition that minimizes R/W, and assigning the network its score. Unfortunately, this will not be useful, since every full partition ($T = \emptyset$) will set $R = 0$ and the score to 1. However, as the experiments below show, it is still indicative of the presence of modules and can be used to compare NCC networks.

Experiments. We performed a set of simple experiments to test the gap and metastability scores.

Figure 1 demonstrates the behavior of the metastability score on networks with a transition region of 1000 nodes (random Erdős–Rényi (ER) graph with density 0.05) and two modules (random graphs with given density and size). To create the network, we first generated the transition region and modules separately, and then for each module randomly identified a vertex from the module and a vertex from the transition region.

In Figure 1(a) the modules have size 100 each, and they are random graphs with density 0.03 to 1. The metastability score was computed for the planted ground truth partition. As the density of modules increases, the metastability score increases also, as the denser modules become more metastable. The transition region does not change, therefore R is the same and only W changes. The gap score increases as well with the module density, except in the cases where the density is low (< 0.18): The 3rd eigenvalue, corresponding to the transition region, is farther and farther from the 2nd eigenvalue as the density of the modules grows further from the transition region density. The errors are a result of the gap not being clear enough when the module density is close to that of the transition region, being identified between the 1st and 2nd eigenvalues. The same cannot be said when the density is constant but the module size changes: The set of networks whose scores are displayed in Figure 1(b) have modules of changing sizes, from 55 to 400, that are complete graphs. Again we see that the metastability increases with the module size, since the larger modules are more metastable. All scores are high since even the small modules, being large complete graphs, are already metastable. The gap score shows the opposite trend: The gap between the 2nd and 3rd eigenvalues decreases as the module size increases and the module size becomes closer to the transition region size. This demonstrates that the gap score does not always agree with our intuition of a modular network, and underscores the need for a better modularity score.

3 Algorithms

We select three clustering algorithms from literature, which we partially adapt for our setting. We choose standard parameters for all algorithms.

SCAN. The SCAN algorithm [26] clusters vertices together based on neighborhood similarity and reachability. It can identify vertices as hubs or outliers; we interpret both as the transition region. SCAN requires a user-defined parameter μ that determines the minimum size of a module. This we set to 10, 1% of the network size of most of the NCC networks we use for evaluation.

Markov Clustering. The idea of the MCL (Markov Clustering) algorithm [5] is to simulate random walks on the network and identify modules as regions where the random walker stays for a prolonged time. MCL always returns a full partition. It has been demonstrated (e. g. [21]) that MCL tends to produce imbalanced clusterings, consisting of a few large clusters and many small clusters of size two or three and singletons. Usually viewed as a shortcoming of the algorithm, we now interpret this tendency to our advantage: We introduce a parameter μ similar to SCAN to set a minimum size for a module. All modules with less than μ nodes are assigned to the transition region.

Markov State Model. The MSM (Markov State Model) algorithm [20] first tries to identify the modular region as the region where a random walker spends the most time. The rest is classified as transition region, and the modular region is clustered with a simple heuristic.

Ground-Truth-Based Evaluation. We evaluate the algorithms by comparing their output to the known clustering using the adjusted Rand index [16, 18], which measures how well two partitions match. We propose three versions of the score:

ρ_{MT} to measure how well the modular region and transition region are distinguished, ρ_M to measure the quality of the clustering within the modular region, and ρ_c as a combined score.

3.1   Experiments on Random Networks

We now test the algorithms on random NCC networks, constructed as follows: Each node belongs either to exactly one of the modules, or to the transition region. We use a similar random graph model as in Section 2, with each module and the transition region being a random ER graph. In addition, each module and the transition region are connected by adding a random spanning tree. Then, from the transition region and from each module a node is chosen at random, and these nodes are connected by a random spanning tree to ensure that the whole graph is connected. Finally, each possible edge between a vertex from a module and a vertex from the transition region is added with a small probability.

This class has the following parameters: **Network size** (total number of nodes) N, **number of modules** M, **total number of nodes in modules** N_{mtot}, **module density** p_m, **transition region density** p_t and **interconnection density** p_i . The inter-connection density is an indicator for the number of edges between modules and the transition region. We also define the number of nodes in transition region $N_t := N - N_{mtot}$ and number of nodes per module $N_m := N_{mtot}/M$.

The standard parameter values are as follows: $N = 1000$ nodes, $M = 5$ modules, $p_m = 0.6$ module density, $p_t = 0.01$ transition region density, and $p_i = 0.01$ interconnection density. The minimal module size under these constraints is then 10 nodes.

Since in practice the running time of the algorithms depends on implementation, and in every case the running time was < 1 minute, we focus here on the accuracy of the algorithm as determined by our evaluation measure.

Experiment 1: Varying transition region size. In this experiment, our goal is to evaluate the behavior of the different algorithms on networks where the transition region comprises between 0% and 90% of the network. In the case of 0% transition region, the modular region occupies the entire network, and the problem will again be that of full partitioning. We hypothesize that the algorithms should perform better on networks with a small transition region, as they are closer to the full partition case: SCAN looks for hubs and outliers but those are usually single nodes, not entire regions; MCL was originally designed for full partitions. Since MSM does not make assumptions about the size of the transition region, it is possible that this algorithm performs the same on the networks regardless of the transition region size.

Indeed, our experiments show that for 80% or less transition region, all algorithms perform optimally ($\rho_c = 1$ for MSM) or close to optimally ($\rho_c > 0.92$). For larger transition regions, all algorithms perform progressively worse.

Figure 2 shows the performance of the algorithms, giving the ρ_c score averaged over 5 networks for each transition region size. SCAN and MCL both identify

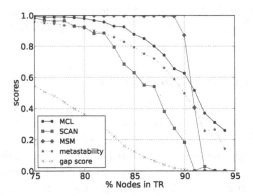

Fig. 2. Comparing the ρ_c score for SCAN, MCL, and MSM on networks with varying transition region size. All networks were generated with 1000 nodes and 5 modules, having the default densities.

only two or three modules, assigning the rest to the transition region. SCAN additionally identifies no hubs or outliers, thus the transition region is a result of small clusters, just as in the case of MCL. MSM separates the modular and transition region well ($\rho_{MT} > 0.95$), identifies five modules in the modular region, but partitions it less than optimally (average $\rho_M = 0.77$). MSM begins to deteriorate a little later than the others, at 89%, but the score decreases fast, with a score of 0 (all nodes are identified as transition region nodes) from 92%. Therefore, MSM is clearly the choice in case the transition region is large, but not too large.

We additionally plot the metastability index of these networks using the ground truth partition. This score also decreases with the size of the modular region, since there are less nodes in modules. The gap scores decreases more quickly, reaching 0.5 when the transition region comprises 75% of the network, but being close to 1 when the transition region is 10% or less.

Experiment 2: Varying module size. As we increase the size of the transition region in Experiment 1, the size of a module decreases automatically, since fewer nodes are now divided into a constant number of 5 modules. Specifically, for a transition region which covers 80% of the network, the corresponding module size is 40, and for 90% it is already 20. To test whether the difference in scores in Experiment 1 is a result of varying the transition region size or of varying the module size, we run a set of experiment where we directly vary the module size. The module size is between 20 and 200, there are 5 modules as before, and the transition region comprises 50% of the network, a value for which all algorithms in Experiment 1 performed perfectly ($\rho_c = 1$). Naturally, to preserve the same proportion of transition region to modular region while varying the total size of the modular region, the overall network size has to change as well, varying from 200 to 2000, respectively. All 3 algorithms performed perfectly for all module sizes: All three ρ scores were 1 or > 0.99. We additionally tested module size < 20, but the results were unstable due to the small difference between the true and the minimal module size: in some cases the modules were detected correctly, in others, only 9 nodes from a 10-node module were detected, and were assigned to the transition region, causing low scores. Therefore, while a very small module size can negatively influence the algorithm, this effect disappears for slightly larger module sizes, and the low scores of Experiment 1 cannot be fully attributed to the module size, but must be due to the proportion of the transition region.

In the next set of experiments we keep the proportions between the network components constant but vary their densities.

Experiment 3: Varying module and transition region densities. In this experiment, we test combinations of the module density p_m and the transition region density p_t. We take as before networks with 1000 nodes and 5 modules, with the transition region comprising 50% of the network. We additionally set $p_i = 0.01$. For these parameters and the default densities $p_t = 0.01, p_m = 0.6$ all three algorithms performed optimally in the previous experiments.

We set $p_m = 0.1, 0.2, \ldots, 0.9, 1$, and $p_t = 0.01, 0.06, 0.11, \ldots, 0.81$. Figure 3 shows a heatmap for each of the algorithms, giving the ρ_c score for each combination of transition region density and module density. Intuitively, we expect the algorithms to do well when the module density is high and the transition density is low. Indeed, we see that this is the case for all algorithms. SCAN performs the best, erring only when $p_t > 0.45$. The other two algorithms perform optimally when $p_t < 0.06$ and $p_m = 0.8$, and performance quickly deteriorates. Looking more closely at the ρ_{MT} and ρ_M scores, we see that the ρ_M score is perfect while ρ_{MT} is low: the entire transition region is detected as a single module in all these cases. The gap score follows this intuition as well, giving a low score only to networks where the module density is much lower than that of the transition region.

The poor performance of MSM could perhaps be attributed to the fact that the algorithm tends to reward (with a high waiting time) those nodes that have a relatively high degree. Those nodes end up being assigned to modules more often. As the density of the transition region increases, so does the average degree. Since we have fixed p_i at 0.01, and as the modules are smaller than the transition region (each module has size 100, compared to 500 for the transition region), the average degree of nodes in the module is also bounded, and for some values of p_m and p_t, the degrees are about the same, and thus MSM cannot tell them apart as well.

Discussion. Unfortunately, no algorithm comes out the clear leader in every case. MSM identifies modules even when the transition region is large, but does not perform so well when the average degree in the transition region is high. While SCAN performs better than the other algorithms whenever the densities of the transition region and modules are close, in many cases it too identifies the transition region as a module.

With regards to the different steps of module identification, we first note that MSM performs best the task of guessing the correct number of modules. SCAN

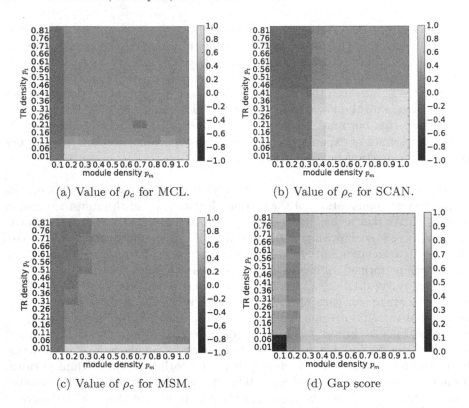

(a) Value of ρ_c for MCL.   (b) Value of ρ_c for SCAN.

(c) Value of ρ_c for MSM.   (d) Gap score

Fig. 3. Plotting the ρ_c score for the three algorithms for different combinations of p_m and p_t. All networks have 1000 nodes and 5 modules with 100 nodes each.

and MCL both under-estimate the module number, identifying modules that are too small and are therefore assigned to the transition region. No algorithm over-estimated the number of modules throughout our experiments. On the task of separating the transition region and the modular region (assessed with the ρ_{MT} measure), the three algorithms had successes and shortcomings: In Experiments 1 and 2 the errors were a result of nodes from the modular region being assigned to the transition region. In Experiment 3, the error resulted from the transition region being identified as a single module.

3.2   Experiments on Real-World Networks

We now apply MSM to a real biological network, the well-known FYI network from [7], in order to test whether the results obtained can provide insight about biological truth. The PPI network of *Saccharomyces cerevisiae* was constructed by integrating the results of several large-scale experiments. The outcome is a network whose nodes represent proteins and an edge between two nodes exists if the interaction between the corresponding proteins has been verified by multiple experiments. We note that we chose to analyze this particular network despite

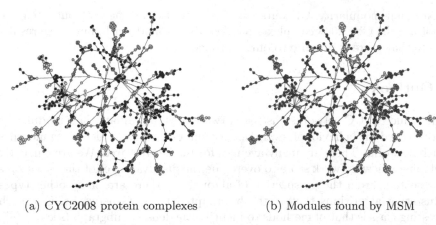

(a) CYC2008 protein complexes   (b) Modules found by MSM

Fig. 4. Modules in the yeast protein interaction network FYI

the abundance of more modern and complete yeast protein interaction networks, since it is unweighted and simpler. In fact, the gap score of this network is 0.17, in contrast to the Newman modularity score [12] of 0.90, suggesting that the partition with the highest Newman modularity could contain many nodes not belonging to modules.

We analyze the largest component of the FYI network, containing 778 nodes. We run the MSM algorithm on this network with the same parameters as for the benchmark networks above. Figure 4(b) shows the FYI network with the modules found by MSM in different colors. The black nodes comprise the transition region, with 556 nodes that do not belong to any modules. We identify 21 modules. The largest module contains 58 nodes, and the smallest 14 nodes.

It is a common approach in the study of PPI networks to equate network modules in PPI networks with putative proteins complexes [1, 4]. This approach can be useful for identifying previously unknown complexes, as well as in assigning previously unknown function to proteins: if a particular protein can be grouped together with a set of other proteins, it can be assumed that it has similar properties or functions to those already known about the protein set.

We therefore compare the modules we identified with the protein complexes listed in the CYC2008 [15] dataset. Figure 4 shows our modules side-by-side with the CYC2008 complexes. For this comparison we projected the complexes on the network, including very small complexes with only two proteins and also complexes comprised partially of proteins that are not a part of our network. We find that large complexes such as the 19/22S regulator complex (far left in the figure) with its 17 protein and the cytoplasmic ribosomal small subunit complex (23 proteins, far right) are identified. Many smaller complexes such as the Cytoplasmic exosome complex with 9 proteins are almost completely identified (MSM finds 8 of the proteins). We observe 38.8% of the nodes do not belong to any CYC2008 complex, and thus indeed the gap score could be said to better capture the modularity of the network than the high score given by the Newman modularity. Of course, as the network is somewhat outdated, we cannot use the CYC2008 complexes as a reliable ground truth, but these results indicate that there is promise to our approach.

4   Outlook

We introduced NCC networks, discussed two scores to evaluate their modularity, and compared the behavior of several algorithms on the task of detecting modules in such networks. There are many avenues for further research. We are currently developing a new network score to overcome the disadvantages of the two scores we presented. From the perspective of algorithms, there are many other types of clustering algorithms that might be adapted to NCC networks. One such interesting class is that of methods to identify the densest subgraph (see e.g. [8]), where it could be possible to run the algorithm repeatedly until all modules are identified. The MSM algorithm can be further improved to avoid the pitfalls of a transition region with a high average degree, as seen in Experiment 3. Finally, our random graph model is quite simple. It would be interesting to apply the three algorithms we employed and the scoring function to a richer set of networks, including more real-world examples.

References

[1] Bader, G.D., Hogue, C.W.: An automated method for finding molecular complexes in large protein interaction networks. BMC Bioinformatics 4 (2003)

[2] Bagrow, J.P.: Communities and bottlenecks: Trees and treelike networks have high modularity. Phys. Rev. E 85, 066118 (2012)

[3] Bovier, A., Eckhoff, M., Gayrard, V., Klein, M.: Metastability and low lying spectra in reversible markov chains. Comm. Math. Phys. 228, 219–255 (2002)

[4] Brohée, S., Van Helden, J.: Evaluation of clustering algorithms for protein-protein interaction networks. BMC Bioinformatics 7, 488 (2006)

[5] Enright, A.J., Van Dongen, S., Ouzounis, C.A.: An efficient algorithm for large-scale detection of protein families. Nucleic Acids Research 30(7), 1575–1584 (2002)

[6] Feng, Z., Xu, X., Yuruk, N., Schweiger, T.A.J.: A novel similarity-based modularity function for graph partitioning. In: Song, I.-Y., Eder, J., Nguyen, T.M. (eds.) DaWaK 2007. LNCS, vol. 4654, pp. 385–396. Springer, Heidelberg (2007)

[7] Han, J.-D.J., Bertin, N., Hao, T., et al.: Evidence for dynamically organized modularity in the yeast protein-protein interaction network. Nature 430, 88–93 (2004)

[8] Khuller, S., Saha, B.: On finding dense subgraphs. In: Albers, S., Marchetti-Spaccamela, A., Matias, Y., Nikoletseas, S., Thomas, W. (eds.) ICALP 2009, Part I. LNCS, vol. 5555, pp. 597–608. Springer, Heidelberg (2009)

[9] King, A.D., Pržulj, N., Jurisica, I.: Protein complex prediction via cost-based clustering. Bioinformatics 20(17), 3013–3020 (2004)

[10] Krause, R., von Mering, C., Bork, P., Dandekar, T.: Shared components of protein complexes–versatile building blocks or biochemical artefacts? BioEssays 26(12), 1333–1343 (2004)

[11] Luxburg, U.: A tutorial on spectral clustering. Statistics and Computing 17(4), 395–416 (2007)

[12] Newman, M.E.J., Girvan, M.: Finding and evaluating community structure in networks. Pattern Recognition Letters 69(2), 413–421 (2004)

[13] Nicosia, V., Mangioni, G., Carchiolo, V., Malgeri, M.: Extending the definition of modularity to directed graphs with overlapping communities. Journal of Statistical Mechanics: Theory and Experiment 2009, 22 (2008)

[14] Palla, G., et al.: Uncovering the overlapping community structure of complex networks in nature and society. Nature 435(7043), 1–10 (2005)

[15] Pu, S., Wong, J., Turner, B., Cho, E., Wodak, S.J.: Up-to-date catalogues of yeast protein complexes. Nucleic Acids Research 37(3), 825–831 (2009)

[16] Rand, W.M.: Objective criteria for the evaluation of clustering methods. Journal of the American Statistical Association 66(336), 846–850 (1971)

[17] Santo, F.: Community detection in graphs. Physics Reports 486(3-5), 75–174 (2010) ISSN 0370-1573

[18] Santos, J.M., Embrechts, M.: On the use of the Adjusted Rand Index as a metric for evaluating supervised classification. In: Alippi, C., Polycarpou, M., Panayiotou, C., Ellinas, G. (eds.) ICANN 2009, Part II. LNCS, vol. 5769, pp. 175–184. Springer, Heidelberg (2009)

[19] Sarich, M., Schuette, C., Vanden-Eijnden, E.: Optimal fuzzy aggregation of networks. Multiscale Modeling and Simulation 8(4), 1535–1561 (2010)

[20] Sarich, M., Djurdjevac, N., Bruckner, S., Conrad, T.O.F., Schütte, C.: Modularity revisited: A novel dynamics-based concept for decomposing complex networks. To Appear, Journal of Computational Dynamics (2012), http://publications.mi.fu-berlin.de/1127/

[21] Satuluri, V., Parthasarathy, S., Ucar, D.: Markov clustering of protein interaction networks with improved balance and scalability. In: Proceedings of the First ACM International Conference on Bioinformatics and Computational Biology, BCB 2010, pp. 247–256. ACM (2010)

[22] Sawardecker, E.N., Sales-Pardo, M., Amaral, L.A.N.: Detection of node group membership in networks with group overlap. EPJ B 67, 277 (2009)

[23] Schütte, C., Sarich, M.: Metastability and Markov State Models in Molecular Dynamics. Submitted to Courant Lecture Notes (2013)

[24] Sprinzak, E., Altuvia, Y., Margalit, H.: Characterization and prediction of protein-protein interactions within and between complexes. PNAS 103(40), 14718–14723 (2006)

[25] Weber, M., Rungsarityotin, W., Schliep, A.: Perron cluster analysis and its connection to graph partitioning for noisy data. ZIB Report, 04-39 (2004)

[26] Xu, X., Yuruk, N., Feng, Z., Schweiger, T.A.J.: SCAN: a structural clustering algorithm for networks. In: Proceedings of the 13th ACM SIGKDD International Conference on Knowledge Discovery and Data Mining, KDD 2007, pp. 824–833. ACM, New York (2007)

[27] Yu, L., Gao, L., Sun, P.G.: A hybrid clustering algorithm for identifying modules in protein protein interaction networks. Int. J. Data Min. Bioinformatics 4(5), 600–615 (2010)

[28] Zhang, S., Wang, R., Zhang, X.: Identification of overlapping community structure in complex networks using fuzzy cc-means clustering. Physica A: Statistical Mechanics and its Applications 374(1), 483–490 (2007)

Telling Stories Fast
Via Linear-Time Delay Pitch Enumeration

Michele Borassi[1,2], Pierluigi Crescenzi[3], Vincent Lacroix[4],
Andrea Marino[3], Marie-France Sagot[4], and Paulo Vieira Milreu[4]

[1] Scuola Normale Superiore, 56126 Pisa, Italy
[2] Università di Pisa, Dipartimento di Matematica, 56127 Pisa, Italy
[3] Università di Firenze, Dipartimento di Sistemi e Informatica, 50134 Firenze, Italy
[4] Inria Rhône-Alpes & Université de Lyon, F-69000 Lyon; Université Lyon 1; CNRS,
UMR5558, Laboratoire de Biométrie et Biologie Évolutive,
69622 Villeurbanne, France

Abstract. This paper presents a linear-time delay algorithm for enumerating all directed acyclic subgraphs of a directed graph $G(V, E)$ that have their sources and targets included in two subsets S and T of V, respectively. From these subgraphs, called pitches, the maximal ones, called stories, may be extracted in a dramatically more efficient way in relation to a previous story telling algorithm. The improvement may even increase if a pruning technique is further applied that avoids generating many pitches which have no chance to lead to a story. We experimentally demonstrate these statements by making use of a quite large dataset of real metabolic pathways and networks.

1 Introduction

Directed graphs are a widely used model in computational biology, notably to represent metabolism, which is the set of chemical transformations that sustain life. If an organism is exposed to a given condition (for instance, some kind of stress), the vertices of the directed graph may be colored depending on whether the quantity of the chemical the vertex represents changed (one color, say black) or remained the same (another color, say white) in relation to what may be defined as the organism's "normal state". Data such as these may be obtained through a technique called metabolomics [9] whose need for analytical methods is giving rise to new research topics. One question of interest then is to understand which subparts of the graph are affected by the condition change. One biologically pertinent definition for such subparts is as follows [7]: a maximal directed acyclic subgraph whose sets of sources and targets are blacks (note that black vertices may also be internal, that is neither sources nor targets, but white vertices can only be internal). In [1], these subgraphs have been called *metabolic stories*, or *stories* for short. Stories are a novel object for the analysis of metabolomics data, but we believe that they may also be useful in other domains. In this paper, we are interested in efficiently enumerating all the stories included in a directed graph.

V. Bonifaci et al. (Eds.): SEA 2013, LNCS 7933, pp. 200–211, 2013.

Enumerating maximal directed acyclic subgraphs of a given directed graph G, without any constraint on their sources and targets, is equivalent to enumerating all feedback arc sets of G, which is itself a classical problem in computer science. An elegant polynomial-time delay algorithm for solving this problem was proposed by Schwikowski and Speckenmeyer [2]. In [1], however, it was shown that the constraint on the sets of sources and targets is enough to drastically change the nature of the problem. Although the complexity of enumerating stories remains open, in [1] the authors proposed an algorithm that is able to go to completion for small enough graphs, and that can be used in a randomized fashion in the case of larger graphs, in order to produce a large sample of stories (as far as we know, this is the only known algorithm for enumerating stories). This algorithm is based on the notion of *pitch*, which is defined as a story without the maximality constraint, and on the following fact: any permutation π of the vertices of G can be transformed in polynomial time to a pitch P_π so that, for any story $\mathcal{S}$, there exists a permutation π of the vertices in G such that P_π can be "completed" in polynomial time in order to obtain $\mathcal{S}$. The algorithm for enumerating stories then proceeds by enumerating all permutations, transforming each of them into the corresponding pitch, and completing this pitch into a story. Unfortunately, this algorithm, called GOBBOLINO, is not polynomial-time delay, that is, the time between the generation of two distinct stories can be exponential in the number of vertices of G (for definitions concerning enumeration algorithms and complexity we refer the reader to the seminal paper [3]). For this reason, in [1] a randomized implementation of the algorithm has been suggested, which simply generates permutations uniformly at random: a biological application of GOBBOLINO and of its randomized version is described in [6].

The main contribution of this paper is twofold. From a theoretical point of view, we show that pitches can be enumerated in linear-time delay. In particular, we show how pitches can be sorted in a rooted tree $\mathcal{T}$ and how a depth-first search of $\mathcal{T}$ can be performed while ensuring the linear-time delay constraint, by applying the so-called *reverse search* technique [4]. From a practical point of view, we propose a new algorithm for enumerating stories, called TOUCHE, which is based on the linear-time delay pitch enumeration and on the pitch completion mechanism introduced in [1]. In particular, we first show how the depth-first search of $\mathcal{T}$ can be made more efficient by using a pruning technique which allows us to avoid visiting parts of the tree that certainly do not contain any story, and we then experimentally compare the GOBBOLINO algorithm with the TOUCHE algorithm, on a large dataset of metabolic networks. Our experiments show that TOUCHE always significantly outperforms GOBBOLINO, and that it is able to enumerate *all* stories in the case of bigger networks for which GOBBOLINO is not even able to produce a significant fraction of them.

1.1 Preliminaries

Let $G(V, E, S, T)$ be a directed graph, where V is the set of all vertices of G, E the set of arcs, and S and T two subsets of V. A vertex u is said to be a *source* if its out-degree is greater than 0 and its in-degree is 0, and it is said to be a

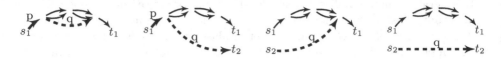

Fig. 1. The visualization of a pitch and a child obtained by adding the dashed path q. The path p is outlined when it is not empty.

target if its in-degree is greater than 0 and its out-degree is 0. A *pitch* P of G is a set of arcs $E' \subseteq E$, such that the subgraph $G' = (V', E')$ of G, where $V' \subseteq V$ is the set of vertices of G having at least one out-going or in-coming arc in E', is acyclic and for each vertex $w \in V' - S$, w is not a source in G', and for each vertex $w \in V' - T$, w is not a target in G'. We say that a pitch P is a story if it is maximal. A vertex w is said to belong to P if it belongs to V'. A path p is simple by definition and is denoted by $p_0, \ldots, p_{|p|}$. We refer to a path p by its natural sequence of vertices or set of edges. We will assume without loss of generality that, for each vertex v, there is a path from a source to v and from v to a target (otherwise we may remove v from the graph). The vertices in S and T are said to be black, while the vertices in $(V - S) - T$ are said to be white. It is worth observing that, besides the fact that a pitch may contain a subset of the black vertices instead of all of them, we work in this paper with a generalization of the definition of pitch introduced in [1] in the sense that here, instead of considering one set of black vertices, we distinguish between black source vertices (they form the set S) and black target vertices (they form the set T). The problem treated in [1] corresponds to the case in which all black vertices are both in S and in T. Finally, we refer to $|V| + |E|$ as the size of the graph $|G|$.

2   Enumerating Pitches

In order to enumerate all the pitches contained in a graph $G = (V, E, S, T)$, we first sort them in a rooted tree $\mathcal{T}$, and then we perform a depth-first search of $\mathcal{T}$. In order to construct $\mathcal{T}$, we introduce an appropriately defined child relationship, such that, for every pitch Q, there exists one and only one pitch P such that Q is a child of P: P is said to be *the father* of Q, and it can be computed starting from Q via a linear-time computable function **father**. A child of a pitch P is always obtained by attaching to P a path q "outside P", that is, such that each internal vertex of q is outside P and each arc of q is not in P (see Fig. 1). We will also impose that in P there is a (possibly empty) path p such that each path in $P \cup q$ from a source to a target is in P or starts by pq. In the following, we will associate to the i-th child of a pitch P the two corresponding paths p_i and q_i (which are uniquely determined).

In order to visit the search tree $\mathcal{T}$ in a depth-first fashion without storing its nodes in a stack (and thus saving space), we also define a linear-time computable function **next**, that allows us to jump from a child Q_i of a pitch P, corresponding to the paths (p_i, q_i), to the next child Q_{i+1}: in particular, $\texttt{next}(P, p_i, q_i) =$

(p_{i+1}, q_{i+1}). If Q_i is the last child of P, then the function returns the empty pair. Moreover, if the function **next** is invoked with arguments P and the two paths p and q that make P the child of its father, then it returns the pairs (p_1, q_1) corresponding to the first child of P (if it exists). By using the **father** and the **next** functions, we can then implement a depth-first search of the pitch tree $\mathcal{T}$ as shown in Fig. 2, where the dotted arcs denote a child relation. The rest of this section is devoted to the definition of the two functions **father** and **next** and to the proof of their time and space complexity.

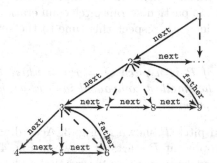

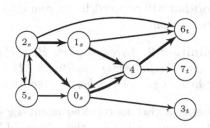

Fig. 2. The depth-first visit of the pitch tree

Fig. 3. An example of a pitch (in bold): the subscript indicates if a vertex is a source or a target

2.1 The father Function

The idea behind the definition of the child relationship (and thus of the pitch tree $\mathcal{T}$) is to build all pitches (starting from the empty one, which is the root of $\mathcal{T}$) by repeatedly adding paths from a source in S to a target in T (in short, *st*-paths). Since we want $\mathcal{T}$ to be a tree, we have to specify in which order *st*-paths are added, so that each pitch has a unique father. To this aim, we fix an arbitrary ordering of V, we lexicographically sort *st*-paths (which are ordered sequences of vertices), and impose that each new (explicitly or implicitly) added *st*-path is bigger than all *st*-paths included in the current pitch. In order to make this approach work, we need to overcome some problems, as shown in the following example.

Example 1. Let us consider the bold pitch shown in Fig. 3, which is formed by the *st*-paths $(0, 4, 6)$, $(1, 4, 6)$, $(2, 0, 4, 6)$, and $(2, 1, 4, 6)$ (listed in lexicographic order). If we allow any *st*-path bigger than $(2, 1, 4, 6)$ to be added to the pitch, then some problems might arise.

1. If we add $(2, 1, 4, 7)$, we are implicitly adding $(0, 4, 7)$, which is smaller than $(2, 1, 4, 6)$, contradicting the uniqueness of the father relationship, since the same pitch will also be reached through the explicit addition of $(0, 4, 7)$.
2. Sometimes it is necessary to implicitly add an *st*-path "much bigger" than the ones already present in the current pitch: adding $(2, 5, 0, 4, 6)$, we also add $(5, 0, 4, 6)$, thus eliminating the possibility of subsequently adding $(2, 6)$. Conversely, if we add first $(2, 6)$, then $(2, 5, 0, 4, 6)$ cannot be added. Therefore a pitch containing both $(2, 6)$ and $(2, 5, 0, 4, 6)$ will be missed.

In order to deal with the second problem, we need to specify that the st-paths that can be added to a pitch satisfy the following definition.

Definition 1. *A path $(p_0, \ldots, p_k)$ of a pitch P is a component of P if it satisfies the following conditions: (1) $p_0 \in S$ is a source and $p_k \in T$ is a target, and (2) p_0 is not reachable in P from a smaller source in P (i.e., there is no $s \in P \cap S$ such that p_0 is reachable in P from s).*

Note that every arc in a pitch belongs to a component. Hence, a pitch can be specified by listing the set of its components: in particular, our pitch enumeration algorithm will proceed in lexicographic order with respect this time to the set of components.

Definition 2. *Given a pitch P, a child of P is a pitch $Q = P \cup c$ where c satisfies the following conditions: (1) c is the smallest component in Q which is not in P; and (2) c is bigger than any component in P.*

Example 2. Let us consider again the bold pitch P shown in Fig. 3. According to the above definition, the subtree of $\mathcal{T}$ rooted at P starts as shown in Fig. 4, where the labels of the edges denote the added component c. Observe that adding a component can cause the implicit addition of other components: for example, adding $(5, 2, 0, 4, 6)$ results also in adding $(5, 2, 1, 4, 6)$, which is however greater than $(5, 2, 0, 4, 6)$. Note also that $(2, 1, 4, 7)$ cannot be added to P, since it would not be the smallest new component not in P.

The following lemma shows that the child relationship defined above sorts all pitches in a tree with root the empty pitch.

Lemma 1. *Every pitch Q, apart from the empty one, has a unique father P.*

Proof. We start with the uniqueness: let us suppose $Q = P \cup c$ where P and c satisfy the conditions of Def. 2. We now show how c can be split into three paths, $c = pqp'$, where p is contained in P, q is "outside" of P (in the sense that it has no arc and no internal vertex in P), and p' is the remaining part of c (we denote by v and w respectively the start and the end of q). By the first condition of Def. 2, p' must be the smallest path in P from w to a target vertex in $T \cap P$: hence, $P = Q - q$ (see Fig. 1). This means that each internal vertex

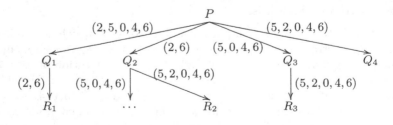

Fig. 4. A fragment of a pitch tree

in q has in-degree and out-degree equal to 1 in Q. Moreover, p is the only path from a source in $S \cap P$ to v: indeed, a smaller one would contradict the first condition of Def. 2, while a bigger one would contradict the second condition. This means that pq is an initial segment of the smallest component of Q which is not in P and that no vertex x in pq before w verifies any of the following conditions: (1) x has at least two incoming edges in Q; (2) x is a source smaller than the first vertex of p; and (3) x has no outgoing edge in Q. Since $Q - q$ is a pitch, w has to satisfy one of the conditions above and this characterizes w. We now need to characterize v. Since no internal vertex in q is in P, v must be the first vertex before w verifying one of the following conditions: (a) v has at least two outgoing edges in Q; (b) v is a target; and (c) v has no incoming edge in Q. Since both v and w are uniquely determined, we have that also q is uniquely determined: hence, the uniqueness is proved. In order to prove the existence, it is enough to show that $P \cup c$ is a child of P, where $c = pqp'$ and p, q and p' are the paths determined by the previous conditions. We have that c is bigger than every component in P because pq is a prefix of the last component of Q and q is outside P. Moreover, c satisfies the first condition of Def. 2 because each component of Q not in P contains an arc in q. This means that it contains the whole pq because of the conditions on v and w. $\qquad\square$

From the proof of the previous lemma, the next result immediately follows.

Corollary 1. *It is possible to find in linear time the father of a pitch.*

Instead, the next consequence of the above lemma will yield a more "algorithmic" definition of the child relationship (see Fig. 1).

Corollary 2. *Let P and Q be two pitches. Q is a child of P if and only if $Q = P \cup pq$ where:*

- *pq is bigger than the last path in P;*
- *$p \subseteq P$;*
- *q is "outside" $P \cup T$, i.e. q is disjoint from P and $q_1, \ldots, q_{|q|-1}$ are not in $P \cup T$;*
- *no vertex in p satisfies Conditions 1-3 in the proof of Lemma 1.*

Moreover, p and q are uniquely determined.

Proof. If Q is a child of P, the proof of Lemma 1 implies all the conditions required and that p and q are uniquely determined. For the other direction, let p' be the first path in P from the end of q to a target. The path pqp' satisfies all conditions required in the child definition. $\qquad\square$

2.2   The next Function

As we already said before, the **next** function should allow us to compute the first child Q_1 of P (if it exists) and the next child Q_{k+1} from the child Q_k (if it exists). In order to define this function well, we first prove the following result.

Lemma 2. *For any pitch P, the function*

$$\Phi_P : \{ children\ of\ P \} \to \{ paths\ in\ G\ satisfying\ Corollary\ 2 \}$$
$$P \cup q \qquad \mapsto \qquad\qquad (p, q) \tag{1}$$

is an order-preserving bijection (pitches are sorted lexicographically as sets of components).

Proof. The function is well defined because of Corollary 2 and it is a bijection because the inverse function is $\Psi_P(p, q) := P \cup q$. It preserves the order because pq is a prefix of the first component Q not in P and a path satisfying Corollary 2 is never a prefix of another component (by the 3rd condition in Corollary 2). □

The function **next** is then defined as follows. Given a graph $G = (V, E, S, T)$, a pitch P, and a path r of G starting from a source which is a prefix of the last path in P or is bigger than every component of P, the function **next** returns the smallest path pq such that P and pq satisfy all conditions in Corollary 2, and pq is strictly bigger than r.

Theorem 1. *The* **next** *function is computable in time* $\mathcal{O}(|G|)$.

2.3   Complexity Analysis

The pitch enumeration algorithm is based on a depth-first search of the pitch tree $\mathcal{T}$, which uses the two functions **father** and **next**. Because of Corollary 1 and Theorem 1, every node can be visited in linear time. By the well-known *alternative output* technique [5, Theorem 1], it is possible to output a solution every time two nodes are visited, to obtain linear delay. To do so, all solutions with even depth in $\mathcal{T}$ must be output as soon as they are found, while solutions with odd depth must be output before computing their father. Since the depth changes by 1 every time a node is visited, the previous condition is accomplished. It is also easy to show that only a linear amount of space to store G, P and $r = (p, q)$ is required.

3   Enumerating Stories

In order to enumerate all the stories contained in a graph $G = (V, E, S, T)$, the approach described in [1] was based on generating all permutations of the vertices, and on cleaning and completing the corresponding DAG in order to turn it into a story. By using the results in the previous section instead, all the stories can be enumerated by enumerating all the pitches and outputting only the maximal ones. However, even if, in many real cases, this approach already outperforms the method proposed in [1], the method itself fails in enumerating, within a reasonable amount of time, all the stories in the case of large graphs. Indeed, usually an exponential number of pitches that are not stories can be generated. In order to avoid the computation of many useless pitches, we will now show how very often it is possible to verify *a priori* whether a pitch can lead to a story, thus performing a *pruning* of the pitch tree $\mathcal{T}$. In order to explain the pruning process, we introduce the following definitions.

Definition 3. *A pitch is a successor of $P \cup r$ if it is a descendant of P bigger than $P \cup r$ (pitches are sorted lexicographically as sets of components).*

Definition 4. *Given a pitch P and a path r, a vertex $v \in V$ is (P,r)-open if it belongs to a successor of $P \cup r$. A vertex is (P,r)-closed if it is not (P,r)-open.*

For example, the vertex 3 of Fig. 3 is $(P, (2,5,0,4,6))$-closed.

Lemma 3. *Let G be strongly connected and let P be a non-empty pitch. If $(S \cup T) - P \neq \emptyset$, then P is not a story.*

Proof. Assume there exists $s \in S - P$ and let p be a shortest path from s to any vertex in P (this path exists since G is strongly connected). Then $P \cup p$ is a pitch strictly containing P: this proves that P is not a story. Analogously, we can prove that if there exists $t \in T - P$, then P is not a story. □

Corollary 3. *Given a pitch P and a path r, if a source or a target is (P,r)-closed, no successor of $P \cup r$ is a story.*

For example, in the case of the fragment of a pitch tree shown in Fig. 4, we have that the subtrees rooted at Q_2, Q_3, and Q_4 do not contain any story (since the vertex 3 of Fig. 3 is $(P, (2,5,0,4,6))$-closed). Actually, vertex 3 is closed with respect to the empty pitch (which is the root of $\mathcal{T}$), and the path $(0,4,6)$, hence P will not even be reached since the pruning will be effective on the very first branch from $\emptyset$ to P. We may now state the main theorem used to prune the tree of all pitches.

Theorem 2. *Given a pitch P and a path r such that there exists a story which is a successor of $P \cup r$, there is no path p that verifies the following conditions: (1) $P \cup p$ is a pitch; (2) the last vertex of p is not in r; and (3) p is "outside" any successor of $P \cup r$, that is, no arc of p is in a successor of $P \cup r$ and all internal vertices of p are (P,r)-closed.*

Proof. Let Q be a successor of $P \cup r$ which is a story (Q exists by hypothesis). By the third condition on p, it follows that p is outside Q. Moreover, $Q \cup p$ is not a pitch (since Q is maximal): hence, there must be a path q in Q from the last vertex of p to the first one. By the first condition on p, it follows that q is not in P (since $P \cup p$ is acyclic). Consider now a path in P from a source to the last vertex of p (this path exists because of the second condition on p) and link this path to q: this can be extended to a component of Q. By the second condition on p, this component is smaller than r: this is a contradiction because this component is not in P (since it contains q which is not in P). □

The above theorem gives us a powerful tool to prune the tree of all pitches. Indeed, given a pitch P and a path r, if we can find a path p satisfying the three conditions of the theorem, we can then conclude that there is no story which is a successor of $P \cup r$. However, in order to apply this pruning criterion, we should be able to compute the set of vertices which are (P,r)-closed (or, equivalently,

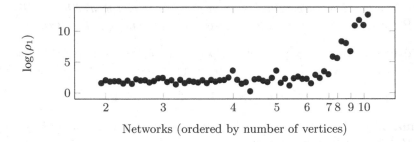

Fig. 5. Ratio between time consumed by GOBBOLINO and TOUCHE to compute all stories in input graphs with 2 to 10 vertices (logarithmic scale)

the set of vertices which are (P, r)-open). So far, we have not been able to solve this latter problem (indeed, we conjecture it is NP-hard), but we can efficiently "approximate from above" the set of (P, r)-open vertices, that is, we can compute in linear time a superset of this set, which in practice is not too much bigger. Thanks to this result and to Theorem 2, we obtain an algorithm that decides if it is possible to prune the pitch tree in time $\mathcal{O}(|V||G|)$. The efficiency of this pruning process will be experimentally validated in the next section.

4   Experimental Results

In order to evaluate the efficiency of the new algorithm for the enumeration of stories, called TOUCHE, we performed three experiments, two of them comparing with the previous algorithm proposed in [1], called GOBBOLINO, and the third one to evaluate the effect of the pruning approach (the entire dataset, the Java code, and the detailed experimental results are available starting from `amici.dsi.unifi.it/lasagne/`).

Enumerating All Stories

Our first experiment consisted in the enumeration of the whole set of stories using both GOBBOLINO and TOUCHE (with the pruning approach implemented) and the comparison of their running time. GOBBOLINO is guaranteed to find all stories only if all permutation orderings of the vertices of the input graph are inspected, which limits its application to small input graphs. In [6], GOBBOLINO was applied in order to automatically recover the so-called *metabolic pathways* in a dataset consisting of 69 such pathways, among which 62 represented an input graph with no more than 10 vertices. For this subset, we obtained the results summarized in Figure 5. Let $t_G(G)$ (resp., $t_T(G)$) denote the time consumed by GOBBOLINO (resp., TOUCHE) to compute all stories in the graph G, and let $\rho_1(G) = t_G(G)/t_T(G)$. In the figure we show the logarithm of ρ_1 for all the 62 graphs, ordered in increasing order with respect to their number of vertices. As it can be seen from the figure, TOUCHE performs better than GOBBOLINO

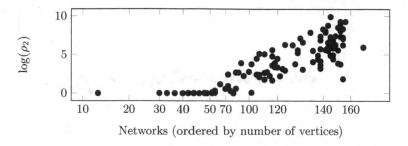

Fig. 6. Ratio between number of stories produced by GOBBOLINO and TOUCHE after 1 minute of computation (logarithmic scale)

for the whole dataset (even if the size of the instances is very small). Clearly, GOBBOLINO consumes more time as the size of the input increases, since it has to check all orderings of the vertices. For inputs with up to 7 vertices, both algorithms finish the enumeration process in less than 1 second. For the three inputs of size 8, GOBBOLINO consumes between 3.8 and 4.7 seconds, while TOUCHE never uses more than 0.05 seconds of computation. The result is even more impressive when we look at the inputs of size 9 and 10. GOBBOLINO takes around 1 minute for the three inputs of size 9 and more than 15 minutes for the input with 10 vertices, while TOUCHE finishes processing them in no more than 0.14 seconds. Indeed, the figure suggests that the ratio ρ_1 increases as an exponential with respect to the number of vertices, in the case of networks with at least 6 vertices (in the case of smaller networks, file management overhead has to be taken into account).

Sampling Stories

One approach used in [6] in order to apply GOBBOLINO for bigger inputs was to use random permutations of the orderings of the vertices to sample the space of pitches and, therefore, the space of solutions (*i.e.*, stories). Our second experiment consisted in comparing this randomized approach of GOBBOLINO to TOUCHE (with the pruning approach implemented), giving a fixed amount of time for both algorithms (1 minute, in our experiment) and checking how many stories each method produced. For this experiment, we selected 118 metabolic networks of various sizes. The dataset may be divided as follows: 8 networks (with size greater than or equal to 10) come from the same metabolic pathways considered in the first experiment; 4 networks are inputs for some experiments also performed in the context of [6] and for which the set of black vertices came from biological experiments; the remaining 106 were metabolic networks downloaded from the public database MetExplore ([8]) and with a random set of black vertices (5% of the vertices of the graph were considered to be black). For this dataset, we obtained the results summarized in Figure 6. Let $s_G(G)$ (resp., $s_T(G)$) denote the number of stories produced by GOBBOLINO (resp., TOUCHE) with input the graph G after 1 minute of computation, and let $\rho_2(G) = s_T(G)/s_G(G)$.

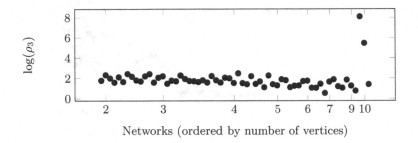

Networks (ordered by number of vertices)

Fig. 7. Ratio between time consumed by TOUCHE to compute all stories without and with pruning

In the figure we show the logarithm of ρ_2 for all the 118 graphs, ordered in increasing order with respect to their number of vertices. The first outcome of this experiment is that TOUCHE always computed a number of stories bigger than or equal to the number of stories computed by GOBBOLINO. For 19 of them (mostly small size ones), the number is the same, but the time spent by TOUCHE is smaller than the limit of 1 minute, which indicates that the number of stories computed is in fact the total number of stories: this highlights another advantage of TOUCHE over the randomized version of GOBBOLINO, that continues exploring permutations of pitches even if it has already computed the whole set of stories. Moreover, note that TOUCHE produces the entire set of stories in the case of 10 other networks. In the case of bigger instances, the number of stories found by TOUCHE could be up to 950 times the number of stories found by GOBBOLINO. Indeed, the figure suggests that the ratio ρ_2 increases as an exponential with respect to the number of vertices. The extreme case is the YERYP364 network for which GOBBOLINO found 4 stories while TOUCHE found 3815 stories: this result strongly suggests that the correspondence between permutation of the vertices and stories is highly biased and that there might be stories corresponding to very few permutations and hence unlikely to be produced by GOBBOLINO.

Evaluating the Pruning Methods

Our third experiment was designed in order to evaluate how effective is the pruning approach described in the previous section. By referring to the dataset used in the first experiment, for each network we collected the running time of TOUCHE with and without the pruning. The results are summarized in Figure 7. Let $t_{T,n}(G)$ (resp., $t_{T,y}(G)$) denote the time consumed by TOUCHE without (resp., with) pruning to compute all stories in the graph G, and let $\rho_3(G) = t_{T,n}(G)/t_{T,y}(G)$. In the figure we show the logarithm of ρ_3 for all the networks, ordered in increasing order with respect to their number of vertices. As it can be seen from the figure, TOUCHE with pruning always performs better than TOUCHE without pruning (even if the size of the instances is very small). The improvement seems to remain constant, even though in the case of two networks (that is, PRPP-PWY and THREOCAT2-PWY) it is quite impressive: the pruning improves the computational time by a factor of 275 in the first case and 43 in the second case.

Finally, we repeat this experiment in the case of two further networks analyzed in [6]: the first contains 10 vertices (8 black) and 222 stories and TOUCHE with pruning computed them about 5 times faster, the second contains 35 vertices (21 black) and TOUCHE with pruning computed its 3,934,160 stories in about three hours while TOUCHE without pruning did not finish after one day.

5 Conclusion

We presented a linear-time delay enumeration algorithm for pitches, that allowed us to enumerate all stories more efficiently than the previous known method [1]. The main question left open by our paper is to determine the complexity of the story enumeration problem.

Acknowledgements. The research leading to these results was funded by: the European Research Council under the European Community's Seventh Framework Programme (FP7 / 2007-2013) / ERC grant agreement n [247073]10; the French project ANR MIRI BLAN08-1335497; and the ANR funded LabEx ECOFECT.

References

1. Acuña, V., Birmelé, E., Cottret, L., Crescenzi, P., Jourdan, F., Lacroix, V., Marchetti-Spaccamela, A., Marino, A., Milreu, P.V., Sagot, M.F., Stougie, L.: Telling stories: Enumerating maximal directed acyclic graphs with a constrained set of sources and targets. Theor. Comput. Sci. 457, 1–9 (2012)
2. Schwikowski, B., Speckenmeyer, E.: On enumerating all minimal solutions of feedback problems. Discrete Applied Mathematics 117(1-3), 253–265 (2002)
3. Johnson, D.S., Papadimitriou, C.H., Yannakakis, M.: On Generating All Maximal Independent Sets. Inf. Process. Lett. 27(3), 119–123 (1988)
4. Avis, D., Fukuda, K.: Reverse Search for Enumeration. Discrete Applied Mathematics 65, 21–46 (1993)
5. Uno, T.: Two general methods to reduce delay and change of enumeration algorithms. NII Technical Report (2003)
6. Milreu, P.V.: Enumerating Functional Substructures of Genome-Scale Metabolic Networks: Stories, Precursors and Organisations. PhD thesis, Université Claude Bernard, Lyon 1, France (2012)
7. Milreu, P.V., Acuña, V., Birmelé, E., Borassi, M., Cottret, L., Junot, C., Klein, C., Marchetti-Spaccamela, A., Marino, A., Stougie, L., Jourdan, F., Lacroix, V., Crescenzi, P., Sagot, M.-F.: Metabolic stories: exploring all possible scenarios for metabolomics data analysi (in preparation, 2013)
8. Cottret, L., Wildridge, D., Vinson, F., Barrett, M.P., Charles, H., Sagot, M.F., Jourdan, F.: Metexplore: a web server to link metabolomic experiments and genome-scale metabolic networks. Nucleic Acids Research 38(Web-Server-Issue), 132–137 (2010)
9. Johnson, C.H., Gonzalez, F.J.: Challenges and opportunities of metabolomics. Journal of Cellular Physiology 227(8), 2975–2981 (2012)

Undercover Branching*

Timo Berthold and Ambros M. Gleixner

Zuse Institute Berlin, Takustr. 7, 14195 Berlin, Germany
{berthold,gleixner}@zib.de

Abstract. In this paper, we present a new branching strategy for non-convex MINLP that aims at driving the created subproblems towards linearity. It exploits the structure of a *minimum cover* of an MINLP, a smallest set of variables that, when fixed, render the remaining system linear: whenever possible, branching candidates in the cover are preferred.

Unlike most branching strategies for MINLP, Undercover branching is not an extension of an existing MIP branching rule. It explicitly regards the nonlinearity of the problem while branching on integer variables with a fractional relaxation solution. Undercover branching can be naturally combined with any variable-based branching rule.

We present computational results on a test set of general MINLPs from MINLPLib, using the new strategy in combination with reliability branching and pseudocost branching. The computational cost of Undercover branching itself proves negligible. While it turns out that it can influence the variable selection only on a smaller set of instances, for those that are affected, significant improvements in performance are achieved.

1 Introduction

State-of-the-art solvers for generic mixed integer linear programs (MIPs) and mixed integer nonlinear programs (MINLPs) are based on the branch-and-bound paradigm [1]. The question of how to split a given MIP or MINLP into subproblems, commonly referred to as the *branching* step, lies at the heart of any branch-and-bound algorithm. Its main purpose is to improve the *dual bound* by, e.g., eliminating fractionality of the integer variables and, for MINLP, reducing the convexification gap between the nonconvex constraint functions and the relaxation. In MIP solving, typically an LP relaxation is solved for the *bounding* step. For MINLP, although an NLP relaxation is a natural choice, most state-of-the-art solvers also rely on an LP relaxation.

The branching rule is one of the components with highest impact on the overall performance of MIP solvers [2,3]. Consequently, the literature has seen many publications on efficient branching rules, which will be reviewed in the next paragraphs. For MINLP, up to now, research has mainly focused on adopting MIP branching rules [4,5].

---

* The authors gratefully acknowledge the support of the DFG Research Center MATHEON *Mathematics for key technologies* in Berlin and the Berlin Mathematical School.

V. Bonifaci et al. (Eds.): SEA 2013, LNCS 7933, pp. 212–223, 2013.

In mixed integer programming, the most common methodology is variable-based branching (an exception being [6]), i.e., considering integer variables with a fractional LP solution value as *branching candidates*. State-of-the-art branching rules, sometimes also called variable selection heuristics, estimate the impact that splitting a variable's domain has on the dual bound and the solvability of the created subproblems. A very prominent approach is the usage of so-called *pseudocosts* [7], an estimate of the increase that branching on a variable has on the optimum of the LP relaxation.

In [8], it is shown that initializing pseudocosts by strong branching [9,10] is beneficial, an approach further refined in reliability branching [11]. Hybrid branching [12] combines reliability branching with VSIDS [13] and inference values [14], two common branching dichotomies in satisfiability testing and constraint programming, respectively. Methods that combine pseudocost and strong branching information can be considered to be the state-of-the-art for MIP solvers.

In recent years several publications have investigated new paradigms for variable-based branching schemes that show superior performance on important classes of hard MIPs. Kılınc et. al. [15] suggest to use conflict learning information for branching on 0-1 integer programs. To this end, they run a sampling phase of 500 branch-and-bound nodes during which they collect conflict constraints, restart the solution process, and prefer branching on variables that appear in short conflict constraints during the second phase.

Backdoor branching [16] goes one step further: it applies multiple restarts, attempting to find a good approximation of a *backdoor*. Here, a backdoor is a (preferably small) set of variables such that, whenever these variables get assigned integer values, solving an LP on the remaining variables gives a proof of feasibility or infeasibility. After each restart, the approximated backdoor is computed by solving a set covering problem. Branching is exclusively performed on backdoor variables until all of them are fixed. *Non-chimerical branching* [17] is a criterion to rule out candidates for strong branching which are not promising.

For nonconvex MINLP, it is possible that the LP relaxation is integral and cannot be strengthened further by gradient cuts (see Footnote 3), while some of the nonconvex constraints are still violated. In this case, spatial branching can be applied, i.e., branching on variables contained in violated nonconvex constraints, including continuous variables. Subsequently, the relaxation can be tightened in the created subproblems; thereby, the infeasible relaxation solution is cut off.

To select a branching variable for spatial branching, Tawarmalani and Sahinidis [18] suggest performing a so-called *violation transfer*. This estimates the impact of each variable on the problem by minimizing and maximizing a Lagrangian function over a neighborhood of the current relaxation solution when holding all other variables fixed. For a linear relaxation, this is similar to selecting variables with large reduced cost.

In [5], the concept of pseudocosts has been extended to continuous variables by investigating suitable counterparts for the violation of integrality, which is used in pseudocost formulas for MIP. Their computational analysis suggests that pseudocost-based branching is superior for hard MINLPs, while for easy instances and nonconvex NLPs it is outperformed by violation transfer or even simpler violation-based rules.

In this paper, we suggest a branching strategy that aims at driving the sub-problems towards linearity. To this end, *Undercover branching* restricts the set of branching candidates to a *minimum cover* [19] of an MINLP, i.e., a smallest set of variables that, when fixed, linearizes all constraints. It builds on the ideas of the Undercover heuristic [19,20], which computes feasible solutions for MINLPs by solving a sub-MIP defined via a minimum cover.

Whereas many branching rules are history-based and share heuristic components, Undercover branching exploits structural information of the problem in an exact manner. In the spirit of backdoor branching, it features a pre-selection rule for branching candidates: independent of the current subproblem, Under-cover branching globally separates a set of variables with a certain predicate from others. Consequently, it can be combined with any variable-based branching rule.

A major characteristic of Undercover branching is that it respects information on the nonlinearity of the problem already in the branching decisions for fractional integer variables, not only during spatial branching. From a computational point of view, Undercover branching has the benefit that it costs little additional time. In the way that we suggest, a minimum cover has to be computed only once in the beginning of the solution process. Our experiments show this to be computationally cheap in practice. In contrast to backdoor branching, Undercover branching does not require repeated restarts of the main solution procedure.

The remainder of the article is organized as follows. Section 2 states a formal definition of a minimum cover, explains how it can be computed, and analyzes minimum cover sizes of the test problems in MINLPLib [21]. In Section 3, we present the general idea and implementational details of the newly proposed branching strategy. In Section 4, we evaluate the applicability and the impact of Undercover branching on instances from MINLPLib. Finally, we discuss the results and give an outlook on future work in Section 5.

2   Covers of Mixed Integer Nonlinear Programs

The branching strategy investigated in this paper relies on the concept of a minimum cover, a structural feature of an MINLP that is a measure for its "grade of nonlinearity". This notion has been introduced in [19] and utilized for the design of a primal heuristic. In the following, we give a brief summary of the main results from [19,20].

Definition 1 (cover of an MINLP). *Let P be an MINLP of form*

$$
\begin{aligned}
\min \quad & c^\mathsf{T} x \\
\text{s.t.} \quad & g_k(x) \leqslant 0 && \textit{for } k = 1, \dots, m, \\
& \ell_i \leqslant x_i \leqslant u_i && \textit{for } i = 1, \dots, n, \\
& x_i \in \mathbb{Z} && \textit{for } i \in \mathcal{I},
\end{aligned}
\tag{1}
$$

where $c \in \mathbb{R}^n$, $g_k : \mathbb{R}^n \to \mathbb{R}$, $\ell_i \in \mathbb{R} \cup \{-\infty\}$, $u_i \in \mathbb{R} \cup \{+\infty\}$, $\ell_i < u_i$, *and* $\mathcal{I} \subseteq \{1, \ldots, n\}$.[1] *We call a set of variable indices* $\mathcal{C} \subseteq \{1, \ldots, n\}$ *a cover of the function* g_k *if and only if for all* $x^* \in [\ell, u]$ *the set*

$$\{(x, g_k(x)) : x \in [\ell, u], x_i = x_i^* \text{ for all } i \in \mathcal{C}\} \tag{2}$$

is an affine set intersected with $[\ell, u] \times \mathbb{R}$. *We call* $\mathcal{C}$ *a cover of* P *if and only if* $\mathcal{C}$ *is a cover of all constraint functions* $g_1, \ldots, g_m$.

Trivial examples of covers are the set of all variables or the set of all variables appearing in nonlinear terms. As will be shown at the end of this section, however, many instances of practical interest allow for significantly smaller covers. Minimum covers can be computed generically by solving a vertex covering problem. This is a crucial observation for exploiting them in an MINLP solver.

Definition 2 (co-occurrence graph). *Let* P *be an MINLP of form* (1) *with* $g_1, \ldots, g_m$ *twice continuously differentiable on the interior of* $[\ell, u]$. *We call* $G_P = (V_P, E_P)$ *the* co-occurrence graph *of* P *with node set* $V_P = \{1, \ldots, n\}$ *given by the variable indices of* P *and edge set*

$$E_P = \{ij : i, j \in V, \exists k \in \{1, \ldots, m\} : \frac{\partial^2}{\partial x_i \partial x_j} g_k(x) \not\equiv 0\},$$

i.e., an edge connects nodes i *and* j *if and only if the Hessian matrix of some constraint has a structurally nonzero entry* (i, j).

This leads to the following result:

Theorem 1. *Let* P *be an MINLP of form* (1) *with* $g_1, \ldots, g_m$ *twice continuously differentiable on the interior of* $[\ell, u]$. *Then* $\mathcal{C} \subseteq \{1, \ldots, n\}$ *is a cover of* P *if and only if it is a vertex cover of the co-occurrence graph* G_P.

Proof. See [20]. □

Since vertex covering is $\mathcal{NP}$-hard [22] and any graph can be interpreted as the co-occurrence graph of a suitably constructed MINLP, we obtain

Corollary 1. *Computing a minimum cover of an MINLP is* $\mathcal{NP}$-*hard.*

In practice, however, minimum covers can be computed rapidly by solving a binary programming formulation of the vertex covering problem with a state-of-the-art MIP solver as has been argued already in [20]. As an example, we have computed minimum covers for 255 instances[2] from MINLPLib [21] for the present paper. Using the MINLP solver SCIP 3.0 [23] and the expression interpreter CppAD 20120101.3 [24] for obtaining the sparsity patterns of the Hessians, the binary programs were all solved within the root node and took at most 0.2 seconds on the hardware described in Sec. 4.

---

[1] W.l.o.g. we may assume a linear objective, because for a nonlinear objective $f(x)$, we can always append a constraint $f(x) \leqslant x_0$ and minimize the auxiliary variable x_0.

[2] This excludes 18 instances that cannot be handled by SCIP 3.0, e.g., because they contain trigonometric functions: `blendgap`, `deb{6,7,8,9,10}`, `dosemin{2,3}d`, `prob10`, `var_con{5,10}`, `water{3,ful2,s,sbp,sym1,sym2}`, and `windfac`.

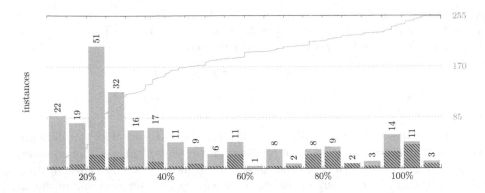

Fig. 1. Distribution of the sizes of a minimum cover relative to the total number of variables over 255 instances from MINLPLib. Numbers above the bars state how many instances fall in the corresponding 5% interval. Shaded bars indicate the proportion of minimum covers with integer variables only. The cumulative distribution function refers to the right-hand scale.

The distribution of the sizes of minimum covers is depicted in Fig. 1. One third of the instances allows for covers consisting of less than 14% of the variables and another third of the instances has covers with less than 36% of the variables. As indicated by the shaded bars, 65 instances have a minimum cover with only integer variables. For the vast majority of 163 instances it contains only continuous variables. The minimum covers for the remaining 27 instances are formed by continuous variables complemented by a small fraction of less than 1% integer variables.

To summarize, we observe that the majority of problems from MINLPLib features small covers. Note that this even holds for many instances that have almost all variables contained in nonlinear terms. The latter underlines that the size of a minimum cover valuably complements other measures of nonlinearity such as number of nonlinear nonzeros, constraints, or variables appearing in nonlinear terms.

3   Using MINLP Covers for Branching

Although MIP and MINLP are both $\mathcal{NP}$-hard, arguably MIPs are computationally easier than MINLPs. For MIP, it is possible to compute a relaxation solution in polynomial time that only drops the integrality requirements, but respects all constraint functions. For MINLP, solving a (nonconvex) NLP relaxation is already $\mathcal{NP}$-hard. Also, generic cutting plane algorithms, which contribute a lot to the practical success of MIP solvers, do not have a direct equivalent in MINLP. Of course, they can be used to strengthen a MIP relaxation, but they do not yield a finite algorithm. Last, but not least, considering today's state-of-the-art in optimization software, MIP codes have reached an impressive maturity and have become a standard industry tool, whereas MINLP software has just recently evolved and only few codes are available by now.

From this point of view it is an important observation that a cover of an MINLP presents a structure that turns an MINLP into a MIP for any assignment of the variables in the cover. Branching shrinks variables' domains, ideally fixes them, and therefore, branching on cover variables offers itself as a promising strategy to drive an MINLP towards linearity. If a pure branch-and-bound algorithm is applied without domain propagation techniques, the size of the cover corresponds to the minimum number of branching decisions that have to be taken before obtaining a linear subproblem. In particular, the following observation holds:

Lemma 1. *Let P be an MINLP of form* (1) *and $C \subseteq \mathcal{I}$ a cover of P with $\ell_i, u_i \in \mathbb{Z}$ for all $i \in C$. Then, P can be solved by solving a sequence of at most $\prod_{i \in C} (u_i - \ell_i + 1)$ MIPs.*

In the case of variables with infinite domain, i.e., continuous or unbounded integer variables, branching on cover variables does not necessarily enforce linearity in a bounded number of steps. Nevertheless, branching on such a variable, and thereby tightening its domain, is likely to produce better underestimators than, e.g., branching on an integer variable which is not even part of a nonlinear expression. Better underestimators lead to better relaxation bounds which lead to earlier pruning (or feasibility) of the created subproblems.

In particular, Undercover branching explicitly regards the nonlinearity of the problem also when branching on integer variables with a fractional relaxation solution.

We therefore suggest to use a branching strategy that prefers cover variables over others as depicted in Fig. 2. The methods branch_int and branch_spat in lines 7 and 10, respectively, are black box methods for which any standard variable-based rule for branching on fractional integer variables and for spatial branching can be used.

```
 1  input MINLP P as in (1) with cover C
 2         set of fractional variables F
 3         set of candidates for spatial branching S
 4  begin
 5  |   if F ≠ ∅ then
 6  |   |   if F ∩ C ≠ ∅ then  F ← F ∩ C ;
 7  |   |   branch_int (P, F) ;
 8  |   else
 9  |   |   if S ∩ C ≠ ∅ then  S ← S ∩ C ;
10  |   |   branch_spat (P, S) ;
11  end
```

Fig. 2. Undercover branching algorithm

To perform Undercover branching, a cover of the MINLP P has to be computed once before the branch-and-bound process starts. This global structure can be exploited also by other solver components, e.g. an Undercover heuristic as in [19,20].

As a distinguishing feature, the structure used for branching is computed exactly, at negligible cost (see previous section), and no sampling phase as, e.g., in [15] or [16], is required. Furthermore, we do not enforce branching on cover variables via strict branching priorities: if the candidate set lies completely outside of the cover, we do not continue branching on unfixed cover variables, but stick with the candidates proposed by the solver.

4   Experimental Results

In this section we investigate the computational impact of Undercover branching when combined with standard branching rules implemented in the MINLP solver SCIP 3.0 [3,23].

SCIP implements a branch-and-bound algorithm based on an LP relaxation that is constructed via gradient cuts[3] for convex constraints and linear over- and underestimators of the nonconvex terms. Further algorithmic components comprise primal heuristics, cutting planes applied to the MIP relaxation, an extensive presolving and propagation engine, conflict analysis, and several reformulation steps to detect convex or convexifiable constraints at the beginning. For details, we refer to [4,25].

By default, SCIP applies binary branching, i.e., it splits the current node into two subproblems. Branching on integer variables is prefered, however, not categorically: if all integer variables have integral value in the LP solution (and nonlinearities are still violated), SCIP continues with spatial branching even if not all of the integer variables are fixed.

For branching on integer variables with a fractional LP value, the default variable selection rule is *hybrid reliability branching* [12], which uses pseudocosts that are initialized by multiple strong branches per variable as in reliability branching [11]. VSIDS [13] and inference values [14,26], two scores from satisfiability testing and constraint programming, are taken into account for tie breaking. For spatial branching, SCIP implements the pseudocost strategy "rb-int-br"[4] from [5] weighted by the violations of the constraints in which a variable appears.

All experiments were conducted on a cluster of 64bit Intel Xeon X5672 CPUs at 3.2 GHz with 12 MB cache and 48 GB main memory and a time limit of one hour. Hyperthreading and Turboboost were disabled. For the latter experiment, we ran only one job per node to avoid random noise in the measured running time that might be caused by cache-misses if multiple processes share common resources. As subroutines, SCIP was linked to the LP solver CPLEX 12.4 [27], the expression interpreter CppAD 20120101.3 [24], and the NLP solver Ipopt 3.10.2 [28].

---

[3] If a convex constraint $g(x) \leqslant 0$, $g \in \mathcal{C}^1([\ell, u], \mathbb{R})$, is violated at some x^*, then x^* can be cut off by the *gradient cut* $\nabla g(x^*)^\top(x - x^*) + g(x^*) \leqslant 0$.

[4] For this strategy, the pseudocosts are multiplied by the distance of the current relaxation solution to the bounds of a variable when computing the branching score.

To avoid interactions, we deactivated the Undercover heuristic. To measure tree sizes accurately, we deactivated restarts.

As test set we chose the MINLPLib [21,29] featuring 273 instances. We excluded 18 instances that cannot be parsed or handled by SCIP 3.0, see Footnote 2. 13 instances were linearized during presolving; 42 further instances could be solved already during root node processing, hence no branching was applied; for two instances, branching had not started after one hour. We also removed three instances for which SCIP 3.0 suffers from numerical inaccuracies which lead to inconsistent solution values (independent of applying Undercover branching). All in all, this leaves a test set of 195 instances. Note that in MINLP, even more than in MIP, standard test sets often decompose into very easy and extremely difficult instances, with very few medium hard problems. Also, they are typically not as heterogeneous as, e.g., the MIPLIBs.

How often? As an initial experiment, we analyzed how often the Undercover pre-selection can be applied to reduce the candidate set. Note that there are two general cases when Undercover branching does not make a difference. If all candidates lie outside the cover (e.g., when the cover is entirely continuous, but there are integer variables with fractional LP solution), the algorithm will select a non-cover variable; if all candidates lie inside the cover (e.g., when the set of fractional variables is a subset of the cover), our algorithm does not yield any impact at that node of the tree.

To this end, we ran SCIP with its default branching rules, and enforced the pre-selection of cover variables as in Fig. 2 whenever possible. At each branching decision taken, we recorded whether all candidates were inside the cover, all candidates were outside, or whether the intersection was nontrivial.

It turned out that on 23 of the instances Undercover branching actually affected the branching decisions. For a further 26 instances the candidates from $\mathcal{F}$ were always included in the cover; all of these had a cover of at least 40% of the variables. For the majority of instances, no candidates were contained in the minimum cover used.[5]

This result is not overly surprising: it is explained by the fact that we are employing *minimum* covers and consequently increase the likelihood that the branching candidates are all outside the cover. However, since the computational overhead of Undercover branching is negligible, it does not degrade the performance on the unaffected instances. It remains to be analyzed whether it is helpful for the set of instances for which it actually affects branching decisions.

How good? The goal of our second experiment was to compare the performance impact of Undercover branching on the 23 instances that are affected. Before discussing the complete results, consider the small instance ex1264a as an illustrative example, modelling a nonconvex trim-loss problem from [30]. Figure 3

---

[5] For 6 instances, SCIP did not branch on integer variables at all; for 140 instances, $\mathcal{F} \cap \mathcal{C}$ was always empty. Spatial branching did not get performed for 61 instances, 111 times $\mathcal{S} \cap \mathcal{C}$ was always empty.

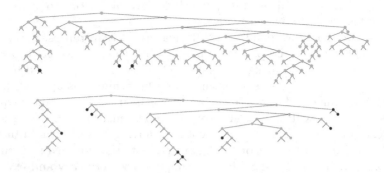

Fig. 3. Search trees explored by SCIP 3.0 for instance ex1264a with default branching reliability/pseudocost (top, 111 nodes processed) and Undercover branching (bottom, 52 nodes processed). White nodes pruned unprocessed. Linear nodes marked black.

shows the actual search trees explored by SCIP 3.0 default and SCIP with Undercover branching added. As can be seen, both trees have similar structure, but the number of nodes is reduced significantly because subtrees can be pruned earlier. This instance indeed confirms our hope that Undercover branching helps to drive the subproblems towards linearity faster: while without, only three nodes are linear, with Undercover branching eleven nodes become linear. Without Undercover branching, linear nodes only appear in depth twelve and below, whereas with Undercover branching, linear nodes can be observed from depth four onwards.

Our main hope is to reduce the number of branch-and-bound nodes processed by exploiting the global perspective provided by a cover in addition to the local perspective of a branching rule at a specific node. However, note that if the base rule employed in the branch_int procedure applies strong branching on its candidates such as the hybrid reliability rule in SCIP, then the restriction of the candidate set affects the solving process beyond the branching variable selection. On the one hand, computation time for solving strong branching LPs on the excluded candidates is saved; on the other hand, variable fixings that can be learned from strong branching might be lost.

Hence, we evaluated the impact of Undercover branching w.r.t. two base rules: SCIP's default reliability/pseudocost as described above, and pseudocost/pseudocost, i.e., exchanging the hybrid reliability rule for branching on integer variables by a pure pseudocost rule without any strong branching. The results for those instances that could be solved by at least one variant can be seen in Tab. 1 and Tab. 2, respectively. Using Undercover branching, for both cases two more instances could be solved as compared to not using it.

On four instances in Tab. 1 solved by both variants, Undercover branching increases the number of branch-and-bound nodes. This might be due to the side effect on strong branching described above. For the majority of cases, however, it reduces the number of branch-and-bound nodes significantly. The impact is even more visible on the number of performed strong branches, which is decreased ten times and only increased once. Since the main goal of Undercover branching is to

Table 1. Impact of Undercover branching on number of nodes, time, and strong branches performed for affected instances solved to optimality within one hour

instance	reliability/pseudocost			with undercover			relative [%]		
	nodes	strbrs	time [s]	nodes	strbrs	time [s]	nodes	strbrs	time
ex1263a	273	292	0.56	132	135	0.17	−52	−54	−70
ex1264a	111	218	0.33	52	120	0.08	−53	−45	−76
ex1265a	135	171	0.18	75	109	0.11	−44	−36	−39
ex1266a	59	220	0.38	101	177	0.35	+71	−20	−8
fac1	5	6	0.04	5	2	0.06	0	−67	+50
fac3	15	54	0.30	9	5	0.24	−40	−91	−20
nvs15	4	5	0.02	4	2	0.02	0	−60	0
pump	509	110	2.96	773	113	3.57	+52	+3	+21
st_e36	206	0	0.79	200	0	0.77	−3	0	−3
st_e40	19	22	0.08	25	22	0.10	+32	0	+25
tln4	2518	291	1.99	171	41	0.55	−93	−86	−72
tln5	500254	2621	373.73	7977	463	5.89	−98	−82	−98
tloss	78	227	0.19	77	166	0.12	−1	−27	−37
tltr	4	57	0.18	17	88	0.20	+325	+54	+11
shifted mean	291	144	3.68	141	79	0.77	−52	−45	−79
timed out:									
tln6	4101772	4007	*58% gap*	17005	471	22.28	−99	−88	.
tln7	2609508	3946	*124% gap*	9703667	5296	3156.38	+272	+34	.

Table 2. Impact of Undercover branching on number of nodes and time for affected instances solved to optimality within one hour

instance	pseudocost/pseudocost			with undercover			relative [%]		
	nodes	strbrs	time [s]	nodes	strbrs	time [s]	nodes	strbrs	time
ex1263a	166	.	0.20	163	.	0.18	−2	.	−10
ex1264a	38	.	0.20	85	.	0.18	+124	.	−10
ex1265a	138	.	0.21	104	.	0.40	−25	.	+90
ex1266a	70	.	0.08	141	.	0.20	+101	.	+150
fac1	7	.	0.05	5	.	0.05	−29	.	0
fac3	23	.	0.52	9	.	0.23	−61	.	−56
nvs15	4	.	0.02	4	.	0.03	0	.	+50
pump	1007	.	4.38	827	.	3.62	−18	.	−17
st_e36	206	.	0.79	200	.	0.73	−3	.	−8
st_e40	23	.	0.07	25	.	0.10	+9	.	+43
tln4	1454	.	1.49	202	.	0.67	−86	.	−55
tln5	622318	.	446.86	11013	.	8.39	−98	.	−98
tloss	73	.	0.09	77	.	0.11	+5	.	+22
tltr	18	.	0.17	4	.	0.16	−78	.	−6
shifted mean	287	.	3.85	159	.	0.91	−45	.	−76
timed out:									
tln6	4963908	.	*79% gap*	16382	.	21.54	−99	.	.
tln7	3207274	.	*226% gap*	468231	.	593.74	−85	.	.

restrict the set of branching candidates, this could be expected. Finally, considering computation time, Undercover branching helps nine times, four times the performance deteriorates. Regarding the shifted geometric means,[6] Undercover

---

[6] We used a shift of 100 nodes, 100 strong branchings, and 10s to compute the means.

branching reduces the computation time by 79% and the number of branch-and-bound nodes by 45%. Note that also when excluding the two positive outliers tln4 and tln5 and one negative outlier tltr, Undercover branching still yields an overall reduction of all considered performance measures.

For pure pseudocost branching, see Tab. 2, we observe a similar behavior: Undercover branching shows an improvement in the performance measures significantly more often than a deterioration, in shifted geometric mean we see improvements of 76% w.r.t. computation time and 45% w.r.t. the number of branch-and-bound nodes.

5   Conclusion and Outlook

In this paper, we have introduced Undercover branching, a new branching strategy for MINLP that exploits minimum covers to drive the subproblems created faster towards linearity. We showed that a combination of Undercover branching with either hybrid reliability branching or pseudocost branching outperforms the corresponding branching rules without Undercover information, yielding savings of 45% w.r.t. the number of branch-and-bound nodes and more than 70% w.r.t. running time in geometric mean for affected instances. The time consumed by the branching rule itself proved negligible.

Currently, the main limitation of Undercover branching is that the fraction of affected instances is relatively small (23 out of 195 instances). Therefore, the main goals of our future research are to employ alternative (not minimum) covers and to investigate the trade-off between cover size and number of affected instances. Furthermore, we work on identifying linear subproblems with large sub-trees, which could then be solved more efficiently using a pure MIP solver.

References

1. Land, A.H., Doig, A.G.: An automatic method of solving discrete programming problems. Econometrica 28(3), 497–520 (1960)
2. Bixby, R., Fenelon, M., Gu, Z., Rothberg, E., Wunderling, R.: MIP: Theory and practice – closing the gap. In: Powell, M., Scholtes, S. (eds.) Systems Modelling and Optimization: Methods, Theory, and Applications, pp. 19–49. Kluwer Academic Publisher (2000)
3. Achterberg, T.: Constraint Integer Programming. PhD thesis, TU Berlin (2007)
4. Vigerske, S.: Decomposition in Multistage Stochastic Programming and a Constraint Integer Programming Approach to MINLP. PhD thesis, HU Berlin (2012)
5. Belotti, P., Lee, J., Liberti, L., Margot, F., Wächter, A.: Branching and bounds tightening techniques for non-convex MINLP. Optimization Methods & Software 24, 597–634 (2009)
6. Karamanov, M., Cornuéjols, G.: Branching on general disjunctions. Math. Prog. 128(1-2), 403–436 (2011)
7. Benichou, M., Gauthier, J., Girodet, P., Hentges, G., Ribiere, G., Vincent, O.: Experiments in mixed-integer programming. Math. Prog. 1, 76–94 (1971)
8. Linderoth, J.T., Savelsbergh, M.W.P.: A computational study of search strategies for mixed integer programming. INFORMS J. Comput. 11, 173–187 (1999)

9. Applegate, D.L., Bixby, R.E., Chvátal, V., Cook, W.J.: Finding cuts in the TSP (A preliminary report). Technical Report 95-05, DIMACS (1995)

10. Applegate, D.L., Bixby, R.E., Chvátal, V., Cook, W.J.: The Traveling Salesman Problem: A Computational Study. Princeton University Press, USA (2007)

11. Achterberg, T., Koch, T., Martin, A.: Branching rules revisited. Operations Research Letters 33, 42–54 (2005)

12. Achterberg, T., Berthold, T.: Hybrid branching. In: van Hoeve, W.-J., Hooker, J.N. (eds.) CPAIOR 2009. LNCS, vol. 5547, pp. 309–311. Springer, Heidelberg (2009)

13. Moskewicz, M.W., Madigan, C.F., Zhao, Y., Zhang, L., Malik, S.: Chaff: Engineering an efficient SAT solver. In: Proc. of the DAC (July 2001)

14. Li, C.M., Anbulagan: Look-ahead versus look-back for satisfiability problems. In: Smolka, G. (ed.) CP 1997. LNCS, vol. 1330, pp. 342–356. Springer, Heidelberg (1997)

15. Kılınç Karzan, F., Nemhauser, G.L., Savelsbergh, M.W.P.: Information-based branching schemes for binary linear mixed-integer programs. Math. Prog. Computation 1(4), 249–293 (2009)

16. Fischetti, M., Monaci, M.: Backdoor branching. In: Günlük, O., Woeginger, G.J. (eds.) IPCO 2011. LNCS, vol. 6655, pp. 183–191. Springer, Heidelberg (2011)

17. Fischetti, M., Monaci, M.: Branching on nonchimerical fractionalities. OR Letters 40(3), 159–164 (2012)

18. Tawarmalani, M., Sahinidis, N.V.: Global optimization of mixed-integer nonlinear programs: A theoretical and computational study. Math. Prog. 99, 563–591 (2004)

19. Berthold, T., Gleixner, A.M.: Undercover – a primal heuristic for MINLP based on sub-MIPs generated by set covering. In: Bonami, P., Liberti, L., Miller, A.J., Sartenaer, A. (eds.) Proc. of the EWMINLP, pp. 103–112 (April 2010)

20. Berthold, T., Gleixner, A.M.: Undercover: a primal MINLP heuristic exploring a largest sub-MIP. Math. Prog. (2013) doi:10.1007/s10107-013-0635-2

21. Bussieck, M.R., Drud, A.S., Meeraus, A.: MINLPLib – a collection of test models for mixed-integer nonlinear programming. INFORMS J. Comput. 15(1), 114–119 (2003)

22. Garey, M.R., Johnson, D.S.: Computers and Intractability: A Guide to the Theory of NP-Completeness. W. H. Freeman & Co., New York (1979)

23. SCIP: Solving Constraint Integer Programs, http://scip.zib.de

24. CppAD: A Package for Differentiation of C++ Algorithms, http://www.coin-or.org/CppAD

25. Berthold, T., Heinz, S., Vigerske, S.: Extending a CIP framework to solve MIQCPs. In: Lee, J., Leyffer, S. (eds.) Mixed Integer Nonlinear Programming. The IMA Volumes in Mathematics and its Applications, vol. 154, pp. 427–444. Springer (2012)

26. Achterberg, T.: Conflict analysis in mixed integer programming. Discrete Optimization 4(1), 4–20 (2007)

27. IBM: CPLEX Optimizer, http://www-01.ibm.com/software/integration/optimization/cplex-optimizer/

28. Wächter, A., Biegler, L.T.: On the implementation of a primal-dual interior point filter line search algorithm for large-scale nonlinear programming. Math. Prog. 106(1), 25–57 (2006)

29. GAMS: MINLP Library, http://www.gamsworld.org/minlp/minlplib.html

30. Harjunkoski, I., Westerlund, T., Pörn, R., Skrifvars, H.: Different transformations for solving non-convex trim-loss problems by MINLP. Eur. J. Oper. Res. 105(3), 594–603 (1998)

Quadratic Outer Approximation
for Convex Integer Programming
with Box Constraints

Christoph Buchheim and Long Trieu

Fakultät für Mathematik, Technische Universität Dortmund
{christoph.buchheim, long.trieu}@math.tu-dortmund.de

Abstract. We present a quadratic outer approximation scheme for solving general convex integer programs, where suitable quadratic approximations are used to underestimate the objective function instead of classical linear approximations. As a resulting surrogate problem we consider the problem of minimizing a function given as the maximum of finitely many convex quadratic functions having the same Hessian matrix. A fast algorithm for minimizing such functions over integer variables is presented. Our algorithm is based on a fast branch-and-bound approach for convex quadratic integer programming proposed by Buchheim, Caprara and Lodi [5]. The main feature of the latter approach consists in a fast incremental computation of continuous global minima, which are used as lower bounds. We generalize this idea to the case of k convex quadratic functions, implicitly reducing the problem to $2^k - 1$ convex quadratic integer programs. Each node of the branch-and-bound algorithm can be processed in $O(2^k n)$ time. Experimental results for a class of convex integer problems with exponential objective functions are presented. Compared with Bonmin's outer approximation algorithm B-OA and branch-and-bound algorithm B-BB, running times for both ternary and unbounded instances turn out to be very competitive.

1   Introduction

Many optimization problems arising in real world applications can be formulated as convex mixed-integer nonlinear programs (MINLP) of the form

$$\min \quad f(x)$$
$$\text{s.t.} \quad g_j(x) \le 0 \ \forall j = 1, \ldots, m$$
$$x_i \in \mathbb{Z} \ \forall i \in \mathcal{I} \,,$$

where $f, g_1, \ldots, g_m : \mathbb{R}^n \to \mathbb{R}$ are convex functions and $\mathcal{I} \subseteq \{1, \ldots, n\}$ is the set of indices of integer variables. Allowing both integrality and nonlinearity makes this class of problems extremely hard. In fact, MINLP comprises the NP-hard subclasses of mixed-integer linear programming (MILP) and general nonlinear programming (NLP). The restriction to convex MINLP preserves NP-hardness, as MILP is still contained as a special case. The most important exact

V. Bonifaci et al. (Eds.): SEA 2013, LNCS 7933, pp. 224–235, 2013.

approaches applied to convex MINLP are branch-and-bound by Dakin [7], generalized Benders decomposition by Geoffrion [10], outer approximation by Duran and Grossmann [8] and Fletcher and Leyffer [9], branch-and-cut by Quesada and Grossmann [12], and the extended cutting plane method by Westerlund and Pettersson [15]. A detailed survey of algorithms and software for solving convex MINLP is given by Bonami, Kilinç and Linderoth [3].

In the further course of this paper, for simplicity, we focus on the pure integer case with box-constraints, thus assuming $\mathcal{I} = \{1, \ldots n\}$ and $l \leq x \leq u$ for some fixed lower and upper bound vectors $l \in (\mathbb{R} \cup \{-\infty\})^n$ and $u \in (\mathbb{R} \cup \{\infty\})^n$. Note that the resulting box-constrained convex integer nonlinear program (INLP) of the form

$$\begin{aligned} \min \quad & f(x) \\ \text{s.t.} \quad & x \in \mathcal{B}\,, \end{aligned} \tag{1}$$

where $\mathcal{B} := \{x \in \mathbb{Z}^n \mid l \leq x \leq u\}$, still belongs to the class of NP-hard problems, since minimizing a convex quadratic function over the integers is equivalent to the Closest Vector Problem, which is known to be NP-hard [14]. The algorithm for solving convex INLP presented in the following is based on the outer approximation scheme.

1.1   Organization of the Paper

After giving a short recapitulation of the standard linear outer approximation scheme, we describe our quadratic outer approximation scheme in Section 2. Section 3 presents our approach for solving a convex piecewise quadratic integer program with constant Hessian matrix, which occurs as a surrogate problem in every iteration of our extended outer approximation scheme. In Section 4, we present computational results and compare the effectiveness of the proposed algorithm, applied to a special class of convex integer nonlinear optimization problems, with existing state-of-the-art software. Finally, we summarize our main results and give a short outlook in Section 5.

2   Outer Approximation

2.1   Linear Outer Approximation

The main idea of the classical linear outer approximation approach is to equivalently transform the original integer nonlinear problem into an integer linear problem by iteratively adding linearizations to the objective function. As soon as we obtain an iterate that has already been computed in an earlier iteration, we have reached optimality. Applied to (1), we get the simplified scheme described in Algorithm 1. The main computational effort of the presented approach lies in the computation of the integer minimizer in Step 3 of each iteration, by solving a box-constrained convex piecewise linear integer program, which can be formulated as an integer linear program (ILP). The approach is illustrated in Fig. 1.

Algorithm 1. Linear Outer Approximation Scheme

---

input : convex and continuously differentiable function f
output: integer minimizer $x^* \in \mathcal{B}$ of f

1. set $k := 1$ and choose any $x^1 \in \mathcal{B}$
2. compute supporting hyperplane for f in x^k:

$$f(x) \geq f(x^k) + \nabla f(x^k)^\top (x - x^k)$$

3. compute $x^{k+1} \in \mathcal{B}$ as an integer minimizer of

$$\max_{i=1,\dots,k} \{f(x^i) + \nabla f(x^i)^\top (x - x^i)\}$$

4. if $x^{k+1} \neq x^i$ for all $i \leq k$, set $k := k+1$ and go to 2, otherwise, x^{k+1} is optimal

---

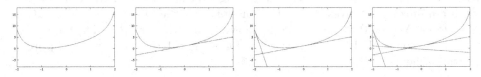

Fig. 1. Linear Outer Approximation applied to $f\colon [-2,2] \to \mathbb{R}$, $f(x) = (x+1)^2 + e^{x^2-2}$. Iterates are $x^1 = 0$, $x^2 = -2$, $x^3 = -1$, and $x^4 = -1$

2.2   Quadratic Outer Approximation

The main drawback of Linear Outer Approximation is that in general many iterations are necessary to obtain an appropriate approximation of the original objective function. The basic idea of our approach is to modify Step 2 of Algorithm 1 by replacing the linearizations by appropriate quadratic underestimators. Unfortunately, the second-order Taylor approximation is not necessarily a global underestimator of the original function. One important challenge therefore is to find a suitable quadratic underestimator. The following observation gives a sufficient condition for an underestimator to be feasible.

Theorem 1. *Let f be twice continuously differentiable and let $Q \in \mathbb{R}^{n \times n}$ such that $Q \preccurlyeq \nabla^2 f(x)$ for all $x \in \mathbb{R}^n$, i.e. $Q - \nabla^2 f(x)$ is negative semidefinite for all $x \in \mathbb{R}^n$. Consider the supporting quadratic function*

$$T(x) := f(x^k) + \nabla f(x^k)^\top (x - x^k) + \tfrac{1}{2}(x - x^k)^\top Q(x - x^k).$$

Then $f(x) \geq T(x)$ for all $x \in \mathbb{R}^n$.

Proof. By construction, we have $\nabla^2(f - T)(x) = \nabla^2 f(x) - Q \succcurlyeq 0$ for all $x \in \mathbb{R}^n$ and $\nabla(f - T)(x^k) = \nabla f(x^k) - \nabla f(x^k) = 0$. This implies that $f - T$ is convex with minimizer x^k, yielding $f(x) - T(x) \geq f(x^k) - T(x^k) = 0$ for all $x \in \mathbb{R}^n$. $\square$

The entire quadratic outer approximation scheme is described in Algorithm 2.

Algorithm 2. Quadratic Outer Approximation Scheme

input : convex and twice continuously differentiable function f,
      matrix Q s.t. $0 \preccurlyeq Q \preccurlyeq \nabla^2 f(x)$ for all $x \in \mathbb{R}^n$
output: integer minimizer $x^* \in \mathcal{B}$ of f

1. set $k := 1$ and choose any $x^1 \in \mathcal{B}$
2. compute supporting quadratic underestimator for f in x^k:

$$f(x) \geq f(x^k) + \nabla f(x^k)^\top (x - x^k) + \tfrac{1}{2} (x - x^k)^\top Q(x - x^k)$$

3. compute $x^{k+1} \in \mathcal{B}$ as an integer minimizer of

$$\max_{i=1,\ldots,k} \{ f(x^i) + \nabla f(x^i)^\top (x - x^i) + \tfrac{1}{2} (x - x^i)^\top Q(x - x^i) \}$$

4. if $x^{k+1} \neq x_i$ for all $i \leq k$, set $k := k + 1$ and go to 2, otherwise, x^{k+1} is optimal

In Step 2 of Algorithm 2, the new underestimator for a given iterate x^k is computed as follows:

$$f(x^k) + \nabla f(x^k)^\top (x - x^k) + \tfrac{1}{2} (x - x^k)^\top Q(x - x^k)$$
$$= \underbrace{f(x^k) - \nabla f(x^k)^\top x^k + \tfrac{1}{2} {x^k}^\top Q x^k}_{=:c_{k+1}} + \underbrace{(\nabla f(x^k)^\top - {x^k}^\top Q)}_{=:L_{k+1}^\top} x + \tfrac{1}{2} x^\top Q x \,,$$

the new quadratic underestimator

$$\tfrac{1}{2} x^\top Q x + L_{k+1}^\top x + c_{k+1}$$

is a convex quadratic function with Hessian $Q \succcurlyeq 0$ not depending on k.

Step 3 of Algorithm 2 requires to compute an integer minimizer of a quadratic program instead of a linear program, as was the case in Algorithm 1. Although the hardness of this surrogate problem increases from a practical point of view, this approach might pay off if the number of iterations decreases significantly with respect to linear approximation. In fact, we observed in our experiments that the number of iterations stays very small in general, even for problems in higher dimensions; see Section 4.

However, the surrogate problem of solving a convex box-constrained piecewise quadratic integer program, being the most expensive ingredient in Algorithm 2, requires an effective solution method to keep the whole algorithm fast. The surrogate problem can be formulated as an integer quadratic program (IQP), it is of the form

$$\min_{x \in \mathcal{B}} \max_{i=1,\ldots,k} (\tfrac{1}{2} x^\top Q x + L_i^\top x + c_i), \tag{2}$$

where $Q \succcurlyeq 0$, $L_1, \ldots, L_k \in \mathbb{R}^n$, and $c_1, \ldots, c_k \in \mathbb{R}$. At this point it is crucial to underline that the Hessian Q of each quadratic function is the same, so that we can rewrite (2) as

$$\min_{x \in \mathcal{B}} \left(\tfrac{1}{2} x^\top Q x + \max_{i=1,\dots,k} (L_i^\top x + c_i) \right) ,$$

which in turn can be reformulated as an integer quadratic program using a dummy variable $\alpha \in \mathbb{R}$:

$$
\begin{aligned}
\min \quad & \tfrac{1}{2} x^\top Q x + \alpha \\
\text{s.t.} \quad & L_i^\top x + c_i \le \alpha \quad \forall i = 1,\dots,k \\
& x \in \mathcal{B} \\
& \alpha \in \mathbb{R} .
\end{aligned}
\tag{3}
$$

The quadratic outer approximation scheme is illustrated in Fig. 2. In this example, the algorithm terminates after two iterations.

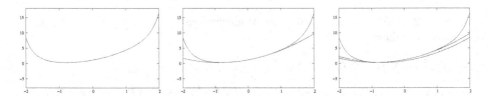

Fig. 2. Quadratic Outer Approximation for $f \colon [-2,2] \to \mathbb{R}$, $f(x) = (x+1)^2 + e^{x^2-2}$, using $Q = 2 + 2e^{-2}$. The iterates are $x^1 = 0$, $x^2 = -1$, and $x^3 = -1$.

3   Convex Piecewise Quadratic Integer Programming

Solving the convex piecewise quadratic integer program (2) in each iteration is the core task of the quadratic outer approximation scheme. We implicitly reduce this problem, in iteration k, to at most $2^k - 1$ convex quadratic integer programs, which are solved by a fast branch-and-bound algorithm proposed by Buchheim, Caprara and Lodi [5]. Our computational results in Section 4 show that only few iterations of Algorithm 2 are necessary in general to solve an instance to optimality, so that the number $2^k - 1$, though being exponential, remains reasonably small in practice.

3.1   Branch-and-Bound for Convex Quadratic Integer Programming

For a better understanding of the branch-and-bound algorithm, we shortly summarize its key ingredients; all results can be found in [5]. For given $Q \succcurlyeq 0$, $L \in \mathbb{R}^n$, and $c \in \mathbb{R}$, consider the convex quadratic integer program

$$\min_{x \in \mathbb{Z}^n} f(x) = \tfrac{1}{2} x^\top Q x + L^\top x + c . \tag{4}$$

For simplicity, assume Q to be positive definite. Otherwise, we can transform the objective function of (4) into a strictly convex function having the same integer minimizer. If Q is positive definite, we can easily determine the unique global minimizer $\bar{x}$ of f over $\mathbb{R}^n$ by solving a system of linear equations, as

$$\bar{x} = -Q^{-1}L \quad \text{with} \quad f(\bar{x}) = c - \tfrac{1}{2}L^\top Q^{-1}L,$$

which we can use as a lower bound for f over $\mathbb{Z}^n$. Simple rounding of the continuous minimizer

$$x_j := \lfloor \bar{x}_j \rceil, \; j = 1, \dots, n$$

to the next integer yields a trivial upper bound $f(x)$ for f over $\mathbb{Z}^n$.

In our branch-and-bound scheme, we branch by fixing the variables in increasing distance to their values in the continuous relaxation. By exploiting the convexity of f and its symmetry with respect to $\bar{x}$, we can cut off the current node of the tree and all its siblings as soon as we fix a variable to some value for which the resulting lower bound exceeds the current best known upper bound. Using these ingredients we get a straightforward branch-and-bound algorithm with running time $O(n^3)$ per node, mainly for computing the continuous minimizer by solving a linear system of equations. However, the running time of this computation can be improved to even linear running time per node, in two steps.

First, note that after fixing the first d variables, the problem reduces to the minimization of

$$\bar{f} \colon \mathbb{Z}^{n-d} \to \mathbb{R}, \; x \mapsto \tfrac{1}{2}x^\top \bar{Q}_d x + \bar{L}^\top x + \bar{c}$$

where $\bar{Q}_d \succ 0$ is obtained by deleting all rows and columns of Q corresponding to fixed variables and $\bar{L}$ and $\bar{c}$ are adapted properly. The main idea is to fix the variables in a predetermined order. Following this approach, the reduced matrices $\bar{Q}_d$ only depend on the depth d, but not on specific fixings. This implies that only n different matrices $\bar{Q}_d$ appear in the entire branch-and-bound tree, so that their inverse matrices can be predetermined in a preprocessing phase. The resulting running time reduces to $O((n-d)^2)$ per node.

The second improvement is a consequence of the following observation [5]:

Theorem 2. *For each $d \in \{0, \dots, n-1\}$, there exist vectors $z^d, v^d \in \mathbb{R}^{n-d}$ and a scalar $s_d \in \mathbb{R}$, only depending on Q and the chosen order of variables, such that the following holds: if $\bar{x}^{old} \in \mathbb{R}^{n-d}$ denotes the minimizer of $\bar{f}$ after fixing variables $x_1, \dots, x_d$, and x_{d+1} is fixed to $r_{d+1} \in \mathbb{Z}$, then the resulting continuous minimizer and associated minimum can be computed incrementally by*

$$\bar{x}^{new} := \bar{x}^{old} + \alpha z^d$$

and

$$\bar{f}(\bar{x}^{new}) := \bar{f}(\bar{x}^{old}) + \alpha\big((\bar{x}^{old})^\top v^d + \bar{L}^\top z^d\big) + \alpha^2 s_d$$

where $\alpha = r_{d+1} - \bar{x}_1^{old}$.

As a conclusion, if z^d, v^d and s_d are computed in a preprocessing phase for all $d = 0, \dots, n-1$, the resulting total running time per node is $O(n-d)$.

3.2   Lower Bound Computation for the Surrogate Problem

To solve the surrogate problem (2) with a branch-and-bound algorithm, we compute a lower bound at every node by solving its continuous relaxation

$$\min_{x \in \mathbb{R}^{n-d}} \max_{i=1,\ldots,k} \left(\tfrac{1}{2} x^T \bar{Q}_d x + \bar{L}_i^T x + \bar{c}_i \right) .$$

The main idea is to decompose this problem into subproblems, namely the minimization of several auxiliary quadratic functions defined on affine subspaces of $\mathbb{R}^{n-d}$, and finally make use of the incremental computation technique described in the last subsection. To describe this decomposition procedure, we need to introduce some definitions. First, we define

$$\bar{f}(x) := \max_{i=1,\ldots,k} \bar{f}_i(x)$$

as the maximum of the reduced functions

$$\bar{f}_i(x) := \tfrac{1}{2} x^T \bar{Q}_d x + \bar{L}_i^T x + \bar{c}_i, \ i = 1, \ldots, k .$$

For all $J \subseteq \{1, \ldots, k\}$, $J \neq \emptyset$, we define

$$U_J := \{x \in \mathbb{R}^{n-d} \mid \bar{f}_i(x) = \bar{f}_j(x) \ \forall i, j \in J\}$$

and consider the auxiliary function

$$\bar{f}_J(x) : U_J \to \mathbb{R}, \quad \bar{f}_J(x) := \bar{f}_i(x), \ i \in J .$$

As all functions $\bar{f}_i$ have the same Hessian matrix $\bar{Q}_d$, each set U_J is an affine subspace of $\mathbb{R}^{n-d}$. In particular, we can compute the minimizers x_J^* of all $\bar{f}_J$ incrementally as described in Section 3.1.

Theorem 3. *For each $J \subseteq \{1, \ldots, k\}$ with $J \neq \emptyset$, let x_J^* be a minimizer of $\bar{f}_J$ over U_J. Then the global minimum of $\bar{f}$ is*

$$\min \ \{\bar{f}_J(x_J^*) \mid \emptyset \neq J \subseteq \{1, \ldots, k\}, \ \bar{f}(x_J^*) = \bar{f}_J(x_J^*)\}.$$

Proof. We clearly have "$\leq$". To show "$\geq$", let $x^* \in \mathbb{R}^{n-d}$ be the global minimizer of $\bar{f}$ and define

$$J^* := \{i \mid \bar{f}_i(x^*) = \bar{f}(x^*)\} \neq \emptyset.$$

Then it follows that $x^* \in U_{J^*}$ and $\bar{f}(x^*) = \bar{f}_{J^*}(x^*)$. Moreover, x^* minimizes $\bar{f}_{J^*}$ over U_{J^*}, hence $x^* = x_{J^*}^*$ by strict convexity. In summary,

$$\bar{f}(x_{J^*}^*) = \bar{f}(x^*) = \bar{f}_{J^*}(x^*) = \bar{f}_{J^*}(x_{J^*}^*) ,$$

from which the result follows. $\qquad\square$

Corollary 1. *The running time of the modified branch-and-bound algorithm for solving the surrogate problem (2) is $O(2^k \cdot (n - d))$ per node.*

Note that the index set J does not need to be considered any more in the current node and its branch-and-bound subtree as soon as the value of $\bar{f}_J(x_J^*)$ exceeds the current upper bound.

4   Experimental Results

To show the potential of our approach, we carried out two types of experiments. First, we compared our branch-and-bound algorithm for solving the surrogate problems (2), called CPQIP, to CPLEX 12.4 [1]. Second, we created a class of hard convex integer programs, to illustrate the effectiveness of our quadratic outer approximation algorithm, called QOA, and to compare its performance to that of Bonmin-OA and Bonmin-BB 1.5.1 [13] using Cbc 2.7.2 and Ipopt 3.10.

4.1   Implementation Details

We implemented our algorithm in C++. To speedup our algorithm, we used a straightforward local search heuristic to determine a good starting point. We start by taking the origin $x = (0, \ldots, 0)$ and continue to increase the first variable x_1 until no further improvement can be found. In the same way we test if decreasing the variable leads to a better solution. We repeat this procedure for every consecutive variable $x_2, \ldots, x_n$. The improvement in running time can be seen in Table 2. Another small improvement in running time can be achieved by using the optimal solution x_k^* of the surrogate problem in iteration k to get an improved global upper bound UB for the next iteration, i.e.

$$UB := \max_{i=1,\ldots,k+1} \tfrac{1}{2} x_k^{*\top} Q x_k^* + L_i^\top x_k^* + c \, .$$

4.2   Surrogate Problem

We randomly generated 160 instances for the surrogate problem (2), 10 for each combination of $n \in \{20, 30, 40, 50\}$ and $k \in \{2, \ldots, 5\}$. We chose $\mathcal{B} = \mathbb{Z}^n$, i.e., we consider unbounded instances. For generating the positive semidefinite matrix Q, we chose n eigenvalues λ_i uniformly at random from $[0, 1]$ and orthonormalize n random vectors v_i, where all entries are chosen uniformly at random from $[-10, 10]$, then we set $Q = \sum_{i=1}^n \lambda_i v_i v_i^\top$. The entries of all L_i and c_i, $i = 1, \ldots, k$, were chosen uniformly at random from $[-10, 10]$.

We compared our algorithm CPQIP with the Mixed-Integer Quadratic Programming (MIQP) solver of CPLEX 12.4 [1] applied to the QP model (3). For both approaches, we used a time limit of 3 hours and an absolute optimality tolerance of 10^{-6}. The relative optimality tolerance of CPLEX was set to 10^{-10}.

The results are summarized in Table 1. Running times are measured in cpu seconds and the numbers of nodes explored in the branch-and-bound trees are given in the corresponding column. First, we can observe that our algorithm could solve 158 out of 160 instances in total within the given time limit, while CPLEX 12.4 managed to solve only 154 instances. Second, for instances with a large number n of variables but small k, our approach turns out to be significantly faster. Instances of this type are the most relevant instances in a quadratic outer approximation scheme, as shown in Section 4.3.

As expected, CPQIP tends to be slower than CPLEX on most of the instances with $k = 5$, since the running time per node in our branch-and-bound algorithm is exponential in k.

Table 1. Average running times, numbers of instances solved and average numbers of branch-and-bound nodes for CPQIP and CPLEX 12.4 on randomly generated unbounded instances for the surrogate problem of type (2)

		CPQIP			CPLEX 12.4		
n	k	solved	time	nodes	solved	time	nodes
20	2	10/10	0.02	1.39e+4	10/10	0.24	2.52e+3
20	3	10/10	0.03	9.84e+3	10/10	0.18	1.68e+3
20	4	10/10	0.06	9.51e+3	10/10	0.21	1.59e+3
20	5	10/10	0.05	2.91e+3	10/10	0.10	1.14e+3
30	2	10/10	0.07	4.23e+4	10/10	0.90	1.33e+4
30	3	10/10	0.40	1.09e+5	10/10	1.94	2.81e+4
30	4	10/10	4.48	4.20e+5	10/10	7.75	9.06e+4
30	5	10/10	4.57	2.63e+5	10/10	2.90	3.92e+5
40	2	10/10	10.41	5.35e+6	10/10	138.77	1.31e+6
40	3	10/10	30.70	7.18e+6	10/10	69.58	7.91e+5
40	4	10/10	66.13	7.15e+6	10/10	73.32	8.28e+5
40	5	10/10	141.53	7.21e+6	10/10	92.80	1.00e+6
50	2	10/10	320.30	1.45e+8	10/10	2337.91	1.81e+8
50	3	10/10	844.18	1.72e+8	8/10	646.15	5.41e+7
50	4	10/10	1925.06	1.79e+8	8/10	864.92	6.95e+7
50	5	8/10	1650.26	6.90e+7	8/10	931.00	7.50e+7

4.3   Quadratic Outer Approximation Scheme

In order to evaluate the entire quadratic outer approximation scheme, we consider the following class of problems:

$$\min_{x \in \mathbb{Z}^n} f(x), \ f : D \subseteq \mathbb{R}^n \to \mathbb{R}, \ f(x) = \sum_{i=1}^{n} \exp(q_i(x)) \tag{5}$$

where $q_i(x) = x^\top Q_i x + L_i^\top x + c_i$. We assume $Q_i \succ 0$ for all $i = 1, \ldots n$, so that f is a strictly convex function.

In order to determine a feasible matrix Q according to Theorem 1, we first compute $m_i := \min_{x \in D} q_i(x)$ for all $i = 1, \ldots, n$ and then choose

$$Q := \sum_{i=1}^{n} 2Q_i \exp(m_i) \succ 0 .$$

It is easy to see that $\nabla^2 f(x) - Q \succeq 0$ for all $x \in \mathbb{R}^n$, so that Q can be used in Algorithm 2. The quality of Q and therefore of the quadratic underestimator strongly depends on D: the smaller the set D is, the larger are the m_i, and the better is the global underestimator.

To test the quadratic outer approximation scheme, we randomly generated ternary and unbounded instances of type (5), i.e., we consider both $D = [-1, 1]$ and $D = \mathbb{R}^n$. All data were generated in the same way as in Section 4.2, except that all coefficients of Q and L_i, c_i, $i = 1, \ldots, n$, were scaled by 10^{-5}, to avoid problems with the function evaluations.

We tested our algorithm against Bonmin-OA and Bonmin-BB [13, 2], which are state-of-the-art solvers for convex mixed-integer nonlinear programming. While Bonmin-OA is a decomposition approach based on outer approximation [8, 9], Bonmin-BB is a simple branch-and-bound algorithm based on solving a continuous nonlinear program at each node of the search tree and branching on the integer variables [11]. Again the time limit per instance was set to 3 hours.

Table 2. Running times, number of instances solved, average and maximum number of iterations of QOA(h), QOA compared to Bonmin-OA and Bomin-BB for randomly generated ternary instances of problem type (5)

	QOA(h)			QOA				Bonmin-OA		Bonmin-BB	
n	solved	ø it	mx	time (s)	solved	ø it	mx	time (s)	solved	time (s)	solved	time (s)
20	10/10	1.00	1	0.03	10/10	2.00	2	0.07	10/10	91.53	10/10	2.12
30	10/10	1.50	4	5.87	10/10	2.50	4	28.46	5/10	1017.99	10/10	148.85
40	10/10	1.40	3	20.19	10/10	2.40	4	65.42	0/10	—	9/10	4573.80
50	10/10	2.00	3	69.54	10/10	3.10	4	151.88	0/10	—	0/10	—
60	9/10	2.11	4	1154.66	9/10	3.00	5	1692.76	0/10	—	0/10	—
70	5/10	3.80	6	3363.11	4/10	3.75	5	2916.21	0/10	—	0/10	—

Table 3. Running times, number of instances solved, average and maximum number of iterations of QOA(h) compared to Bonmin-BB for randomly generated unbounded instances of problem type (5)

	QOA(h)				Bonmin-BB	
n	solved	ø it	mx	time (s)	solved	time (s)
20	10/10	2.30	3	0.12	10/10	113.13
30	8/10	3.12	5	26.38	1/10	4897.10
40	6/10	2.83	3	134.42	0/10	—
50	2/10	3.00	3	7630.12	0/10	—

Results for ternary and unbounded instances are shown in Table 2 and 3, respectively. "QOA(h)" denotes our quadratic approximation scheme using the local search heuristic, while in "QOA" the heuristic was turned off. Running times are again measured in seconds. The columns named "ø it" and "mx" show the average and maximum number of iterations in our approach, respectively.

Overall our quadratic outer approximation approach could solve more instances and seems to be considerably faster for the ternary instances as well as the unbounded instances. For the unbounded instances, we unfortunately experienced some numerical issues, because the function evaluations seem to cause problems. While Bonmin-OA did not converge properly, the branch-and-bound algorithm Bonmin-BB, which seems to be faster than B-OA in all cases, sometimes computed solutions which were slightly worse than ours and hence not always optimal. In the ternary case, no such problems occured in any of the approaches tested.

An important observation in all our experiments is that both the average and maximum number of iterations in our outer approximation scheme tend to be small, here up to 4 for the instance sizes we could solve to optimality within the given time limit. In particular, the number of iterations does not seem to increase significantly with the number of variables n, contrary to the standard linear outer approximation approach.

5   Conclusion

We proposed a quadratic outer approximation scheme for solving convex integer nonlinear programs, based on the classical linear outer approximation scheme. From our computational results, we can conclude that quadratic underestimators have the potential to yield significantly better approximations, which might lead to considerably fewer iterations of the entire algorithm. While the standard linear outer approximation scheme requires to solve an integer linear program in each iteration, our method requires the solution of integer quadratic programs with linear constraints. Therefore we proposed an algorithm which is based on the reduction of the surrogate problems to a set of unconstrained convex quadratic integer programs, which are effectively solved by a branch-and-bound algorithm introduced by Buchheim et al. [5].

For future work it remains to study possible and good choices of Q for other classes of problems. Moreover, the running time could be further reduced by, e.g., trying to eliminate non-active underestimators or approximately solving the surrogate problem instead of solving it to optimality, in order to obtain additional quadratic underestimators quickly. Furthermore, our approach could be extended to constrained convex integer programs by using penalty functions. For this, note that a feasible Q stays feasible if adding a convex penalty function to the objective function.

Finally, one could also consider using non-convex quadratic underestimators for non-convex nonlinear integer optimization within our framework, assuming that a fast algorithm for solving non-convex quadratic integer programs is at hand; potential candidates are the algorithms proposed by Buchheim and Wiegele [4] or by Buchheim, De Santis, Palagi and Piacentini [6]. Such an extension would allow our approach to be applied to a much wider class of problems than classical linear outer approximation.

References

[1] IBM ILOG CPLEX Optimizer 12.4 (2013),
    www.ibm.com/software/integration/optimization/cplex-optimizer
[2] Bonami, P., Biegler, L.T., Conn, A.R., Cornuéjols, G., Grossmann, I.E., Laird, C.D., Lee, J., Lodi, A., Margot, F., Sawaya, N., Wächter, A.: An algorithmic framework for convex mixed integer nonlinear programs. Discrete Optimization 5(2), 186–204 (2008)

[3] Bonami, P., Kilinç, M., Linderoth, J.T.: Part I: Convex MINLP. In: Algorithms and Software for Solving Convex Mixed Integer Nonlinear Programs. IMA Volumes in Mathematics and its Applications: Mixed Integer Nonlinear Programming, vol. 154, pp. 1–39. Springer (2012)

[4] Buchheim, C., Wiegele, A.: Semidefinite relaxations for non-convex quadratic mixed-integer programming. Mathematical Programming (2012) (to appear)

[5] Buchheim, C., Caprara, A., Lodi, A.: An effective branch-and-bound algorithm for convex quadratic integer programming. Mathematical Programming 135, 369–395 (2012)

[6] Buchheim, C., De Santis, M., Palagi, L., Piacentini, M.: An exact algorithm for quadratic integer minimization using ellipsoidal relaxations. Technical report, Optimization Online (2012)

[7] Dakin, R.J.: A tree-search algorithm for mixed integer programming problems. The Computer Journal 8, 250–255 (1965)

[8] Duran, M.A., Grossmann, I.E.: An outer-approximation algorithm for a class of mixed-integer nonlinear programs. Mathematical Programming 36, 307–339 (1986)

[9] Fletcher, R., Leyffer, S.: Solving mixed integer nonlinear programs by outer approximation. Mathematical Programming 66, 327–349 (1994)

[10] Geoffrion, A.: Generalized Benders Decomposition. Journal of Optimization 10, 237–260 (1972)

[11] Gupta, O.K., Ravindran, A.: Branch and bound experiments in convex nonlinear integer programming. Management Science 31(12), 1533–1546 (1985)

[12] Quesada, I., Grossmann, I.E.: An LP/NLP based branch-and-bound algorithm for convex MINLP. Computers and Chemical Engineering 16, 937–947 (1992)

[13] Bonmin 1.5.1: Basic Open source Nonlinear Mixed INteger programming (2013), www.coin-or.org/Bonmin

[14] Van Emde Boas, P.: Another NP-complete problem and the complexity of computing short vectors in a lattice. Technical Report 81-04, University of Amsterdam, Department of Mathematics (1981)

[15] Westerlund, T., Pettersson, F.: A cutting plane method for solving convex MINLP problems. Computers and Chemical Engineering 19, 131–136 (1995)

Separable Non-convex Underestimators
for Binary Quadratic Programming

Christoph Buchheim and Emiliano Traversi

Fakultät für Mathematik, Technische Universität Dortmund
Vogelpothsweg 87, 44227 Dortmund, Germany
{christoph.buchheim,emiliano.traversi}@tu-dortmund.de

Abstract. We present a new approach to constrained quadratic binary programming. Dual bounds are computed by choosing appropriate global underestimators of the objective function that are separable but not necessarily convex. Using the binary constraint on the variables, the minimization of this separable underestimator can be reduced to a linear minimization problem over the same set of feasible vectors. For most combinatorial optimization problems, the linear version is considerably easier than the quadratic version. We explain how to embed this approach into a branch-and-bound algorithm and present experimental results.

1 Introduction

Many combinatorial optimization problems admit natural formulations as binary quadratic optimization problems. Such problems take the form

$$\min \quad f(x) := x^\top Q x + L^\top x$$
$$\text{s.t.} \quad x \in X , \tag{1}$$

where $Q \in \mathbb{R}^{n \times n}$ is a symmetric matrix, $L \in \mathbb{R}^n$ is a vector and $X \subseteq \{0,1\}^n$ is the set of feasible binary vectors. In this paper, we consider problems where the linear counterpart of Problem (1),

$$\min \quad c^\top x$$
$$\text{s.t.} \quad x \in X , \tag{2}$$

can be solved efficiently for any vector $c \in \mathbb{R}^n$. We do not make any assumptions on how Problem (2) is solved. In particular, any combinatorial algorithm can be used, a compact linear description (or polynomial-time separation algorithm) for $\text{conv}(X)$ is not required.

Even under this assumption, the quadratic problem (1) is usually NP-hard. This is true, e.g., for the unconstrained case $X = \{0,1\}^n$, where Problem (1) is equivalent to unconstrained quadratic binary optimization and hence to the max-cut problem. To give another example, the quadratic spanning tree problem is NP-hard [1], while the linear counterpart can be solved very quickly, e.g., by Kruskal's algorithm.

V. Bonifaci et al. (Eds.): SEA 2013, LNCS 7933, pp. 236–247, 2013.
© Springer-Verlag Berlin Heidelberg 2013

The standard approach for solving problems of type (1) is based on linearization. In a first step, a new variable y_{ij} representing the product $x_i x_j$ is introduced for each pair i, j. Then the convex hull of feasible solutions in the extended space is usually approximated either by a polyhedral relaxation or by semidefinite programming (SDP) models, or by a combination of both. The main focus lies on enforcing the connection between x- and y-variables. For the unconstrained case, we point the reader to [5] and the reference therein. In the constrained case, most approaches presented in the literature are highly problem-specific.

A different approach to binary optimization is the QCR technique [2]. Instead of linearizing the problem, it is reformulated as an equivalent binary optimization problem with a *convex* quadratic objective function. This allows to apply more powerful software tailored for convex problems. In particular, it is now possible to solve the continuous relaxation of the problem efficiently. The QCR approach is designed such that this relaxation yields as tight lower bounds as possible.

In this paper, we propose a different approach. It is based on computing underestimators g of the quadratic objective function f. A lower bound on Problem (1) can then be computed by minimizing $g(x)$ over $x \in X$. Unlike most other approaches based on underestimators, we however do not use convex functions in general, but separable non-convex functions. The main idea of our approach is to determine a good separable underestimator g of f in the first step; in the second step we can reduce the separable quadratic function to a linear function exploiting the binarity of all variables. The minimization of $g(x)$ over $x \in X$ can thus be performed by solving Problem (2). Convexity is not required for this approach. The resulting lower bounds are embedded into a branch-and-bound scheme for solving Problem (1) to optimality.

Compared with linearization, the advantage of our approach lies in the fact that we do not need to add any additional variables. Moreover, we do not require any polyhedral knowledge about conv (X) and do not use any LP solver at all. At the same time, any algorithmic knowledge about the linear problem (2) is exploited directly. Compared with the convexification approach, we have a chance to obtain better lower bounds, since we do not require convexity of the underestimator.

An important question in our approach is how to compute the separable underestimator g. We first fix a point $z \in \mathbb{R}^n$ where $f(z) = g(z)$, i.e., where the underestimator touches the original objective function. Reasonable choices discussed in this paper are the stationary point $\bar{x}$ of f, the origin, and the center of the box $\frac{1}{2}\mathbf{1}$. Under this restriction, we compute a separable quadratic function g that is a global underestimator for f and that maximizes the minimum of $g(x)$ over X, i.e., that yields a best possible lower bound. We show that this task can be accomplished efficiently either by solving a semidefinite program or by applying a subgradient method, depending on z and X.

This paper is organized as follows. In the next section, we formalize the main ideas of our approach. In Section 3, we present strategies to determine separable underestimators yielding best possible lower bounds. In Section 4, we discuss how lower bounds can be improved by taking valid linear equations into account.

Details of our branch-and-bound algorithm are given in Section 5. In Section 6, we evaluate our approach computationally, applying it to unconstrained problems and to instances of the quadratic spanning tree problem. It turns out that the new algorithm, though being very general, can solve problems of medium size in reasonable running time.

2   Notation and Basic Idea

We consider Problem (1) and assume that its linear counterpart, Problem (2), can be solved efficiently for any vector $c \in \mathbb{R}^n$. We will use Problem (2) as a black box in the following. Our main idea is to derive a lower bound for Problem (1) by globally underestimating f by a separable but not necessarily convex function g and then using Problem (2) to compute the bound.

For an arbitrary point $z \in \mathbb{R}^n$, we can rewrite $f(x)$ as

$$f(x) = (x - z)^\top Q(x - z) + (L + 2Qz)^\top x - z^\top Qz . \tag{3}$$

Now define

$$
\begin{aligned}
g_z^{(t)}(x) &:= (x - z)^\top \mathrm{Diag}(t)(x - z) + (L + 2Qz)^\top x - z^\top Qz \\
&= \sum_{i=1}^n t_i x_i^2 + \sum_{i=1}^n (-2z_i t_i + l_i + 2q_i^\top z)x_i + \sum_{i=1}^n z_i^2 t_i - z^\top Qz
\end{aligned}
$$

for $t \in \mathbb{R}^n$, where q_i denotes the i-th row of Q. Then $g_z^{(t)}(z) = f(z)$, i.e., the function $g_z^{(t)}$ touches f in the point z. By (3), it is easy to see that the function $g_z^{(t)}$ is a global underestimator of f if and only if $Q \succeq \mathrm{Diag}(t)$. In this case, the desired lower bound can be obtained as

$$\min \ g_z^{(t)}(x) \quad \text{s.t. } x \in X . \tag{4}$$

As $X \subseteq \{0,1\}^n$, we can replace Problem (4) by the equivalent problem

$$\min \ l_z^{(t)}(x) \quad \text{s.t. } x \in X \tag{5}$$

where the function

$$
\begin{aligned}
l_z^{(t)}(x) &:= \sum_{i=1}^n t_i x_i + \sum_{i=1}^n (-2z_i t_i + l_i + 2q_i^\top z)x_i + \sum_{i=1}^n z_i^2 t_i - z^\top Qz \\
&= ((1 - 2z) \cdot t + L + 2Qz)^\top x + z^2 \cdot t - z^\top Qz
\end{aligned}
$$

is bilinear in $x, t \in \mathbb{R}^n$. Here we use $\cdot$ to denote entrywise multiplication and define $z^2 := z \cdot z$. Note that Problem (5) is of type (2) and can thus be solved efficiently by our assumption.

This approach is feasible for each touching point $z \in \mathbb{R}^n$. Throughout this paper, we concentrate on three different choices of z: the origin, the point $\frac{1}{2}\mathbf{1}$, and the stationary point $\bar{x} := -\frac{1}{2}Q^{-1}L$ of f (if Q is a regular matrix). In the respective special cases, the function $l_z^{(t)}$ can be simplified as follows:

$$l_z^{(t)}(x) = \begin{cases} x^\top t + L^\top x & \text{if } z = 0 \\ \frac{1}{4}\mathbf{1}^\top t + (L + Q\mathbf{1})^\top x - \frac{1}{4}\mathbf{1}^\top Q\mathbf{1} & \text{if } z = \frac{1}{2}\mathbf{1} \\ ((1 - 2\bar{x}) \cdot x + \bar{x}^2)^\top t - \bar{x}^\top Q\bar{x} & \text{if } z = \bar{x}. \end{cases}$$

3   Optimal Separable Underestimators

The choice of t is crucial for the strength of the lower bound resulting from (5). As discussed above, this lower bound is valid for each $t \in \mathbb{R}^n$ with $Q \succeq \text{Diag}(t)$. Our objective is to maximize the lower bound induced by t. In other words, our aim is to solve the problem

$$\begin{aligned} \max \quad & \min_{x \in X} l_z^{(t)}(x) \\ \text{s.t.} \quad & Q \succeq \text{Diag}(t). \end{aligned} \tag{6}$$

In the easiest case $X = \{0,1\}^n$, we have

$$\min_{x \in \{0,1\}^n} l_z^{(t)}(x) = \min_{x \in \{0,1\}^n} ((1 - 2z) \cdot t + L + 2Qz)^\top x + z^2 \cdot t - z^\top Qz$$

$$= z^2 \cdot t - z^\top Qz + \sum_{i=1}^n \min\{0, (1 - 2z_i)t_i + l_i + 2q_i^\top z\}$$

so that Problem (6) reduces to solving the semidefinite program

$$\begin{aligned} \max \quad & \sum_{i=1}^n z_i^2 t_i + y_i \\ \text{s.t.} \quad & y_i \le 0 \\ & y_i \le (1 - 2z_i)t_i + l_i + 2q_i^\top z \\ & Q \succeq \text{Diag}(t). \end{aligned}$$

For general X, Problem (6) can be solved by a subgradient approach; this is discussed in Section 3.1. However, if the chosen touching point is $z = \frac{1}{2}\mathbf{1}$, the problem can again be reduced to solving a single semidefinite program, as explained in Section 3.2.

3.1   Subgradient Method

For general $X \subseteq \{0,1\}^n$, Problem (6) can be solved by a subgradient method. For this, we can model the constraint $Q \succeq \text{Diag}(t)$ by a penalty function, and obtain the following problem:

$$\begin{aligned} \max \quad & \min_{x \in X} l_z^{(t)}(x) + \mu \min\{0, \lambda_{\min}(Q - \text{Diag}(t))\} \\ \text{s.t.} \quad & t \in \mathbb{R}^n, \end{aligned} \tag{7}$$

where μ is a suitably large non-negative number. The objective function of (7) is concave, so that a subgradient approach can be used to solve the problem efficiently. The supergradient of

$$\min_{x \in X} l_z^{(t)}(x)$$

at a given point t^k can be computed by using the black box (2), as $l_z^{(t^k)}(x)$ is a linear function in x. Given the optimal solution $\hat{x}^k$, the desired supergradient is the gradient of $l_z^{(t)}(\hat{x}^k)$, which is easily computed since $l_z^{(t)}(\hat{x}^k)$ is a linear function also in t. If $\lambda_{\min}(Q - \mathrm{Diag}(t^k)) < 0$, the supergradient of the penalty term can be obtained as $-\mu v^2$, where v is a normalized eigenvector corresponding to the eigenvalue $\lambda_{\min}(Q - \mathrm{Diag}(t^k))$.

The resulting subgradient approach is sketched in Algorithm 1. Note that Algorithm 1 can be stopped at any time. Let t^k be the best solution to Problem (7) obtained so far. If $Q - \mathrm{Diag}(t^k) \succeq 0$, then t^k is also feasible for (6). If $\lambda_{\min}(Q - \mathrm{Diag}(t^*)) < 0$, then a new solution $\bar{t}$ can be obtained by

$$\bar{t} := t^k + \lambda_{\min}(Q - \mathrm{Diag}(t^k))\mathbf{1}$$

and $\bar{t}$ is a feasible solution for (6) by construction.

Algorithm 1. computation of optimal underestimator

---

input : function f, set X, touching point z, penalty parameter μ,
            procedure for solving Problem (2)
output: a (near-)optimal solution to Problem (6)

$t^0 \leftarrow \lambda_{\min}(Q)\mathbf{1}$;
$k \leftarrow 0$, STOP$\leftarrow$ false;

while *STOP= false* **do**

    solve $\min_{x \in X} l_z^{(t^k)}(x)$, let $\hat{x}^k$ be the optimal inner solution;
    // using black box to solve the inner problem

    $\Delta t^k \leftarrow \left(\nabla_t l_z^{(t)}(\hat{x}^k)\right)(t^k)$; $\lambda \leftarrow \lambda_{\min}(Q - \mathrm{Diag}(t^k))$;
    if $\lambda < 0$ **then**
        choose normalized eigenvector v of $Q - \mathrm{Diag}(t^k)$ to eigenvalue λ;
        $\Delta t^k \leftarrow \Delta t^k - \mu v^2$;
    end
    // computing a supergradient

    if $\Delta t^k \approx 0$ **then**
        STOP$\leftarrow$true; *// t^k is (near-)optimal*
    end
    else
        $t^{k+1} \leftarrow t^k + \Delta t^k$; $k \leftarrow k + 1$;
    end

end

3.2   Box Center as Touching Point

If the touching point z is chosen as $\frac{1}{2}\mathbf{1}$, the optimization problem (6) can be solved more efficiently. In this case, the function

$$l_z^{(t)}(x) = \frac{1}{4}\mathbf{1}^\top t + (L + Q\mathbf{1})^\top x - \frac{1}{4}\mathbf{1}^\top Q\mathbf{1}$$

does not contain any product between x and t. Problem (6) can thus be decomposed as follows:

$$\max \ \tfrac{1}{4}\mathbf{1}^\top t \qquad + \qquad \min \ (L + Q\mathbf{1})^\top x \ - \ \tfrac{1}{4}\mathbf{1}^\top Q\mathbf{1}$$
$$\text{s.t. } Q \succeq \mathrm{Diag}(t) \qquad\qquad \text{s.t. } x \in X$$

The first problem is an SDP, while the second problem can be solved by calling the oracle (2) once. In particular, the optimal underestimator only depends on Q in this case, but not on L. This fact can be exploited in our branch-and-bound algorithm, as explained in Section 5.1.

4   Taking Valid Equations into Account

So far, we assumed that we can access the set X of feasible solutions only via the linear optimization oracle (2). This oracle is used in the second step of the lower bound computation, the minimization of the underestimator. In particular, this step implicitly exploits full knowledge about X.

On the other hand, the computation of an underestimator does not exploit any properties of X, we require that $g_z^{(t)}$ globally underestimates f. If it is known that the set X satisfies certain linear equations $Ax = b$, this information can be used to improve the lower bounds significantly: it is enough to require that the function $g_z^{(t)}$ is an underestimator of f on the affine subspace given by $Ax = b$. This leads to a weaker condition on t, which can still be handled efficiently.

More precisely, let $H = \{x \in \mathbb{R}^n \mid Ax = b\}$ be nonempty and choose $w \in H$. Let $v_1, \ldots, v_k$ be an orthonormal basis of the kernel of A and set $V = (v_1|\ldots|v_k)$, so that $H = \{w + Vy \mid y \in \mathbb{R}^k\}$. We first assume that the touching point z belongs to H, e.g., by defining it as the orthogonal projection of $\frac{1}{2}\mathbf{1}$ to H:

$$z := w + VV^\top(\tfrac{1}{2}\mathbf{1} - w)$$

Now $g_z^{(t)}|_H$ is an underestimator of $f|_H$ if and only if

$$(x - z)^\top Q(x - z) \geq (x - z)^\top \mathrm{Diag}(t)(x - z) \ \forall x \in H$$
$$\Leftrightarrow \qquad y^\top V^\top QVy \geq y^\top V^\top \mathrm{Diag}(t)Vy \qquad \forall y \in \mathbb{R}^k$$
$$\Leftrightarrow \qquad V^\top QV \succeq V^\top \mathrm{Diag}(t)V .$$

The latter constraint can be used to replace the stronger constraint $Q \succeq \mathrm{Diag}(t)$ both in the subgradient approach and in the SDP based computation of t, potentially yielding tighter lower bounds in both approaches. In the former approach, the penalty term can be replaced by $\min\{0, \lambda_{\min}(V^\top(Q - \mathrm{Diag}(t))V)\}$.

The corresponding supergradient is $-(Vv)^2$, where v is a normalized eigenvector of $V^\top(Q - \mathrm{Diag}(t))V$ corresponding to its smallest eigenvalue.

In the latter approach, the constraint $Q - \mathrm{Diag}(t) \succeq 0$ can be replaced by $V^\top(Q - \mathrm{Diag}(t))V \succeq 0$ and the resulting problem remains a semidefinite program. Note that the dimension of this SDP decreases by $n - k = \mathrm{rk}(A)$. In other words, a bigger rank of A implies a smaller number of variables in the semidefinite program.

5   Branch-and-bound Algorithm

In order to solve Problem (1) exactly, we embed the lower bounds derived in Section 3 into a branch-and-bound framework. We thus need to compute the lower bounds as quickly as possible and for many related problems. In the following, we describe how the ideas presented in the previous sections can be adapted to this situation.

5.1   Branching Strategy

In order to compute lower bounds as quickly as possible, we restrict ourselves in two different ways:

1. We determine an order of variables at the beginning and fix variables always in this order. More precisely, if $x_1, \ldots, x_n$ is the chosen order, the next variable to be fixed is the free variable with smallest index. The same idea has been used in [3] and [4].
2. We do not call the subgradient method to compute an optimal t in every node, but try to find one fixed t for each level of the enumeration tree that yields strong lower bounds on average. This reduces the number of oracle calls to one per node.

The reason for accepting Restriction 1 is as follows: by this branching strategy, the reduced matrices Q in the nodes of the enumeration tree only depend on the depth of the node but not on the specific subproblem. Consequently, only n such matrices can appear in the enumeration tree, instead of 2^n when applying other branching strategies. All time-consuming computations concerning this matrix can now be performed in a preprocessing phase. In particular, in combination with Restriction 2, we can now determine one feasible t for all nodes on a given depth in the preprocessing.

More precisely, consider a subproblem on depth d of the enumeration tree. This means that variables $x_1, \ldots, x_d$ have been fixed to some values $\alpha \in \{0,1\}^d$ and the resulting objective function in the given node becomes

$$f_\alpha \colon \mathbb{R}^{n-d} \to \mathbb{R}, \quad f_\alpha(x) = x^\top Q_\alpha x + L_\alpha^\top x + c_\alpha,$$

where Q_α is obtained from Q by deleting the first d rows and columns and

$$(L_\alpha)_j := L_{d+j} + 2\sum_{i=1}^{d} \alpha_i Q_{d+j,i}, \quad c_\alpha := \sum_{i,j=1}^{d} \alpha_i \alpha_j Q_{i,j} + \sum_{i=1}^{d} \alpha_i L_i.$$

As Q_α only depends on the depth d but not on α, we may denote it by Q_d. The coefficients of L_α and c_α can be computed incrementally in $O(n - d)$ time per node using the recursive formulae

$$(L_\alpha)_j = (L_{(\alpha_1,\ldots,\alpha_{d-1})})_{j+1} + 2\alpha_d Q_{d+j,d}$$
$$c_\alpha = c_{(\alpha_1,\ldots,\alpha_{d-1})} + \alpha_d^2 Q_{d,d} + \alpha_d L_d .$$

In Section 3 we showed that for the special touching point $\frac{1}{2}\mathbf{1}$ the optimal lower bound can be computed as

$$\max \ \tfrac{1}{4}\mathbf{1}^\top t \qquad + \quad \min \ (L + Q\mathbf{1})^\top x \quad - \quad \tfrac{1}{4}\mathbf{1}^\top Q\mathbf{1} .$$
$$\text{s.t.} \ Q \succeq \mathrm{Diag}(t) \qquad\qquad \text{s.t.} \ x \in X$$

The first problem is a semidefinite program that does not depend on L_α or c_α. In other words, the optimal t for all nodes on depth d is the same and can be computed in the preprocessing.

In order to accelerate the computation of lower bounds, we can apply this approach for any choice of a touching point z. Any solution of the SDP

$$\max \ \mathbf{1}^\top t \tag{8}$$
$$\text{s.t.} \ Q \succeq \mathrm{Diag}(t)$$

yields feasible lower bounds. In general, the resulting lower bounds are weaker than the bounds obtained from the subgradient method presented in Section 3, but in terms of total running time this approach outperforms the subgradient approach, as only one oracle call per node is necessary.

5.2   Incremental Update for Valid Equations

The fixed order of variables can also be exploited to accelerate the computation of data necessary to handle valid equations. It implies that the induced constraint matrix in a given node again only depends on its depth d in the enumeration tree, it results from deleting the first d columns from A; denote the resulting matrix by A_d. Consequently, the kernel vectors V_d of A_d can be computed in a preprocessing phase again, and the same is true for the matrices $V_d^\top Q_d V_d$ needed in the computation of lower bounds.

On contrary, the induced right hand side of the set of valid equations depends on the fixings applied so far, it turns out to be

$$b_\alpha := b - \sum_{i=1}^{d} \alpha_i A_{\bullet,i} \in \mathbb{R}^m .$$

This implies that the projection z_α of some touching point $z_d \in \mathbb{R}^{n-d}$ to the subspace given by $A_d x = b_\alpha$ depends on the specific node and cannot be computed in the preprocessing. However, it can be calculated incrementally, thus avoiding to solve a linear system of equations in every node of the enumeration tree: in the preprocessing phase we determine vectors $w_0 \in \mathbb{R}^n$ and $y_d \in \mathbb{R}^{n-d}$ for $d = 1, \ldots, n$ satisfying

$$Aw_0 = b \text{ and } A_{\bullet, d+1 \ldots n} y_d = A_{\bullet, d} \text{ for all } d = 1, \ldots, n .$$

When enumerating the branch-and-bound nodes, we incrementally compute a vector w_α satisfying $A_d x = b_\alpha$ as follows: for $d = 0$, we can use $w_\alpha = w_0$. For $d \geq 1$, we set

$$w_\alpha := (w_{(\alpha_1, \ldots, \alpha_{d-1})})_{2 \ldots n-d+1} + ((w_{(\alpha_1, \ldots, \alpha_{d-1})})_1 - \alpha_d)y_d \in \mathbb{R}^{n-d} .$$

Then

$$\begin{aligned} A_d w_\alpha &= A_d (w_{(\alpha_1, \ldots, \alpha_{d-1})})_{2 \ldots n-d+1} + ((w_{(\alpha_1, \ldots, \alpha_{d-1})})_1 - \alpha_d) A_d y_d \\ &= A_{d-1} w_{(\alpha_1, \ldots, \alpha_{d-1})} - \alpha_d A_{\bullet, d} \end{aligned}$$

so that by recursion we obtain

$$A_d w_\alpha = A_0 w_0 - \sum_{i=1}^{d} \alpha_i A_{\bullet, i} = b_\alpha$$

as desired. The projected touching point can now be computed using the formula

$$z_\alpha := w_\alpha + V_d V_d^\top (z_d - w_\alpha)$$

given in Section 4, where $V_d V_d^\top$ can again be computed in the preprocessing. The total running time for computing z_α in a node on depth d using this approach is $O((n - d)^2)$. The time spent in the preprocessing is dominated by the time needed to solve the $n + 1$ systems of linear equations determining w_0 and $y_1, \ldots, y_n$.

5.3   Application of the Subgradient Method

Inside a branch-and-bound framework, the running time of the subgradient method for computing a vector t, as presented in Section 3, can be cut in several ways. As with any subgradient method, a careful tuning of parameters, such as step length, is important for obtaining a decent rate of convergence. Moreover, in a given node on depth $d \geq 1$, we use the best solution $t^*_{(\alpha_1, \ldots, \alpha_{d-1})}$ of the parent node for warmstarting. More precisely, we use the last $n - d$ entries of this solution as initial solution for t^0_α and choose an initial step length that is decreasing with increasing depth d in the enumeration tree.

Furthermore, as every feasible iterate in the subgradient method yields a valid lower bound for our primal problem, we can stop Algorithm 1 as soon as the current lower bound given by (5) exceeds the primal bound, i.e., the objective value of the best known solution of Problem (1).

From a practical point of view, a good strategy is to perform a few re-optimization iterations of Algorithm 1 in every node. An even more restricted approach is to determine the best possible t in the root node and then keep the corresponding underestimator throughout the entire branch-and-bound algorithm, with the necessary adaptions. We will compare these choices in the numerical experiments in Section 6.

6   Experiments

The aim of this section is to determine which variant of our approach yields the most effective underestimator t. As benchmark we use two sets of instances:

– *Unconstrained BQP.*
  We generated a set of random binary instances with $n = 20, 30$. Ten possible levels of convexity of Q are tested: from 10% to 100% of negative eigenvalues. For a given concavity, we randomly generate three different instances for a total of 60 instances.

– *Quadratic Spanning Tree Problem.*
  We generated a set of random graphs $G = (V, E)$ and associated linear and quadratic costs L and Q with uniformly distributed random integer entries, with absolute value in the interval $[1, 100]$. A given instance is characterized by (1) the number of nodes $|V| = 15, 20$, (2) the density $d = 25\%, 50\%, 75\%$ of G, and (3) the percentage of positive coefficients $p = 25\%, 50\%, 75\%$; the matrix Q is dense in all instances. For each combination of parameters we randomly generate three different instances for a total of 54 instances.

For all tests, we use an Intel Xeon E5-2670 processor, running at 2.60 GHz with 64 GB of RAM. Running times are stated in CPU seconds.

As first step, we test different touching points z as explained in Section 2. For each candidate, we solve Problem (8) in the preprocessing phase. As test bed we use the *Unconstrained BQP*. In Figure 1 we present the average number of branch-and-bound nodes for a given percentage of negative eigenvalues. In addition to the results obtained by using the optimal t of Problem (8) (*SDP*), we also report those obtained by using the trivial underestimator $t = -\lambda_{\min}(Q)\mathbf{1}$ (*Triv*). For each policy for t we report the results obtained by fixing the touching point to the origin (0), $\frac{1}{2}\mathbf{1}$ (0.5) or $\bar{x}$ (*stat*). It is obvious from these results that the best choice is $z = \frac{1}{2}\mathbf{1}$ (yellow and green columns): the total number of explored nodes is 10 times and 100 times less than the number of nodes needed with touching point $z = 0$ (blue and red columns) and $z = \bar{x}$ (brown and light blue columns) respectively.

As second step we want to test how t is improved by taking valid inequalities into account; see Section 4. The set *Quadratic Spanning Tree Problem* is used and the (only) valid equation is $\sum_{e \in E} x_e = |V| - 1$. In Table 1 we show how even one single equation is improving the behaviour of the corresponding t. Every line is reporting the number of nodes and computing time (corresponding to an average of three instances). The dimension n of the instances is stated in the second column. With Eq we indicate that t is obtained considering equations and with $NoEq$ the opposite, moreover we report the ratio r between these values. Also in this case the answer is clear: considering equations decreases significantly the number of nodes. E.g., for the larger instances this decreases the number of nodes by a factor of 50 and the solution time by a factor of 20. We finally remark that Table 1 only reports results for smaller instances, because only 15 out of 21 large instances were solved within our time limit of four hours by $NoEq$. Also in

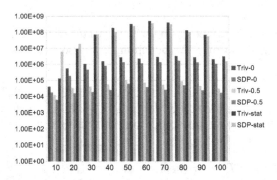

Fig. 1. Touching points comparison

Table 1. Effect of taking equations into account

instances	n	Nodes			Time		
		NoEq	Eq	ratio	NoEq	Eq	ratio
qstp_15_25_25	26	2,714.3	2,685.0	1.0	0.4	1.1	0.4
qstp_15_25_50	26	7,514.3	6,747.0	1.1	0.4	1.1	0.4
qstp_15_25_75	26	5,545.7	5,141.7	1.1	0.4	1.0	0.4
qstp_15_50_25	52	456,420.3	138,324.3	3.3	24.0	66.4	0.4
qstp_15_50_50	52	29,846,421.7	1,203,823.0	24.8	163.0	70.3	2.3
qstp_15_50_75	52	983,578,822.3	18,793,411.7	52.3	4452.3	207.0	21.5

this case, using *Eq* improved the performance, allowing to solve 5 out of the 6 instances unsolved by *NoEq*.

Finally, we test whether updating t during the exploration of the branch-and-bound tree applying Algorithm 1 is better than solving a series of SDPs in the preprocessing. Using the warmstart described in Section 5.3, the values of t in the non-root nodes of the tree are computed by k rounds of Algorithm 1. We tried different settings $k = 1, 2, 5, 10, 100$, but none of them succeeded in improving the overall computation time. In Figure 2, we show how the increase in k affects the total number of nodes and the running time for the set of instances qstp_15_50_50. The red line represents the values obtained by computing t in the preprocessing and the blue line represents the subgradient evolution. As we can see, almost 100 iterations of Algorithm 1 per node are needed in order to improve at least the number of nodes.

The results presented clearly indicate that the best setting is using $\frac{1}{2}1$ as touching point, fixing the best underestimators t levelwise from the beginning and taking into account valid equations. Obtaining a good solution to Problem (6) is a crucial aspect, additional tests showed also that increasing the tolerance in the SDP solver (and hence allowing worse solutions) provides significantly worse underestimators. This consideration, together with the other results

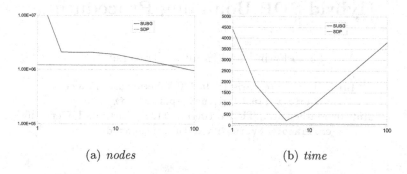

(a) *nodes*   (b) *time*

Fig. 2. Results of the subgradient method for qstp_15_50_50

provided in this section, gives an idea about the strong sensitivity of the overall algorithm to the chosen vector t. The importance of valid inequalities makes problems such as the quadratic assignment problem or the quadratic shortest path problem particularly appealing for future applications of our approach.

Acknowledgments. The authors would like to thank Antonio Frangioni for fruitful discussions and suggestions that improved the present paper significantly.

References

1. Assad, A., Xu, W.: The quadratic minimum spanning tree problem. Naval Research Logistics 39(3), 399–417 (1992)
2. Billionnet, A., Elloumi, S., Plateau, M.-C.: Improving the performance of standard solvers for quadratic 0–1 programs by a tight convex reformulation: The QCR method. Discrete Applied Mathematics 157(6), 1185–1197 (2009)
3. Buchheim, C., Caprara, A., Lodi, A.: An effective branch-and-bound algorithm for convex quadratic integer programming. Mathematical Programming (Series A) 135(1-2), 369–395 (2012)
4. Buchheim, C., De Santis, M., Palagi, L., Piacentini, M.: An exact algorithm for quadratic integer minimization using nonconvex relaxations. Technical report, Optimization Online (2012)
5. Palagi, L., Piccialli, V., Rendl, F., Rinaldi, G., Wiegele, A.: Computational approaches to Max-Cut. In: Handbook on Semidefinite, Conic and Polynomial Optimization, pp. 821–849. Springer (2012)

Hybrid SDP Bounding Procedure

Fabio Furini[1] and Emiliano Traversi[2]

[1] LIPN, Université Paris 13, 93430 Villetaneuse, France
fabio.furini@lipn.univ-paris13.fr
[2] Fakultät für Mathematik, TU Dortmund, 44227 Dortmund, Germany
emiliano.traversi@math.tu-dortmund.de

Abstract. The principal idea of this paper is to exploit Semidefinite Programming (SDP) relaxation within the framework provided by Mixed Integer Nonlinear Programming (MINLP) solvers when tackling Binary Quadratic Problems. We included the SDP relaxation in a state-of-the-art MINLP solver as an additional bounding technique and demonstrated that this idea could be computationally useful. The Quadratic Stable Set Problem is adopted as the case study. The tests indicate that the Hybrid SDP Bounding Procedure allows an average 50% cut of the overall computing time and a cut of more than one order of magnitude for the branching nodes.

Keywords: Binary Quadratic Problems, Semidefinite Relaxation, Branch and Cut, Quadratic Stable Set Problem.

1 Introduction

There are two main classical approaches present in the literature for solving Binary Quadratic Problems (BQP). The first one is directly using a Mixed Integer Nonlinear Programming (MINLP) solver to tackle a mathematical formulation (possibly linearizing the quadratic terms). The second approach uses Branch and Bound techniques which rely on the Semidefinite Programming (SDP) relaxation. The advantage of using MINLP solvers is that they have been strongly developed for decades. To mention just a few examples, we can cite some commercial software like BARON [2], CPLEX [7] and Gurobi [9]; as well as non commercial, for instance Bonmin [4]. They rely on sophisticated Branch and Cut (BC) algorithms based on a smart implicit enumeration of the branching tree. The bounding procedure typically makes use of Continuous Relaxation (CR) in order to prune the branching nodes. This relaxation can be efficiently computed but often, to be effective, it must be strengthened by adding families of valid inequalities. The second approach relies instead on the SDP relaxation. This relaxation is typically stronger than CR but generally it is computationally heavy. To the best of our knowledge, there are not many generic SDP-based solvers, BiqCrunch [3] to name one. On the other hand, there are many problem-oriented SDP algorithms, and we refer the interested reader for instance to [6], [12] or [14].

V. Bonifaci et al. (Eds.): SEA 2013, LNCS 7933, pp. 248–259, 2013.

Within this context, we mention some works (see for example [1]), where the authors compare SDP relaxation and CR proving that stronger bounds can be achieved.

The principal contribution of this paper is to combine the strengths of both approaches, exploiting the SDP relaxation when enhancing the pruning strategies and thus the overall performances of a MINLP solver. To summarize, our goal boils down to addressing the following research questions:

- Is it worth exploiting the strength of bounds provided by SDP relaxation within a well tuned BC framework?
- Which mathematical formulation will profit more from the addition of the SDP bounding techniques?
- Which SDP relaxation has the best trade-off between the time saved by pruning nodes and the time needed for computing the SDP relaxation?
- What additional policies are necessary in order to improve the overall computational performances?

The presentation of the theoretical part of the paper is done for the generic case of the BQP. After each technique is presented, we apply it to the specific case of the Quadratic Stable Set Problem (QSSP). Recalling that a stable set in a graph is a subset of vertices such that for every two vertices selected in the solution, there is no edge connecting the two, the formal definition of the QSSP is the following: given an undirected graph $G = (V, E)$, with V ($n = |V|$) the set of vertices and E the set of edges, a vector of linear profit $L \in \mathbb{R}^n$ and a symmetric matrix of quadratic profit $Q \in \mathbb{R}^{n \times n}$ (possibly negative), the Quadratic Stable Set Problem (QSSP) searches for a stable set of G with maximum profit. In other words, if vertices i and j are in the solution, not only the linear profits are collected but also an additional profit equal to Q_{ij}. This quadratic counterpart of the Linear Stable Set Problem has not received much attention in the literature (we refer the interested reader for the linear case to [8] and [13]). Furthermore, to the best of our knowledge, the only papers that address the QSSP are [10] and [11]. In these works it appears as the sub-problem of a Column Generation algorithm and little computational analysis is presented.

2   Different Mathematical Formulations

The first step of the present work is to introduce the mathematical formulations that can be handled by a generic MINLP solver, i.e. the Quadratic Formulation and the Linear Formulation.

2.1   Quadratic Formulation

The Quadratic Formulation (QF) of the BQP, with n variables and p constraints, is defined as follows:

$$\text{(QF)} \quad \max \sum_{i=1}^{n}\sum_{j=1}^{n} Q_{ij}x_ix_j + \sum_{i=1}^{n} L_ix_i$$

$$x \in K$$

$$x \in \{0,1\}^n,$$

with $Q \in \mathbb{R}^{n\times n}$, $L \in \mathbb{R}^n$, $K = \{x \in \mathbb{R}^n : Ax \geq b\}$, $A \in \mathbb{R}^{p\times n}$ and $b \in \mathbb{R}^p$. Q is a generic symmetric matrix, not restricted to being convex.

In the case of QSSP on a given undirected graph $G = (V,E)$, with V the set of vertices and E the set of edges, we have:

$$K = \{x \in \mathbb{R}^n : x_i + x_j \leq 1, \forall\{i,j\} \in E\}.$$

In order to solve the QF, many generic MINLP solvers are available. In addition, other solvers, explicitly defined for BQP can be used, such as Cplex and GloMIQO.

2.2 Linear Formulation

Another option for modelling BQP is to linearise the quadratic terms (we refer the interested reader to [16]) and obtain the following Linear Formulation (LF):

$$\text{(LF)} \quad \max \sum_{i=1}^{n}\sum_{j=i}^{n} Q_{ij}y_{ij} + \sum_{i=1}^{n} L_ix_i$$

$$\left.\begin{array}{l} y_{ij} \leq x_i \\ y_{ij} \leq x_j \\ y_{ij} \geq x_i + x_j - 1 \\ y_{ij} \geq 0 \end{array}\right\} i,j = 1,\dots,n \qquad (1)$$

$$x \in K$$

$$x \in \{0,1\}^n.$$

This linearization increases the size of the problem adding at most n^2 non-negative variables and $3n^2$ constraints.

3 SDP Formulation

BQP can also be formulated as follows:

$$\max \langle \tilde{Q}, Y \rangle$$

$$Y = \begin{pmatrix} 1 \\ \bar{x} \end{pmatrix}\begin{pmatrix} 1 \\ \bar{x} \end{pmatrix}^\top \qquad (2)$$

$$\bar{x} \in \bar{K} \qquad (3)$$

$$\bar{x} \in \{-1,1\}^n \qquad (4)$$

obtained after applying the linear transformation $\bar{x}_i = 2x_i - 1$, imposing

$$\tilde{Q} = \begin{pmatrix} \frac{1}{2}\sum_{j=1}^{n} L_i + \frac{1}{4}\sum_{i=1}^{n}\sum_{j=1}^{n} Q_{ij} & \frac{1}{4}(L + \frac{1}{2}\sum_{j=1}^{n} Q_j)^\top \\ \frac{1}{4}(L + \frac{1}{2}\sum_{j=1}^{n} Q_j) & \frac{1}{4}Q \end{pmatrix}$$

where Q_j is the j-th column of Q, and K is modified to $\bar{K}$ accordingly.

For the QSSP we have:

$$\bar{K} = \{x \in R^n : x_i + x_j \le 0, \forall\{i,j\} \in E\}.$$

Alternatively, we can exploit the fact that the quadratic constraints $x_i x_j = 0$ for every edge $\{i,j\} \in E$ (valid for the stable set) becomes linear when rewritten in the Y space, we can hence instead of $\bar{K}$ add the following set of valid equations:

$$Y_{ij} + Y_{0i} + Y_{0j} + 1 = 0, \forall\{i,j\} \in E. \tag{5}$$

Constraints (2) and (4) together can be rewritten as $Y \succeq 0$, $diag(Y) = e$ and $rank(Y) = 1$, leading to the following equivalent formulation:

$$\max \langle \tilde{Q}, Y \rangle$$
$$Y \succeq 0$$
$$rank(Y) = 1 \tag{6}$$
$$diag(Y) = e$$
$$Y_0 \in \bar{K} \tag{7}$$

with e being the all-ones vector and Y_0 being the first row of Y without the first element (note that $Y_0 = \bar{x}$). To the best of our knowledge, few solvers are able to deal directly with this formulation, BiqCrunch is one of them. Normally these solvers rely on solving special kinds of SDP relaxations. By relaxing the rank constraints (6) we obtain the classic SDP relaxation which provides valid bounds for all the BQP formulations. Note that this relaxation can be weakened by eliminating constraints (7) or strengthened by substituting them by the equations (5). Moreover, any valid inequalities for the LF, for the convex hull of $\bar{K}$ and for the convex hull of their intersection can be added in order to improve the bound further. An important class of valid inequalities for the SDP relaxation are the so-called triangle inequalities, see for example in [14], where the authors suggest how to carefully separate them and keep the overall computational time under control. Finally, in order to tackle the SDP relaxation, we recall that there are different solvers available in the literature, for instance [5] and [15].

4   Hybrid Bounding Procedure

The Hybrid Bounding Procedure is the idea of mixing two different relaxations, i.e. the basic relaxation used by a generic BQP solver and the SDP relaxation.

As previously underlined there are many different possible ways of deriving an SDP relaxation for a BQP. In the following we will define with Ψ a generic version of the SDP relaxation. Furthermore we introduce another function called Ω which controls the fact of performing function Ψ in a specific node or not. This is done because potentially the computation of Ψ can be time-consuming and hence it has to be used ideally only when worth it, in other words, when it prunes the node. Accordingly we need two functions available at each node of the BC tree:

− Function Ψ is the SDP relaxation that takes as input Q, L and a partial fixing of the variables and returns a bound (UB^{SDP}) on the original objective function.
− Function Ω is an oracle that takes as input all the information about the current node and returns a binary variable indicating whether Ψ should be used or not.

The node processing is represented in Algorithm 1, where LB is the incumbent best feasible solution.

Input: best incumbent solution of value LB and current variable fixing.
Output: 1 if the current node can be fathomed, 0 otherwise.

Solve the continuous relaxation and get the bound UB;
if $(UB \leq LB)$ **then return** 1.
OK $\leftarrow \Omega$.
if (OK = FALSE) **then return** 0.
solve Ψ and get the bound UB^{SDP}.
if $(UB^{SDP} \leq LB)$ **then return** 1.
else return 0.

Algorithm 1. Processing at each decision node

In the following computational section we measure the effectiveness of different options of Ψ and Ω in order to improve the overall efficiency.

5   Computational Experiments

The experiments are divided into two steps: first we start with a wide test-bed of instances of small sizes in order to test different possible options and strategies for the Hybrid SDP bounding procedure described in Section 4, once the best settings have been identified, we test bigger instances to evaluate the practical impact of our procedure. All algorithms were coded in C, and run on a PC with an Intel(R) Core2 Duo CPU E6550 at 2.33GHz and 2 GB RAM memory, under Linux Ubuntu 12 64-bit. The optimization software used in our test was Cplex 12.4 single thread. The SDP relaxation solver used is CSDP (described in [5]) and it was inserted in the Cplex framework using the `callBack` functions of the

Callable Library of Cplex. Moreover, we stop the SDP solver as soon as we find a dual bound able to prune the current node, checking the corresponding dual feasibility.

5.1   Testbed Description

As a first test-bed we randomly generated a set of QSSP instances; the goal was to have a statistical relevance (several instances with the same features) and to have the complete control of the characteristics of the different classes of instance proposed. The instance generator produces random graphs according to the desired number of vertices n and density μ (which implies a number of edges equal to $\lfloor \mu \times \frac{n(n-1)}{200} \rfloor$). The linear and quadratic profits take a uniformly random integer value in the interval $[-100, 100]$, a third parameter ν represents the percentage of positive profits. We generated 27 classes of instances by considering all combinations of:

- number of vertices: $n \in \{50, 60, 70\}$;
- density of edges: $\mu \in \{25\%, 50\%, 75\%\}$;
- percentage of positive costs: $\nu \in \{25\%, 50\%, 75\%\}$.

In addition we created 10 instances for each class using different random seeds, thus obtaining in total 270 QSSP instances. As a second test-bed we focus on instances with 100 vertices with the same range of densities and percentages of positive costs. The whole set of instances is available upon request to the authors. In the next sections we discuss the computational outcome of the experiments.

5.2   Identifying the Best Mathematical Formulations

The goal of this section is to computationally evaluate the different formulations, i.e. QF or LF. In order to do that, we used Cplex with default parameter settings. In Figure 1(a) and 1(b) the computing time (seconds) and number of nodes (logarithmic scale) required for the optimization are reported, dividing the instances (from bottom to top) by vertex number (n), density (μ) and percentage of positive profits (ν). As far as the computing time is concerned, the best mathematical formulation is always QF. It is interesting to stress that the behaviour in terms of number of nodes is exactly the opposite: QF performs a number of nodes on average which is at least one order of magnitude bigger than LF. The tests also confirm the tendency for instances with low density to be more difficult than instances with high density. On the other hand, instances with a high percentage of positive profits ν are more difficult than the ones with a low value. These experiments allow us to conclude that QF is better than LF. Moreover, the fact that QF explores a larger amount of nodes makes it a more promising candidate for testing the addition of a second bounding procedure. For these reasons, in the rest of the tests we will focus only on QF.

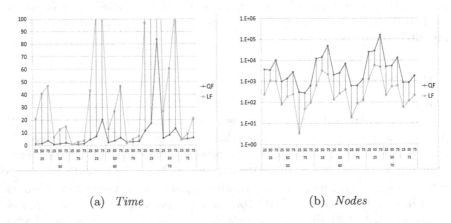

(a) *Time*   (b) *Nodes*

Fig. 1. Performance comparison of different mathematical formulations (QF-LF)

5.3 Different Strategies for Ψ

We propose the following three different options for Ψ:

- **Unconstrained**: SDP relaxation without constraints.
- **Constrained**: SDP relaxation with the family of constraints (7).
- **Constrained2**: SDP relaxation with the family of constraints (5).

In order to access the performance of the different options, six different bounds are computed and compared in Table 1. QF is the root bound of the Cplex BQP solver, LF is the root bound of the continuous relaxation of LF and LF_c is like LF but with the addition of the cuts separated by Cplex in the root node. Finally Uncon, Con1 and Con2 are respectively the bounds of the three different strategies (Ψ). For each option we provide two values: the ratio between the bound obtained and the optimal solution (where a ratio of 1 indicates that the relaxation has no gap with the optimal solution) and the average time needed to compute it. The information concerning QF and LF_c confirms the results obtained in Section 5.2: a BC based on LF is not competitive. It is also interesting to notice that Unconstrained presents a worse gap than QF, although this does not imply that the addition of Unconstrained is not helpful because the SDP relaxation is computationally sensitive to variable fixing during the BC tree

Table 1. Bound Comparison

	Ratios						Times					
n	QF	LF	LF_c	Uncon	Con1	Con2	QF	LF	LF_c	Uncon	Con1	Con2
50	4.0	8.9	2.1	7.5	4.8	1.1	0.1	0.0	7.5	0.0	1.8	1.3
60	4.8	11.8	2.6	9.7	6.0	1.1	0.2	0.1	14.7	0.0	4.6	3.1
70	5.2	14.3	2.9	11.1	6.9	1.1	0.2	0.2	29.2	0.1	10.0	6.9
	4.7	11.7	2.5	9.4	5.9	1.1	0.2	0.1	17.1	0.0	5.5	3.8

explorations and the bound provided becomes stronger in lower levels of the BC tree. If we consider the SDP based bounds we see that `Constrained1` is dominated by `Constrained2` and that there is an interesting trade-off between `Constrained2` and `Unconstrained`. Between the three configurations proposed we hence decided to discard `Constrained` and to focus on `Unconstrained` and `Constrained2`.

5.4 Different Strategies for the Oracle Ω

Ideally the SDP relaxation should be performed only in the cases in which it helps in pruning. To cope with this problem, we tried seven different strategies for the Oracle Ω. The strategies used are:

- 1 `Always`. The SDP relaxation Ψ is triggered at each node of the BC tree.
- 2 `OnOne`. Ψ is triggered every time we branch on one.
- 3 `UnderAverage`. Ψ is triggered if the current integrality gap is lower than the average integrality gap of the nodes explored so far.
- 4 `OverAverage`. Ψ is triggered if the current integrality gap is bigger than the average.
- 5 `SmallGap`. Ψ is triggered only when the integrality gap is within $[0\%, 5\%]$.
- 6 `MediumGap`. Ψ is triggered only when the integrality gap is within $[5\%, 30\%]$.
- 7 `Random`. Ψ is triggered with a random 50% probability.

Strategy `OnOne` exploits the fact that usually when we branch on one the solution in the child nodes will probably change more significantly than when we branch on zero, increasing the probability of pruning. Strategies `OverAverage` and `MediumGap` tend to prune the branch and bound tree at the first levels and they are effective in the case in which the SDP relaxation is much stronger than the continuous relaxation, and/or they tend not to prune in the same points of the branching tree. Strategies `UnderAverage` and `SmallGap` tend to prune the nodes at the final levels and are effective in the cases in which the LP relaxation and the SDP relaxation are strong in the same nodes. Finally, strategies `Always` and `Random` serve as terms of comparison.

5.5 Identifying the Best Ω

Let QF be the basic formulations and QF(Ψ,Ω) be the same formulation using a given couple of SDP relaxation (Ψ) and Oracle strategy (Ω). Four performance indices are collected to assess the computational impact:

$$\delta = \frac{100 \cdot nod_{QF(\Psi,\Omega)}}{nod_{QF}}, \ \tau = \frac{100 \cdot t_{QF(\Psi,\Omega)}}{t_{QF}}, \ \pi = \frac{100 \cdot \hat{t}_{QF(\Psi,\Omega)}}{t_{QF}}, \ \beta = \frac{100 \cdot t_{SDP}}{t_{total}}$$

with t being the total optimization time and nod the number of nodes in the BC tree. Time $\hat{t}$ represents the "useful" computation time, in other words the total time minus the time used for non-pruning SDP relaxation Ψ. t_{SDP} is the total time spent solving SDP in the bounding procedure. Values of δ and τ lower than

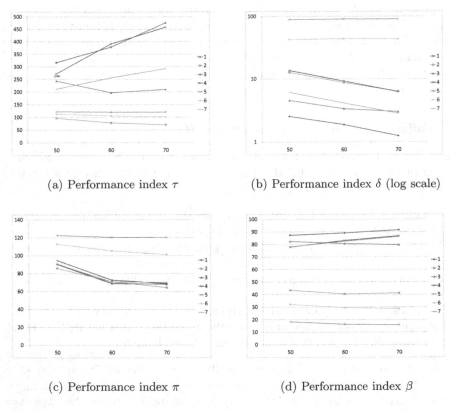

(a) Performance index τ (b) Performance index δ (log scale)

(c) Performance index π (d) Performance index β

Fig. 2. Comparison of the different strategies (Ω)

one hundred correspond to an improvement in the performances of the BC. The percentage π gives what in statistics is called "gold standard", in other words it gives an idea about what the results could be in presence of an "exact" Ω. β is useful in order to establish how much an increase in the speed of the SDP solver would affect the overall performances.

In Figure 2 we compare how these indices change with the increase of the instance size when we use `Constrained2` and all the Ω strategies described. In particular, from Figure 2(a) we see that the Strategies `OnOne`, `SmallGap` and `MediumGap` are promising because of the decreasing trend of τ that goes below the threshold value of 100. The `UnderAverage` trend seems roughly constant and the remaining three strategies seem unpromising. From Figure 2(b) we see that with `SmallGap` and `MediumGap` the pruning index δ is high, for these strategies the SDP relaxation is not often computed. The overall good performance of `OnOne` can be explained by comparing the performance indices δ and β in Figures 2(b) and 2(d): on one side the trend of δ is going below 10, in other words ensuring a decrease of the number of processed BC nodes greater than one order of magnitude; on the other, the fraction of time spent in solving the SDP (β) is significantly lower when compared to other strategies (but still relevant in comparison to SmallGap and MediumGap). Those two points together make OnOne the best candidate for Ω. Finally, as one could expect, the majority of the trends of the "gold standard" π (represented in Figure 2(c)) present values lower than one hundred.

In order to study the interaction between Ψ and Ω we introduce Table 2. The upper part of the table concerns the Unconstrained option and the lower part the Constrained2 option. The structure of both parts is identical: vertically the instances are divided first by density and then by percentage of positive weights and horizontally, the values of δ and τ concerning the 7 different strategies are reported. Each entry in the table is the average value over 10 instances of identical features. Each entry of τ is in bold if the index is less than one hundred (i.e. if the addition of (Ψ,Ω) improved the performances). The instances considered have 70 vertices. Concerning the node reduction, strategy Always gives the best insight about the maximum reduction achievable. With respect to the SDP strategy Ψ, as expected, Constrained2 dominates Unconstrained node-wise, leading to a reduction of almost two orders of magnitude in terms of nodes explored when the edge constraints are also taken into account. If we observe the same entries for the Unconstrained mode we see a decrease of slightly less then one order of magnitude. If we consider the time index τ, the best behaviour is given by the couple (Constrained2, OnOne), that guarantees on average an improvement of about 30% over the total running time.

Table 2. QF + SDP Unconstrained and SDP Constrained2 (Ψ)

strategy			δ								τ				
			1	2	3	4	5	6	7	1	2	3	4	5	6	7
n	μ	ν				SDP Unconstrained										
70	25	25	1.5	8.3	1.6	9.6	83.7	20.4	4.7	**78.6**	**65.3**	**65.9**	**63.4**	144.4	**99.5**	**61.3**
		50	13.8	31.9	13.9	57.9	80.4	19.2	29.5	201.3	125.8	149.2	185.9	134.9	113.6	153.7
		75	21.9	43.0	22.8	83.6	79.4	29.2	44.7	227.3	131.2	159.3	201.9	124.1	136.8	177.8
	50	25	4.0	15.4	4.1	17.6	96.4	50.7	9.9	**96.9**	**68.1**	**82.1**	**93.9**	102.5	**88.2**	**83.5**
		50	13.6	31.7	14.8	46.6	92.3	43.5	28.1	108.9	**80.2**	**89.4**	111.9	100.5	**86.6**	**97.1**
		75	16.4	36.7	16.9	56.3	90.0	47.2	33.4	105.0	**74.2**	**80.2**	116.2	**97.8**	**80.5**	**93.3**
	75	25	10.0	23.3	14.3	31.1	97.1	63.0	20.7	111.6	**90.7**	**95.7**	113.7	**99.9**	**96.3**	103.0
		50	20.8	40.5	22.2	55.8	98.5	65.5	37.9	111.3	**94.9**	**97.1**	113.8	100.3	**97.9**	104.8
		75	18.7	40.0	20.6	57.2	94.6	66.6	36.5	109.6	**89.4**	**91.3**	114.3	**99.5**	**95.7**	101.7
	avg.		13.4	30.1	14.6	46.2	90.2	45.0	27.3	127.8	**91.1**	101.1	123.9	111.5	**99.5**	108.5
n	μ	ν				SDP Constrained2										
70	25	25	0.4	2.1	0.7	3.8	83.7	20.4	1.1	120.3	**65.8**	109.9	**76.4**	186.9	174.0	**83.4**
		50	0.4	1.8	0.8	4.0	80.4	13.9	0.9	**71.8**	**49.4**	**61.2**	**54.1**	160.5	103.5	**46.8**
		75	0.1	0.4	0.2	1.1	79.9	17.7	0.3	**16.6**	**13.4**	**16.1**	**15.4**	133.4	**76.1**	**13.4**
	50	25	1.2	6.4	2.7	5.8	96.4	50.7	3.0	765.7	**95.6**	404.3	623.8	103.4	**94.3**	446.8
		50	0.7	5.2	1.8	4.9	92.3	43.9	2.0	232.5	**82.4**	113.7	262.4	101.6	**91.6**	169.3
		75	0.5	3.1	1.4	3.5	90.0	47.0	1.3	207.0	**55.2**	**97.4**	213.5	**98.1**	**82.0**	146.4
	75	25	2.9	15.4	8.7	10.1	97.1	63.0	6.0	1033.3	**94.0**	371.7	1139.7	**99.9**	**96.8**	655.1
		50	3.4	13.9	6.6	14.3	98.5	65.4	6.0	796.6	**91.4**	319.9	660.0	100.4	**98.3**	420.9
		75	1.7	8.6	4.3	8.8	94.6	66.6	5.0	1038.2	**85.9**	398.1	1079.1	**99.6**	**95.8**	654.8
	avg.		1.3	6.3	3.0	6.2	90.3	43.2	2.8	475.8	**70.4**	210.3	458.3	120.4	101.4	293.0

5.6   Results for Instances with 100 Nodes

One main issue when dealing with a SDP-based approach is the scalability of the method. In order to answer to this question, we keep the best settings obtained from the first set of tests and we use them to solve bigger instances, i.e., strategy `Constrained2` (Ψ) and `OnOne` (Ω). We also decided to keep strategy `Always` (Ω) in the analysis for comparison. In Table 3 we report the results concerning the four parameters introduced in Section 5.5 subdivided horizontally by μ and ν and vertically by Ω strategy. In Columns *avg. times* we report the average times (in seconds) needed for solving the instances without (column QF) and with (column QF(Ψ,Ω)) the additional bounding procedure `Constrained2` using `OnOne` as oracle. The strategy (`Constrained2`, `OnOne`) is able to reduce the overall computational time by 50%. If we consider only the non dense instances ($\mu = 25$) the overall computational time is reduced by 80%. The results concerning `Always` shows how the idea of running the additional bounding procedure without an oracle is useless in terms of overall computational time.

Table 3. QF + `Constrained2` SDP (Ψ) - Instances with 100 nodes

			δ		τ		π		β		avg. time	
		str	1	2	1	2	1	2	1	2	QF	QF(Ψ,Ω)
n	μ	ν										
100	25	25	0.1	0.7	**89.5**	**25.0**	13.9	18.3	99.0	95.5	345.4	85.2
		50	0.1	1.0	**49.1**	**24.0**	11.5	17.9	90.9	84.6	419.2	86.9
		75	0.0	0.2	**5.0**	**3.1**	1.3	2.4	90.1	86.4	4378.0	133.7
	50	25	0.3	2.5	**679.3**	**67.0**	57.9	57.4	97.5	50.3	41.6	27.9
		50	0.2	1.9	**489.7**	**54.9**	42.7	48.9	96.7	41.5	54.5	30.0
		75	0.2	2.2	**274.8**	**31.0**	22.9	23.6	95.8	48.8	120.6	37.6
	75	25	1.5	7.2	**3449.9**	**80.9**	172.6	79.7	99.3	9.1	20.3	16.4
		50	1.4	11.3	**2490.0**	**83.1**	164.6	79.7	98.9	9.2	22.7	18.8
		75	0.9	6.4	**2269.8**	**72.2**	123.9	69.3	98.2	10.3	27.0	19.5
		avg.	0.5	3.7	1088.5	49.0	67.9	44.1	96.3	48.4	603.3	50.7

6   Conclusions

In this paper we have explored the use of SDP-based bounding procedures within the BC framework provided by MINLP solvers. In order to do that, we performed an extensive computational analysis on the QSSP that allows us to conclude that the Hybrid SDP Bounding Procedure allows a noticeable reduction of computing time and BC nodes. The SDP bounds help in pruning but are heavy to compute, and thus in this optic we proposed different strategies in order to make the Hybrid bounding procedure more efficient. The addition of these strategies is crucial for the improvement of the performances over the standard BC.

Finally, the SDP relaxation can be used to enhance any of the essential ingredients of a generic purpose MINLP solver, which are: bounding techniques, primal heuristics and branching strategies. Specifically the information provided by the solution of the SDP relaxation at each branching node can be used either to derive alternative branching strategies, to compute different heuristic solutions or to strengthen LP relaxation for pruning the node. In this paper we have focused on this last aspect, deriving effective Hybrid Bounding Procedures. We leave the other two aspects for further development.

References

1. Anstreicher, K.M.: Semidefinite programming versus the reformulation-linearization technique for nonconvex quadratically constrained quadratic programming. J. Global Optim. 43(2-3), 471–484 (2009)
2. BARON (2012), http://archimedes.cheme.cmu.edu/?q=baron
3. BiqCrunch (2012), http://www-lipn.univ-paris13.fr/BiqCrunch/
4. Bonmin (2012), https://projects.coin-or.org/Bonmin
5. Borchers, B.: CSDP, A C library for semidefinite programming. Optim. Methods Softw. 11, 613–623 (1999)
6. Burer, S., Vandenbussche, D.: Globally solving box-constrained nonconvex quadratic programs with semidefinite-based finite branch-and-bound. Comp. Optim. Appl. 43, 181–195 (2009)
7. Cplex (2012), http://www-01.ibm.com/software/integration/optimization/cplex-optimizer/
8. Giandomenico, M., Letchford, A.N., Rossi, F., Smriglio, S.: A new approach to the stable set problem based on ellipsoids. In: Günlük, O., Woeginger, G.J. (eds.) IPCO 2011. LNCS, vol. 6655, pp. 223–234. Springer, Heidelberg (2011)
9. Gurobi (2012), http://www.gurobi.com/
10. Jaumard, B., Marcotte, O., Meyer, C.: Estimation of the Quality of Cellular Networks Using Column Generation Techniques. Cahiers du GÉRAD (1998)
11. Jaumard, B., Marcotte, O., Meyer, C., Vovor, T.: Comparison of column generation models for channel assignment in cellular networks. Discrete Appl. Math. 112(1-3), 217–240 (2001)
12. Krislock, N., Malick, J., Roupin, F.: Improved semidefinite bounding procedure for solving max-cut problems to optimality. To appear in Math. Prog. (2013)
13. Mahdavi Pajouh, F., Balasundaram, B., Prokopyev, O.: On characterization of maximal independent sets via quadratic optimization. J. Heuristics, 1–16 (2011)
14. Rendl, F., Rinaldi, G., Wiegele, A.: Solving max-cut to optimality by intersecting semidefinite and polyhedral relaxations. Math. Prog. 121, 307–335 (2010)
15. SeDuMi (2012), http://sedumi.ie.lehigh.edu/
16. Sherali, H.D., Adams, W.P.: A Reformulation-Linearization Technique for Solving Discrete and Continuous Nonconvex Problems. Springer (1998)

Computing Multimodal Journeys in Practice[*]

Daniel Delling[1], Julian Dibbelt[2], Thomas Pajor[2],
Dorothea Wagner[2], and Renato F. Werneck[1]

[1] Microsoft Research Silicon Valley, Mountain View, CA 94043, USA
{dadellin,renatow}@microsoft.com
[2] Karlsruhe Institute of Technology (KIT), 76128 Karlsruhe, Germany
{dibbelt,pajor,wagner}@kit.edu

Abstract. We study the problem of finding multimodal journeys in
transportation networks, including unrestricted walking, driving, cycling,
and schedule-based public transportation. A natural solution to this
problem is to use multicriteria search, but it tends to be slow and to
produce too many journeys, several of which are of little value. We propose algorithms to compute a full Pareto set and then score the solutions in a postprocessing step using techniques from fuzzy logic, quickly
identifying the most significant journeys. We also propose several (still
multicriteria) heuristics to find similar journeys much faster, making the
approach practical even for large metropolitan areas.

1 Introduction

Efficiently computing good journeys in transportation networks has been an
active area of research in recent years, with focus on the computation of routes
in both road networks [11] and schedule-based public transit [2,5], but these
are often considered separately. In practice, users want an integrated solution
to find the "best" journey considering all available modes of transportation.
Within a metropolitan area, this includes buses, trains, driving, cycling, taxis,
and walking. We refer to this as the *multimodal route planning* problem.

In fact, any public transportation network has a multimodal component,
since journeys require some amount of walking. To handle this, existing solutions [4,10,14] predefine transfer arcs between nearby stations, then run a search
algorithm on the public transit network to find the "best" journey. Unlike in
road networks, however, defining "best" is not straightforward. For example,
while some people want to arrive as early as possible, others are willing to spend
a little more time to avoid extra transfers. Most recent approaches therefore
compute the *Pareto set* of non-dominating journeys optimizing multiple criteria, which is practical even for large metropolitan areas [10].

Extending public transportation solutions to a full multimodal scenario (with
unrestricted walking, biking, and taxis) may seem trivial: one could just incorporate routing techniques for road networks [9,17] to solve the new subproblems.

---

[*] Partial support by DFG grant WA654/16-1 and EU grant 288094 (eCOMPASS).

V. Bonifaci et al. (Eds.): SEA 2013, LNCS 7933, pp. 260–271, 2013.

Unfortunately, meaningful multimodal optimization must take more criteria into account, such as walking duration and costs. Some people are happy to walk 10 minutes to avoid an extra transfer, while others are not. In fact, some will walk half an hour to avoid using public transportation at all. Taking a taxi to the airport is a good solution for some; users on a budget may prefer cheaper alternatives. Considering more criteria leads to much larger Pareto sets, however, with many of the additional journeys looking unreasonable (see full paper [8]).

Previous research thus tends to avoid multicriteria search altogether [3], looking for reasonable routes by other means. A natural approach is to work with a weighted combination of all criteria, transforming the search into a single-criterion problem [19]. When extended to find the k-shortest paths [6], this method can even take user preferences into account. Unfortunately, linear combination may miss Pareto-optimal journeys [7] (also see full paper [8]). To avoid such issues, another line of multimodal single-criterion research considers label-constrained quickest journeys [1]. Here, journeys are required to obey a user-defined pattern, typically enforcing a hierarchy of modes [6] (such as "no car travel between trains"). Although this approach can be quite fast when using preprocessing techniques for road networks [12], it has a fundamental conceptual problem: it relies on the user to know her options before planning the journey.

Given the limitations of current approaches, we revisit the problem of finding multicriteria multimodal journeys on a metropolitan scale. Instead of optimizing each mode of transportation independently [15], we argue in Section 2 that most users optimize three criteria: travel time, convenience, and costs. As this produces a large Pareto set, we propose using fuzzy logic [20] to identify, in a principled way, a modest-sized subset of representative journeys. This postprocessing step is very quick and can incorporate personal preferences. As Section 3 shows, we can use recent algorithmic developments [10,12,17] to answer exact queries optimizing time and convenience in less than two seconds within a large metropolitan area, for the simpler scenario of walking, cycling, and public transit. Unfortunately, this is not enough for interactive applications and becomes much slower when more criteria, such as costs, are incorporated. Section 4 proposes heuristics (still multicriteria) that are significantly faster and closely match the representative journeys in the actual Pareto set. Section 5 presents a thorough experimental evaluation of all algorithms in terms of both solution quality and performance and shows that our approach can be fast enough for interactive applications. Moreover, since it does not rely on heavy preprocessing, it can be used in dynamic scenarios.

2   Problem Statement

We want to find journeys in a network built from several *partial networks*. The first is a *public transportation network* representing all available schedule-based means of transportation, such as trains, buses, rail, or ferries. We can specify this network in terms of its timetable, which is defined as follows. A *stop* is a location in the network (such as a train platform or a bus stop) in which a user can board or leave a particular vehicle. A *route* is a fixed sequence of stops for which there is scheduled service during the day; a typical example is a bus or subway line. A route is served by one or more distinct *trips* during the day; each trip is associated with a unique vehicle, with fixed (scheduled) arrival and departure times for every stop in the route. Each stop may also keep a *minimum change time*, which must be obeyed when changing trips. Besides the public transportation network, we also take as input several *unrestricted networks*, with no associated timetable. Walking, cycling, and driving are modeled as distinct unrestricted networks, each represented as a directed graph $G = (V, A)$. Each vertex $v \in V$ represents an intersection and has associated coordinates (latitude and longitude). Each arc $(v, w) \in A$ represents a (directed) road segment and has an associated *duration* $\text{dur}(v, w)$, which corresponds to the (constant) time to traverse it. The *integrated transportation network* is the union of these partial networks with appropriate *link vertices*, i.e., vertices (or stops) in different networks are identified with one another to allow for changes in modes of transportation. Note that, unlike previous work [18], we do not necessarily require explicit *footpaths* in the public transportation networks (to walk between nearby stops). A query takes as input a *source location* s, a *target location* t, and a *departure time* τ, and it produces *journeys* that leave s no earlier than τ and arrive at t. A *journey* is a valid path in the integrated transportation network that obeys all timetable constraints.

We still have to define *which* journeys the query should return. We argue that users optimize three natural criteria in multimodal networks: arrival time, costs, and "convenience". For our first (simplified) scenario (with public transit, cycling, and walking, but no taxi), we work with three criteria. Besides arrival time, we use number of trips and walking duration as proxies for convenience. We add cost for the scenario that includes taxi. Given this setup, a first natural problem we need to solve is the *full multicriteria problem*, which must return a full (maximal) Pareto set of journeys. We say that a journey J_1 *dominates* J_2 if J_1 is strictly better than J_2 according to at least one criterion and no worse according to all other criteria. A *Pareto set* is a set of pairwise nondominating journeys. If two journeys have equal values in all criteria, we only keep one.

Solving the full multicriteria problem, however, can lead to solution sets that are too large for most users. Moreover, many solutions provide undesirable trade-offs, such as journeys that arrive much later to save a few seconds of walking (or walk much longer to save a few seconds in arrival time). Intuitively, most criteria are diffuse to the user, and only large enough differences are significant. Pareto optimality fails to capture this. To formalize the notion of significance, we propose to *score* the journeys in the Pareto set in a post-processing step using concepts from fuzzy logic [20]. Loosely speaking, fuzzy logic generalizes Boolean logic to handle (continuous) degrees of truth. For example, the statement "60 and 61 seconds of walking are equal" is false in classical logic, but "almost true" in fuzzy logic. Formally, a *fuzzy set* is a tuple $\mathcal{S} = (\mathcal{U}, \mu)$, where $\mathcal{U}$ is a set and $\mu \colon \mathcal{U} \to [0, 1]$ a *membership function* that defines "how much" each element in $\mathcal{U}$ is contained in $\mathcal{S}$. Mostly, we use μ to refer to $\mathcal{S}$. Our application requires fuzzy relational operators $\mu_<$, $\mu_=$, and $\mu_>$. For any

$x, y \in \mathbb{R}$, they are evaluated by $\mu_<(x - y)$, $\mu_>(y - x)$, and $\mu_=(x - y)$. We use the well-known [20] exponential membership functions for the operators: $\mu_=(x) := \exp(\frac{\ln(\chi)}{\varepsilon^2} x^2)$, where $0 < \chi < 1$ and $\varepsilon > 0$ control the degree of fuzziness. The other two operators are derived by $\mu_<(x) := 1 - \mu_=(x)$ if $x < 0$ (0 otherwise) and $\mu_> := 1 - \mu_=(x)$ if $x > 0$ (0 otherwise). Moreover, we require binary operators (norms) $T, S : [0,1]^2 \to [0,1]$ to represent fuzzy (logical) disjunction (T) and conjunction (S). We use the *maximum/minimum norms*, i.e., $T = \max$ and $S = \min$. Note that $S(x, y) = 1 - T(1 - x, 1 - x)$ holds, which is important for consistency. Other norms are evaluated in the full paper [8].

We now recap the concept of fuzzy dominance in multicriteria optimization, which is introduced by Farina and Amato [16]. Given journeys J_1 and J_2 with M optimization criteria, we denote by $n_b(J_1, J_2)$ the (fuzzy) number of criteria in which J_1 is better than J_2. More formally $n_b(J_1, J_2) := \sum_{i=1}^{M} \mu_<^i(\kappa^i(J_1), \kappa^i(J_2))$, where $\kappa^i(J)$ evaluates the i-th criterion of J and $\mu_<^i$ is the i-th fuzzy less-than operator. (Note that each criterion may use different fuzzy operators.) Analogously, we define $n_e(J_1, J_2)$ for equality and $n_w(J_1, J_2)$ for greater-than. By definition, $n_b + n_e + n_w = M$. Hence the Pareto dominance can be generalized to obtain a *degree of domination* $d(J_1, J_2) \in [0, 1]$, defined as $(2n_b + n_e - M)/n_b$ if $n_b > (M - n_e)/2$ (and 0 otherwise). Here, $d(J_1, J_2) = 0$ means that J_1 does not dominate J_2, while a value of 1 indicates that J_1 Pareto-dominates J_2. Otherwise, we say J_1 *fuzzy-dominates* J_2 by degree $d(J_1, J_2)$. Now, given a (Pareto) set $\mathcal{J}$ of n journeys $J_1, \ldots, J_n$, we define a *score function* sc: $\mathcal{J} \to [0, 1]$ that computes the degree of domination by the whole set for each J_i. More precisely, sc(J) := $1 - \max(J_1, \ldots, J_n)$, i.e., the value sc($J$) is determined by the (one) journey that dominates J most. See the full paper [8] for more details, including an illustration of the fuzzy dominance function d. We finally use the score to order the journeys by significance. One may then decide to only show the k journeys with highest score to the user.

3   Exact Algorithms

We now study exact algorithms for the multicriteria multimodal problem. We first propose two solutions (building on different methods for multicriteria optimization on public transportation networks), then describe an acceleration technique that applies to both. For simplicity, we describe the algorithms considering only the (schedule-based) public transit network and the (unrestricted) walking network. We later deal with cycling and taxis, which are unrestricted but have special properties.

Multi-label-correcting Algorithm. Traditional solutions to the multicriteria problem on public transportation networks typically model the timetable as a graph. A particularly effective approach is to use the *time-dependent route model* [18]. For each stop p, we create a single *stop vertex* linked by time-independent *transfer edges* to multiple *route vertices*, one for each route serving p. We also add *route edges* between route vertices associated to consecutive stops within the same route. To model the trips along a route, the cost of a route edge is given by a function reflecting the traversal time (including waiting for the next departure).

A journey in the public transportation network corresponds to a path in this graph. The *multi-label-correcting* (MLC) [18] algorithm uses this to find full Pareto sets for arbitrary criteria that can be modeled as edge costs. MLC extends Dijkstra's algorithm [13] by operating on labels that have multiple values, one per criterion. Each vertex v maintains a *bag* $B(v)$ of nondominated labels. In each iteration, MLC extracts from a priority queue the minimum (in lexicographic order) unprocessed label $L(u)$. For each arc (u, v) out of the associated vertex u, MLC creates a new label $L(v)$ (by extending $L(u)$ in the natural way) and inserts it into $B(v)$; newly-dominated labels (possibly including $L(v)$ itself) are discarded, and the priority queue is updated if needed. MLC can be sped up with target pruning and by avoiding unnecessary domination checks [14].

To solve the multimodal problem, we extend MLC by augmenting its input graph to include the walking network, creating an integrated network. The MLC query remains essentially unchanged. Although labels can now be associated to vertices in different networks, they can all share the same priority queue.

Round-based Algorithm. A drawback of MLC (even restricted to public transportation networks) is that it can be quite slow: unlike Dijkstra's algorithm, MLC may scan the same vertex multiple times (the exact number depends on the criteria being optimized), and domination checks make each such scan quite costly. Delling et al. [10] have recently introduced RAPTOR (*Round bAsed Public Transit Optimized Router*) as a faster alternative. The simplest version of the algorithm optimizes two criteria: arrival time and number of transfers. Unlike MLC, which searches a graph, RAPTOR uses dynamic programming to operate directly on the timetable. It works in rounds, with round i processing all relevant journeys with exactly $i - 1$ transfers. It maintains one label per round i and stop p representing the best known arrival time at p for up to i trips. During round i, the algorithm processes each *route* once. It reads arrival times from round $i - 1$ to determine relevant trips (on the route) and updates the labels of round i at every stop along the way. Once all routes are processed, the algorithm considers potential transfers to nearby (predefined) stops in a second phase. Simpler data structures and better locality make RAPTOR an order of magnitude faster than MLC. Delling et al. [10] have also proposed McRAPTOR, which extends RAPTOR to handle more criteria (besides arrival times and number of transfers). It maintains a *bag* (set) of labels with each stop and round.

Even with multiple modes of transport available, one trip always consists of a single mode. This motivates adapting the round-based paradigm to our scenario. We propose MCR (*multimodal multicriteria RAPTOR*), which extends McRAPTOR to handle multimodal queries. As in McRAPTOR, each round has two phases: the first processes trips in the public transportation network, while the second considers arbitrary paths in the unrestricted networks. We use a standard McRAPTOR round for the first phase (on the timetable network) and MLC for the second (on the walking network). Labels generated by one phase are naturally used as input to the other. During the second phase, MLC extends bags instead of individual labels. To ensure that each label is processed at most once, we keep track of which labels (in a bag) have already been extended. The initialization routine (before the first round) runs Dijkstra's algorithm on the walking network from the source s to determine the fastest walking path to each stop in the public transportation network (and to t), thus creating the initial labels used by MCR. During round i, the McRAPTOR subroutine reads labels from round $i - 1$ and writes to round i. In contrast, the MLC subroutine may read and write labels of the same round if walking is not regarded as a trip.

Contracting Unrestricted Networks. As our experiments will show, the bottleneck of the multimodal algorithms is processing the walking network $G = (V, A)$. We improve performance using a quick preprocessing technique [12]. For any journey involving public transportation, walking between trips always begins and ends at the restricted set $K \subset V$ of link vertices. During queries, we must only be able to compute the pairwise distances between these vertices. We therefore use preprocessing to compute a smaller *core graph* that preserves these distances. More precisely, we start from the original graph and iteratively *contract* [17] each vertex in $V \setminus K$ in the order given by a rank function r. Each contraction step (temporarily) removes a vertex and adds shortcuts between its uncontracted neighbors to maintain shortest path distances (if necessary). It is usually advantageous to first contract vertices with relatively small degrees that are evenly distributed across the network [17]. We stop contraction when the average degree in the core graph reaches some threshold (we use 12 in our experiments) [12].

To run a faster multimodal s–t query, we use essentially the same algorithm as before (based on either MLC or RAPTOR), but replacing the full walking network with the (smaller) core graph. Since the source s and the target t may not be in the core, we handle them during initialization. It works on the graph $G^+ = (V, A \cup A^+)$ containing all original arcs A as well as all shortcuts A^+ added during the contraction process. We run upward searches (only following arcs (u, v) such that $r(u) > r(w)$) in G^+ from s (scanning forward arcs) and t (scanning reverse arcs); they reach all potential entry and exit points of the core, but arcs within the core are not processed [12]. These core vertices (and their respective distances) are used as input to MCR's (or MLC's) standard initialization, which can operate on the core from this point on. The main loop works as before, with one minor adjustment. Whenever MLC extracts a label $L(v)$ for a scanned core vertex v, we check if it has been reached by the reverse search during initialization. If so, we create a temporary label $L'(t)$ by extending $L(v)$ with the (already computed) walking path to t and add it to $B(t)$ if needed. MCR is adjusted similarly, with bags instead of labels.

Beyond Walking. We now consider other unrestricted networks (besides walking). In particular, our experiments include a bicycle rental scheme, which can be seen as a hybrid network: it does not have a fixed schedule (and is thus unrestricted), but bicycles can only be picked up and dropped off at designated *cycling stations*. Picking a bike from its station counts as a trip. To handle cycling within MCR, we consider it during the first stage of each round (together with

RAPTOR and before walking). Because bicycles have no schedule, we process them independently (from RAPTOR) by running MLC on the bicycle network. To do so, we initialize MLC with labels from round $i - 1$ for all relevant bicycle stations and, during the algorithm, we update labels of (the current) round i.

We consider a taxi ride to be a trip as well, since we board a vehicle. Moreover, we also optimize a separate criterion reflecting the (monetary) *cost* of taxi rides. If taxis were not penalized in any way, an all-taxi journey would almost always dominate all other alternatives (even sensible ones), since it is fast and has no walking. Our round-based algorithms handle taxis as they do walking, except that in the taxi stage labels are read from round $i - 1$ and written into round i. Note that we link the taxi network to public transit stops and bicycle stations.

Dealing with personal cars or bicycles is simpler. Assuming that they are only available for the first or last legs of the journey, we must only consider them during initialization. Initialization can also handle other special cases, such as allowing rented bicycles to be ridden to the destination (to be returned later).

Note that contraction can be used for cycling and driving. For every unrestricted network (walking, cycling, driving), we keep the link vertices (stops and bicycle stations) in one common core and contract (up to) all other nodes. As before, queries start with upward searches in each relevant unrestricted network.

4   Heuristics

Even with all accelerations, the exact algorithms proposed in Section 3 are not fast enough for interactive applications. This section proposes quick heuristics aimed at finding a set of journeys that is similar to the exact solution, which we take as ground truth. We consider three approaches: weakening the dominance rules, restricting the amount of walking, and reducing the number of criteria. We also discuss how to measure the quality of the heuristic solutions we find.

Weak Dominance. The first strategy we consider is to weaken the domination rules during the algorithm, reducing the number of labels pushed through the network. We test four implementations of this strategy. The first, MCR-hf, uses fuzzy dominance (instead of strict dominance) when comparing labels during the algorithm: for labels L_1 and L_2, we compute the fuzzy dominance value $d(L_1, L_2)$ (cf. Section 2) and dominate L_2 if d exceeds a given threshold (we use 0.9). The second, MCR-hb(κ), uses strict dominance, but discretizes criterion κ: before comparing labels L_1 and L_2, we first round $\kappa(L_1)$ and $\kappa(L_2)$ to predefined discrete values (*buckets*); this can be extended to use buckets for several criteria. The third heuristic, MCR-hs(κ), uses strict dominance but adds a slack of x units to κ. More precisely, L_1 already dominates L_2 if $\kappa(L_1) \leq \kappa(L_2) + x$ and L_1 is at least as good L_2 in all other criteria. The last heuristic, MCR-ht, weakens the domination rule by trading off two or more criteria. More concretely, consider the case in which walking (walk) and arrival time (arr) are criteria. Then, L_1 already dominates L_2 if arr(L_1) $\leq$ arr(L_2)$+a\cdot$(walk(L_1)$-$walk(L_2)), walk(L_1) $\leq$ walk(L_2) $+ a \cdot$ (arr(L_1) $-$ arr(L_2)), and L_1 is at least as good as L_2 in all other criteria, for a tradeoff parameter a (we use $a = 0.3$).

Restricting Walking. Consider our simple scenario of walking and public transit. Intuitively, most journeys start with a walk to a nearby stop, followed by one or more trips (with short transfers) within the public transit system, and finally a short walk from the final stop to the actual destination. This motivates a second class of heuristics, MCR-tx. It still runs three-criterion search (walking, arrival, and trips), but limits walking transfers between stops to x minutes; in this case we precompute these transfers. MCR-tx-ry also limits walking in the beginning and end to y minutes. Note that existing solutions often use such restrictions [4].

Fewer Criteria. The last strategy we study is reducing the number of criteria considered during the algorithm. As already mentioned, this is a common approach in practice. We propose MR-x, which still works in rounds, but optimizes only the number of trips and arrival times explicitly (as criteria). To account for walking duration, we count every x minutes of a walking segment (transfer) as a trip; the first x minutes are free. With this approach, we can run plain Dijkstra to compute transfers, since link vertices no longer need to keep bags. The round index to which labels are written then depends on the walking duration (of the current segment) of the considered label. A special case is $x = \infty$, where a transfer is never a trip. Another variant is to always count a transfer as a single trip, regardless of duration; we abuse notation and call this variant MR-0. We also consider MR-∞-tx: walking duration is not an explicit criterion and transfers do not count as trips, but are limited to x minutes.

For scenarios that include cost as a criterion (for taxis), we consider variants of the MCR-hb and MCR-hf heuristics. In both cases, we drop walking as an independent criterion, leaving only arrival time, number of trips, and costs to optimize. We account for walking by making it a (cheap) component of the costs.

Quality Evaluation. To measure the quality of a heuristic, we compare the set of journeys it produces to the *ground truth*, which we define as the solution found by MCR. To do so, we first compute the score of each journey with respect to the Pareto set that contains it (cf. Section 2). Then, for a given parameter k, we measure the similarity between the top k scored journeys returned by the heuristics and the top k scored journeys in the ground truth. Note that the score depends only on the algorithm itself and does not assume knowledge of the ground truth, which is consistent with a real-world deployment. To compare two sets of k journeys, we run a greedy maximum matching algorithm. First, we compute a $k \times k$ matrix where entry (i, j) represents the similarity between the i-th journey in the first set and the j-th in the second. Given two journeys J_1 and J_2, the similarity with respect to the i-th criterion is given by $c^i := \mu^i_=(\kappa^i(J_1) - \kappa^i(J_2))$, where κ^i is the value of this criterion and $\mu^i_=$ is the corresponding fuzzy equality relation. Then, we define the total similarity between J_1 and J_2 as $\min(c^1, c^2, \ldots, c^M)$. After computing the pairwise similarities, we greedily select the unmatched pairs with highest similarity (by picking the highest entry in the matrix that does not share a row or column with a previously picked entry). The similarity of the whole matching is the average similarity of its pairs, weighted by the fuzzy score of the reference journey. This means that matching the highest-scored reference journey is more important than matching the k-th one.

5   Experiments

All algorithms from Sections 3 and 4 were implemented in C++ and compiled with g++ 4.6.2 (64 bits, flag -O3). We ran our experiments on one core of a dual 8-core Intel Xeon E5-2670 clocked at 2.6 GHz, with 64 GiB of DDR3-1600 RAM.

We focus on the transportation network of London (England); results for other instances (available in the full paper [8]) are similar. We use the timetable information made available by Transport for London (TfL), from which we extracted a Tuesday in the periodic summer schedule of 2011. The data includes subway (tube), buses, tram, and light rail (DLR), as well as bicycle station locations. To model the underlying road network, we use data provided by PTV AG from 2006, which explicitly indicates whether each road segment is open for driving, cycling and/or walking. We set the walking speed to 5 km/h and the cycling speed to 12 km/h, and we assume driving at free-flow speeds. We do not consider turn costs, which are not defined in the data. The resulting combined network has 564 cycle stations and about 20 k stops, 5 M departure events, and 259 k vertices in the walking network.

Recall that we specify the fuzziness of each criterion by a pair (χ, ε), roughly meaning that the corresponding Gaussian (centered at $x = 0$) has value χ for $x = \varepsilon$. We set these pairs to $(0.8, 5)$ for walking, $(0.8, 1)$ for arrival time, $(0.1, 1)$ for trips, and $(0.8, 5)$ for costs (given in pounds; times are in minutes). Note that the number of trips is sharper than the other criteria. Our approach is robust to small variations in these parameters, but they can be tuned to account for user-dependent preferences. We run *location-to-location* queries, with sources, targets, and departure times picked uniformly at random (from the walking network and during the day, respectively).

For our first experiment, we use walking, cycling, and the public transportation network and consider three criteria: arrival time, number of trips, and walking duration. We ran 1 000 queries for each algorithm. Table 1 summarizes the results (the full paper [8] has additional statistics). For each algorithm, the table first shows which criteria are explicitly taken into account. The next five columns show the average values observed for the number of rounds, scans per entity (stop/vertex), label comparisons per entity, journeys found, and running time (in milliseconds). The last four columns evaluate the quality of the top 3 and 6 journeys found by our heuristics, as explained in Section 4. We show both averages and standard deviations.

The methods in Table 1 are grouped in blocks. Those in the first block compute the full Pareto set considering all three criteria (arrival time, number of trips, and walking). MCR, our reference algorithm, is round-based and uses contraction in the unrestricted networks. As anticipated, it is faster (by a factor of about three) than MCR-nc (which does not use the core) and MLC (which uses the core but is not round-based). Accordingly, all heuristics we test are round-based and use the core.

Table 1. Performance and solution quality on journeys considering walking, cycling, and public transit. Bullets (•) indicate the criteria taken into account by the algorithm.

Algorithm	Arr. Trp. Wlk	Scans Rnd.	/Ent.	Comp. /Ent.	Jn.	Time [ms]	Quality-3 Avg.	Sd.	Quality-6 Avg.	Sd.
MCR-nc	• • •	13.8	13.8	168.2	29.1	4634.0	100 %	0 %	100 %	0 %
MCR	• • •	13.8	3.4	158.7	29.1	1438.7	100 %	0 %	100 %	0 %
MLC	• • •	—	10.6	1246.7	29.1	4543.0	100 %	0 %	100 %	0 %
MCR-hf	• • •	15.6	2.9	14.3	10.9	699.4	89 %	15 %	89 %	11 %
MCR-hb	• • •	10.2	2.1	12.7	9.0	456.7	91 %	12 %	91 %	10 %
MCR-hs	• • •	14.7	2.6	11.1	8.6	466.1	67 %	28 %	69 %	23 %
MCR-ht	• • •	10.5	2.0	6.4	8.6	373.6	84 %	22 %	82 %	20 %
MCR-t10	• • •	13.8	2.7	132.7	29.0	1467.6	97 %	10 %	95 %	10 %
MCR-t10-r15	• • •	10.7	1.7	73.3	13.2	885.0	38 %	40 %	30 %	31 %
MCR-t5	• • •	13.8	2.7	126.6	28.9	891.9	93 %	16 %	92 %	15 %
MR-∞	• • ○	7.6	1.4	4.8	4.5	44.4	63 %	28 %	63 %	24 %
MR-0	• • ○	13.7	2.1	6.9	5.4	61.5	63 %	28 %	63 %	24 %
MR-10	• • ○	20.0	1.1	4.8	4.3	39.4	51 %	33 %	45 %	29 %
MR-∞-t10	• • ○	7.6	1.1	4.8	4.5	22.2	63 %	28 %	62 %	24 %

The second block contains heuristics that accelerate MCR by weakening the domination rules, causing more labels to be pruned (and losing optimality guarantees). As explained in Section 4, MCR-hf uses fuzzy dominance during the algorithm, MCR-hb uses walking *buckets* (discretizing walking by steps of 5 minutes for domination), MCR-hs uses a slack of 5 minutes on the walking criterion when evaluating domination, and MCR-ht considers a tradeoff parameter of $a = 0.3$ between walking and arrival time. All heuristics are faster than pure MCR, and MCR-hb gives the best quality at a reasonable running time.

The third block has algorithms with restrictions on walking duration. Limiting transfers to 10 minutes (as MCR-t10 does) has almost no effect on solution quality (which is expected in a well-designed public transportation network). Moreover, adding precomputed footpaths of 10 minutes is not faster than using the core for unlimited walking (as MCR does). Additionally limiting the walking range from s or t (MCR-t10-r15) improves speed, but the quality becomes unacceptably low: the algorithm misses good journeys (including all-walk) quite often. If instead we allow even more restricted transfers (with MCR-t5), we get a similar speedup with much better quality (comparable to MCR-hb).

The MR-x algorithms (fourth block) reduce the number of criteria considered by combining trips and walking. The fastest variant is MR-∞-t10, which drops walking duration as a criterion but limits the amount of walking at transfers to 10 minutes, making it essentially the same as RAPTOR, with a different initialization. As expected, however, quality is much lower than for MCR-tx, confirming that considering the walking duration explicitly during the algorithm is

Table 2. Performance on our London instance when taking taxi into account

Algorithm	Arr. Trip Walk Cost	Rnd.	Scans / Ent.	Comp. / Ent.	Jn.	Time [ms]	Quality-3 Avg.	Sd.	Quality-6 Avg.	Sd.
MCR	● ● ● ●	16.3	3.1	369 606.0	1 666.0	1 960 234.0	100 %	0 %	100 %	0 %
MCR-hf	● ● ● ●	17.1	2.1	137.1	35.2	6 451.6	92 %	12 %	92 %	6 %
MCR-hb	● ● ● ●	9.9	1.3	86.8	27.6	2 807.7	96 %	8 %	92 %	6 %
MCR	● ● ○ ●	14.6	2.4	7 901.4	250.9	25 945.8	98 %	6 %	97 %	5 %
MCR-hf	● ● ○ ●	12.0	1.4	33.6	17.6	2 246.3	87 %	12 %	74 %	12 %
MCR-hb	● ● ○ ●	9.0	1.0	20.0	11.6	996.4	86 %	12 %	74 %	12 %

important to obtain a full range of solutions. MR-10 attempts to improve quality by transforming long walks into extra trips, but is not particularly successful.

Summing up, MCR-hb should be the preferred choice for high-quality solutions, while MR-∞-t10 can support interactive queries with reasonable quality.

Our second experiment considers the full multimodal problem, including taxis. We add *cost* as fourth criterion (at 2.40 pounds per taxi-trip plus 60 pence per minute). We do not consider the cost of public transit, since it is significantly cheaper. Table 2 presents the average performance of some of our algorithms over 1 000 random queries in London. The first block includes algorithms that optimize all four criteria (arrival time, walking duration, number of trips, and costs). While exact MCR is impractical, fuzzy domination (MCR-hf) makes the problem tractable with little loss in quality. Using 5-minute buckets for walking and 5-pound buckets for costs (MCR-hb) is even faster, though queries still take more than two seconds. The second block shows that we can reduce running times by dropping walking duration as a criterion (we incorporate it into the cost function at 3 pence per minute, instead), with almost no loss in solution quality. This is still not fast enough, though. Using 5-pound buckets (MCR-hb) reduces the average query time to about 1 second, with reasonable quality.

6   Final Remarks

We have studied multicriteria journey planning in multimodal networks. We argued that users optimize three criteria: arrival time, costs, and convenience. Although the corresponding full Pareto set is large and has many unnatural journeys, fuzzy set theory can extract the relevant journeys and rank them. Since exact algorithms are too slow, we have introduced several heuristics that closely match the best journeys in the Pareto set. Our experiments show that our approach enables efficient realistic multimodal journey planning in large metropolitan areas. A natural avenue for future research is accelerating our approach further to enable interactive queries with an even richer set of criteria in dynamic scenarios, handling delay and traffic information. The ultimate goal is to compute multicriteria multimodal journeys on a global scale in real time.

References

1. Barrett, C., Bisset, K., Holzer, M., Konjevod, G., Marathe, M.V., Wagner, D.: Engineering Label-Constrained Shortest-Path Algorithms. In: The Shortest Path Problem: 9th DIMACS Impl. Challenge. DIMACS, vol. 74, pp. 309–319. AMS (2009)
2. Bast, H.: Car or Public Transport – Two Worlds. In: Albers, S., Alt, H., Näher, S. (eds.) Efficient Algorithms. LNCS, vol. 5760, pp. 355–367. Springer, Heidelberg (2009)
3. Bast, H.: Next-Generation Route Planning: Multi-Modal, Real-Time, Personalized (2012) Talk given at ISMP
4. Bast, H., Carlsson, E., Eigenwillig, A., Geisberger, R., Harrelson, C., Raychev, V., Viger, F.: Fast Routing in Very Large Public Transportation Networks using Transfer Patterns. In: de Berg, M., Meyer, U. (eds.) ESA 2010, Part I. LNCS, vol. 6346, pp. 290–301. Springer, Heidelberg (2010)
5. Bauer, R., Delling, D., Wagner, D.: Experimental Study on Speed-Up Techniques for Timetable Information Systems. Networks 57(1), 38–52 (2011)
6. Bielli, M., Boulmakoul, A., Mouncif, H.: Object modeling and path computation for multimodal travel systems. EJOR 175(3), 1705–1730 (2006)
7. Corne, D., Deb, K., Fleming, P., Knowles, J.: The Good of the Many Outweighs the Good of the One: Evolutionary Multi-Objective Optimization. Connections 1(1), 9–13 (2003)
8. Delling, D., Dibbelt, J., Pajor, T., Wagner, D., Werneck, R.F.: Computing and Evaluating Multimodal Journeys. Technical Report 2012-20, Faculty of Informatics, Karlsruhe Institute of Technology (2012)
9. Delling, D., Goldberg, A.V., Pajor, T., Werneck, R.F.: Customizable Route Planning. In: Pardalos, P.M., Rebennack, S. (eds.) SEA 2011. LNCS, vol. 6630, pp. 376–387. Springer, Heidelberg (2011)
10. Delling, D., Pajor, T., Werneck, R.F.: Round-Based Public Transit Routing. In: ALENEX, pp. 130–140. SIAM (2012)
11. Delling, D., Sanders, P., Schultes, D., Wagner, D.: Engineering Route Planning Algorithms. In: Lerner, J., Wagner, D., Zweig, K.A. (eds.) Algorithmics. LNCS, vol. 5515, pp. 117–139. Springer, Heidelberg (2009)
12. Dibbelt, J., Pajor, T., Wagner, D.: User-Constrained Multi-Modal Route Planning. In: ALENEX, pp. 118–129. SIAM (2012)
13. Dijkstra, E.W.: A Note on Two Problems in Connexion with Graphs. Numerische Mathematik 1, 269–271 (1959)
14. Disser, Y., Müller–Hannemann, M., Schnee, M.: Multi-Criteria Shortest Paths in Time-Dependent Train Networks. In: McGeoch, C.C. (ed.) WEA 2008. LNCS, vol. 5038, pp. 347–361. Springer, Heidelberg (2008)
15. Ensor, A., Lillo, F.: Partial order approach to compute shortest paths in multimodal networks. Technical report (2011), http://arxiv.org/abs/1112.3366v1
16. Farina, M., Amato, P.: A Fuzzy Definition of "Optimality" for Many-Criteria Optimization Problems. IEEE Tr. Syst., Man, and Cyb. A 34(3), 315–326 (2004)
17. Geisberger, R., Sanders, P., Schultes, D., Vetter, C.: Exact Routing in Large Road Networks Using Contraction Hierarchies. Transp. Sci. 46(3), 388–404 (2012)
18. Müller-Hannemann, M., Schulz, F., Wagner, D., Zaroliagis, C.: Timetable Information: Models and Algorithms. In: Geraets, F., Kroon, L.G., Schoebel, A., Wagner, D., Zaroliagis, C.D. (eds.) Railway Optimization 2004. LNCS, vol. 4359, pp. 67–90. Springer, Heidelberg (2007)
19. Modesti, P., Sciomachen, A.: A Utility Measure for Finding Multiobjective Shortest Paths in Urban Multimodal Transportation Networks. EJOR 111(3), 495–508 (1998)
20. Zadeh, L.A.: Fuzzy Logic. IEEE Computer 21(4), 83–93 (1988)

Efficient Computation of Jogging Routes*

Andreas Gemsa, Thomas Pajor, Dorothea Wagner, and Tobias Zündorf

Department of Computer Science, Karlsruhe Institute of Technology (KIT)
{gemsa,pajor,dorothea.wagner}@kit.edu, tobias.zuendorf@student.kit.edu

Abstract. We study the problem of computing jogging (running) routes
in pedestrian networks: Given source vertex s and length L, it asks for a
cycle (containing s) that approximates L while considering niceness criteria such as the surrounding area, shape of the route, and its complexity.
Unfortunately, computing such routes is NP-hard, even if the only optimization goal is length. We therefore propose two heuristic solutions:
The first incrementally extends the route by joining adjacent faces of
the network. The other builds on partial shortest paths and is even able
to compute sensible alternative routes. Our experimental study indicates
that on realistic inputs we can compute jogging routes of excellent quality
fast enough for interactive applications.

1 Introduction

We study the problem of computing *jogging routes* in pedestrian networks. Given
a source vertex s (the user's starting point), and a desired length L (in kilome-
ters), the problem asks for a cycle of length (approximately) L that contains the
vertex s. A "good" jogging route is, however, not only determined by its length;
other criteria are just as important. An ideal route might follow paths through
nice areas of the map (e. g., forests, parks, etc.), has rather circular shape, and
not too many intersection at which the user is required to turn. A practical
algorithm must, therefore, take all of these criteria into account.

Much research focused on efficient methods for the related, but simpler,
problem of computing point-to-point (shortest) paths. In fact, a plethora of
algorithms exist, many of which are surveyed in [3,9]. They usually employ sophisticated preprocessing to speed up query performance. In contrast, much less
practical work exists for computing cycles. Graphs may contain exponentially
many (in the number of vertices) cycles, even if they are planar [1]. If the length
of the cycles is restricted by L, they can be enumerated in time $O((n+m)(c+1))$,
where c is the number of cycles of length at most L [8]. If one is interested in
computing cycles with exactly k edges, the problem can be solved in $O(2^k m)$
expected time [10]. Unfortunately, none of these methods seem practical in our
scenario. To the best of our knowledge, no efficient algorithms that quickly compute sensible jogging routes exist.

This work introduces the JOGGING PROBLEM. It turns out to be NP-hard,
hence, we propose two heuristic approaches. The first, *Greedy Faces* is based

---

* Partially supported by DFG grant WA 654/16-1.

on building the route by successively joining adjacent faces of the network. The second, *Partial Shortest Paths*, exploits the intuition of constructing equilateral polygons via shortest paths. The latter can be easily parallelized and has the inherent property of providing sensible alternative routes. The result of our algorithms are routes of length within $(1 \pm \varepsilon)L$, but also consider other important criteria that optimize the surrounding area, shape, and route complexity. An experimental study justifies our approaches: Using OpenStreetMap data, we are able to compute jogging routes of excellent quality in under 200 ms time, which is fast enough for interactive applications.

The paper is organized as follows. Section 2 defines variants of the problem and shows NP-hardness. Section 3 introduces our two algorithmic approaches. Section 4 presents experiments, and Section 5 contains concluding remarks.

2 Problems

Before we formally define the considered problems, we need to develop some notation. We model pedestrian networks as *undirected graphs* $G = (V, E)$ with nonnegative integral *edge costs* $\ell \colon E \to \mathbb{Z}_{\geq 0}$. Usually, vertices correspond to intersections and edges to walkable segments. Also, we assume that our graphs admit straight-line embeddings, since vertices have associated latitude/longitude coordinates. For simplicity, our graphs are always connected. A *path* P is a sequence of vertices $P = [u_1, \ldots, u_k]$ for which $u_i u_{i+1} \in E$ must hold. Note that we sometimes just write u_1-u_k-path or P_{u_1, u_k} for short. If the first and last vertices coincide, we call P a *cycle*. The *cost* of a path, denoted by $\ell(P)$, is the sum of its edge-costs. A *shortest path* between two vertices u_1 and u_2 is a u_1-u_2-path with minimum cost. At some places we require intervals around a value $x \in \mathbb{Z}_{\geq 0}$ with error $\varepsilon \in \mathbb{R}_{\geq 0}$. We define them by $I(x, \varepsilon) = [\lfloor (1 - \varepsilon)x \rfloor, \lceil (1 + \varepsilon)x \rceil]$.

Simple Jogging Problem. The first problem we consider is the SIMPLE JOGGING PROBLEM (SJP): We are given a graph G, source vertex $s \in V$, and a targeted cost $L \in \mathbb{Z}_{\geq 0}$ as input. The goal is to compute a cycle P through s with cost $\ell(P) = L$. In practical scenarios, cost usually represent geographical length. It turns out that SJP is NP-hard by reduction from HAMILTONIAN CYCLE. Note that from this, NP-hardness follows for the respective optimization problem, i.e., finding a cycle that *minimizes* $|\ell(P) - L|$.

If we allow running time in the order of L, one can solve SJP by a dynamic program, similarly as it is known for the SUBSET SUM PROBLEM [5]. The algorithm maintains a boolean matrix $Q \colon V \times \mathbb{Z}_{\geq 0} \to \{0, 1\}$ of size $|V| \times L$, which indicates whether a path to vertex u with cost ℓ exists. Initially, Q is set to all-zero, except for the entry $Q(s, 0)$, which is set to 1. It then considers subsequent cost values ℓ in increasing order (beginning at 0). In each step, the algorithm checks for all edges $uv \in E$ if an existing path can be extended to v with cost ℓ. It does so by looking if $Q(u, \ell - \ell(uv))$ is set to 1, updating $Q(v, \ell)$ accordingly. The algorithm stops as soon as ℓ exceeds the input cost L. Then, the requested jogging route exists iff $Q(s, L) = 1$ holds. The running time of the algorithm is $O(L|E|)$, and thus we conclude that the SJP is weakly NP-hard [5].

Relaxed Jogging Problem. In practice, solely optimizing length (or cost) may result in undesirable routes. Jogging is a recreational activity, therefore, one usually also considers the surrounding area (parks and forests), the shape (preferably edge-disjoint), and the complexity of the route (small number of turns). We argue that the primary goal remains geographical length. However, we allow some (user-specified) slack on the length to take the aforementioned criteria into account. This motivates the RELAXED JOGGING PROBLEM (RJP): Given a graph G, a source vertex $s \in V$, input length $L \in \mathbb{Z}_{\geq 0}$, and a parameter $\varepsilon \in [0, 1]$, the goal is to compute a cycle P through s with cost $\ell(P) \in I(L, \varepsilon)$ while optimizing a set of *soft criteria*. We identify three important criteria in the following.

To account for the surrounding area, we introduce *badness* as a mapping on the edges bad: $E \to [0, 1]$. Smaller values indicate "nicer" areas (e. g., parks). Badness values on the edges are provided by the input data. To extend badness to paths, we combine it with the path's length. (Note that we assume costs to represent geographical length for the remainder of the paper.) That is, for a path $P = [u_1, \ldots, u_k]$ its badness is defined by $\mathrm{bad}(P) = \sum \mathrm{bad}(u_i u_{i+1}) \ell(u_i u_{i+1}) / \ell(P)$. By these means, badness values are scaled by their edge lengths, but are still in the interval $[0, 1]$. This enables comparing paths (wrt. badness) of different lengths.

To optimize edge-disjointness of paths, we consider *sharing*. It counts edges that appear at least twice on P, scaled by their length. Formally, it first accumulates into a set D all indices i, j for which either $u_i u_{i+1} = u_j u_{j+1}$ or $u_i u_{i+1} = u_j u_{j-1}$ hold. (Note that edges are undirected.) The sharing of path P is then $\mathrm{sh}(P) = \sum_{i \in D} \ell(u_i u_{i+1}) / \ell(P)$. Sharing values are also in $[0, 1]$.

To evaluate route complexity, we consider *turns*. For two edges a and b, we measure their angle $\angle(a, b)$, and regard them as a turn, iff $\angle(a, b) \notin I(180°, \alpha)$ holds. We usually set α to $15\,\%$.

3   Algorithms

We now introduce our two approaches for the RELAXED JOGGING PROBLEM: *Greedy Faces* and *Partial Shortest Paths*. We present each approach in turn, starting with a basic version, then, proposing optimizations along the way.

3.1   Greedy Faces

Assume that we are already given a tentative jogging route (i. e., a cycle in G that contains s). A natural way to extend it, is to attach one of its adjacent "blocks" that lie on the "outer" side of the route. Then, repeat this step, until a route of desired size and shape has been grown. In a planar graph, blocks correspond to faces. But our inputs may contain intersecting edges (such as bridges and tunnels), albeit only few in practice. We, therefore, propose preprocessing G to identify blocks (we still call them faces). These are used by our greedy faces algorithm. Finally, we present smoothening techniques to reduce route complexity in a quick postprocessing step.

Identifying Faces. For our algorithm to work, we must precompute a set F of *faces* in G. We identify each face $f \in F$ with its *enclosing* path P_f. Our pre-processing involves several steps. First, we delete the *1-shell* of G by iteratively removing vertices (and their incident edges) from G that have degree one. The resulting graph is 2-connected and no longer contains dead-end streets (which we want to avoid, anyway). Next, we consider all remaining edges $uv \in E$. For each, we perform a *right-first search*, thereby, constructing an enclosing path P_f for a new face f. More precisely, we run a depth-first search, beginning at uv. Whenever it reaches a vertex x (via an edge a), it identifies the unique edge b that follows a in the (counterclockwise) circular edge ordering at x. (Note that this ordering is always defined for embedded graphs.) It adds b to P_f. If $b = uv$, the algorithm stops, and adds f to F, discarding duplicates. However, since G is not necessarily planar, the edge b might intersect with one of its preceding edges on P_f. In this case, it removes b from P_f, and considers the next edge (after b) in the circular order at x for expansion. While constructing F, the algorithm remembers for each edge a list of its incident faces. It uses them to build a *dual graph* $G^* = (V^*, E^*)$: Vertices correspond to faces (of G), and two faces are connected in G^*, iff they share at least one edge in G. This definition of G^* extends the well-known graph duality for planar graphs, however, as G may not be planar, so may not be G^*. The running time of the preprocessing is dominated by the face-detection step. For every edge it runs a right-first search, each in time $O(|E|)$. Whenever it expands an edge, it must perform intersection tests with up to $O(|V|)$ preceding edges. This results in a total running time of $O(|V||E|^2)$. Note that we expect much better running times in practice: On realistic inputs we may assume faces to have constant size.

Greedy Faces Algorithm. Our greedy faces algorithm, short GF, now uses G^* as input. Its basic idea is to run a (modified) breadth first search (BFS) on G^*. It starts by selecting an arbitrary face $f \in V^*$ that contains the source vertex s, i. e., where $s \in P_f$ holds. It then grows a BFS-tree T (rooted at f), until a stopping condition is met. When it stops, the jogging route P is retrieved by looking at the set of *cut edges* that separate T from $V^* - T$: Their corresponding edges in G constitute a cycle. (Note that this is a well-known property on planar graphs, but carries over to our definition of G^*.) However, to make P a feasible jogging route, we must ensure two properties: The cycle must be (a) simple, and (b) still contain s. We ensure both while growing T. Regarding (a), we know that the corresponding cycle P in G is simple iff the subgraph induced by $V^* - T$ is connected. We check this condition when expanding an edge $fg \in E^*$ during the BFS, discarding fg if adding g to T would disconnect $V^* - T$. Regarding (b), The vertex s is still part of the jogging route as long as at least one incident face of s remains in $V^* - T$. We also perform this check while expanding edges, discarding them whenever necessary. The result of every iteration of the BFS is a potential jogging route P. The algorithm stops as soon as the cost of P exceeds $(1 + \varepsilon)L$. It then returns, among all discovered routes whose length is in $I(L, \varepsilon)$, the one with minimum total badness.

However, up to now, GF does not *optimize* badness. To guide the search towards "nice" areas of the graph, we propose a force-directed approach. Therefore, consider a face f and the *geometric center* $C(f)$ of its enclosing path. Inspired by Newton's law of gravity, we define a force vector $\phi(f, p)$ acting upon a point p of the map by $\phi(f, p) = (\text{bad}(f) - 0.5)\ell(f)/|\boldsymbol{d}|^2 \cdot \boldsymbol{d}/|\boldsymbol{d}|$, where $\boldsymbol{d} = p - C(f)$. Note that, depending on $\text{bad}(f)$, the force is repelling/attracting. Also, the vector $\phi(f, p)$ is directed, and its intensity decreases with the distance squared. Now, the force that acts upon a face g is the sum of the forces over all (other) faces in the graph (toward g). More precisely, $\phi(g) = \sum_{f \in V^*} \phi(f, C(g))$. In practice, we quickly precompute these values restricted to reachable faces (i. e., faces within a radius of $L/2$ from s). The BFS in our algorithm now extends the edge $fg \in E^*$ next, for which g has the highest force in *direction of extension*. More precisely, it extends fg, iff g maximizes the term $\phi(g) \cos(\angle(\phi(g), C(f) - C(P)))$. Note that $C(P)$ is the geometric center of the current (tentative) jogging route P in the algorithm, and $\angle(\cdot, \cdot)$ measures the angle of two vectors. In principle, further criteria can be added to the BFS (e. g., via linear combinations): The *roundness* considers the ratio of the route's perimeter to its area (lower values are better); *convexity* takes the distance between a candidate face and the current route into account (higher values are better). However, preliminary experiments showed that (on realistic inputs) the effect of these criteria is limited. The running time of GF is bounded by the BFS on G^*. In the worst case, it scans $O(|V^*|)$ faces. The next face it expands to can be determined in time $O(|V^*|)$, yielding a total running time of $O(|V^*|^2)$. Finally, recall that our preprocessing removes the 1-shell of G. For the case that the source vertex s is part of the 1-shell, we quickly find the (unique) path P' to the first vertex s' that is not in the 1-shell. We then run our algorithm, but initialized with s' and $L' = L - 2\ell(P)$, simply attaching P' to the route afterward. Also note that routes obtained by GF are optimal with respect to sharing: The only (unavoidable) place it may occur is on P' (in case s is in the 1-shell).

Route Smoothening. By default, GF provides no guarantee on route *complexity* (i. e., on the number of turns). We, therefore, propose reducing it by *smoothing* the route in a postprocessing step. To do so, we first select a small subsequence $P' \subset P$ of the route's vertices. (Note that s must be part of P'.) Then, for each two subsequent vertices $uv \in P'$, we compute a shortest u-v-path (e. g., by Dijkstra's algorithm [4]). Finally, concatenating these paths produces the smoothened route. To also take badness into account, we use a custom metric $\omega: E \to \mathbb{Z}_{\geq 0}$, defined by $\omega(a) = \text{bad}(a)\ell(a)$, when computing shortest paths.

It remains to discuss how we choose the subsequence P' from P. We propose three rules. The first, called *equidistant rule* (es), simply selects the k (an input parameter) vertices from P, which are distributed equally regarding their subsequent distances. More precisely, vertex $u \in P$ is selected as the i-th vertex on P' if it minimizes $\ell(P)i/k - \ell(P_{s,u})$ (here, $P_{s,u}$ denotes the subpath of P up to vertex u). Unfortunately, this rule may select vertices at arbitrary (with respect to the route's shape) positions. Therefore, our second rule, called *convex rule* (cs), obtains P' by computing the *convex hull* of P, e. g., by running Graham's Scan

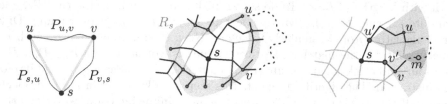

Fig. 1. Left: Intuition of constructing 2-via-routes. Middle: Shortest path tree rooted at s and ring R_s with candidate vertices u, v forming a feasible route (dotted). Right: Selecting middle vertices m that lie "behind" u, v in the shortest path trees of u', v'.

algorithm [6] on P. In case the source vertex s is not part of the convex hull, we must still add it to P': We set its position next to the first vertex of P that is contained in P's convex hull. Finally, the third rule, called *important vertex rule* (ivs), tries to identify k (again, an input parameter) "important" vertices of P: At first, it slices P into k subpaths of equal length. From each, it then selects the vertex u whose incident edges have lowest total badness (i.e., $\prod_{uv \in E} \mathrm{bad}(uv)$ is minimized). This rule follows the intuition that vertices that share many edges of low badness are more likely in "nicer" areas. Note that while smoothening helps to reduce route complexity, its drawback is that the route's length may change arbitrarily. We address this issue by our next approach.

3.2 Partial Shortest Paths

As discussed, GF provides no guarantee on the deviation from the requested route length, if they are smoothened. We, therefore, propose a second approach: It directly computes a set of *via vertices*, connected by shortest paths, but such that the length of the resulting routes is guaranteed to be in $I(L, \varepsilon)$. In the following, we refer to jogging routes that use k via vertices by k-*via-routes*.

2-via-routes. For our basic version, we exploit the intuition of constructing equilateral triangles (see Fig. 1, left), thus, obtaining 2-via-routes. We know that s must be part of the route. Therefore, we choose s as one of the triangle's vertices. It now remains to compute two vertices u, v (and related paths), such that $\ell(P_{s,u}), \ell(P_{u,v}), \ell(P_{v,s}) \in I(L/3, \varepsilon)$. From this, we obtain the required total length of $I(L, \varepsilon)$. To select u and v, we, at first, define a metric on the edges $\omega \colon E \to \mathbb{Z}_{\geq 0}$ that takes the edge's badness into account. As in Section 3.1, we set $\omega(a) = \mathrm{bad}(a)\ell(a)$. We now run a shortest path computation on G from s using this metric with Dijkstra's algorithm [4]. To limit the search, we do not relax edges out of vertices x for whom $\ell(P_{s,x})$ exceeds $(1 + \varepsilon)L/3$. (Note that $\ell(P_{x,s})$ can be stored with x during the algorithm with negligible overhead.) The resulting shortest path tree T_s (rooted at s) accounts for "nice" paths by optimizing ω, and provably contains all feasible candidate vertices u (and v). We refer to this subset of candidate vertices as *ring* around s with distance $I(L/3, \varepsilon)$, in short R_s. We must now find two vertices of the ring that have a connecting path with length $I(L/3, \varepsilon)$. To do so, we pick a vertex u from the ring R_s, and, compute *its* ring R_u (also with respect to length $I(L/3, \varepsilon)$) by running Dijkstra's algorithm from u, similarly to before. Now, the intersection of R_s with R_u exactly contains the matching vertices v, that is, concatenating $P_{s,u}, P_{u,v}, P_{v,s}$ yields an admissible jogging route (i. e., of length $I(L, \varepsilon)$). See Fig. 1 (middle) for an illustration. The algorithm repeats this step for all vertices in R_s, and selects among all admissible routes it discovers the one minimizing badness. We call this algorithm PSP2 (partial shortest paths with two vias). We remark that distances other than $L/3$ are possible when computing rings. This varies the route's shape, and corresponds to constructing "triangles" with nonuniform side lengths. The running time of PSP2 is dominated by up to $O(|V|)$ shortest path computations, thus, it is bounded by $O(|V|^2 \log |V| + |V||E|)$. Note that we expect much better performance in practice, as the shortest path computations are local.

We now propose two optimizations for PSP2. First, the algorithm can be sped up by a *stopping criterion*. For it to work, it must pick vertices u from R_s in order of increasing value $\omega(P_{s,u})$. Note that this order is automatically provided by Dijkstra's algorithm. It then only needs to consider paths $P_{v,s}$ as third leg of the route, for whom $\omega(P_{v,s}) \geq \omega(P_{s,u})$ holds (all others have been evaluated earlier). By this, the total badness of any route P the algorithm may still find is lower-bound by $\text{bad}_{\text{lb}} = 2\omega(P_{s,u})/(1+\varepsilon)L$. If we keep track of the route P_{opt} minimizing badness, the algorithm may stop as soon as bad_{lb} exceeds $\text{bad}(P_{\text{opt}})$— it will provably not find any route with lower badness. Up to now, PSP2 has no guarantee on the sharing of P. In fact, it can be up to 100 % in extreme cases, thus, we propose the following optimization. When the algorithm computes R_u for a vertex $u \in R_s$, we forbid it to relax any edges from $P_{s,u}$. This ensures that $P_{s,u}$ and $P_{u,v}$ are edge-disjoint. To also make $P_{u,v}$ and $P_{v,s}$ edge-disjoint, we disregard routes whose last edges of $P_{u,v}$ and $P_{v,s}$ coincide. Note that we still allow sharing wrt. to the first and last legs of the route (around s).

3-via-routes. Jogging routes obtained by PSP2 follow shortest paths for each of its three legs $P_{s,u}$, $P_{u,v}$, and $P_{v,s}$. However, no such guarantee exists around u and v, which might be undesirable. We now propose an optimized variant of our algorithm, PSP3. It aims to smooth the route around u and v. Moreover, it uses three via-vertices, which, in general, produces more circular shaped routes.

The algorithm follows the intuition of constructing regular *quadrilaterals*. Taking the source vertex s as one of the quadrilateral's vertices, it must therefore compute vertices u, m, and v, connected by paths $P_{s,u}, P_{u,m}, P_{m,v}$, and $P_{v,s}$, each with length $I(L/4, \varepsilon)$. We refer to m as *middle vertex*. The algorithm starts, again, by first computing a ring R_s of vertices from s, but now with distance $I(L/4, \varepsilon)$. (It does so by using Dijkstra's algorithm with metric ω.) To smoothen the route around u and v, we do not use u and v directly as sources for the subsequent shortest path computations (like we did with PSP2). Instead, we consider the (tighter) ring R'_s of vertices around s with distance $I(\alpha L/4, \varepsilon)$. Here, the parameter α takes values from $[0.5, 1]$, and controls smoothness around u and v. We obtain the ring R'_s by traversing the shortest path tree from each vertex $u \in R_s$ upward, until the distance condition is met. Moreover, the vertex u remembers which vertex u' it created in R'_s (this is required later). Next, the algorithm picks vertices u' from R'_s (in any order), and computes, for each, a ring $R_{u'}$ around u'. To account for α, we set the distance of $R_{u'}$ to $I((2-\alpha)L/4, \varepsilon)$. It follows that vertices in $R_{u'}$ have distance $I(L/2, \varepsilon)$ from s, containing potential middle vertices. Having computed all rings, we then consider for each pair of vertices u', v' in R'_s the intersection M of their rings, i.e., $M = R_{u'} \cap R_{v'}$. The algorithm now selects only such middle vertices $m \in M$ that result in smooth paths around u and v. More precisely, a vertex $m \in M$ is selected, iff the *smoothing condition* holds, i.e., the path $P_{u',m}$ contains u and the path $P_{v',m}$ contains v. Intuitively, we are only interested in the part of M that lies "behind" u (resp. v) on the shortest path tree of $R_{u'}$ ($R_{v'}$). See Fig. 1 (right) for an illustration. Each vertex m that fulfills the smoothing condition represents an admissible jogging route by concatenating $P_{s,u}$, $P_{u,m}$, $P_{m,v}$, and $P_{v,s}$. The algorithm returns, among those, the one with minimum badness. With PSP3, the only vertex around which sharing may occur is m (besides s). We avoid it by discarding middle vertices m, for which the last edges of $P_{u',m}$ and $P_{v',m}$ coincide. This can be efficiently checked during the algorithm.

We now propose two optimizations to speed up PSP3. The first avoids the costly computation of set-intersections: Instead of storing (and intersecting) rings $R_{u'}$, the algorithm maintains a vertex-set M_m at each vertex m of the graph. Whenever Dijkstra's algorithm scans a potential middle vertex m, it adds u to M_m (iff the smoothing condition holds). Moreover, it suffices to keep the (at most) two vertices u, v with lowest associated badness values in each set M_m. As a result, managing middle vertices is a constant time operation. The second optimization avoids some calls to Dijkstra's algorithm: If the ring R'_s contains vertices u' and v' for which u' is an ancestor of v' in the shortest path tree, a single Dijkstra run from u' suffices to handle both u' and v'. Including these optimizations, PSP3 essentially runs $O(|V|)$ times Dijkstra's algorithm. Its total running time is thus $O(|V|^2 \log |V| + |V||E|)$, as well as PSP2's.

Bidirectional Search. To allow more flexibility for selecting the middle vertex, we propose the algorithm PSP3-Bi which is an extension of PSP3 using bidirectional search [2]. As PSP3, it starts by computing R_s, and from that, R'_s. However, it now runs (in turn) for each *pair* of vertices u', v' a bidirectional search. Whenever it scans a vertex m that has already been scanned by the opposite direction, it checks (a) whether u (resp. v) are ancestors of m in the forward (resp. backward) shortest path tree, and (b) if the total length of the combined route is in $I(L, \varepsilon)$. If both hold true, it stops, and considers the just-found jogging route as output (it keeps track of the one that minimizes badness). Note that by design, sharing around m cannot occur. Since PSP3-Bi must run a bidirectional search for each pair of vertices in R'_s, its running time is bounded by $O(|V|^3 \log |V| + |V|^2 |E|)$.

Parallelization. All PSP-based algorithms can be parallelized quite easily in a shared memory setup: They, first, sequentially compute the ring R_s (resp. R'_s). Subsequent Dijkstra runs may then be distributed among the available processors. Each processor computes its locally optimal route, and the globally optimal route is selected in a sequential postprocessing step. To avoid race conditions, we use locking as synchronization primitive, whenever necessary.

Alternative Routes. All PSP-based algorithms provide *alternative routes* without significant computational overhead. Instead of just outputting the route with minimum badness, we may output the k best routes. However, these routes tend to be too similar. We, therefore, only consider routes as alternatives that are pairwise different in their via-vertices u and v from R_s (still selecting the k best regarding badness). By these means, we obtain jogging routes that cover different regions of the graph around the source vertex s.

4   Experiments

We implemented all algorithms from Section 3 in C++ compiled with GCC 4.7.1 and flag -O3. Experiments were run on one core of a dual 8-core Intel Xeon E5-2670 clocked at 2.6 GHz with 64 GiB of DDR3-1600 RAM. We focus on the pedestrian network of the greater Karlsruhe region in Germany. We extracted data from a snapshot of the freely available OpenStreetMap[1] (OSM) on 5 August 2012. We only keep walkable street segments and use OSM's `highway` and `landuse` (of the surrounding polygon, if available) tags to define sensible badness values (see [11] for details). The resulting graph has 104 759 vertices and 118 671 edges.

Our first experiment evaluates quality and performance of our algorithms. For each, we ran (the same) 1 000 queries with source vertex s chosen at random. We request routes of 10 km length and ε set to 10 %. Results are summarized in Table 1. We report the average length (in km) of the computed routes, the standard deviation (Std.-Dev.) of their length, their average badness values (Bad.), their average amount of sharing (Sh.), the number of turns on them (No. Trn.), and the average running time of the algorithm on one, and where applicable, also on four and eight processors (Time-x). Sometimes our algorithms may not find any feasible solution. Therefore, we also report their success rates (Succ. Rate).

Algorithms in Table 1 are grouped into blocks. The first evaluates the greedy faces approach from Section 3.1. We observe that GF succeeds in approximating the required route length of 10 km with very little error. However, for 7 % of our queries no solution was found. One reason is that GF is unable to recover from local optima. However, sharing is almost nonexistent with an average value of 0.2 %. This is expected, since by design sharing for GF only occurs around s, iff it lies in a dead-end street. On the downside, route complexity is quite high with 51 turns on average. This justifies our smoothening rules by shortest paths. We set the number of selected vertices to 6 for GF-es and to 9 for GF-ivs. Interestingly, figures are quite similar for all rules: They reduce route complexity by a factor of almost two, which comes with little increase in sharing (up to 6.9 %). Recall that smoothening may arbitrarily change route lengths. Our experiments indicate that the average route length deviates little (it is still 9.5–9.7 km, depending on the specific rule). However, the figure is much less stable: The mean error (Std.-Dev.) increases to

---

[1] http://openstreetmap.org

Table 1. Solution quality and performance on our Karlsruhe input for both the Greedy Faces (GF) and Partial Shortest Paths (PSP) algorithms. For smoothening, we apply the equidistant rule (es), convex hull rule (cs), and important vertex rule (ivs) to GF.

Algorithm	Length [km]	Std.-Dev.	Bad. [%]	Sh. [%]	No. Trn.	Succ. Rate	Time-1 [ms]	Time-4 [ms]	Time-8 [ms]
GF	9.89	0.58	48.7	0.2	51	93 %	285	—	—
GF-es	9.61	2.07	43.8	6.5	28	93 %	289	—	—
GF-cs	9.73	2.23	43.0	6.9	29	93 %	296	—	—
GF-ivs	9.48	1.98	41.7	6.0	30	93 %	293	—	—
PSP2	9.99	0.58	27.3	52.5	16	98 %	179	84	63
PSP3	10.14	0.41	31.0	23.6	20	98 %	155	78	72
PSP3-Bi	10.06	0.53	33.4	13.9	21	98 %	446	177	140

around 2 km. Regarding running times, GF runs in 285 ms on average, with a mild increase up to 296 ms ($\approx 4\,\%$), if we enable smoothening.

The second block evaluates the PSP approach from Section 3.2 (we set α to 0.6, where applicable). Again, we succeed approximating the required route length of 10 km with little error (≈ 0.5 km on average for all algorithms). Because PSP considers more route combinations than GF, it is more likely to find a feasible solution. This is reflected by the excellent success rate of 98 % (for all PSP algorithms). Regarding badness, PSP finds "nicer" routes (lower average badness) than any of the GF algorithms. However, their sharing (still only possible around s) is much higher. On average, sharing is 52 % for PSP2's, though, we are able to reduce it to 14 % with PSP3-Bi. This is well acceptable in practice. An important advantage of PSP over GF is route complexity: With 16–21 turns on average, this figure is lower than *any* of the GF algorithms, even with applied smoothening. Enabling the stopping criterion decreases running times from 3 579 ms (not reported in the table) to 179 ms, a factor of 20. The fastest algorithm is PSP3 with 155 ms on average. PSP3-Bi is slower by a factor of 2.9. (Recall that it must run a bidirectional search for every *pair* of vertices from R_s; cf. Section 3.2.) Regarding parallelism, we observe speedups of factor 2.1 (PSP2) and 1.9 (PSP3) on four processors over a sequential execution. As expected, with a speedup of 2.5, PSP3-Bi benefits most from parallelization. Increasing the number of processors to eight, improves little. Still, PSP3-Bi benefits most, with a total speedup of 3.1.

We now present two detailed experiments. The first concerns our smoothening rules, the second evaluates variations of the input parameter ε. Each datapoint is based on (the same) 1 000 queries with s selected at random, and L set to 10 km. Fig. 2 shows results of our first experiment. We set ε to 10 %, and vary (on the abscissa) the number of vertices between which the smoothening process computes shortest paths. The left plot reports, for each smoothening rule, how much it affects the length of the routes. We report the average amount (in percent) it changes. The right plot shows the same figure, but for badness. We observe that our routes tend to get shorter after smoothening. This is expected,

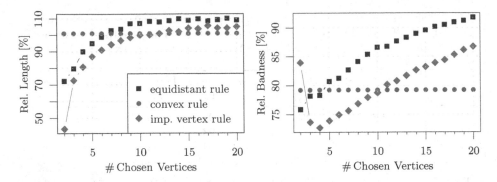

Fig. 2. Evaluating the effect of the smoothening rules on GF. We report the relative amount by which the route's length (left) and badness (right) change while varying the number of vertices the algorithm selects to compute shortest paths (cf. Section 3.1). The legend of the left figure also applies to the right.

since we rebuild routes using shortest paths. Selecting too few vertices shortens routes severely (to below 50 %). Their length eventually stabilizes above 90 % for six vertices and more. Badness generally improves when using smoothening, but continuously increases with more vertices. Interestingly, the convex rule (which is independent of the number of vertices) seems good regarding both length and badness, which makes it the preferred rule in practice.

Our final experiment evaluates all algorithms for varying input parameter ε. Results are summarized in Fig. 3, which evaluates, for each ε, the average success rate (left plot) and the resulting route's badness (right plot). Note that applying smoothening to GF does not affect the success rate, therefore, we do not enumerate smoothening rules in the left figure. We observe that too much restriction on the allowed length (small ε-values), may result in a low success rate (down to 75 %) and high badness values (more than 50 % for GF). Setting $\varepsilon > 0.07$ already significantly improves the success rate. Unsurprisingly, badness values gradually improve with increasing ε, as this gives the algorithms more room for optimization. Here, a good tradeoff seems setting ε to 0.1. Interestingly, PSP3-Bi's success rate is almost unaffected by ε, even for tiny values below 0.07.

5   Conclusion

In this work, we introduced the NP-hard JOGGING PROBLEM. To compute useful jogging routes, we presented two novel algorithmic approaches that solve a relaxed variant of the problem. Besides length, both explicitly optimize two important criteria: Badness (i. e., surrounding area) and sharing (i. e., shape of the route). The methods are based on different intuitions. The first incrementally extends routes by carefully joining adjacent faces of the graph, possibly smoothened by a quick postprocessing step. The second computes sets of alternative routes that resemble equilateral polygons via shortest path computations. Experiments on real-world data reveal that our algorithms are indeed practical: They compute

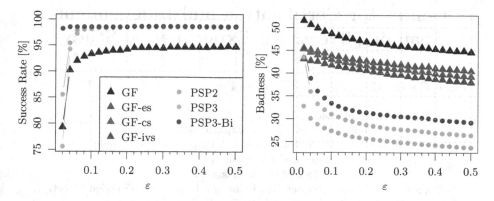

Fig. 3. Evaluating success rate and badness on all algorithms for varying ε. The legend of the left figure also applies to the right. Note that, regarding the greedy faces approach, smoothening does not affect the success rate, hence, we only report it for GF.

jogging routes of excellent quality in under 200 ms time, which is fast enough for interactive applications. Future work includes comparing our algorithms to exact solutions, and better methods for selecting via vertices—either as smoothening rules, or for computing routes directly. Also, providing via vertices (or "areas") as input is an interesting scenario. Finally, we like to accelerate our algorithms further. Especially, PSP may benefit from speedup techniques [3,9]. This, however, requires adapting them to compute rings instead of point-to-point paths.

References

1. Buchin, K., Knauer, C., Kriegel, K., Schulz, A., Seidel, R.: On the number of cycles in planar graphs. In: Lin, G. (ed.) COCOON 2007. LNCS, vol. 4598, pp. 97–107. Springer, Heidelberg (2007)
2. Dantzig, G.: Linear Programming and Extensions. Princeton University Press (1962)
3. Delling, D., Sanders, P., Schultes, D., Wagner, D.: Engineering Route Planning Algorithms. In: Lerner, J., Wagner, D., Zweig, K.A. (eds.) Algorithmics. LNCS, vol. 5515, pp. 117–139. Springer, Heidelberg (2009)
4. Dijkstra, E.W.: A Note on Two Problems in Connexion with Graphs. Numerische Mathematik 1, 269–271 (1959)
5. Garey, M.R., Johnson, D.S.: Computers and Intractability: A Guide to the Theory of NP-Completeness. W. H. Freeman & Co., New York (1979)
6. Graham, R.L.: An efficient algorithm for determining the convex hull of a finite planar set. Information Processing Letters 1(4), 132–133 (1972)
7. Karp, R.M.: Reducibility among Combinatorial Problems. In: Complexity of Computer Computations, pp. 85–103. Plenum Press (1972)
8. Liu, H., Wang, J.: A new way to enumerate cycles in graph. In: AICT and ICIW 2006, pp. 57–60. IEEE Computer Society (2006)
9. Sommer, C.: Shortest-Path Queries in Static Networks (2012) (submitted), Preprint available at http://www.sommer.jp/spq-survey.html
10. Yuster, R., Zwick, U.: Color-coding. Journal of the ACM 42(4), 844–856 (1995)
11. Zündorf, T.: Effiziente Berechnung guter Joggingrouten. Bachelor thesis, Karlsruhe Institute of Technology (October 2012)

Dominator Certification and Independent Spanning Trees: An Experimental Study

Loukas Georgiadis[1], Luigi Laura[2], Nikos Parotsidis[1], and Robert E. Tarjan[3]

[1] Department of Computer Science, University of Ioannina, Greece
loukas@cs.uoi.gr, nickparo1@gmail.com
[2] Dip. di Ingegneria Informatica, Automatica e Gestionale
"Sapienza" Università di Roma
laura@dis.uniroma1.it
[3] Department of Computer Science, Princeton University, 35 Olden Street,
Princeton, NJ, 08540, and Hewlett-Packard Laboratories
ret@cs.princeton.edu

Abstract. We present the first implementations of certified algorithms for computing dominators, and exhibit their efficiency experimentally on graphs taken from a variety of applications areas. The certified algorithms are obtained by augmenting dominator-finding algorithms to compute a certificate of correctness that is easy to verify. A suitable certificate for dominators is obtained from the concepts of low-high orders and independent spanning trees. Therefore, our implementations provide efficient constructions of these concepts as well, which are interesting in their own right. Furthermore, we present an experimental study of efficient algorithms for computing dominators on large graphs.

1 Introduction

A *flow graph* is a directed graph with a distinguished *start* vertex s such that every vertex is reachable from s. Throughout this paper $G = (V, A, s)$ is a flow graph with vertex set V, arc set A, start vertex s, and no arc entering s. (Arcs entering s can be deleted without affecting any of the concepts we study.) We denote the number of vertices by n and the number of arcs by m ($m \geq n - 1$). A fundamental concept in flow graphs is that of *dominators*. A vertex u is a *dominator* of a vertex v (u *dominates* v) if every path from s to v contains u; u is a *proper dominator* of v if u dominates v and $u \neq v$. The dominator relation is reflexive and transitive. Its transitive reduction is a rooted tree, the *dominator tree* D: v dominates w if and only if v is an ancestor of w in D. If $v \neq s$, $d(v)$, the parent of v in D, is the *immediate dominator* of v: it is the unique proper dominator of v that is dominated by all proper dominators of v. Dominators have applications in diverse areas including program optimization and code generation [11], constraint programming [32], circuit testing [4], theoretical biology [2], memory profiling [27], connectivity and path-determination problems [15,16,24], and the analysis of diffusion networks [22]. Allen and Cocke showed that the dominance relation can be computed iteratively from a set of

V. Bonifaci et al. (Eds.): SEA 2013, LNCS 7933, pp. 284–295, 2013.
© Springer-Verlag Berlin Heidelberg 2013

data-flow equations [1]. A direct implementation of this method has an $O(mn^2)$ worst-case time bound, for a flowgraph with n vertices and m edges. Cooper, Harvey, and Kennedy [10] presented a clever tree-based space-efficient implementation of the iterative algorithm. Although it does not improve the $O(mn^2)$ worst-case time bound, the tree-based version is much more efficient in practice. Purdom and Moore [31] gave an algorithm, based on reachability, with complexity $O(mn)$. Improving on previous work by Tarjan [34], Lengauer and Tarjan [25] gave two near-linear-time algorithms for computing D that run fast in practice and have been used in many of these applications. The simpler of these runs in $O(m \log_{(m/n+1)} n)$ time. The other runs in $O(m\alpha(m,n))$ time, where α is a functional inverse of Ackermann's function [35]. Subsequently, more-complicated but truly linear-time algorithms were discovered [3,6,7,17].

In [18,19,20] the problem of verifying the dominator tree of a flow graph was considered: we wish for a simple way to verify that the tree produced by one of the fast but complicated dominator-finding algorithms is in fact the dominator tree. The correctness of a simpler but less efficient algorithm for computing dominators has been mechanically verified [37], but to our knowledge none of the fast algorithms has had its correctness mechanically verified. The approach in [19,20] was to augment the dominator-finding algorithm to compute additional information, a *certificate of correctness*. The verifier uses the certificate to make dominator verification easier. This makes the dominator-finding algorithm a *certifying algorithm* [28]. A suitable certificate for dominators is that of a *low-high order* of a flow graph and a rooted tree. Let T be a rooted tree. We denote by $t(v)$ the parent of vertex v; $t(v) = null$ if v is the root of T. If v is an ancestor of w, $T[v,w]$ is the path from v to w. Tree T is *flat* if its root is the parent of every other vertex. Given a tree T rooted at s with vertex set V (not necessarily a spanning tree of G), a preorder of T is *low-high* on G if, for all $v \neq s$, $(t(v), v) \in A$ or there are two arcs $(u, v) \in A$, $(w, v) \in A$ such that u is less than v, v is less than w, and w is not a descendant of v. See Figure 1.

Low-high orders are related to the notion of *independent spanning trees*. Two spanning trees B and R rooted at s are *independent* if for all v, $B[s,v]$ and $R[s,v]$ share only the dominators of v; B and R are *strongly independent* if for every pair of vertices v and w, either $B[s,v]$ and $R[s,w]$ share only the common dominators of v and w, or $B[s,w]$ and $R[s,v]$ share only the common dominators of v and w. Given a low-high order it is easy to construct in $O(n)$ time two strongly independent spanning trees, and, conversely, given two independent spanning trees we can construct a low-high order in $O(n)$ time [19,20]. These three definitions are interesting in their own right, and have applications in other graph problems [19]. Previously, they were considered only for flow graphs with flat dominator trees [9,23,30,36].

In this work we present efficient implementations of certified, near-linear-time algorithms for computing dominators, based on low-high orders, as described in [19]. [1] This way we also obtain efficient implementations of near-linear-time algorithms for computing a low-high order and two strongly independent spanning

---

[1] We refer to [19] for the complete description of the algorithms and proofs of correctness.

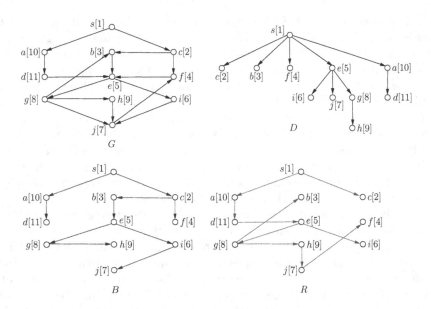

Fig. 1. A flow graph, its dominator tree with vertices numbered in low-high order (numbers in brackets), and two strongly independent spanning trees B and R.

trees of a given flow graph.[2] We exhibit the efficiency of our algorithms experimentally on graphs taken from a variety of applications areas. Furthermore, we present experimental results for various efficient algorithms for computing dominators. A previous experimental study of algorithms for computing dominators was presented in [21], where careful implementations of both versions of the Lengauer-Tarjan algorithm, the iterative algorithm of Cooper, Harvey, and Kennedy, and a new hybrid algorithm (SNCA) were given. In these experimental results the performance of all these algorithms was similar, but the simple version of the Lengauer-Tarjan algorithm and the hybrid algorithm were most consistently fast, and their advantage increased as the input graph got bigger or more complicated. The graphs used in [21] have moderate size (at most a few thousand vertices and edges) and simple enough structure that they can be efficiently processed by the iterative algorithm. In our experiments we deal with larger and more complicated graphs for which simple iterative algorithms are not competitive with the more sophisticated algorithms based on Lengauer-Tarjan.

2   Dominator Certification

Let T be a rooted tree whose vertex set is V. Tree T has the *parent property* if for all $(v, w) \in A$, $t(w)$ (the parent of w in T) is an ancestor of v in T. Since

---

[2] Similarly, we can obtain truly linear-time algorithms by augmenting the linear-time algorithm for computing dominators in [6].

s has no entering arcs, but every other vertex has at least one entering arc (all vertices are reachable from s), the parent property implies that T is rooted at s. Tree T has the *sibling property* if v does not dominate w for all siblings v and w. The parent and sibling properties are necessary and sufficient for a tree to be the dominator tree.

Theorem 1. [19,20] *A tree T has the parent and sibling properties if and only if $T = D$.*

By Theorem 1, to verify that a tree T is the dominator tree, it suffices to show (1) T is a rooted tree, (2) T has the parent property, and (3) T has the sibling property. It is straightforward to verify (1) and (2) in $O(m + n)$ time as follows. We execute a preorder traversal of T, assigning a preorder number $pre(v)$ from 1 to n to each vertex v, and computing the number of descendants $size(v)$ of v. To verify (1) we simply check if $size(s) = n$. To verify (2), we use the fact that v is an ancestor of w if and only if $pre(v) \leq pre(w) < pre(v) + size(v)$ [33]. This takes $O(1)$ time per arc, for a total of $O(m)$ time. If T is given by its parent function, we can number the vertices and compute their sizes by first building a list of children for each vertex and then doing a depth-first traversal, all of which take $O(n)$ time. The hardest step in verification is to show (3), but the use of a low-high order gives a straightforward test.

Theorem 2. [19,20] *A tree with the parent property has the sibling property, and hence is the dominator tree, if and only if it has a low-high order with respect to G.*

Given a tree T with the parent property and a preorder ord, we can test the sibling property by performing the following steps. Construct lists of the children of each vertex in T in increasing order with respect to ord. Do a depth-first traversal of T to verify that the preorder generated by the search is the same as ord and to compute the size of each vertex. Check that each vertex has the one or two entering arcs needed to make ord a low-high order, using the numbers and sizes to test the ancestor-descendant relation in $O(1)$ time.

Although it is easy to test if a given order is low-high, it is not so easy to test for the existence of a low-high order. Thus we place the burden of constructing such an order on the algorithm that computes the dominator tree, not on the verification algorithm: the order certifies the correctness of the tree. Furthermore, an algorithm that computes the dominator tree and a low-high order does not need to output the low-high order explicitly. It suffices to output the edges of the dominator tree in an order such that an edge (u, v) precedes (u, w) if and only if v is less than w in the low-high order. A verification algorithm can then perform the verification steps (1)-(3) concurrently. In the following section we will refer to this verification algorithm as VERIFY.

2.1 Low-High Orders and Independent Spanning Trees

An efficient way to construct a low-high order of G is via two independent spanning trees. In [18,19] it is shown that two such spanning trees can be computed by a simple extension of the fast algorithms for finding dominators [3,6,25]. This extension requires additional $O(n)$ time and space computations.

Let F be a depth-first spanning tree of G rooted at s, with vertices numbered from 1 to n as they are first visited by the search. Identify vertices by number. A path from u to v is *high* if all its vertices other than u and v are higher than both u and v. If $v \neq s$, the *semi-dominator* of v, $sd(v)$, is the minimum vertex u such that there is a high path from u to v. Vertex $sd(v)$ is the ancestor u of v in F closest to s such that there is a path from u to v avoiding all other vertices on $F[u, v]$ (the path in F from u to v) [25]. Since there is a path from s to v avoiding all vertices on $F[sd(v), v]$ except $sd(v)$ and v, $d(v) \leq sd(v)$. The *relative dominator* $rd(v)$ is the vertex x on $F(sd(v), v]$ (the path in F from a child of $sd(v)$ to v) such that $sd(x)$ is minimum, with a tie broken in favor of the smallest x. The algorithm to construct two independent spanning trees, B and R, processes the vertices in increasing order. For each vertex $v \neq s$ it chooses one of $f(v)$ and $g(v)$ to be the parent $b(v)$ of v in B and the other to be the parent $r(v)$ of v in R, as follows: if $sd(v) = sd(rd(v))$ or $b(rd(v)) = f(rd(v))$, set $b(v) = g(v)$ and $r(v) = f(v)$; otherwise, set $b(v) = f(v)$ and $r(v) = g(v)$. We will refer to this algorithm in the next section as SLTIST.

For the computation of a low-high order it is convenient to modify the above construction as follows. For each vertex $v \neq s$, if $(d(v), v)$ is an arc of G then replace $b(v)$ and $r(v)$ by $d(v)$. Clearly, B and R remain independent spanning trees of G after this modification.

2.2 Derived Arcs

The next step of the construction is to compute the *derived arcs* of B and R. Let (v, w) be an arc of G. By the parent property, $d(w)$ is an ancestor of v in D. The *derived arc* of (v, w) is null if w is an ancestor of v in D, (v', w) otherwise, where $v' = v$ if $v = d(w)$, v' is the sibling of w that is an ancestor of v if $v \neq d(w)$. Given a list of arcs L we can compute in $O(n + |L|)$ time the derived arcs of all arcs in L using a three-pass radix sort. See [19] for the details.

2.3 Construction of a Low-High Order

Let B' and R' be the graphs formed, respectively, by the derived arcs of B and R. Let G' be the union of B' and R'. Then, G and G' have the same dominator tree, and B' and R' are independent spanning trees of G'.

Given B', R' and G' we can compute a low-high order of D as follows. For each v, initialize its list of children $C(v)$ to be empty. Then apply the following reduction step. If G' contains only one vertex $v \neq s$, insert v anywhere in $C(s)$. Otherwise, let v be a vertex whose in-degree in G' exceeds its number of children in B' plus its number of children in R'. Assume v is a leaf in R'; proceed symmetrically if v is a leaf in B'. If v is not a leaf in B', let w be its child in B', and replace $b'(w)$ by $b'(v)$. Delete v, and apply the reduction step recursively to insert the remaining vertices other than s into lists of children. If $b'(v) = d(v)$, insert v anywhere in $C(d(v))$; otherwise, insert v just before $b'(v)$ in $C(d(v))$ if

$r'(v)$ is before $b'(v)$ in $C(d(v))$, just after $b'(v)$ otherwise. After processing all vertices, do a depth-first traversal of D, visiting the children of each vertex v in their order in $C(v)$. Number the vertices from 1 to n as they are visited. The resulting order is low-high on G.

In order to implement this algorithm we need to specify how to: (1) perform the insert operations in each list of children, and (2) maintain G', B' and R' as the algorithm removes vertices during the reduction step. The first problem is a special case of the *dynamic list maintenance* problem [5,13], for which a simple solution that supports insertions and order tests in $O(1)$ time is given in [19]. For problem (2), we can maintain the parents in B' and R' easily using two arrays that store the corresponding parent functions. Also, instead of storing the adjacency lists of G' it is more convenient to store the adjacency lists of B' and R' using three integer arrays, each of size n. Consider the structure for B'. The first array, $list_{B'}$, stores the vertices sorted in increasing order by parent in B', the second array, $first_{B'}$, stores the position in $list_{B'}$ of the first child of each vertex, and the third array, $position_{B'}$, stores the position of each vertex in $list_{B'}$. Consider vertex v that is removed in an application of the reduction step. If v is a leaf in B' then we set $list_{B'}[position_{B'}[v]] = null$. Otherwise, let w be the (unique) child of v in B'. To find w, we search for the first (and only) child of v in $first_{B'}$ that is not null. Then we set $list_{B'}[position_{B'}[v]] = w$ and $position_{B'}[w] = position_{B'}[v]$. The structure for R' is updated similarly. Since we search the list of children of a vertex v only once (when v is removed) the total running time for updating B' and R' is $O(n)$. In the next section, we will refer to this algorithm as SLTCERT.

2.4   Alternative Construction of a Low-High Order

In [19] it is conjectured that the spanning trees B and R computed by the algorithm described in Section 2.1 can be generated from some low-high order of D. That is, there is an order of the vertices such that for each vertex v, $b(v) < v < r(v)$ if $b(v) \neq r(v)$, and $b(v) = r(v) = d(v) < v$ otherwise. This conjecture is supported by our experimental results, but we have no proof. If the conjecture holds, then we can compute a low-high order with the following simpler alternative to the algorithm of Section 2.3. Construct the independent spanning trees B' and R' of the derived graph, as in Section 2.2. Form the graph K whose arcs are all those in B' and the reversals of all arcs in R' that are not in B'. Find a topological order of K. For each vertex $v \neq s$, arrange its children in D in an order consistent with the topological order of K. The preorder on D corresponding to its ordered lists of children will be a low-high order. We will refer to this algorithm in the next section as SLTCERT-II.

3   Empirical Analysis

We evaluate the performance of eight algorithms, four to compute dominators (the simple (SLT) and the sophisticated (LT) versions of the Lengauer-Tarjan algorithm [25], the hybrid algorithm (SNCA) of [21], and the iterative algorithm of Cooper, Harvey, and Kennedy (CHK) [10]), the simple version of the Lengauer-Tarjan algorithm augmented to compute two independent spanning trees (SLTIST), two versions of a certified simple version of the Lengauer-Tarjan algorithm (SLTCERT and SLTCERT-II), and the low-high verification algorithm (VERIFY). For the first four algorithms we adapted the implementations from [21]. Our implementations of the independent spanning trees constructions and the certified algorithms are based upon SLT which usually performs better than LT in practice and it is easier to code. We made all implementations as efficient and uniform as we could, within reason. (Of course, there might be room for further improvements.) Both certified algorithms compute a low-high order of the dominator tree via two independent spanning trees, which are computed as described in Section 2.1. Then, algorithm SLTCERT computes a low-high order by applying the method of Section 2.3, while algorithm SLTCERT-II applies the method of Section 2.4. The latter construction of a low-high order is simpler than the former, but we have no proof that it is guaranteed to work. The algorithm detects if the construction has succeeded or not; it succeeds if and only if graph K, defined in Section 2.4, is acyclic. Of course, one can combine both methods by applying that of Section 2.3 if the method of Section 2.4 fails. In all the experiments we performed, however, the method of Section 2.4 always succeeded in computing a low-high order. We note that we cannot directly apply these methods to augment SNCA into a certified algorithm. The reason is that the construction of the two independent spanning trees requires the computation of relative dominators, which are not computed by SNCA.

Experimental Setup. Our implementations have been written in C++, and the code was compiled using g++ v. 4.3.4 with full optimization (flag -O4). The source code is available from the authors upon request. We have tested our code under Windows (Seven), Mac OSX (10.8.2 Mountain Lion), and GNU/Linux Debian (6.06); the behaviour of the code was comparable on all the architectures, and therefore in the following, due to space constraints, we report only the results against the GNU/Linux machine: namely an HP Proliant server 64-bit NUMA machine composed by two AMD Opteron 6174 processors and 32GB of RAM memory. Each processor is equipped with 12 cores that share a 12MB L3 cache, and each core has a 512KB private L2 cache and 2200MHz speed. We report CPU times measured with the `getrusage` function. All the running times reported in our experiments were averaged over ten different runs. To minimize fluctuations due to external factors, we used the machine exclusively for tests, took each measurement three times, and picked the best. Running times do not include reading the input file (the programs read the input arcs from a text file and store them in an array), but they do include creating the graph (successor and predecessor lists, as required by each algorithm), and allocating and deallocating the arrays used by each algorithm.

Instances. We conducted experiments for a collection of (mostly) large graphs, detailed in Table 1, from several distinct application domains. This collection includes road networks taken from the 9th DIMACS Implementation

Table 1. Real-world graphs sorted by file size; n is the number of vertices, m the number of edges, and δ_{avg} is the average vertex degree.

Graph	n	m	file size	δ_{avg}	type
rome99	3.3k	8.8k	98k	2.65	road network
s38584	20.7k	34.4k	434k	1.67	circuit
Oracle-16k	15.6k	48.2k	582k	3.08	memory profiling
p2p-gnutella25	22.6k	54.7k	685k	2.41	peer2peer
soc-Epinions1	75.8k	508k	5.9M	6.71	social network
USA-road-NY	264k	733k	11M	2.78	road network
USA-road-BAY	321k	800k	12M	2.49	road network
Amazon0302	262k	1.2M	18M	4.71	product co-purchase
web-NotreDame	325k	1.4M	22M	4.6	web graph
web-Stanford	281k	2.3M	34M	8.2	web graph
Amazon0601	403k	3.4M	49M	8.4	product co-purchase
wiki-Talk	2.3M	5.0M	69M	2.1	social network
web-BerkStan	685k	7.6M	113M	11.09	web graph
SAP-4M	4.1M	12.0M	183M	2.92	memory profiling
Oracle-4M	4.1M	14.6M	246M	3.55	memory profiling
Oracle-11M	10.7M	33.9M	576M	3.18	memory profiling
SAP-11M	11.1M	36.4M	638M	3.27	memory profiling
LiveJournal	4.8M	68.9M	1G	14.23	social network
USA road complete	23.9M	58.3M	1.1G	2.44	road network
SAP-32M	32.3M	81.9M	1.5G	2.53	memory profiling
SAP-47M	47.0M	131.0M	2.2G	2.8	memory profiling
SAP-70M	69.7M	215.7M	3.7G	3.09	memory profiling
SAP-187M	186.9M	556.2M	11G	2.98	memory profiling

Challenge website [12], a circuit from VLSI-testing applications [4] obtained from the ISCAS'89 suite [8], graphs taken from applications of dominators in memory profiling (see e.g., [29]), and graphs taken from the Stanford Large Network Dataset Collection [26]. Almost all the graphs considered were completely reachable from the chosen root vertex; only three of them had a reachable fraction of nodes smaller than 90%: soc-Epinions (63%), web-Stanford (77%), and web-BerkStan (67%).

Evaluation. We first focus on algorithms for computing dominators, in order to later provide a perspective for the other algorithms we implemented.

Dominators computation. In Table 2, we present the results of the four dominator-finding algorithms (SLT, LT, SNCA, and CHK) against small and medium scale graphs. We observe that the iterative algorithm is not competitive with the more sophisticated LT-based algorithms as the graph size increases. Among the LT-based algorithms there is no clear winner. In order to provide a better picture of the overall performances of these three algorithms, Figure 2 gives a plot of their running times in microseconds, normalized to the number of the edges. As we can see, the processing time per edge is almost constant, i.e. the values range in a very small interval. This result is consistent with what observed in [21] and, for large scale graphs, in [14].

Table 2. Running times against small and medium graphs, measured in seconds, of the four dominator computation algorithms considered. The best result in each row is marked in bold.

Graph	n	m	LT	SLT	SNCA	CHK
rome99	3.3k	8.8k	0.002	**0.001**	**0.001**	**0.003**
s38584	20.7k	34.4k	0.008	**0.007**	**0.007**	0.010
Oracle-16k	15.6k	48.2k	0.004	0.006	**0.003**	**0.003**
p2p-gnutella25	22.6k	54.7k	**0.006**	0.007	**0.006**	**0.006**
soc-Epinions1	75.8k	508k	**0.052**	0.057	0.053	0.083
USA-road-NY	264k	733k	0.072	0.066	**0.061**	0.624
USA-road-BAY	321k	800k	0.086	0.076	**0.070**	1.130
Amazon0302	262k	1.2M	**0.35**	**0.35**	0.39	0.96
web-NotreDame	325k	1.4M	0.10	**0.09**	0.10	0.51
web-Stanford	281k	2.3M	0.29	0.28	**0.27**	5.68
Amazon0601	403k	3.4M	0.51	0.53	**0.47**	1.41
wiki-Talk	2.3M	5.0M	1.44	1.38	**1.14**	2.14
web-BerkStan	685k	7.6M	0.36	0.35	**0.33**	15.64
SAP-4M	4.1M	12.0M	1.35	1.22	**1.07**	1199.08
Oracle-4M	4.1M	14.6M	1.77	1.63	**1.45**	6.05
Oracle-11M	10.7M	33.9M	7.29	6.06	**5.16**	4583.74

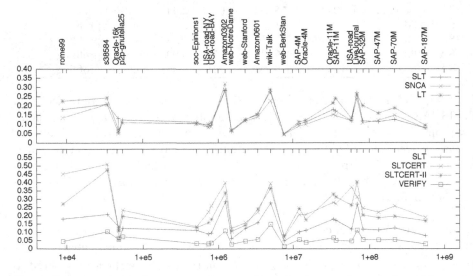

Fig. 2. Running times (in microsecs) normalized to the number of edges (shown in logarithmic scale)

Dominators certification. Table 3 shows the running times against medium and large scale graphs for all algorithms except CHK. We also provide the running times for executing a DFS traversal of the graphs (including the time to build adjacency lists and other auxiliary arrays used by DFS), which we use as a baseline; the first step of all algorithms considered in this study is to execute a DFS on the input graph, with the exception of VERIFY, which executes a DFS

Table 3. Running times against medium and large graphs, measured in seconds. The best result in each row among the three dominator-finding algorithms, and between the two certified algorithms is marked in bold.

Graph	n	m	DFS	LT	SLT	SNCA	SLTCERT	SLTCERT-II	VERIFY	SLTIST
Amazon0302	262k	1.2M	0.10	**0.35**	**0.35**	0.39	0.49	**0.42**	0.13	0.32
web-NotreDame	325k	1.4M	0.03	0.10	**0.09**	0.10	0.18	**0.15**	0.04	0.09
web-Stanford	281k	2.3M	0.10	0.29	0.28	**0.27**	0.36	**0.33**	0.10	0.28
Amazon0601	403k	3.4M	0.13	0.51	0.53	**0.47**	**0.74**	0.81	0.19	0.50
wiki-Talk	2.3M	5.0M	0.32	1.44	1.38	**1.14**	1.97	**1.82**	0.74	1.36
web-BerkStan	685k	7.6M	0.15	0.36	0.35	**0.34**	**0.53**	0.59	0.12	0.36
SAP-4M	4.1M	12.0M	0.38	1.35	1.22	**1.07**	**2.49**	2.92	0.67	1.29
Oracle-4M	4.1M	14.6M	0.54	1.77	1.64	**1.45**	3.03	**2.57**	0.57	1.70
Oracle-11M	10.7M	33.9M	1.42	7.30	6.06	**5.16**	**9.46**	11.13	2.23	6.37
SAP-11M	10.7M	36.4M	1.27	8.73	6.22	**5.30**	**9.86**	11.36	1.81	6.60
USA road complete	23.9M	58.3M	3.13	7.72	**6.88**	**6.88**	21.59	**15.24**	2.68	7.92
LiveJournal	4.8M	68.9M	3.08	18.09	18.52	**16.44**	**21.50**	27.64	7.72	18.74
SAP-32M	32.3M	81.9M	3.13	16.46	9.62	**8.77**	20.01	**16.62**	4.53	11.26
SAP-47M	47.0M	131.0M	5.13	20.91	**14.82**	15.79	28.77	**24.56**	6.99	18.29
SAP-70M	69.7M	215.7M	7.86	40.47	**26.95**	31.81	55.34	**42.11**	11.67	38.77
SAP-187M	186.9M	556.2M	21.44	51.61	43.77	**41.95**	104.77	**93.53**	21.38	46.93

on the input tree. Figure 2 gives the corresponding plot of the running times, normalized to the number of the edges. The results indicate that the certified algorithms perform well in practice even for very large graphs: in all of our tests the running time of SLTCERT and SLTCERT-II is very close to SLT, and only for the largest graph they took slightly more than twice its time. With respect to DFS, we see that all algorithms are slower than DFS by a small constant factor. As already mentioned, the correctness of SLTCERT-II hinges on a conjecture from [19]. We tested it on several synthetic and real world graphs (not reported in the tables) and, so far, we did not find a counterexample.

Dominator tree and low-high order verification. The running time of the verification algorithm, VERIFY, are shown in Table 3, and plotted in Figure 2. As expected, the verification algorithm is very fast, and for some inputs even faster than DFS.

Independent spanning trees computation. Finally, in the last column of Table 3 we can see the running times needed to compute two independent spanning trees: the overhead needed by SLTIST, compared to SLT that is based on, is approximately no more than 15-20%, with only one exception (43% more for the sap-70M graph).

4 Conclusion

In light of our experimental results we conclude that the use of the more elaborate algorithms for computing dominators is necessary for applications that deal with very large graphs. This, in turn, advocates for the use of practical dominator verification methods. Here we presented efficient implementations of certified algorithms for computing dominators that make verification straightforward and very fast. In the process, the algorithms compute two independent spanning trees and a low-high order of the input flow graph, which are useful in other graph algorithms. Our algorithms were based on the simple version of the Lengauer-

Tarjan algorithm, which is easier to implement compared to the sophisticated version, and usually performs better in practice.

Acknowledgements. We thank Bob Foster for bringing to our attention the use of dominators in memory profilers and for providing the "Oracle" graphs, and Andreas Buchen and Krum Tsvetkov for providing the "SAP" graphs. We also thank Alessandro Pellegrini and Francesco Quaglia for providing us the computing infrastructure for our experiments.

References

1. Allen, F.E., Cocke, J.: Graph theoretic constructs for program control flow analysis. Tech. Rep. IBM Res. Rep. RC 3923, IBM T.J. Watson Research Center (1972)
2. Allesina, S., Bodini, A.: Who dominates whom in the ecosystem? Energy flow bottlenecks and cascading extinctions. Journal of Theoretical Biology 230(3), 351–358 (2004)
3. Alstrup, S., Harel, D., Lauridsen, P.W., Thorup, M.: Dominators in linear time. SIAM Journal on Computing 28(6), 2117–2132 (1999)
4. Amyeen, M.E., Fuchs, W.K., Pomeranz, I., Boppana, V.: Fault equivalence identification using redundancy information and static and dynamic extraction. In: Proceedings of the 19th IEEE VLSI Test Symposium (March 2001)
5. Bender, M.A., Cole, R., Demaine, E.D., Farach-Colton, M., Zito, J.: Two simplified algorithms for maintaining order in a list. In: Möhring, R.H., Raman, R. (eds.) ESA 2002. LNCS, vol. 2461, pp. 152–164. Springer, Heidelberg (2002)
6. Buchsbaum, A.L., Georgiadis, L., Kaplan, H., Rogers, A., Tarjan, R.E., Westbrook, J.R.: Linear-time algorithms for dominators and other path-evaluation problems. SIAM Journal on Computing 38(4), 1533–1573 (2008)
7. Buchsbaum, A.L., Kaplan, H., Rogers, A., Westbrook, J.R.: A new, simpler linear-time dominators algorithm. ACM Transactions on Programming Languages and Systems 20(6), 1265–1296 (1998); Corrigendum in 27(3), 383–387 (2005)
8. CAD Benchmarking Lab: ISCAS'89 benchmark information, http://www.cbl.ncsu.edu/www/CBL_Docs/iscas89.html
9. Cheriyan, J., Reif, J.H.: Directed s-t numberings, rubber bands, and testing digraph k-vertex connectivity. Combinatorica, 435–451 (1994), also in SODA 1992
10. Cooper, K.D., Harvey, T.J., Kennedy, K.: A simple, fast dominance algorithm. Software Practice & Experience 4, 110 (2001)
11. Cytron, R., Ferrante, J., Rosen, B.K., Wegman, M.N., Zadeck, F.K.: Efficiently computing static single assignment form and the control dependence graph. ACM Transactions on Programming Languages and Systems 13(4), 451–490 (1991)
12. Demetrescu, C., Goldberg, A., Johnson, D.: 9th DIMACS Implementation Challenge: Shortest Paths (2007), http://www.dis.uniroma1.it/~challenge9/
13. Dietz, P., Sleator, D.: Two algorithms for maintaining order in a list. In: Proc. 19th ACM Symp. on Theory of Computing, pp. 365–372 (1987)
14. Firmani, D., Italiano, G.F., Laura, L., Orlandi, A., Santaroni, F.: Computing strong articulation points and strong bridges in large scale graphs. In: Klasing, R. (ed.) SEA 2012. LNCS, vol. 7276, pp. 195–207. Springer, Heidelberg (2012)
15. Georgiadis, L.: Testing 2-vertex connectivity and computing pairs of vertex-disjoint s-t paths in digraphs. In: Abramsky, S., Gavoille, C., Kirchner, C., Meyer auf der Heide, F., Spirakis, P.G. (eds.) ICALP 2010. LNCS, vol. 6198, pp. 738–749. Springer, Heidelberg (2010)

16. Georgiadis, L.: Approximating the smallest 2-vertex connected spanning subgraph of a directed graph. In: Demetrescu, C., Halldórsson, M.M. (eds.) ESA 2011. LNCS, vol. 6942, pp. 13–24. Springer, Heidelberg (2011)
17. Georgiadis, L., Tarjan, R.E.: Finding dominators revisited. In: Proc. 15th ACM-SIAM Symp. on Discrete Algorithms, pp. 862–871 (2004)
18. Georgiadis, L., Tarjan, R.E.: Dominator tree verification and vertex-disjoint paths. In: Proc. 16th ACM-SIAM Symp. on Discrete Algorithms, pp. 433–442 (2005)
19. Georgiadis, L., Tarjan, R.E.: Dominator tree certification and independent spanning trees. CoRR abs/1210.8303 (2012) (submitted for journal publication)
20. Georgiadis, L., Tarjan, R.E.: Dominators, directed bipolar orders, and independent spanning trees. In: Czumaj, A., Mehlhorn, K., Pitts, A., Wattenhofer, R. (eds.) ICALP 2012, Part I. LNCS, vol. 7391, pp. 375–386. Springer, Heidelberg (2012)
21. Georgiadis, L., Tarjan, R.E., Werneck, R.F.: Finding dominators in practice. Journal of Graph Algorithms and Applications (JGAA) 10(1), 69–94 (2006)
22. Gomez-Rodriguez, M., Schölkopf, B.: Influence maximization in continuous time diffusion networks. In: 29th International Conference on Machine Learning, ICML (2012)
23. Huck, A.: Independent trees in graphs. Graphs and Combinatorics 10, 29–45 (1994)
24. Italiano, G.F., Laura, L., Santaroni, F.: Finding strong bridges and strong articulation points in linear time. Theoretical Computer Science 447, 74–84 (2012)
25. Lengauer, T., Tarjan, R.E.: A fast algorithm for finding dominators in a flowgraph. ACM Transactions on Programming Languages and Systems 1(1), 121–141 (1979)
26. Leskovec, J.: Stanford large network dataset collection (2009), http://snap.stanford.edu
27. Maxwell, E.K., Back, G., Ramakrishnan, N.: Diagnosing memory leaks using graph mining on heap dumps. In: Proceedings of the 16th ACM SIGKDD International Conference on Knowledge Discovery and Data Mining, KDD 2010, pp. 115–124 (2010)
28. McConnell, R.M., Mehlhorn, K., Näher, S., Schweitzer, P.: Certifying algorithms. Computer Science Review 5(2), 119–161 (2011)
29. Mitchell, N.: The runtime structure of object ownership. In: Thomas, D. (ed.) ECOOP 2006. LNCS, vol. 4067, pp. 74–98. Springer, Heidelberg (2006)
30. Plehn, J.: Über die Existenz und das Finden von Subgraphen. Ph.D. thesis, University of Bonn, Germany (May 1991)
31. Purdom Jr., P.W., Moore, E.F.: Algorithm 430: Immediate predominators in a directed graph. Communications of the ACM 15(8), 777–778 (1972)
32. Quesada, L., Van Roy, P., Deville, Y., Collet, R.: Using dominators for solving constrained path problems. In: Van Hentenryck, P. (ed.) PADL 2006. LNCS, vol. 3819, pp. 73–87. Springer, Heidelberg (2006)
33. Tarjan, R.E.: Depth-first search and linear graph algorithms. SIAM Journal on Computing 1(2), 146–159 (1972)
34. Tarjan, R.E.: Finding dominators in directed graphs. SIAM Journal on Computing 3(1), 62–89 (1974)
35. Tarjan, R.E.: Efficiency of a good but not linear set union algorithm. Journal of the ACM 22(2), 215–225 (1975)
36. Whitty, R.W.: Vertex-disjoint paths and edge-disjoint branchings in directed graphs. Journal of Graph Theory 11, 349–358 (1987)
37. Zhao, J., Zdancewic, S.: Mechanized verification of computing dominators for formalizing compilers. In: Hawblitzel, C., Miller, D. (eds.) CPP 2012. LNCS, vol. 7679, pp. 27–42. Springer, Heidelberg (2012)

Novel Techniques for Automorphism Group Computation [*]

José Luis López-Presa[1], Luis Núñez Chiroque[2], and Antonio Fernández Anta[2]

[1] DIATEL-UPM, Madrid, Spain
jllopez@diatel.upm.es
[2] Institute IMDEA Networks, Madrid, Spain
{luisfelipe.nunez,antonio.fernandez}@imdea.org

Abstract. Graph automorphism (GA) is a classical problem, in which the objective is to compute the automorphism group of an input graph. In this work we propose four novel techniques to speed up algorithms that solve the GA problem by exploring a search tree. They increase the performance of the algorithm by allowing to reduce the depth of the search tree, and by effectively pruning it.

We formally prove that a GA algorithm that uses these techniques correctly computes the automorphism group of the input graph. We also describe how the techniques have been incorporated into the GA algorithm conauto, as *conauto-2.03*, with at most an additive polynomial increase in its asymptotic time complexity.

We have experimentally evaluated the impact of each of the above techniques with several graph families. We have observed that each of the techniques by itself significantly reduces the number of processed nodes of the search tree in some subset of graphs, which justifies the use of each of them. Then, when they are applied together, their effect is combined, leading to reductions in the number of processed nodes in most graphs. This is also reflected in a reduction of the running time, which is substantial in some graph families.

1 Introduction

Graph automorphism (GA), graph isomorphism (GI), and finding a canonical labeling (CL) are closely-related classical graph problems that have applications in many fields, ranging from mathematical chemistry [4,20] to computer vision [1]. Their general time-complexity is still an open problem, although there are several cases for which they are known to be solvable in polynomial time. Hence, the construction of tools that are able to solve these problems efficiently for a large variety of problem instances has significant interest. This work focuses on the GA problem, whose objective is to compute the automorphism group of an

[*] Research was supported in part by the Comunidad de Madrid grant S2009TIC-1692, Spanish MINECO/MICINN grant TEC2011-29688-C02-01, Factory Holding Company 25, S.L., grant SOCAM, and National Natural Science Foundation of China grant 61020106002.

V. Bonifaci et al. (Eds.): SEA 2013, LNCS 7933, pp. 296–307, 2013.
© Springer-Verlag Berlin Heidelberg 2013

input graph (e.g., by obtaining a set of generators, the orbits and the size of this group). In this paper, novel techniques to speed up algorithms that solve the GA problem are proposed. Additionally, most of these techniques can be applied to increase the performance of algorithms for solving the other two problems as well.

1.1 Related Work

There are several practical algorithms that solve the GA problem. Most of them can also be used for CL (and consequently, for GI testing). For the last three decades, *nauty* [13,14] has been the most widely used tool to tackle all these problems. Other interesting algorithms that solve GA and CL are *bliss* [6,5], *Traces* [17], and *nishe* [19,18]. Recently, McKay and Piperno have jointly released a new version of both nauty and Traces [15] with significant improvements over their previous versions. Another tool, named *saucy* [3,7,8], which solves GA (but not CL), has the advantage of being the most scalable for many graph families, since it is specially designed to efficiently process big and sparse graphs. Recently, it was shown that the combined use of saucy and bliss improves the running times of bliss for the canonical labeling of graphs from a variety of families [9].

All these tools are based on the same principles, using variants of the Weisfeiler-Lehman individualization-refinement procedure [21]. They explore a search tree, whose nodes are identified by equitable vertex partitions, using a backtracking algorithm to compute the automorphism group of the graph and, optionally, a canonical labeling. The efficiency of an algorithm depends on the speed at which it performs basic operations, like refinement, and, mainly, on the size of the search tree generated (the number of nodes of the search tree which are explored). There are two main ways to reduce the search space: pruning, and choosing a good target cell (and vertex) for individualization.

Miyazaki showed in [16] that it is possible to make nauty choose bad target cells for individualization, so its search space becames exponential in size when computing the automorphism group for a family of colored graphs. This suggests that a rigid criterion cell selector may be easily misled so that many nodes are explored, while choosing the right cells could dramatically reduce the search space. Thus, different colorings of a graph, or just differently labeled instances, may generate radically different search trees. Algorithms for CL use different criteria to choose the target cell for individualization, but these criteria must be isomorphism invariant to ensure that the search tree for isomorphic graphs are isomorphic, what is not necessary for GA. Examples of cell selectors are: the first cell, the maximum nonuniformly joined cell, the cell with more adjacencies to non-singleton cells, etc. A cell selector immune to this dependency on the coloring or the labeling would be desirable.

Pruning the search space may be accomplished using several techniques. Orbit pruning and coset pruning are extensively used by GA and CL algorithms. Perhaps, the most sophisticated pruning based on orbit stabilizer algorithms is that of the latest versions of nauty and Traces [15], that use the random Schreier method. However, when the number of generators grow, the overhead imposed is not negligible. *Conflict propagation* is used by bliss [5] to prune brother nodes when one of them generates a conflict which was not found in the corresponding node of the first path. Conflicts may be detected at the nodes of the search tree, or during the refinement process as done by conauto [12] (for GI) and saucy [8].

Limited early automorphism detection, when a node has exactly the same non-singleton cells (in the same position) as the corresponding (and compatible) node in the first path, is present in all versions of conauto [10]. Recently, this feature has been added to saucy [8] under the name of *matching OPP pruning*. A more ambitious *component detection* was added to bliss [5] for early automorphism detection. However, components are not always easy to discover and keep track of.

1.2   Contributions

In this paper we propose a novel combination of four techniques to speed up GA algorithms, but which can be used in GI and CL algorithms as well. (Such extensions are out of the scope of this work.) These techniques can be used in GA algorithms that follow the individualization-refinement approach. One key concept that we define, and that is used by some of the proposed techniques, is the property of a partition being a *subpartition* of another partition (see the definition in Section 3).

We propose a novel approach to *early automorphism detection* (EAD) without the need of explicitly identifying components, unlike the component recursion of bliss. EAD is based on the concept of subpartition, and its correctness is proved by Theorem 2. This technique is useful, for example, when the graph is built from regularly connected sets of isomorphic components, and components which have automorphisms themselves.

A second technique which, to our knowledge, has never be used in any other GA algorithm is *backjumping* (BJ) in the search tree, under the condition that the partition of the current node is a subpartition of its parent node. In this case, if the current node has been fully explored and no automorphism has been found, instead of backtracking to its parent node, it is possible to backtrack directly to another ancestor. Specifically, to the nearest ancestor of which the current node is not a subpartition. The correctness of BJ is proved by Theorem 3. This technique helps, for example, when there are isomorphic and non-isomorphic components in a graph.

As previously stated, the target cell selector for individualization is key to yield a good search tree. We propose a *dynamic cell selector* (DCS) that tries to generate a tree in which nodes are subpartitions of their parent nodes, so the previous techniques can be applied. If that is not possible, it chooses the vertex to individualize to be the one, among a non isomorphism invariant subset of all the possible candidates, that generates the partition with the largest number of cells. DCS adapts to a large variety of graph families. Since it is not isomorphism invariant, it cannot be applied to CL. However, it can be used for GA, using a different one for CL, once the automorphism group has been computed, in a way similar to the combined use of saucy and bliss for CL proposed in [9].

The last technique proposed is *conflict detection and recording* (CDR), an improvement of the conflict propagation of bliss. Besides recording a hash for each different conflict found exploring branches of the nodes of the first path, the number of times each conflict appeared is counted. Then, if the number of times a certain conflict has been found on a node outside the first path exceeds the number of times it was found in the corresponding node of the first path, then no other branches need to be explored in this node. This technique helps in a large variety of graph families.

We have implemented the four techniques described, and integrated them into our program conauto-2.0,[1] resulting in the new version conauto-2.03. It is worth to mention that all versions of conauto process both directed and undirected graphs (in fact they consider all graphs as directed).

We have performed an analysis of the time complexity of conauto-2.03. It is easy to adapt prior analyses [12] to show that conauto-2.0 has asymptotic time complexity $O(n^3)$ with high probability when processing a random graph $G(n, p)$, for $p \in [\omega(\ln^4 n/n \ln \ln n), 1 - \omega(\ln^4 n/n \ln \ln n)]$ [2]. We then show that, in the worst case, the techniques proposed here increase the asymptotic time complexity of conauto-2.03 by an additive polynomial term with respect to that of conauto-2.0. In particular, DCS can increase the asymptotic time complexity in up to $O(n^5)$, while EAD and BJ in up to $O(n^3)$. Finally, CDR does not increase the asymptotic time complexity. Hence, if conauto-2.0 has polynomial execution time, the execution time of conauto-2.03 does not become superpolynomial. Furthermore, as will be observed experimentally, in some cases the techniques added can drastically reduce the computing time.

We have experimentally evaluated the impact of each of the above techniques for the processing of several graph families, and different graph sizes for each family. To do so, we have compared the number of nodes traversed by conauto-2.0 and the number of nodes traversed when each of the above techniques is applied. Then we have compared the number of nodes traversed, and the running times of conauto-2.0 and conauto-2.03. The improvements are significant as the size of the search tree increases, and the overhead introduced is only noticeable for very small search trees.

1.3   Structure

The next section defines the basic concepts and notation used in the analytical part of the paper. In Section 3 we define the concept of subpartition and state the main theoretical properties, which imply the correctness of EAD and BJ. Then, in Section 4 we describe how these results have been implemented in conauto-2.03 and in Section 5 we evaluate the time complexity of conauto-2.03. Finally, in Section 6 we present the experimental evaluation of conauto-2.03, concluding the paper with Section 7.

---

[1] The original algorithm conauto [12] solves the GI problem but not the GA problem; conauto-2.0 is a modified version that computes automorphism groups and uses limited, though quite effective, coset and orbit pruning.

2 Basic Definitions and Notation

Most of the concepts and notation introduced in this section are of common use. For simplicity of presentation, graphs are considered undirected. However, all the results obtained can be almost directly extended to directed graphs.

2.1 Basic Definitions

A *graph* G is a pair (V, E) where V is a finite set, and E is a binary relation over V. The elements of V are the *vertices* of the graph, and the elements of E are its *edges*. The set of graphs with vertex set V is denoted by $\mathcal{G}(V)$. Let $W \subseteq V$, the subgraph induced by W in G is denoted by G_W. Let $W \subseteq V$ and $v \in V$, we denote by $\delta(G, W, v)$ the number of neighbors of vertex v which belong to W. More formally, $\delta(G, W, v) = |\{(v, w) \in E : w \in W\}|$. If $W = V$, then it denotes the *degree* of the vertex.

Two graphs $G = (V_G, E_G)$ and $H = (V_H, E_H)$ are *isomorphic* if and only if there is a bijection $\gamma : V_G \to V_H$, such that $(v, w) \in E_G \iff (\gamma(v), \gamma(w)) \in E_H$. This bijection γ is an isomorphism of G onto H. An *automorphism* of a graph G is an isomorphism of G onto itself. The *automorphism group* $\mathrm{Aut}(G)$ is the set of all automorphisms of G with respect to the composition operation.

An *ordered partition* (or *partition* for short) of V is a list $\pi = (W_1, ..., W_m)$ of nonempty pairwise disjoint subsets of V whose union is V. The sets W_i are the *cells* of the ordered partition. For each vertex $v \in V$, $\pi(v)$ denotes the index of the cell of π that contains v (i.e., if $v \in W_i$, then $\pi(v) = i$). The number of cells of π is denoted by $|\pi|$. Let $A \subseteq V$, π^A denotes the partition of A obtained by restricting π to A. The set of all partitions of V is denoted by $\Pi(V)$. A partition is *discrete* if all its cells are singletons, and *unit* if it has only one cell. Let $\pi, \rho \in \Pi(V)$, then ρ is *finer* than π, if π can be obtained from ρ by replacing, one or more times, two or more consecutive cells by their union. Let $\pi = (W_1, ..., W_m)$ and $v \in W_i$, the partition obtained by *individualizing* vertex v is $\pi \downarrow v = (W_1, ..., W_{i-1}, \{v\}, W_i \setminus \{v\}, W_{i+1}, ..., W_m)$.

A *colored graph* is an ordered pair $(G, \pi) \in \mathcal{G}(V) \times \Pi(V)$. Partition π assigns color $\pi(v)$ to each vertex $v \in V$. Let $\pi = (W_1, ..., W_m)$, for each vertex $v \in V$, its *color-degree vector* is defined as $d(G, \pi, v) = (\delta(G, W_i, v) : i = 1, ..., m)$. A colored graph (G, π) is *equitable* if for all $v, w \in V$, $\pi(v) = \pi(w)$ implies $d(G, \pi, v) = d(G, \pi, w)$. (I.e., if all vertices of the same color have the same number of adjacent vertices of each color.) The notion of isomorphism and automorphism can be extended to colored graphs as follows. Two colored graphs (G, π) and (H, ρ) are isomorphic if there is an isomorphism γ of G onto H, such that $\gamma(v) = w$ implies $\pi(v) = \rho(w)$.

Two equitable colored graphs $(G, \pi) \in \mathcal{G}(V_G) \times \Pi(V_G)$ and $(H, \rho) \in \mathcal{G}(V_H) \times \Pi(V_H)$ are *compatible* if and only if (1) $|\pi| = |\rho| = m$; (2) let $\pi = (W_1, ..., W_m)$ and $\rho = (W'_1, ..., W'_m)$, then for all $i \in [1, m]$, $|W_i| = |W'_i|$; (3) and for all $v \in V_G$, $w \in V_H$, $\pi(v) = \rho(w)$ implies $d(G, \pi, v) = d(H, \rho, w)$. Note that, if two colored graphs are not compatible, then they can not be isomorphic.

2.2   Individualization-Refinement and Search Trees

Most algorithms for computing GA or CL use variants of the Weisfeiler-Lehman individualization-refinement procedure [21]. This procedure requires two functions: a *cell selector* and a *partition refiner*. A *cell selector* is a function S that, given a colored graph (G, π), returns the index i of a cell $W_i \in \pi$ such that $|W_i| > 1$. A *partition refiner* is an isomorphism-invariant function R that, given a colored graph (G, π), returns (G, π) if it is already equitable. Otherwise, it returns an equitable colored graph (G, ρ) such that ρ is finer than π.

The automorphism group of a graph is usually computed by traversing a search tree in a *depth-first* manner. A *search tree* of a graph $G \in \mathcal{G}(V)$ is a rooted tree $\mathcal{T}(G)$ of colored graphs defined as follows.

1. The root of $\mathcal{T}(G)$ is the colored graph $R(G, (V))^2$.
2. Let (G, π) be a node of $\mathcal{T}(G)$. If π is discrete, it is a leaf node.
3. Otherwise, let $\pi = \{W_1, ..., W_m\}$ and assume $S(G, \pi) = j, j \in [1, m]$, and $W_j = \{v_1, ..., v_k\}$ (recall that $|W_j| > 1$ from the definition of a cell selector). Then, (G, π) has exactly k children, where the ith child is $(G, \pi_i) = R(G, \pi \downarrow v_i)$.

A *path* in $\mathcal{T}(G)$ starts at some internal (non-leaf) node and moves toward a leaf. A path can be denoted as $\pi_0[v_1\rangle\pi_1...[v_k\rangle\pi_k$, indicating that, starting at node (G, π_0) and individualizing vertices $v_1, ..., v_k$, node (G, π_k) is reached. The *depth* (or *level*) of a node in $\mathcal{T}(G)$ is determined by the number of vertices which have been individualized in its path from the root. Thus, if (G, π_0) is the root node, then π_0 is the partition at level 0, and π_k is the partition at level k. The first path traversed in $\mathcal{T}(G)$ is called the *first-path*, and the leaf node of the first-path is called the *first-leaf*.

Theorem 1. *Let $G = (V, E)$ be a graph. Let (G, π) and (G, ρ) be two compatible leaf-nodes in $\mathcal{T}(G)$. Then, mapping $\gamma : V \to V$ such that, for all $v \in V$, $\pi(v) = \rho(\gamma(v))$ is an automorphism of G.*

Proof. Direct from the definition of compatibility among colored graphs, and the fact that, since (G, π) and (G, ρ) are leaf-nodes, all their cells are singleton.

3   Correctness of EAD and BJ

In this section we define specific concepts needed to develop our main results, like the concept of the *kernel* of a partition, and that of a partition being a *subpartition* of another partition. Then, we state theorems that prove the correctness of the EAD and BJ techniques.

We start by defining the *kernel* of a partition, which intuitively is the subset of vertices in non-singleton cells with edges to other vertices in non-singleton cells, but not to all of them. More formally, we can define the kernel as follows.

---

^{2} We write $R(G, (V))$ and $S(G, \pi)$ instead of $R((G, (V)))$ and $S((G, \pi))$ to avoid duplicated parentheses.

Definition 1. *Let $(G, \pi) \in \mathcal{G}(V) \times \Pi(V)$ be an equitable colored graph, $\pi = (W_1, ..., W_m)$ and $W = \bigcup_{i:|W_i|>1} W_i$. Then, the* kernel *of partition π is defined as $\kappa(\pi) = \{v \in W : \delta(G, W \setminus \{v\}, v) \in [1, |W| - 1]\}$. The* kernel complement *of π is defined as $\overline{\kappa}(\pi) = (V \setminus \kappa(\pi))$.*

Now we can define the concept of a subpartition of another partition.

Definition 2. *Let (G, π) and (G, ρ) be two equitable colored graphs such that ρ is finer than π. Then, ρ is a* subpartition *of π if and only if each cell in the kernel of ρ is contained in a different cell of π. (I.e., $\rho^{\kappa(\rho)} = \pi^{\kappa(\rho)}$.)*

The next result allows for *early automorphism detection* (EAD) when, at some node in the search tree, the node's partition is a subpartition of an ancestor's partition. In practice, it limits the maximum depth in the search tree, necessary to determine if a path is automorphic to a previously explored one.

Definition 3. *Let $G \in \mathcal{G}(V)$ and $\mathcal{T}(G)$ its search tree. Let (G, π_k) be a node of $\mathcal{T}(G)$. Let (G, π_l) and (G, ρ_l) be two descendants of (G, π_k) such that (1) they are compatible, and (2) π_l and ρ_l are subpartitions of π_k. Let $\pi_l = (W_1, ..., W_m)$ and $\rho_l = (W'_1, ..., W'_m)$. For all $i \in [1, m]$, let β_i be any bijection from W_i to W'_i. Let us define the function $\alpha : V \to V$ as follows.*
- *For all $v \in \overline{\kappa}(\pi_l)$, $\alpha(v) = \beta_{\pi_l(v)}(v)$.*
- *For all $v \in \kappa(\pi_l)$, $\alpha(v) = f(v)$, where $f(v) = v$ if $v \in \kappa(\rho_l)$, and $f(v) = f(\beta^{-1}(v))$ if $v \in \overline{\kappa}(\rho_l)$.*

Theorem 2. *Let $G \in \mathcal{G}(V)$ and $\mathcal{T}(G)$ its search tree. Let (G, π_k) be a node of $\mathcal{T}(G)$. Let (G, π_l) and (G, ρ_l) be two descendants of (G, π_k) such that (1) they are compatible, and (2) π_l and ρ_l are subpartitions of π_k. Then, (G, π_l) and (G, ρ_l) are isomorphic, and α (as defined in Definition 3) is an automorphism of G.*

Interestingly, some of the properties used for early automorphism detection in other graph automorphism algorithms are special cases of the above theorem. For instance, the early automorphism detection used in saucy-3.0 is limited to the case in which all the non-singleton cells are the same in both partitions. This corresponds to the particular case of Theorem 2 in which $\overline{\kappa}(\pi_l) \cap \kappa(\rho_l) = \emptyset$, and all the cells in $\overline{\kappa}(\pi_l)$ are singleton.

The following theorem shows the correctness of *backjumping* (BJ) when searching for automorphisms. This allows to backtrack various levels in the search tree at once.

Theorem 3. *Let (G, π_k) be a node of $\mathcal{T}(G)$. Let (G, π_l) and (G, ρ_l) be two compatible descendants of (G, π_k). Let (G, π_m) and (G, ρ_m) be two descendants of (G, π_l) and (G, ρ_l) respectively, such that π_m is a subpartition of π_l and ρ_m is a subpartition of ρ_l. If (G, π_m) and (G, ρ_m) are compatible but not isomorphic, then (G, π_l) and (G, ρ_l) are not isomorphic either.*

A direct practical consequence of Theorem 3 is that, when exploring alternative paths at level k, if a level m is reached that satisfies the conditions of the theorem, it is not necessary to explore alternative paths at level l. Instead, it is possible to backjump directly to the closest level $j \in [k, l)$ such that ρ_m is not a subpartition of ρ_j.

4   Implementation of the Techniques in *conauto-2.03*

The starting point is algorithm conauto-2.0, which is the first version of co-nauto that solves GA. It obtains a set of generators, and computes the orbits and the size of the automorphism group using the individualization-refinement approach. Its cell selector chooses a non-singleton cell with the largest number of adjacencies to non-singleton cells, and the one with the smallest size among them. The basic algorithm works in the following way. It starts by generating the first path, recording the positions of the individualized cells at each node of the path, for future use. Then, starting from the leaf parent, it explores each alternative branch. When a leaf node compatible with the leaf of the first path is reached, an automorphism is found and stored. Then, the algorithm moves to the parent node and explores the new branches of its subtree, which will gener-ate paths of length two. This process continues until the root node of the search tree has been explored, using limited coset and orbit pruning.

EAD is implemented as follows. The first path is explored to find, for each non-leaf node (G, π), its nearest successor (G, ρ) which is a subpartition of (G, π). Note that a leaf node is a subpartition of all its ancestors. (G, ρ) is recorded as the search limit for (G, π). Then, when searching for automorphisms from (G, π), if a new node compatible with (G, ρ) is found, an automorphism α is inferred applying Definition 3. This requires a subpartition test which is linear in the number of cells, that will be executed, for each non-leaf node in the first path, at most as many times as the length of the path from that node to the leaf. Every time the search limit is not a leaf, a subtree is pruned.

BJ requires the execution of the subpartition test for the ancestors of each node (G, π) of the first path, until a node of which it is not a partition is found. That will be the backjump point for node (G, π). The point is recorded, and BJ can be subsequently applied with zero overhead.

EAD and BJ can only be applicable if there are nodes in the first path that satisfy the subpartition condition. Without a cell selector that favours subparti-tions, they cannot be expected to be useful in general. Hence, a cell selector like DCS is needed. DCS works in the following way. At node (G, π), it first selects, as candidates, one cell in $\kappa(\pi)$ of each size and number of adjacencies to its kernel. From each such cell, it takes the first vertex v, and computes the corresponding refinement $R(G, \pi{\downarrow}v)$. If it gets a partition which is a subpartition of π, it selects that cell (and vertex) for individualization. If no such cell is found, it selects the cell (and vertex) which produces the partition with the largest number of cells. Observe that this function is not isomorphism-invariant (not all the vertices of a cell will always produce compatible colored graphs), and it has a significant cost in both time and number of additional nodes explored. However, it pays off because the final search tree is drastically reduced for a great variety of graphs, and other techniques compensate the overhead introduced.

Conflict detection and recording (CDR) requires a function to compute the hash of each conflict found, and storing a couple of integers for the hash and the counters. The cost incurred is very limited and there is a large variety of graphs that benefit from this technique.

5 Complexity Analysis

It was shown in [12] that conauto-1.0 is able to solve the GI problem in polynomial time with high probability if at least one of the two input graphs is a
random graph $G(n, p)$ for $p \in [\omega(\ln^4 n/n \ln \ln n), 1 - \omega(\ln^4 n/n \ln \ln n)]$. Using
a similar analysis, it is not hard to show a similar result for the complexity of
conauto-2.0 solving the GA problem. I.e., conauto-2.0 solves the GA problem
in polynomial time with high probability if the input graph is a random graph
$G(n, p)$ for $p \in [\omega(\ln^4 n/n \ln \ln n), 1 - \omega(\ln^4 n/n \ln \ln n)]$.

We argue now that the techniques proposed in this work only increase the
asymptotic time complexity of conauto-2.0 by a polynomial additive term. This
implies that there is no risk that, if a graph is processed in polynomial time by
conauto-2.0, by using these techniques it will require superpolynomial time with
conauto-2.03. Let us consider each of the techniques proposed independently.

DCS only increases the execution time during the computation of the first-
path. This follows since it is only used by the cell selector to choose a cell, and
the cell selector is only used to choose the first-path. (Every time the cell selector
returns a cell index, this index is recorded to be used in the rest of the search
tree exploration.) The cell selector is called at most a linear number of times
in n, where n is the number of vertices of the graph. Then, DCS is applied
a linear number of times. Each time it is applied it may require to explore a
linear number of branches. Each branch is explored with a call to the partition
refiner function, whose time complexity if $O(n^3)$. Therefore, DCS increases the
asymptotic time complexity of the execution by an additive term of $O(n^5)$.

Regarding EAD, like DCS, it requires additional processing while the first-
path is created. In particular, for each partition π in the first-path, the closest
partition down the path which is a subpartition of π is determined. This process
always finishes, since the leaf of the first-path is a trivial subpartition of all the
other partitions in the first-path. There is at most a linear number of partitions
π and, hence, at most a linear number of candidate subpartitions. Moreover,
checking if a partition is a subpartition of another takes at most linear time.
Hence, EAD adds a term $O(n^3)$ to the time complexity of processing the first-
path. On the other hand, when the rest of the search tree is explored, checking the
condition to apply EAD has constant time complexity. If EAD can be applied,
an automorphism is generated in linear time. Observe that if EAD were not
used, then an equivalent automorphism would have been found, but at the cost
of exploring a larger portion of the search tree (which takes at least linear time
and may have up to exponential time complexity). Hence the application of EAD
does not increase the asymptotic time complexity of exploring the rest of the
search tree, and may in fact significantly reduce it.

The time complexity added by BJ to the processing of the first-path is similar
to that of EAD, i.e., $O(n^3)$, since for each partition in π the task is to find
the closest partition up the first-path which is not a subpartition of π (if such
a partition exists). The application of BJ in the exploration of the rest of the
search tree takes constant time to check and to apply, while the time complexity
reduction can be exponential.

CDR on its hand involves no processing during the generation of the first-path. Then, during the exploration of the rest of the search tree, every time a conflict is detected, the hash of that conflict is computed and the corresponding counter has to be updated (see Section 4). This takes in total at most linear time. Observe that conflict detection, which takes at least linear time, has to be done in any case. Hence, CDR does not increase the asymptotic time complexity of the algorithm.

6   Evaluation of the Techniques in *conauto-2.03*

In this section, we evaluate the improvement in performance of conauto-2.0 by adding the proposed techniques. The experiments have been carried out in an Intel(R) Core(TM) i5 750 @2.67GHz, with 16GiB of RAM under Ubuntu Server 9.10. All the programs have been compiled with gcc 4.4.1 and optimization flag '-O2', and all the results have been verified to be correct. First, we evaluate the impact of each of the techniques proposed separately on the number of nodes that are explored during the search. Then, we evaluate the impact of their joint use in conauto-2.03 with respect to conauto-2.0. Finally, we compare the running times of conauto-2.03 vs. conauto-2.0. For the experiments, we have used all the graphs in our benchmark [11], which include a variety of graph families with different characteristics. It includes strongly regular graphs, random graphs, projective planes, Hadamard matrices, multiple variations of Miyazaki's

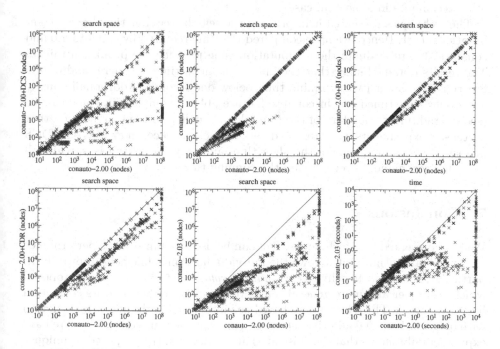

Fig. 1. Performance evaluation for the different techniques in conauto-2.0

construction, different kinds of union graphs, etc. When counting the number of nodes of the search tree explored, each execution was stoped when the count reached 10^8. For the time comparison, a timeout of 5,000 seconds was established. When an execution reached the limit, its corresponding point is placed on boundary of the plotting area. The plots are shown in Figure 1.

As can be observed in the plots, EAD, BJ, and CDR never increase the number of nodes explored. This number slightly increases with DCS in some graphs, but only in a few executions with small search tress, and the benefit attained for most graphs is very noticeable. In fact, many executions that reached the count limit without DCS, lay within the limit when DCS is used (see the rightmost boundary of the plot). In the case of component-based graphs with subsets of isomorphic components, EAD is able to prune many branches, but with other graph families it has no visible effect. That is why the diagonal of the plot is crowded. BJ has a similar effect, but for different classes of graphs. It is mostly useful for component-based graphs which have few automorphisms, so they are complementary. EAD exploits the existence of automorphisms, and BJ exploits the inexistence of automorphisms. CDR is useful with a variety of graphs. It is mostly useful when the target cells used for individualization are big and there are few automorphisms. When DCS and/or BJ are combined with DCS, their effect increases, since DCS favours the subpartition condition, generating more nodes at which EAD and BJ are applicable. When all the techniques proposed are used (in conauto-2.03), the gain is general (big search trees have disapeared from the diagonal), and the overhead generated by DCS is compensated by the other techniques in almost all cases.

The techniques presented help pruning the search tree, but they have a computational cost. Hence, we have compared the time required by conauto-2.0 and conauto-2.03, to evaluate the computation time paid for the pruning attained. The results obtained show that the improvement in time is general and only a few runs are slower (with running time below one second). Additionally, many executions that timed out in conauto-2.0 are able to complete in conauto-2.03 (see the rightmost boundary of the *time* plot). Finally, we want to mention that exetensive experiments, not presented here for lack of space, show that only DCS increases the running time of the algorithm, and only for a few cases, while all the other techniques never increase the running times.

7 Conclusions

We have presented four techniques than can be used to improve the performance of any GA algorithm that follows the individualization-refinement approach. In particular, a new way to achieve *early automorphism detection* has been proposed which is simpler and more general than previous approaches, and its correction has been proved. These techniques have been integrated in the algorithm conauto with only a polynomial additive increase in asymptotic time complexity. We have experimentally shown that, both isolated and combined, the proposed techniques drastically prune the search tree for a large collection of graph instances.

References

1. Conte, D., Foggia, P., Sansone, C., Vento, M.: Graph matching applications in pattern recognition and image processing. In: ICIP, Barcelona, Spain, vol. 2, pp. 21–24 (September 2003)
2. Czajka, T., Pandurangan, G.: Improved random graph isomorphism. Journal of Discrete Algorithms 6(1), 85–92 (2008)
3. Darga, P.T., Liffiton, M.H., Sakallah, K.A., Markov, I.L.: Exploiting structure in symmetry detection for cnf. In: DAC, pp. 530–534 (2004)
4. Faulon, J.-L.: Isomorphism, automorphism partitioning, and canonical labeling can be solved in polynomial–time for molecular graphs. Journal of Chemical Information and Computer Science 38, 432–444 (1998)
5. Junttila, T., Kaski, P.: Conflict propagation and component recursion for canonical labeling. In: Marchetti-Spaccamela, A., Segal, M. (eds.) TAPAS 2011. LNCS, vol. 6595, pp. 151–162. Springer, Heidelberg (2011)
6. Junttila, T.A., Kaski, P.: Engineering an efficient canonical labeling tool for large and sparse graphs. In: ALENEX (2007)
7. Katebi, H., Sakallah, K.A., Markov, I.L.: Symmetry and satisfiability: An update. In: Strichman, O., Szeider, S. (eds.) SAT 2010. LNCS, vol. 6175, pp. 113–127. Springer, Heidelberg (2010)
8. Katebi, H., Sakallah, K.A., Markov, I.L.: Conflict anticipation in the search for graph automorphisms. In: Bjørner, N., Voronkov, A. (eds.) LPAR-18. LNCS, vol. 7180, pp. 243–257. Springer, Heidelberg (2012)
9. Katebi, H., Sakallah, K.A., Markov, I.L.: Graph symmetry detection and canonical labeling: Differences and synergies. In: Turing-100. EPiC Series, vol. 10, pp. 181–195 (2012)
10. Presa, J.L.L.: Efficient Algorithms for Graph Isomorphism Testing. PhD thesis, ETSIT, Universidad Rey Juan Carlos, Madrid, Spain (March 2009)
11. López-Presa, J.L., Chiroque, L.N., Anta, A.F.: Benchmark graphs for evaluating graph isomorphism algorithms. Conauto website by Google sites (2011), http://sites.google.com/site/giconauto/home/benchmarks
12. López-Presa, J.L., Fernández Anta, A.: Fast algorithm for graph isomorphism testing. In: Vahrenhold, J. (ed.) SEA 2009. LNCS, vol. 5526, pp. 221–232. Springer, Heidelberg (2009)
13. McKay, B.D.: Practical graph isomorphism. Congressus Numerantium 30, 45–87 (1981)
14. McKay, B.D.: The nauty page. Computer Science Department, Australian National University (2010), http://cs.anu.edu.au/~bdm/nauty/
15. McKay, B.D., Piperno, A.: Practical graph isomorphism, ii (2013)
16. Miyazaki, T.: The complexity of McKay's canonical labeling algorithm. In: Groups and Computation II, pp. 239–256. American Mathematical Society (1997)
17. Piperno, A.: Search space contraction in canonical labeling of graphs (preliminary version). CoRR, abs/0804.4881 (2008)
18. Tener, G.: Attacks on difficult instances of graph isomorphism: sequential and parallel algorithms. Phd thesis, University of Central Florida (2009)
19. Tener, G., Deo, N.: Attacks on hard instances of graph isomorphism. Journal of Combinatorial Mathematics and Combinatorial Computing 64, 203–226 (2008)
20. Tinhofer, G., Klin, M.: Algebraic combinatorics in mathematical chemistry. Methods and algorithms III. Graph invariants and stabilization methods. Technical Report TUM-M9902, Technische Universität München (March 1999)
21. Weisfeiler, B. (ed.): On construction and identification of graphs. Lecture Notes in Mathematics, vol. 558 (1976)

Blinking Molecule Tracking

Andreas Karrenbauer[1],[*] and Dominik Wöll[2],[**]

[1] Max-Planck-Institute for Informatics, Saarbrücken, Germany
andreas.karrenbauer@mpi-inf.mpg.de
[2] Department of Chemistry, University of Konstanz, Germany
dominik.woell@uni-konstanz.de

Abstract. We discuss a method for tracking individual molecules which globally optimizes the likelihood of the connections between molecule positions fast
and with high reliability even for high spot densities and blinking molecules. Our
method works with cost functions which can be freely chosen to combine costs
for distances between spots in space and time and which can account for the
reliability of positioning a molecule. To this end, we describe a top-down polyhedral approach to the problem of tracking many individual molecules. This immediately yields an effective implementation using standard linear programming
solvers. Our method can be applied to 2D and 3D tracking.

1 Introduction

The possibility to observe single fluorescent molecules in real-time has opened up a
lot of new insights into the dynamics of systems in biology and material sciences. Single molecule microscopy (SMM) allows for the parallel observation of translational
and rotational motion of many single fluorescent molecules beyond the diffraction limit
provided that their concentration is reasonably low. However, tracking of single fluorescent molecules bears the challenge that the fluorescent spots are rather weak with
a low signal/noise-ratio and show significant changes in signal intensity [1,2,3]. In the
extreme case, a fluorescent molecule is dark for several recorded frames, a phenomenon
which is termed blinking [4,5].

The fact that the fluorescence signals of single molecules cannot be classified according to their intensity or shape in different frames and even disappear in some frames
causes severe problems for single molecule tracking. Thus, many tracking algorithms
which have been developed for single particle tracking (e.g. for cells) in video microscopy fail for tracking single molecules.

The path from the recorded single molecule microscopy movies to the results about
their motion includes typically the following steps:

 (i) determination of the positions of each fluorescent molecule,
 (ii) connecting the positions to single molecule tracks, and
(iii) statistical analysis of these tracks.

* Supported by the Max-Planck-Center for Visual Computing and Communication.
** Supported by the Zukunftskolleg, University of Konstanz, Germany.

1.1 Previous Work

Nowadays, single molecule positions are most often determined using center of mass or Gaussian fits, with the latter being the best option for low signal-to-noise ratios of around four [6]. Usually the images are preprocessed by various filters, e.g. *Mexican hat* [7], before the actual localization.

After localization, the positions of subsequent frames have to be connected to tracks [8]. Different approaches have been developed for this purpose, but up to now it remains challenging to improve and develop algorithms not only for special tasks but for a universal set of problems [9].

Single Particle Tracking procedures started with the connection of one point with its closest neighbor in consecutive frames [10]. In 1999, Chetverikov et al. published a new algorithm called IPAN Tracker [11]. Using a competitive linking process that develops as the trajectories grow, this algorithm deals better with incomplete trajectories, high spot densities, faster moving particles and appearing and disappearing spots. Sbalzarini et al. used the same approach, but did not make any assumptions about the smoothness of trajectories [12]. Their algorithm was implemented as ParticleTracker in ImageJ.

The SpotTracker [7] is a very powerful tool to follow single spots throughout one movie, but it can only proceed spot by spot. The algorithm proposed by Bonneau et al. [13] falls in the same category of greedy algorithms that iteratively compute shortest paths in space-time, which are not revised subsequently.

One of the most accurate solutions to single particle tracking is provided by multiple-hypothesis tracking (MHT). This method chooses the largest non-conflicting ensemble of single particle paths simultaneously accounting for all position in each frame. Jaqaman et al. used such an approach where they first linked positions in consecutive frames by solving bipartite matching problems and combined these links into entire trajectories [14] with a post-processing step to account for missing points in a frame. Both steps were optimized independently yielding a very likely solution of the tracking problem. Dynamic multiple-target tracing was used by Sergé et al. to generate dynamic maps of tracked molecules at high density [15]. Subtracting detected peaks from the images allows them for a detection of low intensity peaks which would be otherwise hidden in movies of high particle density. Peak positions were connected using statistical information from past trajectories.

Moreover, manual or semi-automated approaches, which only perform unambiguous choices automatically, are still used though there are cumbersome due to many user interaction at high particle densities.

For the analysis of the tracks, different approaches have been developed [16,9]. The most common approach is the analysis of the mean squared displacement for different time intervals [17,18] which can readily distinguish between different modes of motion such as normal diffusion, anomalous diffusion, confined diffusion, drift and active transport [19]. Alternatively, the empirical distribution of squared displacements [20] and radii of gyration [21,22,16] can be used to analyze single molecule tracks.

In a compagnion paper [23], we report on the implications of our work from a chemical point of view. Whereas in this paper, we highlight the algorithmic aspects of our approach.

1.2   Our Contribution

We present a method for single molecule tracking which globally optimizes the likelihood of the connections between molecule positions fast and with high reliability even for high spot densities. Our method uses cost functions which can be freely chosen to combine costs for distances between spots in space and time and which can account for the reliability of positioning a molecule. Using a suitable positioning procedure, reliable tracking can be performed even for highly mobile, frequently blinking and low intensity fluorescent molecules, cases for which most other tracking algorithms fail.

In the following, we present a top-down approach for modeling molecule tracking. We thereby unify the previous approaches in one framework. By developing a suitable polyhedral model in Sec. 2, we show theoretically that it remains computationally tractable. A major advantage of our method is that we immediately obtain an effective software solution using standard linear programming software. Moreover, we experimentally evaluate our implementation in Sec. 3. To this end, we use real-world data and realistic data, i.e. randomly generated according to a physical model. We qualitatively compare our tracking on the real-world data to tracks obtained by a human expert, whereas we exploit the knowledge about the ground-truth in the realistic data to quantitatively measure the impact of noise and the validity of the parameters that we have chosen. We evaluated our approach for two-dimensional tracking, but the extension to 3D is straight-forward. We provide our software as open source code[1] for MATLAB using CPLEX as LP-solver.

2   A Polyhedral Model for Molecule Tracking

It is easy to see that the number of possible trajectories grows exponentially with the number of points. To tackle this *combinatorial explosion* [13], we consider the a top-down polyhedral approach for a concise representation in this paper. Suppose we are given a set of points $V = \{v_1, \ldots, v_n\}$. Each point has one temporal and d spatial coordinates, say (t_i, x_i, y_i), $i \in V$, with $d = 2$ as used in the following for the sake of presentation. We postulate the following conditions for a track:

- Each point has at most one predecessor.
- Each point has at most one successor.

We model the predecessor/successor relation of two points by ordered pairs. To this end, let $V_{t<t_j} := \{v_i \in V : t_i < t_j\}$ and $A := \{(v_i, v_j) : v_i \in V_{t<t_j}, v_j \in V\}$. We denote the predecessor/successor relation by $f : A \to \{0, 1\}$. Moreover, let $g, h : V \to \{0, 1\}$ denote missing predecessors and successors, respectively. That is, $g(v) = 1$ iff $v \in V$ does not have a predecessor, and $h(v) = 1$ iff it does not have a successor. Let $\delta^{in}(v), \delta^{out}(v) \subseteq A$ denote the sets of possible predecessor/successor relations for point v.

Definition 1. *A **track partition** or **tracking** of V is a collection of disjoint tracks covering V, i.e. each point appears in exactly one track, which might consist of a single point.*

---

[1] http://arxiv.org/src/1212.5877v2/anc/tracking.m

The characteristic vector (f, g, h) of a track partition is a $\{0, 1\}$-vector in which the first $|A|$ entries corresponding to f denote the predecessors/successor relation, the following $|V|$ entries corresponding to g determine the starting points of the tracks, and the last $|V|$ entries corresponding to h define the endpoints of the tracks.

Theorem 1. *The **tracking polytope**, i.e. the convex hull of all track partitions, is given by*

$$P := \{(f, g, h) \in \mathbb{R}_{\geq 0}^{|A|+2|V|} : \forall v \in V : g(v) + f(\delta^{in}(v)) = 1, h(v) + f(\delta^{out}(v)) = 1\}.$$

Proof. It is easy to see that the characteristic vector of a tracking is contained in P. Moreover, each $\{0, 1\}$-vector in P corresponds to a tracking. Hence, it remains to show that these are the only vertices of P. To this end, we prove that the constraint matrix M that defines P in the form $P = \{x : Mx = 1, x \geq 0\}$ is totally unimodular. First, we observe that $M = (M'I)$ where M' corresponds to the f-variables. Hence, it suffices to show total unimodularity for M'. To this end, we consider an auxiliary graph $G' = (V_1 \cup V_2, E')$ at which each of $V_{1,2}$ contains a copy of each point in V and E' is the set of edges that mimics the set A on $V_1 \times V_2$. Note that by this definition G' is bipartite. Moreover, its adjacency matrix is given by M'. Hence, M' and M are totally unimodular.

2.1 Optimization

Based on the compact representation of all possible tracks as described before, we now consider the problem of selecting an appropriate tracking out of all these possibilities. To this end, we leverage the fundamental paradigm of normal diffusion: Tracks are Markov chains, i.e. the transition probability from one state to another does only depend on the current state and not on the history that led to it. Thus, all transitions are independent random events.

Suppose that we are given probabilities $p_1 : A \to [0, 1]$ for the transitions and $p_{2,3} : V \to [0, 1]$ denoting the probability that a point is the beginning or the end of a track, respectively. We wish to find a tracking with maximum likelihood, i.e. a tracking that maximizes the joint probability of the independent random events

$$\mathcal{L}(f, g, h) = \prod_{\substack{a \in A \\ f(a)=1}} p_1(a) \prod_{\substack{v \in V \\ g(v)=1}} p_2(v) \prod_{\substack{v \in V \\ h(v)=1}} p_3(v)$$

or, equivalently,

$$\log \mathcal{L}(f, g, h) = \sum_{\substack{a \in A \\ f(a)=1}} \log p_1(a) + \sum_{\substack{v \in V \\ g(v)=1}} \log p_2(v) + \sum_{\substack{v \in V \\ h(v)=1}} \log p_3(v).$$

Hence, by substituting $c_i = -\log p_i \geq 0$, finding the most likely tracking amounts to solve the linear programming problem

$$\min \left\{ \sum_{a \in A} c_1(a) f(a) + \sum_{v \in V} c_2(v) g(v) + \sum_{v \in V} c_3(v) h(v) : (f, g, h) \in P \right\}, \quad (1)$$

where we exploit the consequence of Thm. 1 that the minimum is attained by a $\{0, 1\}$-solution for (f, g, h). Put differently, it is not necessary to enforce an integer solution by Integer Linear Programming, which is NP-hard in general, but it is sufficient to solve the LP-relaxation (1), which can be done in polynomial time using the ellipsoid method [24] or interior point methods [25,26]. Note that the formulation (1) is general enough to capture arbitrary separable likelihood functions $\mathcal{L}(f, g, h)$.

Lemma 1. *For all optimum solutions* $(f, g, h) \in P$ *and* $a = (v, w) \in A$, *we have*

$$f(a) = 1 \quad \Rightarrow \quad c_1(a) \leq c_2(v) + c_3(w).$$

Proof. By contradiction: We would obtain a better feasible solution by setting $f(a) = 0$ and $g(v) = h(w) = 1$.

Although it is possible to consider different probabilities for appearing and vanishing particles, we choose a constant one, i.e. let $c_2(v) = c_3(v) = C$ for all $v \in V$. Hence, $c(a) \leq 2 \cdot C$ for all $a \in A$ which appear in any optimum solution. This inspires the definition of a **tracking radius** R such that we will only consider predecessors and successors within that range. This dramatically limits the size of an instance and enables us to use space partition techniques to efficiently construct the tracking LP. Put differently, the restriction to a certain tracking radius for efficiency reason is justified by Lem. 1. In the experiments section, we will discuss suitable choices for R. Similarly, it makes sense to limit the number of frames that a molecule might be invisible.

2.2   Dealing with Noise

Since the points are usually detected from noisy images, there are false positive and false negatives. That is, a spot $v \in V$ is a false positive, if it does not correspond to any track. A false negative is a point that does not have a correspondent in V.

Moreover, there might be $v, w \in V$ that correspond to the same point. To deal with these duplicates, we introduce so-called *joins* into A. That is, we allow that two points from the same frame appear in one track. We thereby maintain the integrality of our polyhedron. However, we treat these joins differently w.r.t. the objective function to reflect the special situation. We propose to set the cost of such a link to the ordinary cost of that connection plus the mean of the penalties for not having a successor and a predecessor, respectively. Taking the penalty into account is necessary to avoid 2-cycles.

Since we can only deal with points that are present in V, we shall avoid false negatives in the point detection. However, a low false negative rate often leads to a high false positive rate. Therefore, we utilize the possibility to consider a quality measure of each detected point. That is, we reduce the cost of not tracking a point according to its quality. This can be modeled easily by multiplying $c_{2,3}(v)$ by some quality factor $q(v)$, e.g. proportional to the strength of the signal of this spot.

3   Experiments

So far, we described a generic approach for tracking blinking molecules. In this section, we propose our choice for the cost-function, i.e.

$$c(v_i, v_j) = (x_i - x_j)^2 + (y_i - y_j)^2 + (t_i - t_j)^2,$$

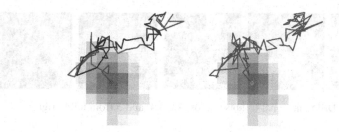

Fig. 1. Tracking with $\Delta x^2 + \Delta y^2$ (left) and $\Delta x^2 + \Delta y^2 + \Delta t^2$ (right). Two interleaved tracks are produced for one molecule on the left, whereas there is exactly one on the right. The background images are inverted for better readability.

which is validated experimentally. The rationale for using this function is based on the following observation: if time is not penalized, then track fragmentation becomes more likely as shown in Fig. 1.

Thus, we introduced the superlinear term Δt^2 such that two time steps of length 1 are cheaper than one time step of length 2. Hence, the spatial distance is mainly responsible for comparing positions within the same frame. We consider closer destinations to be more likely. Therefore, we do not use the time in this part of the objective.

We evaluated our approach w.r.t. efficiency and accuracy (the latter is only discussed briefly in this paper to the extent that is relevant for algorithmic conclusions and a more detailed analysis, in particular w.r.t. chemistry, is presented in [23]). We first consider a controlled testing environment based on the normal diffusion model mentioned above. We thereby obtain realistic randomly generated instances. Though the experiments with the simulated realistic data has the advantage that we know the ground truth and thus we can quantify the deviation of the computed results in certain situations, we shall also validate our approach on real-world instances. To this end, we compare the diffusion coefficients obtained manually by a human expert with our automated approach in the final subsection of this paper.

3.1 Realistic Data

Synthetic trajectories were generated to test our tracking approach. Random walk simulations with 500 steps were performed using four different diffusion coefficients (10^{-15}, 10^{-14}, 10^{-13}, and 10^{-12} m^{2}s^{-1}), five different signal-to-noise ratios (1, 2, 3, 4, and 5, see Fig. 2) and five different particle densities (100, 200, 300, 400 and 500 particles per frame). These values can be found in many tracking challenges of practical importance, but were also chosen to determine the sensitivity of our algorithms.

The initial x- and y-position of each spot was chosen uniformly at random in the range [0;500]. The single molecule trajectories were used to construct movies with 500 frames. Two dimensional Gaussian functions were constructed with their center located at the positions obtained from the random walk simulations and their widths being diffraction limited (ca. 300 nm). We did not explicitly vary the intensities of the spots as it might appear in practice because testing the localization routine is not our primary

Fig. 2. Different signal-to-noise ratios (1,2,3,4, and 5 from left to right)

focus in this paper.[2] However, we implicitly simulated variations in the intensities due to noise and blinking. The time between consecutive frames was chosen as 0.1 s and the resolution as 100 nm per pixel. Gaussian white noise was added to the movies corresponding to the signal-to-noise ratio defined as with the amplitude of the center of the 2D Gaussian and the standard deviation of the Gaussian white noise.

Evaluation of Running Times. The running times were measured on a Dell Precision T7500 with an Intel Xenon CPU X5570 at 2.93 GHz and 24 GB RAM. The memory usage never exceeded 2 GB. In the following, we will discuss the scaling behavior of the running time. That is, the CPU time used for constructing the constraints using a space partition[3] and for solving the LP with CPLEX's barrier interior point method, which turned out to perform best to solve such problems from scratch. In Fig. 3, we present the scaling behavior in dependence on the particle density.

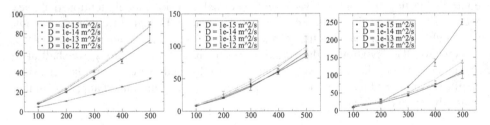

Fig. 3. The running times in seconds depending on the number of particles per frame. The plots correspond to tracking radii 5,10,15 pixels from left to right. The error bars indicate the 95% confidence intervals. Quadratic polynomials are fitted to the data.

The naive hypothesis for the running time coming from the theoretical bound of Ye [26] for log-barrier interior point methods, i.e. cubic in the number of variables, can be clearly dismissed. This is not surprising since the special sparse structure of our constraint matrix is unlikely to serve as a worst-case example. Instead, we found experimental support for the hypothesis that the running times scale quadratically w.r.t. the number of particles per frame. The rationale for this hypothesis is that the degree of the

---

[2] Note that our approach is modular such that any localization procedure may be used.

[3] We use a regular grid with spacing equal to the tracking radius.

node of a point in a frame is proportional to the number of particles per frame. Thus, the number of arcs is quadratic in the number of particles per frame. Hence, the construction of the LP model takes quadratic time, which is the dominating part for low densities and slow molecules. However, with high densities and fast molecules, we see a turn-over to the LP-solver: the faster the molecules, the fewer arcs (because the shorter is the time a molecule stays within a fixed circle), but on the other hand the more connected the graph becomes because each molecule *sees* more other molecules. We believe that the increasing correlation is responsible for the slow-down of the LP-solver and the higher running times for $D = 10^{-12}$ m^{2}s^{-1}. This is supported by the observation that for small tracking radii the computations for fast molecules finish earliest while for the large tracking radius it takes more time than the others (see Fig. 3). Nevertheless, we stress that these running times are negligible w.r.t. the time necessary for preparing and executing such an experiment in reality. Thus, concerning computational resources, our approach is well suited for being applied in the lab.

Determination of Tracking Accuracy for Simulated Data. Tracking procedures have to meet several conditions to be suitable. Apart from practical aspects such as tracking speed and memory consumption, the number of false positives is the key factor which has to be minimized in order to obtain reliable results. Similarly to the definition of false positive and false negative w.r.t. particle locations, false positives in this context are connections between positions in different frames which have been set even though the positions do not belong to the same molecule/particle. They can result in severe errors in single molecule tracking and cause wrong interpretations of collected data. Thus, the number of false positives should be kept as low as possible. False negatives are connections between positions in different frames which are not recognized by the tracking algorithm.

We counted the number of false positives and false negatives for movies of different diffusion coefficients and signal-to-noise-ratios by comparing ground truth and analyzed connections between points. The fraction of false positive connections decreases from 13% to 3% as the S/N-ratio increases from 1 to 5. False negatives particularly occur with fast moving molecules if the tracking radius is not chosen carefully. The reason is that the probability of finding the destination within a radius of R is

$$P(R,t) = \int_0^R \frac{r}{2Dt} \exp\left(-\frac{r^2}{4Dt}\right) dr = 1 - \exp\left(-\frac{R^2}{4Dt}\right).$$

Thus, picking the tracking radius too low yields biased false negatives and hence an underestimation of the diffusion coefficients. However, the tempting choice of an excessive tracking radius does not only require much more computational resources, but may also lead to an overestimation of the diffusion coefficients if there occur leaps in the tracks due to false positives (in particular with high particle densities).

We propose to choose the tracking radius such that a displacement is smaller with a probability of about 99%. Since those leaps are easily determined in a post-processing steps, a repetition of the tracking with different radii in a feedback loop is feasible. In particular, allowing or disallowing connections between spots can be done efficiently with linear programming since the LP remains primal feasible or dual feasible, respectively. Thus, in the former case, we shall use the primal simplex method for reoptimization and the dual simplex method in the latter case.

Positioning and tracking are the key steps in the determination of diffusion coefficients. To distinguish the errors appearing in these steps, we analyzed four different cases shown in Fig. 4.

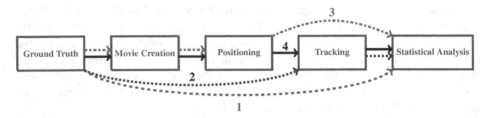

Fig. 4. Steps from the simulated ground truth data to the analysis of the trajectories. To analyze the accuracy of each single step different paths were considered (labeled 1 to 4).

In case (1), the distribution of diffusion coefficient was directly calculated from the ground truth tracks. Though all tracks were created with respect to fixed diffusion coefficients, the calculation yields peaked distributions around the true values because of the finite number of sample points in each track.

In a second set of analysis (case 2) the same ground truth positions were tracked using our polyhedral model solved with CPLEX. Good results were obtained except for fast molecules and high densities. That is, the probability of foreign spots moving into the tracking range of another molecule is too high, and thus the tracking algorithm in general returns diffusion coefficients lower than the real value.

The third set of analysis, case 3, allows for an investigation of the influence of positioning inaccuracies on the distribution of diffusion coefficients. Movies were constructed from the ground truth positions with different S/N-ratios. Our positioning algorithm was applied to these movies and, where possible, the positions matched to the positions of the ground truth tracks. With the determined positions of each track, a diffusion coefficient was obtained. The distribution of diffusion coefficients of these tracks resembles the distributions of the ground truth tracks with the exception of low diffusion coefficients with low S/N-ratios where the poor localization accuracy results in a seemingly higher diffusion coefficient than simulated.

Case 4 describes the procedure which is applied for real movies to determine single molecule diffusion coefficients. Spots in movies are positioned, the positions tracked and a diffusion coefficient calculated from these tracks. For high S/N-ratios, the obtained distributions are similar to the simulated distributions. Two trends can be observed in particular at low S/N- ratios:

(i) molecules with very low diffusion coefficients tend to be analyzed as being faster as they were simulated, and
(ii) analysis of the motion of very fast molecules in average results in a lower diffusion coefficient as the ground truth data.

The former observation can be explained by the poorer localization accuracy at low S/N-ratios. This inaccuracy resembles diffusion and thus a low diffusion coefficient will be assigned even to immobile molecules. The localization accuracy determines the lowest diffusion coefficient which can be determined by the corresponding experimental settings.

Tracking of Blinking Data. In the previous subsection, points were missing due to false negatives in the localization at low S/N-ratios. However, even with perfect localization missing points may occur naturally in real-world experiments because the fluorescence intensity of single molecules is typically not constant, but shows blinking behavior due to photochemical or photophysical quenching processes [1]. The lengths of on- and off-times typically show a power law distribution [1]. In order to simulate blinking behavior, we generated on- and off-times for our simulated tracks using the following procedure. At the beginning a molecule was u.a.r. set as on or off. The number of frames remaining in this state was determined randomly from a probability distribution with $P(t) \sim (t/\tau)^{\alpha}$. We chose realistic values for τ and α, i.e. $\tau = 1s$ and $\alpha = -2$, respectively.

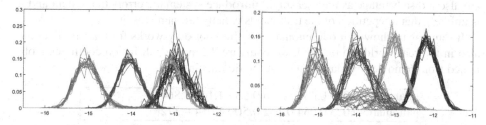

Fig. 5. Distributions of diffusion coefficients obtained after tracking of blinking data for a maximum step length of 5 (left) and 15 (right). The scale for the x-axis is $\log_{10}(D \cdot m^{-2}s)$.

The analyzed distributions of diffusion coefficients for tracking of blinking ground truth data are shown in Fig. 5. For a tracking radius of 5 pixels, the distributions of diffusion coefficients resemble the ground truth data except for fast molecules where a tracking radius of 5 pixels is not sufficient as discussed before. A tracking radius of 15 pixels yields good results for fast diffusing molecules, but for slower ones, the distributions have a long tail or even a second band at higher D values (see Fig. 5 (right)). The deviation of distribution from the ground truth distribution is caused by tracks which include at least one large jump from one molecule to another one which results in a significant increase of the track radius. However, these situations can be recognized easily or even automatically by an outlier detection algorithm. As said before, we propose to integrate such a post-processing in a feedback loop to deal with the such situations especially when heterogeneous ensembles with slow and fast molecules are observed.

3.2 Real-World Data

The real-world data were obtained from single molecule fluorescence widefield experiments during the bulk radical polymerization of styrene to polystyrene. The motion of single perylene diimide fluorophores was observed at various monomer-to-polymer-conversions and thus different viscosities which allowed us to probe a broad range of diffusion coefficients. The interested reader is referred to [27] for more details.

Before this project, the tracks were constructed semi-manually due to the lack of a satisfying alternative. That is, only a simple search in the neighborhood of the points was performed automatically as long as there were no ambiguities, i.e. only one localized point within the tracking radius and no competition among potential predecessors. In the case when the automatic continuation of the tracks fails, the user was presented with 10 consecutive frames of the movie with the options to select a successor among the alternatives, to introduce a new spot that was not detected by the localization, or to end the track. Needless to say that this was a tedious task, which took several working days to complete the tracking of a 5-minute-movie with high particle density. The advantage of this method is that the human expert maintains the full control over the process and the pattern recognition capabilities of the human brain is leveraged to resolve situations in which the image processing tools fail. On the other hand, these possibilities are also a disadvantage as the user might introduce systematic errors in the data and it is unlikely that a repetition of the task yields exactly the same results.

It remains to show that our automatic method not only works for realistic data but also in the real-world. To this end, we compare the average diffusion coefficients obtained from 6 movies by manual and automatic tracking:

manual	0.019 0.053 0.126 0.537 1.166 4.864	$\cdot 10^{-13}$ m^{2}s^{-1}
automatic	0.023 0.054 0.132 0.509 1.054 4.372	

References

1. Cichos, F., von Borczyskowski, C., Orrit, M.: Power-law intermittency of single emitters. Current Opinion in Colloid & Interface Science 12(6), 272–284 (2007)
2. Lippitz, M., Kulzer, F., Orrit, M.: Statistical evaluation of single nano-object fluorescence. Chemphyschem 6(5), 770–789 (2005)
3. Bingemann, D.: Analysis of 'blinking' or 'hopping' single molecule signals with a limited number of transitions. Chemical Physics Letters 433(1-3), 234–238 (2006)
4. Göhde, W., Fischer, U.C., Fuchs, H., Tittel, J., Basché, T., Bräuchle, C., Herrmann, A., Müllen, K.: Fluorescence blinking and photobleaching of single terrylenediimide molecules studied with a confocal microscope. Journal of Physical Chemistry A 102(46), 9109–9116 (1998)
5. Yip, W.T., Hu, D.H., Yu, J., Vanden Bout, D.A., Barbara, P.F.: Classifying the photophysical dynamics of single- and multiple-chromophoric molecules by single molecule spectroscopy. Journal of Physical Chemistry A 102(39), 7564–7575 (1998)
6. Cheezum, M.K., Walker, W.F., Guilford, W.H.: Quantitative comparison of algorithms for tracking single fluorescent particles. Biophysical J. 81(4), 2378–2388 (2001)
7. Sage, D., Neumann, F.R., Hediger, F., Gasser, S.M., Unser, M.: Automatic tracking of individual fluorescence particles: Application to the study of chromosome dynamics. IEEE Transactions on Image Processing 14(9), 1372–1383 (2005)

8. Schmidt, T., Schutz, G.J., Baumgartner, W., Gruber, H.J., Schindler, H.: Imaging of single molecule diffusion. Proceedings of the National Academy of Sciences of the United States of America 93(7), 2926–2929 (1996)
9. Saxton, M.J.: Single-particle tracking: conneting the dots. Nature Methods 5, 671–672 (2008)
10. Ghosh, R.N., Webb, W.W.: Automated detection and tracking of individual and clustered cell-surface low-density-lipoprotein receptor molecules. Biophysical J. 66(5), 1301–1318 (1994)
11. Chetverikov, D., Verestoy, J.: Feature point tracking for incomplete trajectories. Computing 62(4), 321–338 (1999)
12. Sbalzarini, I.F., Koumoutsakos, P.: Feature point tracking and trajectory analysis for video imaging in cell biology. J. Structural Biology 151(2), 182–195 (2005)
13. Bonneau, S., Dahan, M., Cohen, L.D.: Single quantum dot tracking based on perceptual grouping using minimal paths in a spatiotemporal volume. IEEE Transactions on Image Processing 14(9), 1384–1395 (2005)
14. Jaqaman, K., Loerke, D., Mettlen, M., Kuwata, H., Grinstein, S., Schmid, S.L., Danuser, G.: Robust single-particle tracking in live-cell time-lapse sequences. Nature Methods 5(4), 695–702 (2008)
15. Sergé, A., Bertaux, N., Rigneault, H., Marguet, D.: Dynamic multiple-target tracing to probe spatiotemporal cartography of cell membranes. Nature Methods 5, 687–694 (2008)
16. Elliott, L.C.C., Barhoum, M., Harris, J.M., Bohn, P.W.: Trajectory analysis of single molecules exhibiting non-brownian motion. Physical Chemistry Chemical Physics 13, 4326–4334 (2011)
17. Hellriegel, C., Kirstein, J., Bräuchle, C.: Tracking of single molecules as a powerful method to characterize diffusivity of organic species in mesoporous materials. New Journal of Physics 7(1), 1–14 (2005)
18. Schmidt, T., Schutz, G.J., Baumgartner, W., Gruber, H.J., Schindler, H.: Characterization of photophysics and mobility of single molecules in a fluid lipid-membrane. Journal of Physical Chemistry 99(49), 17662–17668 (1995)
19. Saxton, M.J., Jacobson, K.: Single-particle tracking: Applications to membrane dynamics. Annual Rev. of Biophysics and Biomolecular Structure 26, 373–399 (1997)
20. Schütz, G.J., Schindler, H., Schmidt, T.: Single-molecule microscopy on model membranes reveals anomalous diffusion. Biophysical J. 73(2), 1073–1080 (1997)
21. Rudnick, J., Gaspari, G.: The shapes of random-walks. Science 237(4813), 384–389 (1987)
22. Werley, C.A., Moerner, W.E.: Single-molecule nanoprobes explore defects in spin-grown crystals. Journal of Physical Chemistry B 110(38), 18939–18944 (2006)
23. Wöll, D., Kölbl, C., Stempfle, B., Karrenbauer, A.: Novel method for automatic single molecule tracking of blinking molecules at low intensities. Physical Chemistry Chemical Physics (2013)
24. Khachiyan, L.G.: A polynomial algorithm in linear programming. Dokl. Akad. Nauk SSSR 244, 1093–1097 (1979)
25. Karmarkar, N.: A new polynomial-time algorithm for linear programming. Combinatorica 4(4), 373–395 (1984)
26. Ye, Y.: An $O(n^3L)$ potential reduction algorithm for linear programming. Mathematical Programming 50(1-3), 239–258 (1991)
27. Stempfle, B., Dill, M., Winterhalder, M., Müllen, K., Wöll, D.: Single molecule diffusion and its heterogeneity during the bulk radical polymerization of styrene and methyl methacrylate. Polym. Chem. 3, 2456–2463 (2012)

The Quest for Optimal Solutions for the Art Gallery Problem: A Practical Iterative Algorithm⋆

Davi C. Tozoni, Pedro J. de Rezende, and Cid C. de Souza

Institute of Computing, University of Campinas, Campinas, Brazil
www.ic.unicamp.br
davi.tozoni@gmail.com, {rezende,cid}@ic.unicamp.br

Abstract. The general Art Gallery Problem (AGP) consists in finding the minimum number of guards sufficient to ensure the visibility coverage of an art gallery represented by a polygon. The AGP is a well known NP-hard problem and, for this reason, all algorithms proposed so far to solve it are unable to guarantee optimality except in special cases. In this paper, we present a new method for solving the Art Gallery Problem by iteratively generating upper and lower bounds while seeking to reach an exact solution. Notwithstanding that convergence remains an important open question, our algorithm has been successfully tested on a very large collection of instances from publicly available benchmarks. Tests were carried out for several classes of instances totalizing more than a thousand hole-free polygons with sizes ranging from 20 to 1000 vertices. The proposed algorithm showed a remarkable performance, obtaining provably optimal solutions for every instance in a matter of minutes on a standard desktop computer. To our knowledge, despite the AGP having been studied for four decades within the field of computational geometry, this is the first time that an exact algorithm is proposed and extensively tested for this problem. Future research directions to expand the present work are also discussed.

1 Introduction

The Art Gallery Problem (AGP) is one of the most investigated problems in computational geometry. The problem's input consists of a simple polygon, representing the outline of an art gallery, and one seeks to find a smallest set of points where guards should be placed so that the entire gallery is visually covered. Guards are assumed to have 360 degrees of unlimited vision, distance wise, and a polygon is said to be (visually) covered when every point of it is visible by at least one guard. We say that a point is visible by another whenever the line segment connecting them does not intersect the exterior of the polygon. Figure 1 illustrates a simplified floor plan of the Musée du Louvre and an optimal solution consisting of ten guards.

⋆ This research was supported by FAPESP – Fundação de Amparo à Pesquisa do Estado de São Paulo, CNPq – Conselho Nacional de Desenvolvimento Científico e Tecnológico and Faepex/Unicamp.

V. Bonifaci et al. (Eds.): SEA 2013, LNCS 7933, pp. 320–336, 2013.

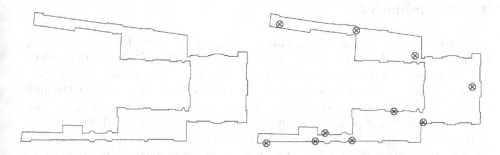

Fig. 1. Louvre polygon representation (left); an optimal guard positioning (right)

In the version of the AGP considered in this work, often referred to as the AGP *with point guards*, there are no constraints on the positions of the guards. Moreover, we assume that the polygon is hole-free, meaning that its complement is connected. Many variations of this problem have been studied in the literature, including generalizations in which one allows for the presence of holes in the interior of the polygon or restrictions where guard placement is limited a priori to some finite set of points (e.g., the set of vertices), see [18,20]. Both of these cases are known to be NP-hard [18,17]. To this date, as no exact algorithm exists except for the latter case [8], a great deal of effort has been placed on the development of heuristics and approximation techniques [1,4,13].

Our Contribution. In this paper, we present a practical iterative algorithm for the Art Gallery Problem with point guards, which finds a sequence of decreasing upper bounds and increasing lower bounds for the optimal value. As evidence of the effectiveness of the proposed algorithm, we also present results showing that for *every one* of more than *1440 benchmark polygons of various classes gathered from the literature* with *up to a thousand vertices*, optimal solutions are attained in just a few minutes of computing time. This work is unprecedented since, despite several decades of extensive investigation on the AGP, all previously published algorithms were unable to handle instances of that size and often failed to prove optimality for a significant fraction of the instances tested. As a matter of fact, as recently as last year, experts have claimed that "practical algorithms to find optimal solutions or almost-optimal bounds are not known" for this problem [15].

Organization of the Text. In the next section, a few basic definitions and notations are presented. Section 3 is devoted to a short survey of the relevant related works. Section 4 describes the main steps of the proposed algorithm. The most relevant implementation details are given in Section 5, while Section 6 discusses the computational results obtained from the experiments. Finally, in Section 7, we draw some conclusions and identify future directions in this research.

2 Preliminaries

We briefly review, in this section, concepts that are relevant for understanding the remaining of the paper. Recall that, in the Art Gallery Problem, the gallery is represented by a simple polygon P, which, in this work, is assumed to contain no holes. We say that two points in P are *visible* from each other if the line segment that joins them does not intersect the exterior of the polygon. The *visibility polygon* of a point $p \in P$, denoted by $\mathrm{Vis}(p)$, is the set of all points in P that are visible from p. The edges of $\mathrm{Vis}(p)$ are called *visibility edges* and they are said to be *proper* for p if and only they are not contained in any edges of P.

Given a finite set S of points in P, a maximal connected region in $\cup_{p \in S} \mathrm{Vis}(p)$ $(P \setminus \cup_{p \in S} \mathrm{Vis}(p))$ is called a *covered* (*uncovered*) region induced by S in P. Moreover, the geometric arrangement defined by the visibility edges of the points in S partitions P into a collection of closed polygonal faces called *Atomic Visibility Polygons* or simply *AVPs*. Clearly, the edges of an AVP are either portions of edges of P or portions of proper visibility edges for points of S. AVPs can be classified according to their visibility properties relative to the points of S. We say that an AVP $\mathcal{F}$ is a *light* (*shadow*) AVP if there exists a subset T of S such that $\mathcal{F}$ is (is not) visible from any point in T and the only proper visibility edges that bounds $\mathcal{F}$ emanate from points in T (see Figure 2).

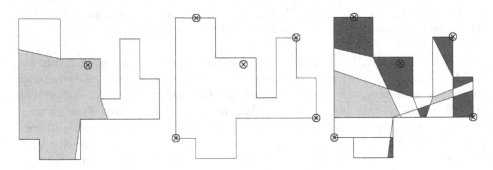

Fig. 2. Basic definitions: Visibility polygon of a point (left); a finite subset of points S (center) and the arrangement induced by S with the *light* (gray) and *shadow* (dark gray) AVPs (right)

We now describe discretized versions of the AGP that are fundamental to our approach. Recollect that in the most general version of the problem, the set of points to be covered and the set of potential guards are both infinite (equal to P). In contrast, the discretizations we consider here aim at reducing the set of points to be covered and/or the set of guard candidates to be finite.

Firstly, in the *Art Gallery Problem With Fixed Guard Candidates* (AGPFC), one is given a finite set of points $C \subset P$, and the question consists of selecting the minimum number of guards in C that are sufficient to cover the entire polygon. A special case of the AGPFC is obtained when the elements of C are restricted to the vertices of P, in which case we call it the *Art Gallery Problem With Vertex Guards* (AGPVG).

In another discrete version, named *Art Gallery Problem With Witnesses* (AGPW), one is given a finite set of points $W \subset P$, and the problem consists in finding the minimum number of guards in P that are sufficient to cover all points in W. Clearly, coverage of W does not ensure that of P. In [7], the authors define a polygon P to be *witnessable* if there exists a finite witness set $W \subset P$ satisfying the property that any set of guards that covers W also covers the entire polygon P. They also show that non-witnessable polygons are rather common. Nonetheless, a simple process for constructing witness sets for witnessable polygons is described. In Section 4.2, we show how the implementation of our algorithm takes advantage of this process.

Finally, a third discretization is introduced when both the witness set and the guard candidate set are required to be finite. This discretization leads to a hybrid of the last two problems which we will denote by AGPWFC. It is worth noting that the latter problem can easily be cast as a *Set Cover Problem* (SCP) in which the elements of W have to be covered using the subsets comprised of the witness points that are covered by the candidate guards. Despite being NP-hard, large instances of the SCP can be solved quite efficiently using modern integer programming solvers. Our algorithm takes advantage of this fact.

Although we tackle the discrete versions of AGP described above, the reader should bear in mind that, under the assumption of convergence, the algorithm presented in Section 4 leads to an optimal solution to the original problem. Later, we shall further elucidate this point. As we close this section, let us evoke that, to avoid ambiguities, unless stated otherwise, the term *Art Gallery Problem* and its acronym AGP are employed throughout this text to refer to the formulation with *point guards*.

3 Related Work

The Art Gallery Problem was initially proposed by Klee in 1973 as the question of determining the minimum number of guards sufficient to watch over an entire art gallery, represented by a simple polygon of n edges [14]. Since then, the AGP became one of the most discussed problem in Computational Geometry and gave rise to several important works including O'Rourke's classical book [18], a recent text by Ghosh on visibility algorithms [12] and surveys by Shermer in 1992 [19] and Urrutia in 2000 [20].

The first significant theoretical result on this problem was due to Chvátal in 1975, when he proved that $\lfloor n/3 \rfloor$ guards are sufficient to visually cover any simple polygon of size n [6]. Among other important theoretical results, Lee and Lin proved, in 1986, the NP-hardness of the AGP when using vertex guards, point guards or edge guards [17].

On the algorithmic side, several techniques have been proposed for different variants of the problem, including approximation algorithms, heuristics and even an exact method. For instance, based on an early work [11], Ghosh [13] recently presented an $O(n^4)$ time approximation algorithm for simple polygons yielding solutions within a $\log n$ factor of the optimal. Further approximation results are also found in Eidenbenz's work [10], which describes algorithms designed for several variations of terrain guarding problems.

On the other hand, in 2007, Amit et al. [1] presented a series of heuristic techniques for the AGP based on greedy strategies and methods that employ polygon partitions. According to the authors, some of these algorithms achieved good results for a large set of instances and, in many cases, optimal solutions were found.

In 2011, Bottino and Laurentini [4] proposed a new heuristic for the AGP based on an incremental algorithm that solves a restricted problem in which the goal is to cover only the edges of the polygon using as few guards as possible. The *edge covering algorithm* from [3] performed quite well in practice and, for this reason, was modified by the authors so as to yield a guard set to ensure the covering of the entire polygon. The heuristic thus obtained was tested and led to high quality solutions, even in comparison with the results of Amit et al. [1].

Finally, in 2011 Couto et al. [8] extended their previous work [9] and presented an *exact algorithm* for the AGPVG (with vertex guards). That algorithm iteratively discretizes a witness set creating a sequence of AGPWFC instances (see Section 2 for notation) which are then modeled as SCPs and solved using integer programming techniques. The experimental tests carried out by the authors confirm the algorithm's efficiency and robustness, showing that it is a viable option for the *exact* computation of AGPVG in practice. Remarkably, the authors also showed how to determine an initial witness set that enables the algorithm to execute in a single iteration, notwithstanding that experiments showed that such decrease in the number iterations does not always pay off in terms of total computing times. The challenges of this approach reside not only on finding an effective way to compute this ideal witness set, but also on coping with the huge set cover instance that ensues. In the research presented here, the algorithm of Couto et al. turned out to be a useful tool in solving the AGP (with point guards) to optimality, as we shall see later.

Kröller et al. [16] developed another approach aiming at solving the AGP in an exact way. The idea of their algorithm is again to discretize not only the witness set but also the guard set and to model the restricted AGP as an SCP. Firstly, lower and upper bounds are computed iteratively from the linear programming relaxation of the SCP formulation. The solutions of the primal and dual linear programs are used to guide the refinement of the witness and the guard candidate sets, giving rise to larger models, in an attempt to continuously reduce the duality gap. Whenever convergence happens and integrality of the primal linear relaxation variables is obtained, an optimal solution is found. The computational results proved the usefulness of the algorithm to generate bounds of high quality, even for polygons with holes. Nevertheless, convergence uncertainty and difficulties in obtaining integral solutions are major drawbacks that do not commend this algorithm as an effective alternative to optimally solve the AGP in practice.

The article by Chwa et al. [7] is also relevant to the work presented here. Therein, the authors study the so-called "Witness Problem" in which one wishes to determine whether a given polygon is *witnessable*. Necessary and sufficient conditions for the occurrence of this property are presented along with a description of how to create a minimum-sized set of witnesses that ensures this attribute. This result will be used later in the development of our algorithm.

4   An Iterative Exact Algorithm for the AGP

Before we describe the algorithm we developed for solving the AGP, some additional notation will be introduced to facilitate the exposition. Let V denote the set of vertices of the input polygon P and assume that $|V| = n$. Given a finite set S of points in P, we denote by $\text{Arr}(S)$ the arrangement defined by the visibility edges of the points in S. Let $C_{\mathcal{U}}(S)$ be a set comprised of one point from the interior of each uncovered region induced by S in P. Denote by $V_{\mathcal{L}}(S)$ the set of vertices of the light AVPs of $\text{Arr}(S)$ and by $C_{\mathcal{S}}(S)$ the set of centroids of the shadow AVPs of this arrangement.

Let D and C denote, respectively, a finite witness set and a finite candidate guard set. Let $\text{AGPW}(D)$ indicate the AGP with witness set D and $\text{AGPFC}(C)$ the AGP with candidate guard set C. Lastly, $\text{AGPWFC}(D, C)$ refers to the AGP with witness set D and candidate guard set C.

4.1   Fundamental Results

At each iteration of the algorithm, lower (dual) and upper (primal) bounds are generated for the optimal value of the AGP. The next two theorems establish basic results needed to compute these bounds and to ensure the correctness of our algorithm.

Theorem 1. *Let D be a finite subset of points in P. Then, there exists an optimal solution for $AGPW(D)$ where each guard belongs to a light AVP of $\text{Arr}(D)$.*

Proof. Let G be an optimal (cardinality-wise) set of guards that covers all points in D. Suppose there is a guard g in G that belongs to a face F of the arrangement $\text{Arr}(D)$ that is not a light AVP. Then, there must exist an edge e of F that belongs to the boundary of the visibility polygon of some point p in D, which is not visible from any point inside F. Let F' be an AVP of $\text{Arr}(D)$ that share e with F. It follows from the construction of $\text{Arr}(D)$ that every point in $D \setminus \{p\}$ visible from F is also visible from F'. If g' is any point in F', then g' sees p along with every point of D seen by g. An inductive argument suffices to show that this process eventually reaches a light AVP wherein lies a point that sees at least as much of D as g does, i.e., g may be replaced by a guard that lies on a light AVP. The Theorem then follows, by induction, on the number of guards of G. □

From Theorem 1, one can conclude that there exists an optimal solution for AGPW(D) for which all the guards are in $V_{\mathcal{L}}(D)$. Therefore, an optimal solution for AGPW(D) can be obtained simply by solving AGPWFC($D, V_{\mathcal{L}}(D)$). As seen before, the latter problem can be modeled as an SCP by an integer program whose number of constraints and variables are polynomial in $|D|$. Moreover, it is important to observe that, since D is a subset of points of P, the optimum of AGPW(D) gives rise to a lower bound for AGP. Later, we will see how a well chosen sequence of increasingly larger sets can be constructed to augment one such lower bound.

Thus, as we now know how to produce dual bounds for the AGP, the next task is to find a way to compute good upper bounds for the problem. To this end, we rely on the following result.

Theorem 2. *Let D and C be two finite subsets of P, so that C covers P. Assume that $G(D, C)$ is an optimal solution for AGPWFC(D, C) and let $z(D, C) = |G(D, C)|$. If $G(D, C)$ covers P, then $G(D, C)$ is also an optimal solution for AGPFC(C).*

Proof. Firstly, assume that $G(D, C)$ covers P, but it is not an optimal solution for AGPFC(C). Then, there exists $G' \subseteq C$ with $|G'| < z(D, C)$ such that G' is a feasible solution for AGPFC(C), i.e., G' covers P. This implies that G' is also a feasible solution for AGPWFC(D, C), contradicting the fact that $G(D, C)$ is optimal for this problem.   □

Notice that, as a corollary of Theorem 2, we also have that $z(D, C)$ is an upper bound for the optimum value of the AGP on P. This result can be explored in practice by applying a strategy analogous to that used in [8] to solve the AGPVG. Below, we describe how this is done.

Let D be a witness set and $C = V_{\mathcal{L}}(D) \cup V$. Assume that $G(D, C)$ is an optimal solution for AGPWFC(D, C) computed, possibly, with the aid of Theorem 1. Suppose, regrettably, that $G(D, C)$ does not cover P lending the conditions of Theorem 2 unfulfilled. Hence, $z(D, C) = |G(D, C)|$ is not a valid upper bound for the AGP. To mend this situation, the witness set D is updated to $D \cup C_{\mathcal{U}}(G(D, C))$. This process is then repeated until $G(D, C)$ covers P and, thus, Theorem 2 is applicable and $z(D, C)$ is an upper bound for the AGP. As in the algorithm given in [8], since the entire set C is a coverage for P, one can prove that this procedure converges in a number of steps that is polynomial in $|C|$ and n. Moreover, a single step actually suffices if one solves the SCP instance corresponding to AGPWFC(D, C) for $D = C_{\mathcal{S}}(C)$ (see [8] for a proof of this).

Basic Steps. Algorithm 1 displays a pseudo-code that summarizes the steps of our method to solve the AGP. Up to the third line, the algorithm only does the basic initializations, including that of the initial discretization of the witness set and of a known solution so far, G^*. Tests show that initializing G^* to V makes our algorithm run faster, in general, than spending $O(n \log n)$ time to compute the placement of $|V|/3$ guards based on the algorithm in [2]. The remaining lines form the main loop. In line 4, AGPW(D) is solved. According to Theorem 1

Algorithm 1. AGP Algorithm

1: $D \leftarrow$ initial witness set   {see Section 4.2}
2: Set: $LB \leftarrow 0$, $UB \leftarrow n$ and $G^* \leftarrow V$
3: **loop**
4:     Solve AGPW(D) : set $G_w \leftarrow$ optimal solution and $z_w \leftarrow |G_w|$
5:     $C \leftarrow V_{\mathcal{L}}(D) \cup V$
6:     **if** G_w is a coverage of P **then**
7:         **return** G_w
8:     **else**
9:         $U \leftarrow C_{\mathcal{U}}(G_w)$
10:         $LB \leftarrow \max\{LB, z_w\}$ {Theorem 1}
11:     **end if**
12:     **if** $LB = UB$ **then**
13:         **return** G^*
14:     **end if**
15:     $D_f \leftarrow D \cup U$
16:     **repeat**
17:         Solve AGPWFC(D_f, C) : set $G_f \leftarrow$ optimal solution and $z_f \leftarrow |\bar{G}_f|$
18:         **if** G_f is a coverage of P **then**
19:             $UB \leftarrow \min\{UB, z_f\}$ and, if $UB = z_f$, set $G^* \leftarrow G_f$ {Theorem 2}
20:         **else**
21:             $D_f \leftarrow D_f \cup C_{\mathcal{U}}(G_f)$
22:         **end if**
23:     **until** G_f is a coverage of P
24:     **if** $LB = UB$ **then**
25:         **return** G_f
26:     **else**
27:         $D \leftarrow D \cup U \cup M$ {M: see Section 4.3}
28:     **end if**
29: **end loop**

---

this can be done via the solution of AGPWFC($D, V_{\mathcal{L}}(D) \cup V$). Also, from this theorem, a lower bound is computed in line 10. The commands in lines 5, 9 and 15 prepare the witness and the candidate guard sets to obtain a (new) upper bound for the AGP, according to Theorem 2 and the subsequent discussion. The actual computation of the primal bound is accomplished in the repeat loop from lines 16 to 23. If the upper and lower bounds do not coincide, a new iteration of the outer loop is set up by redefining the witness set in line 27. Notice that the update of D requires the computation of a subset M, which, together with the choice of the initial witness set (first line), is a crucial issue for the performance of the algorithm. Both of these points are discussed in the Sections that follow. Finally, note that halting the main loop (lines 7, 13 and 25) strictly depends on finding a viable solution with the same cardinality as an AGPW solution. Thus, convergence of the algorithm, which remains an open question, is contingent on the choice of the initial set of witnesses and on how it is incremented.

4.2   Constructing the Initial Witness Set

We implemented and tested several procedures to calculate the initial witness set that best improves the algorithm's performance. The first of them assigned D simply to V, i.e., to the set of all vertices of P. This strategy was named *All-Vertices (AV)*. Preliminary tests indicated, though, that reflex vertices are not very hard to be covered, so little can be gained by including them in D. This led us to a second strategy, *Convex-Vertices (CV)*, where D starts off with only the convex vertices. Recall that the more witnesses there are in D, the more complex $\mathrm{Arr}(D)$ is and, consequently, the larger the set $V_{\mathcal{L}}(D)$ becomes. As a consequence, with the second strategy, the number of constraints and variables in the integer program that models the associated SCP instance tend to be much smaller, leading to the expectation that the algorithm will perform better.

In [7], Chwa et al. presented a theorem stating that if P is witnessable, then a minimum size witness set for P can be constructed by placing a witness anywhere in the interior of every *reflex-reflex* edge of P and on the convex vertices of every *convex-reflex* edge. The terms *convex* and *reflex* here refer to the angles formed at a vertex or at the endpoints of an edge. Based on this result, we devised a third discretization method called *Chwa-Points (CP)*. In our implementation, this discretization is made up of the midpoints of all reflex-reflex edges and all convex vertices from convex-reflex edges. Notice that, when the polygon is witnessable and this initial witness set is used, our algorithm finds an optimal solution in just one iteration, halting on the first execution of line 7.

However, as one should expect, most polygons are not witnessable. Consequently, even with the CP strategy, the algorithm often performs multiple iterations. Nevertheless, as we shall see in Section 6 , this strategy performs well in practice. Still, it inspired us to design a fourth strategy, named *Chwa-Extended (CE)*. In this case, the initial witness set is populated with the same points as in CP plus all reflex vertices from convex-reflex edges. Preliminary experiments showed that this strategy speeds up the algorithm in some cases.

It is worth noticing that the size of the discretization set certainly affects the computation time, since it directly determines the size of the SCP integer programs and, indirectly, it may also have an influence on the number of iterations. Nonetheless, from our experience, the algorithm's performance is not merely dependent on the size of this set. The *quality* of the points chosen to be brought into the witness set is critical in determining for how long the algorithm iterates. Experience shows that points less exposed to surveillance seem to play a better role as witnesses. Moreover, even though a solid formalization of this concept is yet unresolved, heuristically speaking, the earlier such points can be identified, the better.

4.3   Incrementing the Witness Set

If the last test comparing the bounds in line 24 of Algorithm 1 fails, the *else* clause is executed, and the algorithm will iterate again. This brings about an update on the witness set (line 27). Initially, we included in D only the centroids of the regions that remain uncovered by G_w from the solution of $\mathrm{AGPW}(D)$ (line 4), but this did

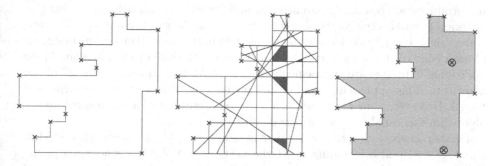

Fig. 3. Solving the AGPW (Lower Bound): The initial witness set D (Chwa-Points) (left); the arrangement $\mathrm{Arr}(D)$ and the light AVPs (center); the solution to $\mathrm{AGPW}(D)$ (right)

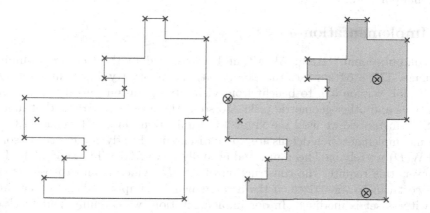

Fig. 4. Solving AGPFC (Upper Bound): Updated witness set D_f (left); the solution to $\mathrm{AGPWFC}(D_f, V_{\mathcal{L}}(D) \cup V)$ and to $\mathrm{AGPFC}(V_{\mathcal{L}}(D) \cup V)$ (right)

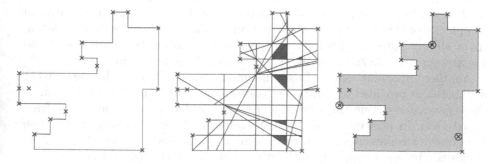

Fig. 5. Solving the AGPW (Lower Bound): The new witness set D (left); the arrangement $\mathrm{Arr}(D)$ and the light AVPs (center); the solution to $\mathrm{AGPW}(D)$ and to AGP (right)

not produce good results. Apparently, the main reason for this behavior is that this criterion precludes the choice of points on the boundary of the polygon. To overcome this drawback, we decided to also include in the discretization the midpoint of each edge of the shadow regions which lies on an edge of the polygon. These additional witnesses are the elements of the set M that appear in line 27 of the algorithm. This approach proved to be very effective for most polygon instances tested. However, some rare cases still required a number of iterations before the algorithm was capable of closing the duality gap.

Further analyzes led to yet another attempt to curb the occurrence of slow convergence by also including in M the vertices of the edges whose midpoints had been inserted as previously mentioned. These additional witnesses proved conducive to a performance improvement, as we will see in Section 6, by boosting the iterative algorithm to find optimal solutions quicker.

Figures 3, 4 and 5 illustrate the execution of the AGP algorithm on an orthogonal polygon from [8].

5   Implementation

For computational testing, Algorithm 1 was coded in the C++ programming language. The program uses the Computational Geometry Algorithms Library (CGAL) [5], version 3.9, to benefit from visibility operations, arrangement constructions and other geometric tasks. To solve the integer programs that model the SCP instances, we used the XPRESS Optimization Suite [21], version 7.0.

Some implementation details are worth discussing. Firstly, notice that to solve $AGPW(D)$ we rely on Theorem 1 and actually solve $AGPWFC(D, V_{\mathcal{L}}(D) \cup V)$. However, this requires the construction of $Arr(D)$ which is an expensive step. This cost can be amortized, if the arrangement is simply updated every time the witness set is modified. In our implementation, we carefully kept track of the arrangement changes, avoiding its recalculation from scratch throughout the iterations. For this task, we used the **Arrangement_2** class from CGAL, which employs the DCEL data structure. Moreover, a hash table containing all the visibility polygons from witnesses and guard candidates already calculated was used to speed up the process. If D_l denotes the final witness set, the overall complexity of the arrangement construction amounts to the time for calculating $|D_l|$ visibility polygons plus the time spent adding each of the $n \times |D_l|$ edges to the arrangement. Asymptotically, this procedure is quadratic in $n \times |D_l|$.

Another important aspect of the implementation amounts to the use of exact arithmetic as provided by CGAL. The coordinates of the input points as well as those generated during the execution of the algorithm are all expressed in fractional form. In principle, both the numerators and the denominators of these fractions are represented by integers with an unlimited number of digits. As mentioned before, the update of the witness set along the iterations of our algorithm avails itself of the computation of a point interior to each uncovered region. A natural choice would be to compute the centroid of any triangle contained in the uncovered region. However, even when the vertices of P have integral coordinates, it is easy to encounter instances for which these calculations lead to numbers whose representations are extremely large. As a consequence, any arithmetic operation with such numbers becomes very time consuming, severely deteriorating the algorithm's overall performance. To circumvent this situation, we initially replaced the computation of this centroid by that of another point in the interior of this triangle with a much shorter representation. For the correctness of the algorithm, it is immaterial which point we choose in the interior of the uncovered region. Notwithstanding, exceedingly long terms of fractions might still be generated. In this case, we perform verifiably valid truncation operations along with a simple trial-and-error procedure on these terms. This produced a dramatic gain in performance when compared with the computation of the centroids. Hence, this method was incorporated to the code and used in all tests, which are reported in the next section.

6   Computational Experiments

In this section, we describe the computational experiments that were carried out, reporting and analyzing the results obtained for a collection of 1440 instances in the public domain containing polygons from a large variety of classes and sizes.

Environment. All tests were conducted using a single desktop PC featuring an Intel® Core™ i7-2600 at 3.40 GHz, 8 GB of RAM and running under GNU/Linux 3.2.0. As described in Section 5, CGAL and XPRESS libraries were used in our implementation. Furthermore, all our tests were run in isolation, meaning that no other processes were executed at the same time on the machine.

Instances. The experiments were conducted on a set of instances in the public domain. This allowed us to make more direct comparisons with other works published earlier on the AGP. The benchmark is composed of instances grouped together according to their distinctive polygonal forms and sizes. This helps in highlighting the quality and robustness of our algorithm.

Firstly, our algorithm was tested with random simple polygon instances obtained from Bottino et al. [4] and Couto et al. [8]. A total of 670 polygons were used in this experiment, 250 of which came from [4], while the remaining 420 were collected from [8]. These instances were divided into groups according to their sizes: those from Bottino et al. contain from 30 to 60 vertices while those from Couto et al. range from 20 to 1000 vertices.

The second category of instances included 500 random orthogonal polygons, which also came from [4] (80 instances) and [8] (420 instances). As in the previous case, they were grouped according to their original benchmarks and their sizes.

Finally, we also tested with random orthogonal von Koch polygons from [8] since, as shown by the authors, such instances represent a challenge to AGP solvers. In total, this group contained 270 instances partitioned into nine smaller subgroups according to their sizes, ranging from 20 to 500 vertices.

Additionally, we experimented on a few singular polygons such as a floor plan of the Musée du Louvre and a simple polygon used in [16] to illustrate a convergence problem occurring with the algorithm proposed in that paper.

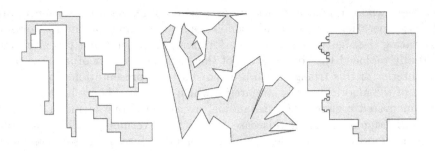

Fig. 6. Sample instances used in experiments obtained from [4] and [8]

Examples of polygons (in the public domain) used in the computational experiments are shown in Figure 6.

Analysis. Let us start off by declaring unambiguously that our algorithm was able to prove optimality for *all* 1440 instances of the benchmark. This unprecedented success for an AGP solver can be better appreciated by analyzing the data on the tables that follow.

In Table 1 we compare the results we have obtained with those reported in [16]. For this comparison, we used the simple and orthogonal polygons of sizes 60, 100, 200 and 500 from Couto et al. [8], as these were the polygon sizes used in the experiments reported in [16]. The most important aspect to analyze, not to say the only one that has significance, refers to the number of instances solved to proven optimality. Recall that the algorithm presented by Kröller et. al. in [16] generates a sequence of primal and dual bounds for AGP which are computed through a sequence of linear programs obtained by incrementing the number of variables and constraints of the model. As mentioned earlier, that method has no proof of convergence and is only able to guarantee optimality if the primal linear program has an integer optimal solution whose value is equal to the known dual bound. As seen in this table, the algorithm often failed to converge and, hence, to provide an optimal solution. In contrast, our algorithm always returned an optimal solution, leading to a substantial increase in the number of instances in the literature with known optima. This can be perceived from the results of optimality rate on Table 1. It should be noticed that in [16] the algorithm was always halted after 20 minutes of execution. Therefore, one can legitimately argue that it might have converged to the optimum in other cases had extra computation time been given. However, as noted by the authors, in the cases of failure, little or no improvement in the duality gap had been observed when the algorithm was halted.

The computation times given in [16] are also displayed in Table 1 even though a fair comparison is not possible since the computational environments for the two experiments are rather different. Therefore, we restrict ourselves to pointing out that our implementation was able to solve each of the 240 instances that comprise the tests displayed in this table within a maximum time of 141 seconds.

Table 1. Comparison between the method of Kröller et al. [16] and ours

Instance Groups	n	Optimality Rates		Instance Groups	n	Average Time (sec)	
		Method [16]	Our Method			Method [16]	Our Method
Simple (30 inst. per size)	60	80%	100%	Simple (30 inst. per size)	60	0.70	0.57
	100	64%	100%		100	29.40	1.72
	200	44%	100%		200	14.90	7.09
	500	4%	100%		500	223.30	65.64
Orthogonal (30 inst. per size)	60	80%	100%	Orthogonal (30 inst. per size)	60	0.40	0.30
	100	54%	100%		100	1.10	0.95
	200	19%	100%		200	4.30	3.95
	500	7%	100%		500	25.30	30.85

Now, we analyze the cardinality of the guard sets produced by the heuristic of Bottino et al. [4] relative to the optima computed with our algorithm. This study involved uniquely the instances treated in [4]. The data are summarized in Table 2. Although an instance-based comparison would be more desirable, this was not possible since in [4] only average values are reported. Nevertheless, one can see that, except for the small random simple polygons with 30 vertices, the heuristic was unable to reach the optimum on all remaining instance subgroups. As a side remark, one may perceive a trend of a growing gap between the results from the heuristic and the optimal values as the number of vertices grow. As in the previous analysis, a direct comparison of execution times would not be adequate as the two algorithms have distinct goals and were executed on different computer systems. As an illustration, the tests in [4] were conducted on an Intel® Core2™ processor at 2.66 GHz and 2 GB of RAM. Despite this observation, it seems remarkable that the average time spent by our code to find a provably optimal solution is orders of magnitude smaller than that consumed by the heuristic of Bottino et al. to generate a suboptimal solution.

Table 2. Comparison between the method of Bottino et al. [4] and ours

Instance Groups	n	Number of Guards (average)		Instance Groups	n	Average Time (sec)	
		Method [4]	Our Method			Method [4]	Our Method
Simple (20 inst. per size)	30	4.20	4.20	Simple (20 inst. per size)	30	1.57	0.17
	40	5.60	5.55		40	2.97	0.23
	50	6.70	6.60		50	221.92	0.42
	60	8.60	8.35		60	271.50	0.54
Orthogonal (20 inst. per size)	30	4.60	4.52	Orthogonal (20 inst. per size)	30	1.08	0.12
	40	6.10	6.00		40	9.30	0.17
	50	7.80	7.70		50	6.41	0.23
	60	9.30	9.10		60	81.95	0.30

Initial Discretizations. As explained in Section 4.2, four different discretization strategies were developed to construct the initial witness set W. Our tests revealed significant changes in the performance of the algorithm when these strategies are adopted. This observation can be better understood from the data exhibited in Tables 3 and 4.

Table 3. Number of iterations (main loop) and average time spent until an optimal solution is found for each initial discretization strategy, using Bottino et al. instances [4]

Instance Groups	n	Iterations			
		AV	CV	CP	CE
50x30 (170 inst.)	30	1.14	1.14	1.10	1.11
Simple (20 inst. per size)	30	1.50	1.55	1.45	1.50
	40	1.25	1.40	1.15	1.10
	50	1.45	1.70	1.55	1.35
	60	1.55	1.80	1.20	1.40
Orthogonal (20 inst. per size)	30	1.38	1.38	1.14	1.10
	40	1.50	1.75	1.45	1.35
	50	1.55	1.65	1.45	1.45
	60	1.80	1.90	1.40	1.55

Instance Groups	n	Time (seconds)			
		AV	CV	CP	CE
50x30 (170 inst.)	30	0.19	0.15	0.16	0.18
Simple (20 inst. per size)	30	0.22	0.17	0.19	0.22
	40	0.32	0.25	0.23	0.29
	50	0.61	0.43	0.42	0.58
	60	0.91	0.79	0.54	0.84
Orthogonal (20 inst. per size)	30	0.14	0.12	0.12	0.13
	40	0.21	0.18	0.17	0.20
	50	0.28	0.26	0.23	0.28
	60	0.41	0.35	0.30	0.38

Table 4. Number of iterations (main loop) and average time spent until an optimal solution is found for each initial discretization strategy, using Couto et al. instances [8]

Instance Groups	n	Iterations			
		AV	CV	CP	CE
Simple (30 inst. per size)	100	2.53	2.63	1.80	1.87
	500	3.83	3.93	3.97	3.80
	1000	4.70	4.67	4.47	4.57
Orthogonal (30 inst. per size)	100	2.80	2.57	2.37	2.33
	500	4.50	4.37	3.73	3.80
	1000	5.40	5.87	5.00	5.43
Von Koch (30 inst. per size)	100	1.57	1.70	1.60	1.77
	500	2.03	2.20	2.43	2.13

Instance Groups	n	Time (seconds)			
		AV	CV	CP	CE
Simple (30 inst. per size)	100	3.03	2.29	1.72	2.12
	500	114.51	78.62	65.64	103.60
	1000	926.39	554.24	408.71	718.93
Orthogonal (30 inst. per size)	100	1.34	1.09	0.95	1.17
	500	68.01	41.31	30.85	42.46
	1000	297.50	233.82	155.00	235.35
Von Koch (30 inst. per size)	100	2.26	1.44	1.62	2.60
	500	1064.08	256.77	595.89	1639.80

Analyzing these tables one can notice that the number of iterations increases slightly as the size of the polygons grows. As an example, in the case of random orthogonal instances from [8], the number of iterations increases by a factor of 2 when the polygon size is multiplied by 10. Regarding the alternative strategies, one can verify that CP and CE lead to the fewest number of iterations in almost all cases, with some advantage to the former.

To analyze computation times attained with each strategy, we initially focus our attention on the tests with Random (Simple and Orthogonal) polygons. It is clear that CP outperforms the other strategies and, hence, should be the preferred one. Although the CE strategy needs almost the same number of iterations as CP, as far as computing times are concerned, it failed to keep up with the latter. This can be explained by the fact that the CE discretization starts with more witnesses, which increases the time needed to calculate the visibility polygons, the arrangement and, as a consequence, the SCP integer model. In this context, one can also notice that the execution time required grows approximately quadratically with the size of the polygon, strongly contrasting with what happens with the number of iterations. An explanation for this behavior is that, at each iteration, the number of witnesses and, therefore, the complexity of the arrangement formed by them increase, leading to extra time spent at each iteration.

We now turn our attention to the results obtained for the random Von Koch polygons. In this case, CV was, surprisingly, the strategy with which the algorithm reached the optima faster. It is also important to note that, for these instances, the execution times seem to grow more rapidly than a quadratic function on the size of the polygon. For both observations, a possible explanation could be the much higher complexity of the arrangements associated to Von Koch polygons when compared to other type of polygons.

7 Concluding Remarks

The absence of algorithms for the Art Gallery Problem with efficiency confirmed in practice has often been mentioned in the literature. This work contributes to fill this gap. We developed an algorithm for the AGP aiming at solving the problem exactly. This algorithm was implemented and the code tested on more than a 1400 instances of polygons in the public domain. Not only the algorithm found provably optimal solutions for all instances, but it also achieved low computation times to accomplish the task. This is particularly remarkable in light of the fact that, in many situations, these times were even smaller than what heuristics published earlier in the literature required.

Despite the excellent experimental results, on the theoretical side, it must be said that to prove that the algorithm always converges remains a challenge. To raise the appreciation for the difficulty of this endeavor, recall that in [16] the authors introduced a carefully crafted instance to illustrate a convergence problem occurring with the algorithm proposed in their own work. Given a particular initial discretization of the witness set and a strategy for choosing the new witnesses to be added along the iterations, one can tweak this instance of Kröller et al.'s to create a pathological example that forces our algorithm to iterate forever. Randomization obviously affords the possibility of dramatically reducing the chance that the algorithm might run into convergence uncertainty. However, the theoretical question remains on whether there exists a strategy to initialize and update the witness set in a way that avoids this bad behavior.

On the other hand, in view of further practical applications, we should divulge that we are presently refining the implementation in order to broaden the classes of polygons that it can address, by allowing for the solution of instances comprised of polygons with holes.

Acknowledgments. We thank A. Kröller for fruitful discussions and A. Bottino for providing us with some polygon instances. We are also grateful to the anonymous referees for their helpful comments.

References

1. Amit, Y., Mitchell, J.S.B., Packer, E.: Locating guards for visibility coverage of polygons. In: ALENEX, New Orleans, Lousiana (January 2007)
2. Avis, D., Toussaint, G.T.: An efficient algorithm for decomposing a polygon into star-shaped polygons. Pattern Recognition 13(6), 395–398 (1981)

3. Bottino, A., Laurentini, A.: A nearly optimal sensor placement algorithm for boundary coverage. Pattern Recognition 41(11), 3343–3355 (2008)
4. Bottino, A., Laurentini, A.: A nearly optimal algorithm for covering the interior of an art gallery. Pattern Recognition 44(5), 1048–1056 (2011)
5. CGAL. Computational Geometry Algorithms Library, www.cgal.org (last access January 2012)
6. Chvátal, V.: A combinatorial theorem in plane geometry. Journ. of Combin. Theory Series B 18, 39–41 (1975)
7. Chwa, K.-Y., Jo, B.-C., Knauer, C., Moet, E., van Oostrum, R., Shin, C.-S.: Guarding art galleries by guarding witnesses. Intern. Journal of Computational Geometry and Applications 16(02n03), 205–226 (2006)
8. Couto, M.C., de Rezende, P.J., de Souza, C.C.: An exact algorithm for minimizing vertex guards on art galleries. International Transactions in Operational Research 18(4), 425–448 (2011)
9. Couto, M.C., de Souza, C.C., de Rezende, P.J.: An exact and efficient algorithm for the orthogonal art gallery problem. In: Proc. of the XX Brazilian Symp. on Comp. Graphics and Image Processing, pp. 87–94. IEEE Computer Society (2007)
10. Eidenbenz, S.: Approximation algorithms for terrain guarding. Inf. Process. Lett. 82(2), 99–105 (2002)
11. Ghosh, S.K.: Approximation algorithms for art gallery problems. In: Proc. Canadian Inform. Process. Soc. Congress (1987)
12. Ghosh, S.K.: Visibility Algorithms in the Plane. Cambridge University Press, New York (2007)
13. Ghosh, S.K.: Approximation algorithms for art gallery problems in polygons. Discrete Applied Mathematics 158(6), 718–722 (2010)
14. Honsberger, R.: Mathematical Gems II. The Dolciani Mathematical Expositions, vol. 2. MAA (1976)
15. Kröller, A., Baumgartner, T., Fekete, S.P., Moeini, M., Schmidt, C.: Practical solutions and bounds for art gallery problems (August 2012), http://ismp2012.mathopt.org/show-abs?abs=1046
16. Kröller, A., Baumgartner, T., Fekete, S.P., Schmidt, C.: Exact solutions and bounds for general art gallery problems. J. Exp. Algorithmics 17(1), 2.3:2.1–2.3:2.23 (2012)
17. Lee, D.T., Lin, A.: Computational complexity of art gallery problems. IEEE Transactions on Information Theory 32(2), 276–282 (1986)
18. O'Rourke, J.: Art Gallery Theorems and Algorithms. Oxford University Press, New York (1987)
19. Shermer, T.: Recent results in art galleries. Proceedings of the IEEE 80(9), 1384–1399 (1992)
20. Urrutia, J.: Art gallery and illumination problems. In: Sack, J.R., Urrutia, J. (eds.) Handbook of Computational Geometry, pp. 973–1027. North-Holland (2000)
21. XPRESS. Xpress Optimization Suite (2009), http://www.fico.com/en/Products/DMTools/Pages/FICO-Xpress-Optimization-Suite.aspx (access January 2012)

An Improved Branching Algorithm
for Two-Layer Planarization Parameterized
by the Feedback Edge Set Number

Mathias Weller*

Institut für Softwaretechnik und Theoretische Informatik, TU Berlin, Germany
mathias.weller@tu-berlin.de

Abstract. Given an undirected graph G and an integer $k \geq 0$, the NP-complete TWO-LAYER PLANARIZATION problem asks whether G can be transformed into a forest of caterpillar trees by removing at most k edges. Since transforming G into a forest of caterpillar trees requires breaking every cycle, the size f of a minimum feedback edge set is a natural parameter with $f \leq k$. We refine and enhance ideas that led to previous algorithms running in $O(3.562^k k + |G|)$ time and $O(6^f f^2 + f \cdot |G|)$ time, respectively, to an improved branching algorithm running in $O(3.8^f f^2 + f \cdot |G|)$ time. Since we expect f to be significantly smaller than k for a wide range of input instances, the presented algorithm can be considered superior to the previous algorithms. We present an empirical study of an implementation of our algorithm and compare it to implementations of previous algorithms. Our experiments show that even large instances can be solved as long as they are sparse.

1 Introduction

A strategy of drawing hierarchical graphs in human readable form is the "Sugiyama approach" [12, 9, 7], an important part of which is finding good 2-layered drawings of graphs. A 2-layered drawing can be understood as an assignment and arrangement of the vertices to two layers such that edges only occur between the layers and edges are drawn as straight lines. Furthermore, "good" means that the number of edges that do not "behave well", that is, the number of edges we have to remove such that no two edges of the drawing cross, is minimum. If a graph has a 2-layered drawing such that no two edges cross, then the graph is called *biplanar*. Given a set of edges such that their removal makes a graph biplanar, then the corresponding drawing can be computed efficiently. Hence, we focus on the problem of finding such an edge set or, more precisely, its decision variant.

TWO-LAYER PLANARIZATION (2LP):
Given: An undirected graph $G = (V, E)$ and an integer $k \geq 0$.
Question: Is there an edge subset $E' \subseteq E$ with $|E'| \leq k$ such that $(V, E \setminus E')$ is biplanar?

* Supported by the DFG, research project DARE (NI 369/11).

V. Bonifaci et al. (Eds.): SEA 2013, LNCS 7933, pp. 337–353, 2013.
© Springer-Verlag Berlin Heidelberg 2013

It has been shown that a graph is biplanar if and only if it consists of disjoint caterpillars[1] [9, 2]. This allows for appropriate alternative formulations of 2LP. Apart from being proposed as an alternative method to minimize crossings [9], solving 2LP is important in DNA mapping [15] and global routing for row-based VLSI layout [8]. Due to the space constraint, proofs are deferred to an appendix.

Previous and Related Work. Two-Layer Planarization is NP-hard even if the input graph is bipartite and one set of the partition contains only vertices of degree at most two [4]. Concerning the parameter k ("number of edge deletions"), Dujmović et al. [2] showed that 2LP can be solved in $O(6^k \cdot k + |G|)$ time by devising a search tree algorithm and several polynomial-time data reduction rules leading to a problem kernel with $O(k)$ vertices and edges. Suderman and Whitesides [11] implemented and tested this algorithm, also in comparison with an ILP formulation developed earlier [6]. Fernau [5] presented a refined search tree for 2LP leading to a running time of $O(5.19276^k \cdot k^2 + |G|)$. Finally, based on a different branching analysis, Suderman [10] developed an $O(3.562^k \cdot k + |G|)$-time algorithm and published running time results. Unfortunately, we were unable to get in contact with Suderman to obtain his implementation. Hence, we will have to compare our results to the ones obtained in 2005 [10].

Recently, we considered 2LP with respect to the parameter "size f of a minimum feedback edge set of G" [14]. We developed data reduction rules that led to a problem kernel of size $O(f)$. We also presented a branching algorithm that solves 2LP in $O(6^f \cdot f + f \cdot |G|)$ time [14].

New Results. In this work, we consider the parameter "feedback edge set number f" of the input graph and develop a branching algorithm running in $O(3.8^f f^2 + f \cdot |G|)$ time. The algorithm refines previous branching algorithms [10, 14] by choosing adequate forbidden subgraphs to branch on and applying further data reduction if no such subgraphs can be found. We make use of our previously shown kernelization algorithm for 2LP [14].

To support the practical relevance of our work, we performed experiments with our algorithm. On the one hand, we used the generated bipartite graphs used by Mutzel [9] and Suderman and Whitesides [11]. On the other hand, we generated treelike graphs since we expect a variety of inputs encountered in the context of drawing hierarchical graphs to be very sparse. Dujmović et al. [2] even pointed out that "instances of Two-Layer Planarization for *dense* graphs are of little interest from a practical point of view" since the resulting drawings are unreadable anyway. Although our algorithm does not perform significantly better than the state-of-the-art branching algorithm [10] on the first type of instances, it shows its strength in the second experiment.

Considering the parameter f has numerous advantages. For example, the number of necessary edge deletions to make a graph biplanar is an upper bound on the feedback edge set number, since we have to destroy all cycles to obtain a forest of caterpillars. In this sense, we improve on previous work [2, 5, 10]. As Dujmović et al. [2] pointed out (see quote above), the solution size can be expected

---

[1] A caterpillar is a tree each of whose vertices is adjacent to at most two non-leaves.

to be small in practice. This is even more plausible for the feedback edge set number of a graph, which is directly linked to the number of edges, and, hence, the sparseness of the graph. The feedback edge set number f is a parameter that can easily be computed in advance and, hence, allows for a meta-algorithm that chooses an algorithm for a given input by computing an estimation on the running time prior to running the algorithm for the problem itself. Since the parameter k ("number of edge deletions") is NP-hard to compute, this is another advantage over previous approaches.

2   Preliminaries

We assume the reader is familiar with general graph notation. For a vertex v of a graph G, we write $G - v$ for $G[V(G) \setminus \{v\}]$. Analogously, for an edge set S we abbreviate $(V(G), E(G) \setminus S)$ to $G - S$. If $G - S$ is acyclic, S is called *feedback edge set* of G. We denote the size of a minimum feedback edge set, the *feedback edge set number* of G (also known as *circuit rank*, *cyclomatic number* or *nullity*), by $f(G)$ or simply f if G is clear from the context. An edge whose removal does not decrease f is called a *bridge*. Note that, for a graph (V, E) with c connected components, $f = |E| - |V| + c$. A tree is a *caterpillar tree* (or *caterpillar* for short) if each of its vertices has at most two non-leaf neighbors (we say its *non-leaf degree* is at most two). Equivalently, a caterpillar is a tree that does not contain a 2-claw [4] (a claw whose edges have been subdivided, that is, replaced by paths of length two (see Figure 2a)). Thus, a graph is a forest of caterpillars if and only if it is acyclic and does not contain a 2-claw as subgraph.

For a graph G, the maximum induced subgraph G^* of G that has minimum degree two is the result of repeatedly removing degree-one vertices and is called the *2-core* of G. Its edgewise complement with respect to G (called *1-shell* of G) is acyclic. For a vertex $v \in V(G^*)$ let T^v denote the tree in $G - E(G^*)$ that contains v and note that no other vertex of G^* is contained in T^v. The tree T^v is called the *pendant tree* of v and v is called its *connection point*. The pendant trees shown in Figure 1 are of particular interest in this work. Herein, the neighbor w of v in a Y-graph Y is called center(Y). We use $G^\bigcirc$ to denote the subgraph of G that results from deleting all bridges from G. Given G, both G^* and $G^\bigcirc$ can be computed in linear time. The next lemma is essential in various proofs throughout this work.

Lemma 1 ([14]). *Let G be a graph that is reduced with respect to the kernelization of Uhlmann and Weller [14] and let T^v be the pendant tree of a vertex v in G^*. Then, T^v is isomorphic to one of the trees shown in Figure 1.*

The exponential part of the running time of our branching algorithm depends only on the feedback edge set number f. In the literature, such an algorithm is called *fixed parameter tractable*, or *fpt* with respect to f. The idea behind a branching algorithm is to find a subgraph H that contains a forbidden subgraph (A 2-claw or a cycle) and then try all feasible ways of destroying all forbidden subgraphs in H. We call H a *branching structure*. We present our algorithm as a

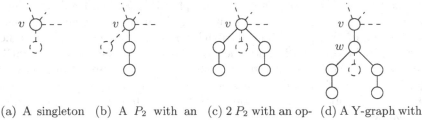

(a) A singleton or a leaf   (b) A P_2 with an optional leaf   (c) 2 P_2 with an optional leaf   (d) A Y-graph with an optional leaf

Fig. 1. In an instance reduced with respect to the known kernelization for parameter f [14], the pendant tree T^v of each vertex $v \in V(G^*)$ is isomorphic to one of the trees shown in Figures 1a–1d

collection of *branching rules*, that is, polynomial-time executable graph modification rules that, given a graph, create "partial solutions" (edge sets whose removal destroys the forbidden subgraphs in H). We call a rule *correct* if one of the created partial solutions can be extended to an optimal solution for the input instance. Each partial solution corresponds to a decrease in the parameter. The vector of these differences for all partial solutions of a branching rule is called its *branching vector*. If only one partial solution is created, then we call the modification rule a *data reduction rule*. If the requirements of a rule are not met by a graph, then we call this graph *reduced* with respect to this rule. A collection of data reduction rules such that the size of graphs reduced with respect to these rules can be bounded in the parameter, is called a *kernelization* with respect to the parameter.

3   An Improved Branching Algorithm for 2LP

Our algorithm is a non-trivial adaptation of the algorithm of Suderman [10], which runs in $O(3.8^k \cdot |G|)$ time.[2] Suderman [10] defined five branching structures and developed a branching rule for each of them (see Figure 2). However, the branching vectors of these rules are with respect to the solution size k and not with respect to the feedback edge set number f. We use the same structures for our branching, but, in order to maintain the branching vectors, we ensure that the feedback edge set number decreases in each branch. To this end, we find specific locations of branching structures in the input graph that allow this kind of branching. However, the price we pay for this advantage is that it becomes possible that none of the branching rules apply to the input graph. To deal with such graphs, we augment the process with a reduction rule that applies to graphs reduced with respect to the branching rules, thereby solving the given instance.

Throughout the section, G denotes the input graph, G^* denotes its 2-core, and $G^\bigcirc$ denotes the result of stripping G of all its bridges.

---

[2] Note that Suderman [10] provided a refined algorithm running in $O(3.562^k \cdot |G|)$ time that we could not adapt for our parameterization.

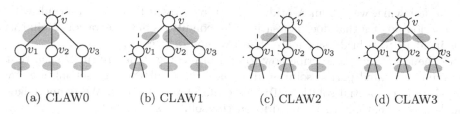

| (a) CLAW0 | (b) CLAW1 | (c) CLAW2 | (d) CLAW3 |

Fig. 2. Schematic view on four of the first five branching rules of Suderman [10] ("3CYC" is omitted since it is equal to Branching Rule 1). Gray ellipses indicate the edges deleted in each branch. The branching vectors are $(1, 1, 1, 2)$, $(1, 1, 1, 2, 2)$, $(1, 1, 1, 2, 2)$, and $(1, 1, 1, 2, 2, 2)$.

3.1 Adapting Previous Branching Rules

The first branching rule of Suderman [10] branches on cycles of length at most three and, thus, already decreases the feedback edge set number in each branch.

Branching Rule 1 *Let u, v, w be vertices in G forming a cycle. Then, create the partial solutions $\{\{u, v\}\}, \{\{v, w\}\}$, and $\{\{w, u\}\}$.*

In the following, we develop a strategy to apply the branching rules of Suderman [10] while avoiding bridges of G. This is necessary since deleting bridges does not decrease the feedback edge set number. Two obstacles arise when trying to maintain the branching vectors of the branching rules of Suderman [10]: First, there may be a bridge of G in a partial solution. Second, a partial solution may contain two non-bridges but deleting one of them makes the other a bridge.

By carefully selecting a branching structure, we can avoid bridges of G. To this end, we differentiate between two kinds of bridges in G. The first kind are edges in $G - E(G^*)$, that is, edges of pendant trees. We call them *A-bridges*. The second kind are the bridges in G^*, called *B-bridges*. The following observation (whose proof is roughly the same as the proof of Lemma 6 in [14]) shows that not all A-bridges pose a problem to the branching rules, allowing us to ignore partial solutions containing these bridges.

Observation 1 *Let G be reduced with respect to the kernelization [14] and let T^v be the pendant tree of a vertex v of G^* such that T^v is not a Y-graph. Then, there is an optimal solution for G that does not delete any edge of T^v.*

With Observation 1, we can simply ignore branches that include A-bridges, unless they belong to Y-graphs. In fact, we can further limit the structure of optimal solutions involving certain Y-graphs.

Observation 2 (consequence of Observation 1 of [14]) *Let Y be a pendant Y-graph with connection point v and center w such that v is incident to exactly two non-bridges e_1 and e_2 and no B-bridge in G. Then, there is an optimal solution S for $G - \{e_1, e_2\}$ such that $S \cup \{\{v, w\}\}$ or $S \cup \{e_1, e_2\}$ is an optimal solution for G.*

In the following, we call an A-bridge $\{v, w\}$ *relevant* if there is a Y-graph with connection point v that does not satisfy the conditions of Observation 2, that is, v is incident to B-bridges or more than two non-bridges.

In the following, we describe how to find branching structures in G such that applying a created partial solution decreases the feedback edge set number by the size of the partial solution. To this end, let $v \in V(G^\bigcirc)$. We define edges incident to v that can be included in partial solutions.

Definition 1. *Let $v \in V(G^\bigcirc)$. We call v* branchable *if there are three vertices $v_1, v_2, v_3 \in N_G(v)$ (called* branching partners *of v) such that*

(1) For each $1 \le i \le 3$, $\{v, v_i\}$ is not a relevant A-bridge.
(2) For each $1 \le i \le 3$, there is no B-bridge incident to v_i.

By finding a vertex with degree at least three in the graph that remains after deleting all relevant A-bridges and all B-bridges from G, a branchable vertex and its branching partners can be found in linear time. Although we need some more properties of branching partners to prove the desired branching vectors, we can show that these properties follow from Definition 1.

Observation 3 *Let $v \in V(G^\bigcirc)$ be branchable and v has three branching partners in $G^\bigcirc$. Then, there are three branching partners $v_1, v_2, v_3 \in V(G^\bigcirc)$ of v such that*

(1) $\deg_G(v_1) = 2 \Rightarrow \deg_G(v_2) = 2$ and $\deg_G(v_2) = 2 \Rightarrow \deg_G(v_3) = 2$.
(2) if $\deg_G(v_1) = 2$, then $\{v, v_2\}$ is not a bridge in $G - \{\{v, v_1\}\}$, and
(3) if $\deg_G(v_1) > \deg_G(v_2) = 2$, then $\{v, v_3\}$ is not a bridge in $G - \{\{v, v_2\}\}$.

For Observation 3(1), we can simply sort v_1, v_2, and v_3 by their degree in G. For Observation 3(2), we can just swap v_2 and v_3 if necessary. For Observation 3(3), note that if deleting $\{v, v_2\}$ makes $\{v, v_3\}$ a bridge, then, deleting both $\{v, v_2\}$ and $\{v, v_3\}$ does not make $\{v, v_1\}$ a bridge. Since $v_1 \in V(G^\bigcirc)$, there is some $v_4 \in N_{G^\bigcirc}(v)$ such that $\{v, v_4\}$ is not a bridge in G. This allows us to replace v_2 by v_4, while maintaining all properties stated before.

In the following, we assume that v is branchable and its branching partners v_1, v_2, and v_3 fulfill all properties of Definition 1 and Observation 3. To complete the branching structure, it remains to select edges incident to v_1, v_2, and v_3.

Definition 2. *For $1 \le i \le 3$, let $e_i = \{v, v_i\}$ and let E_i denote a set of at most two edges of G such that*

(i) all edges in E_i are incident to v_i but not to v,
(ii) if $|E_i| = 1$, then $\deg_G(v_i) = 2$,
(iii) if $\deg_{G^\bigcirc}(v_i) \ge 3$, then there is an edge $\{v_i, u\} \in E_i$ such that u is connected to v via a path in G that avoids v_i, and
(iv) if there is a bridge of G in E_i, then $\deg_{G^\bigcirc}(v_i) = 2$.

An algorithm to compute E_i for given v and v_i can be found in the appendix. We can now state the modified versions of the branching rules of Suderman [10].

Branching Rule 2 (based on "CLAW0" in [10]) *Let* $|E_1| = 1$. *Then, create the partial solutions* E_1, E_2, E_3, *and* $\{e_1, e_2\}$. *Discard all partial solutions containing non-relevant A-bridges.*

Branching Rule 3 (based on "CLAW1" in [10]) *Let* $|E_1| > |E_2| = 1$. *Then, create the partial solutions* E_2, E_3, $\{e_1\}$, *and* $\{e_2, e_3\}$. *If* E_1 *does not contain an A-bridge, then additionally create the partial solution* E_1. *Discard all partial solutions containing non-relevant A-bridges.*

Branching Rule 4 (based on "CLAW2" in [10]) *Let* $|E_2| > |E_3| = 1$. *Then, create the partial solutions* E_3, $\{e_1\}$, *and* $\{e_2\}$. *For each* $1 \leq i \leq 2$, *if* E_i *does not contain an A-bridge, then additionally create the partial solutions* E_i. *Discard all partial solutions containing non-relevant A-bridges.*

Branching Rule 5 (based on "CLAW3" in [10]) *Let* $|E_3| > 1$. *Then, create the partial solutions* $\{e_1\}$, $\{e_2\}$, *and* $\{e_3\}$. *For each* $1 \leq i \leq 3$, *if* E_i *does not contain an A-bridge, then additionally create the partial solutions* E_i. *Discard all partial solutions containing non-relevant A-bridges.*

The correctness proof of Branching Rules 2–5 is based on the correctness of the original rules [10]. We can show that the worst-case branching vector $(1, 1, 1, 2, 2, 2)$ (see Figure 2) is matched by our branching rules.

Lemma 2. *If one of the branching partners of v is not in $G^{\bigcirc}$, then the branching number of Branching Rules 2–5 is at most 2.733. Otherwise, removing the edges of a partial solution created by one of Branching Rules 2–5 decreases the feedback edge set number of G by $|S|$.*

3.2   Reducing the Remaining Graph

A reason why our branching rules could not be applied is that branching partners must not be incident to B-bridges (see Definition 1). Thus, we present a way to deal with B-bridges in reduced graphs. To this end, consider the tree that results from contracting each connected component of $G^{\bigcirc}$ and making two components adjacent if there is a bridge between them in G. We call this tree the "component tree" T^C of G. By considering a leaf in T^C, we can limit the possibilities for branching structures to contain B-bridges. Note that T^C can be computed in linear time.

In the following, we consider graphs G that are reduced with respect to the presented data reduction and branching rules. Consider a leaf L in the component tree T^C of G and let u denote the only vertex in L that is incident to a B-bridge in G. We can observe the following properties of G.

Observation 4 *Let G be reduced with respect to all presented branching and reduction rules. Let L be a leaf in T^C and let $u \in L$ with $N_G(u) \not\subseteq L$. Then,*
(a) $G[L]$ does not contain B-bridges;
(b) $|N_{G^{\bigcirc}}(u)| = 2$ because, otherwise, we could apply a branching rule to u;

(c) each vertex $x \in L \setminus N_{G^{\bigcirc}}[u]$ has at most two non-leaf neighbors in G since, otherwise, we could apply a branching rule to x;

(d) the two vertices in $N_{G^{\bigcirc}}(u)$ have the same degree in $G^{\bigcirc}$, since by (c) and (b) all degree-2 paths starting in one of them must end in the other.

Observation 4 fixes the structure of $G[L]$ which we can exploit with the following data reduction rule. In the following, we call a vertex *dirty* if its pendant tree is not a leaf, a singleton or a Y-graph. Let $\{v, w\} = N_{G^{\bigcirc}}(u)$ such that if w is dirty, then so is v.

Reduction Rule 1 *If* $\deg_{G^{\bigcirc}}(v) = 2$ *and* w *is dirty, then delete an edge of* $G^*[L]$ *with maximum distance to* u. *Otherwise, delete* $\{u, v\}$. *In both cases, decrement* k.

Since Reduction Rule 1 can be applied whenever none of the other reduction or branching rules can be applied, applying all presented rules exhaustively solves the input instance. The worst-case branching vector corresponds to Branching Rule 5 and is $(1, 1, 1, 2, 2, 2)$. This implies a search tree with 3.8^f nodes.

Theorem 1. TWO-LAYER PLANARIZATION *can be solved in* $O(3.8^f \cdot f^2 + f \cdot |G|)$ *time, where* f *denotes the feedback edge set number of the input graph.*

4   Heuristic Speedups and Experimental Results

4.1   Heuristic Speedups

In the following, we describe heuristic tricks that we used in our implementation to speed up the computation of the size of an optimal solution.

Observe that the correctness proofs of Suderman [10] for the branching rules we employ are not limited to $|E_i| \leq 2$. They work just as well if E_i contains all edges incident to v_i except $\{v, v_i\}$. Hence, we extended the sets E_i accordingly. Note that, since the new sets E'_i are supersets of the sets E_i, the branching vectors of Branching Rules 2-5 improve.

In each search-tree node, we are challenged with finding a "good" branching structure to continue our search for an optimal solution. A branching structure is *good* if the smallest branching number of any applicable branching rule is small. Our strategy is to find all reasonable branching structures, sort them by their branching number, and branch using the best possible branching rule.

In each search-tree node, we use a linear-time algorithm of Tarjan [13] to find and mark all bridges in the current graph G. This algorithm is also capable of detecting whether G is disconnected. If G contains multiple connected components, then an optimal solution is split among them, allowing us to return the sum of the sizes of optimal solutions for each component.

We keep track of an optimal solution found by our algorithm so far. If any branch cannot contain a better solution, then we cancel the branching and return failure back to the parent of the search-tree. Lower-bound techniques are used to determine whether a better solution is possible in this branch. We tested different algorithms and found that the best overall performance was delivered by simply using the feedback edge set number f as a lower bound. On the one hand, this is not a good bound, since f can be far from the solution size k. On the other hand, f can be computed very quickly.

If a partial solution contains a single edge e then, after searching the search-subtree corresponding to the deletion of this edge, we can exclude e from further branching, thereby improving the branching vectors.

4.2   Experiments

For comparability of results, we followed the example of Suderman and White-sides [11] and included the size of the search tree in the results, since this value is a measure of speed that depends only on the algorithm, not the hardware.

The tests were run on an Intel(R) Xeon(R) E5-1620 CPU at 3.6GHz without taking advantage of the multiprocessor capabilities. The systems were running Debian Linux 3.2 with GNU libc 2.13 and gcc 4.7.1. The program was compiled with CFLAGS=-march=native -msahf -O3. Each run was canceled after 600s.

Instance Generation. We studied two test-case scenarios. First, we reproduced the generated instances used by Mutzel [9], Suderman and Whitesides [11], and Suderman [10] (where detailed descriptions on reproducing the instances can be found). This test set comprises 1700 "dense" bipartite graphs $(V_1 \uplus V_2, E)$ with $|V_1| = |V_2| = 20$ and $|E|$ between 20 and 100 and 900 "sparse" bipartite graphs with $|V_1| = |V_2|$ between 20 and 100 and $|E| = 2|V_1|$.

The second set of instances comprises 1500 large, sparse graphs that were generated by for each $n \in \{100i \mid 1 \leq i \leq 10\}$ and each $p \in \{3\%, 6\%, 9\%, 12\%, 15\%\}$, constructing a tree on n vertices and adding $p \cdot n$ edges uniformly at random. If some insertion failed because the edge was already present, we repeated the insertion with new random values so that the graphs are guaranteed to contain $n - 1 + p \cdot n$ edges. The results can be found in Table 2 in the appendix.

Results. The results obtained by the branching algorithm of Suderman [10] and the ILP formulation of Jünger and Mutzel [6] (which had a 300s timeout) are compared to the results of our implementation in Table 1. First, consider the set of "dense" graphs (first 12 rows). Although our average running times rival those of Suderman's algorithm, it is important to realize that these results were obtained in 2005 on a 1GHz Pentium III computer [10]. Thus, we can conclude that our algorithm does not perform as well as theory suggests. In the following, we identify possible reasons hopefully leading to future improvements.

1. The tested graphs do not fit well in the picture we painted in the introduction. More precisely, their feedback edge set number f differs from the solution size k by at most 2.

2. Looking at the search-tree sizes for the algorithm of Suderman, it quickly becomes apparent that they differ only marginally for all k between 14 and 61 leading to the conjecture that the search-tree sizes are influenced by some other, hidden factor.

Table 1. Results of the first test. The columns labeled "%" give the percentages of instances solved within the respective timelimit. "∅" indicates an average over *these* instances, while "med." indicates the median over *all* instances. Columns labeled "steps" contain the numbers of explored search-tree nodes. Columns labeled "t" contain running-times in seconds.

$|V_i|$	$|E|$	∅k	∅f	ILP ∅ t	3.562^k algo. [10] ∅ t	∅ steps	%	3.8^f time algorithm ∅ t	med. t	∅ steps	med. steps	%
20	45	11	9	26	0	85	100	0	0	157	80	100
20	50	14	13	100	4	4,694	100	0	0	897	253	100
20	55	18	17	81	1	946	100	0	0	2,417	102	100
20	60	23	22	56	5	6,232	100	1	0	18,596	128	100
20	65	27	27	54	3	3,645	97	8	0	305,501	117	99
20	70	32	31	26	7	8,263	99	14	0	489,962	107	98
20	75	37	36	22	2	2,249	100	4	0	147,080	85	99
20	80	41	41	12	2	2,060	99	1	0	27,630	88	99
20	85	46	46	20	5	5,366	100	2	0	82,563	139	99
20	90	51	51	8	6	6,503	99	2	0	89,623	77	99
20	95	56	55	4	8	8,276	99	3	0	126,372	84	97
20	100	61	60	4	4	5,243	98	1	0	37,733	109	96
20	40	7	6	6	0	95	100	0	0	24	13	100
30	60	11	10	49	0	356	100	0	0	231	73	100
40	80	16	13	150	3	3,002	100	0	0	1,546	212	100
50	100	19	16	-	14	11,876	99	1	0	21,754	403	99
60	120	24	19	-	64	48,240	96	2	0	37,182	3,852	99
70	140	28	23	-	129	91,339	88	6	1	112,015	11,098	99
80	160	31	26	-	-	-	-	22	2	339,282	37,172	90
90	180	35	29	-	-	-	-	44	4	661,619	60,600	91
100	200	38	32	-	-	-	-	74	26	1,097,335	323,228	81

3. Suderman employs a very tight lower bound to cancel branches that cannot yield a better solution than what was already computed. Our lower bound, however, is simply the feedback edge set number of the current graph. If this was indeed the cause of the observed difference in search-tree sizes, then we could just replace our crude lower bound with Suderman's.

4. Finally, Suderman describes a sophisticated divide-and-conquer technique based on "p-components". While a mathematical analysis of this technique is open, Suderman described it as very effective and should be incorporable in our algorithm as well.

On the "sparse" instances (last 9 rows of Table 1), we expected our algorithm to perform better than on the set of dense graphs. In fact, we were able to solve a good portion of the larger instances that could not be solved in the past. On the one hand, this may again be due to our hardware advantage. On the other hand, we observe a larger divergence between the parameters f and k.

A closer inspection of the running times of our implementation reveals that averages do not reflect the behavior of our algorithm very well. Therefore, we also provided median running times and search-tree sizes in Table 1 and draw

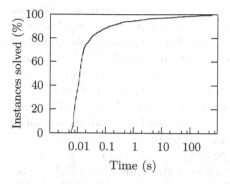
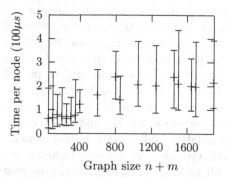

Fig. 3. Cumulative distribution of running times of our algorithm on "dense" graphs (first 12 rows of Table 1). It plots the percentage of instances solved before time x versus the time x.

Fig. 4. Average, minimum, and maximum time spent per search-tree node by our algorithm over all experiments. Only search trees with at least 10 nodes are considered.

the cumulative distribution function of running times on the set of dense graphs in Figure 3. Notice the striking difference between average and median running times and search-tree sizes that differ by a factor of up to $4,500$. Figure 3 shows that, after about one second, more than 90% of all "dense" instances were solved. Table 1 and Figure 3 raise hope that our algorithm will perform well on a wide range of inputs.

While the tested instances are very sparse, we also note that Suderman [10] performed tests on instances with $|E|/|V| = 0.6$. Furthermore, our algorithm is *designed* to run on sparse graphs, making dense graphs an unreasonable input.

Last but not least, we want to get a glimpse of the efficiency of our implementation of the kernelization with respect to the feedback edge set number [14]. To this end, we plotted the time per search-tree node versus the size of the input graph in Figure 4. Although the time per search-tree node is also influenced by our elaborate method of selecting the best possible branching vector first, we estimate that the application of the reduction rules dominates the running time. Although no clear trend can be made out, times between $100\mu s$ and $200\mu s$ per search-tree node can be observed for all input sizes, suggesting a rather slowly growing function.

5   Conclusion

In this work, we presented a branching algorithm solving the Two-Layer Planarization problem in $O(3.8^f f^2 + f \cdot |G|)$ time, where f denotes the feedback edge set number of the input graph. Although the theoretical advantages of our algorithm are apparent, our implementation does, on average, not deliver the desired results on "dense" inputs. However, our algorithm is designed to perform well on *sparse* graphs, which we could demonstrate. Results of our tests indicate that Suderman's heuristic speedups enable his implementation to outperform ours, especially the "p-component" technique which we consider a reasonable future addition to our implementation as well. This can be seen as a general lesson that heuristic speedups are equally important for developing fast solvers as theoretical considerations. We interpreted the slow increase in the time per search-tree node as an indicator for the efficiency of our kernelization implementation. It would be interesting to provide a theoretical analysis thereof.

In further theoretical development, it is desirable to search for fixed-parameter algorithms (and problem kernels) for parameters upper-bounded by f. The feedback vertex set number and the odd cycle transversal number would be canonical candidates. Additionally, it may be interesting to investigate the parameter $k - f$ that represents an "above guarantee" parameter for the problem.

Other interesting problems in the context of Sugiyama's algorithm [12] are the multilayered problem versions [3] and ONE LAYER PLARARIZATION [2, 5]. Are they also fixed-parameter tractable with respect to the feedback edge set number? Another variant of 2LP is obtained by replacing edge deletion as the allowed graph modification operation by the so-called "node duplication" operation[3], yielding the NODE DUPLICATION BASED CROSSING ELIMINATION problem [1].

References

[1] Chaudhary, A., Chen, D.Z., Whitton, K., Niemier, M.T., Ravichandran, R.: Eliminating wire crossings for molecular quantum-dot cellular automata implementation. In: Proc. 2005 ICCAD, pp. 565–571. IEEE Computer Society (2005)

[2] Dujmović, V., Fellows, M., Hallett, M., Kitching, M., McCartin, G.L.C., Nishimura, N., Ragde, P., Rosamond, F., Suderman, M., Whitesides, S., Wood, D.R.: A fixed-parameter approach to 2-layer planarization. Algorithmica 45(2), 159–182 (2006)

[3] Dujmović, V., Fellows, M., Hallett, M., Kitching, M., McCartin, G.L.C., Nishimura, N., Ragde, P., Rosamond, F., Suderman, M., Whitesides, S., Wood, D.R.: On the parameterized complexity of layered graph drawing. Algorithmica 52(2), 267–292 (2008)

[4] Eades, P., Whitesides, S.: Drawing graphs in two layers. Theor. Comp. Sci. 131(2), 361–374 (1994)

[5] Fernau, H.: Two-layer planarization: Improving on parameterized algorithmics. J. Graph Algorithms Appl. 9(2), 205–238 (2005)

[6] Jünger, M., Mutzel, P.: 2-layer straightline crossing minimization: Performance of exact and heuristic algorithms. J. Graph Algorithms Appl. 1 (1997)

[7] Koenig, P.-Y., Melançon, G., Bohan, C., Gautier, B.: Combining DagMaps and Sugiyama layout for the navigation of hierarchical data. In: Proc. 11th IV, pp. 447–452. IEEE Computer Society (2007)

[8] Lengauer, T.: Combinatorial Algorithms for Integrated Circuit Layout. Wiley (1990)

[9] Mutzel, P.: An alternative method to crossing minimization on hierarchical graphs. SIAM J. Optim. 11(4), 1065–1080 (2001)

[10] Suderman, M.: Layered Graph Drawing. PhD thesis, School of Computer Science, McGill University Montréal (2005)

[11] Suderman, M., Whitesides, S.: Experiments with the fixed-parameter approach for two-layer planarization. J. Graph Algorithms Appl. 9(1), 149–163 (2005)

---

[3] Duplicating a vertex v means deleting v and adding u, w with $N(u) \uplus N(w) = N(v)$.

[12] Sugiyama, K., Tagawa, S., Toda, M.: Methods for visual understanding of hierarchical system structures. IEEE Trans. Syst., Man, Cybern. 11(2), 109–125 (1981)
[13] Tarjan, R.E.: A note on finding the bridges of a graph. Inf. Process. Lett. 2(6), 160–161 (1974)
[14] Uhlmann, J., Weller, M.: Two-layer planarization parameterized by feedback edge set. Theor. Comp. Sci. (2013), doi:10.1016/j.tcs.2013.01.029
[15] Waterman, M.S., Griggs, J.R.: Interval graphs and maps of DNA. Bulletin of Mathematical Biology 48(2), 189–195 (1986)

Appendix

Observation 1. *Let G be reduced with respect to the kernelization [14] and let T^v be the pendant tree of a vertex v of G^* such that T^v is not a Y-graph. Then, there is an optimal solution for G that does not delete any edge of T^v.*

Proof (of Observation 1). For the sake of contradiction, assume that there is an optimal solution S^* containing an edge e of T^v. Lemma 1 allows us to assume that e is incident to v. Deleting e splits G into two components, one of which, say T', is a subgraph of T^v. Let $S' := S^* \setminus \{e\}$ and note that S' is not a solution for G, that is, there is a 2-claw centered at a vertex $u \in N_{G-S'}[v]$. With Lemma 1, it is easy to see that $u \notin V(T^v) \setminus \{v\}$. Hence, deleting $\{v, u\}$ from $G - S'$ decreases the non-leaf degree of v. Thus, all vertices whose non-leaf degree in $G - S^*$ is larger than their non-leaf degree in $G - S' - \{v, u\}$ are in T', contradicting that no 2-claw is centered in T'. $\qquad\square$

$N_i \leftarrow$ non-bridges incident to v_i except for $\{v, v_i\}$;
if $\deg_{G\bigcirc}(v_i) = 2$ **then**
   $E_i \leftarrow N_i$;
   if *there is a bridge a incident to v_i* **then** add a to E_i;
else if $\deg_{G\bigcirc}(v_i) \geq 3$ **then**
   $E_i \leftarrow$ a non-bridge in N_i that respects Definition 2(iii);
   add a non-bridge in $N_i \setminus E_i$ to E_i;

Algorithm 1. An algorithm that, given v and v_i, computes E_i

Lemma 3. *Branching rules 2–5 are correct, that is, for each of these branching rules, one of the created partial solutions can be extended to an optimal solution for G.*

Proof (of Lemma 3). For the sake of contradiction, assume that none of the created partial solutions can be extended to an optimal solution for G. Hence, by correctness of the branching rules of Suderman [10], there is some $1 \leq i \leq 3$ such that E_i is not created and E_i can be extended to an optimal solution S^* for G. Since E_i is not created, E_i contains an A-bridge b. Hence, Definition 2(iv) implies $\deg_{G\bigcirc}(v_i) = 2$ and, thus, Observation 2 is applicable. Then, however, we can replace b with e_i in S^*, implying that the partial solution $\{e_i\}$, which is created in all branching rules in question, can be extended to an optimal solution for G. $\qquad\square$

Lemma 2. *If one of the branching partners of v is not in $G^\bigcirc$, then the branching number of Branching Rules 2–5 is at most 2.733. Otherwise, removing the edges of a partial solution created by one of Branching Rules 2–5 decreases the feedback edge set number of G by $|S|$.*

Proof (of Lemma 2). First, consider $V' := \{v_1, v_2, v_3\} \setminus V(G^\bigcirc)$.

By Definition 1, no $\{v, v_i\}$ is a relevant A-bridge or a B-bridge Hence, by the statement of the branching rules, all partial solutions containing edges incident to vertices in V' are discarded. Thus, if $|V'| \geq 2$, then the worst-case branching vector of the branching rules is $(1,1)$, corresponding to a branching number of two. Hence, assume $|V'| = 1$. Consider $|E_i|$ for all $v_i \notin V'$ and note that E_i does not contain a bridge of G because, by Definition 1, branching partners of v are not incident to B-bridges and the branching rules do not create E_i if it contains A-bridges. By Definition 2(iii), deleting E_i decreases the feedback edge set number by $|E_i|$.

In the following, let $V' = \{v_x\}$ and recall that E_x and any partial solution containing e_x contain non-relevant A-bridges and are, therefore, discarded. Consider the branching rules separately:

Case 1. 1 Branching Rule 2 applies to v. Then, by symmetry, we may relabel v_1, v_2, v_3 such that $v_x = v_2$. This implies a branching vector of $(1,1)$ corresponding to a branching number of two.

Case 2. 2 Branching Rule 3 applies to v. Then, either $v_x = v_1$, implying a branching vector of $(1,1,2)$ or $v_x \in \{v_2, v_3\}$, implying a branching vector of $(1,1,2)$. In both cases, the branching number is 2.415.

Case 3. 3 Branching Rule 4 applies to v. Then, either $v_x \in \{v_1, v_2\}$, implying a branching vector of $(1,1,2)$ or $v_x = v_3$, implying a branching vector of $(1,1,2,2)$. In both cases, the branching number does not exceed 2.733.

Case 4. 4 Branching Rule 4 applies to v. Then, the branching vector is $(1,1,2,2)$, corresponding to a branching number of 2.733.

Next, let $v_1, v_2, v_3 \in V(G^\bigcirc)$ and let S be a partial solution created by one of Branching Rules 2–5.

Consider the case that $|S| = 1$, that is, $S = \{e\}$. If e is incident to v, then, since $v_1, v_2, v_3 \in V(G^\bigcirc)$, e is not a bridge. If e is not incident to v, then, $e \in E_i$ for some $1 \leq i \leq 3$. Since partial solutions E_i are only created if they do not contain A-bridges and, by Definition 1, e is not a B-bridge, e is not a bridge. In both cases, removing e decreases the feedback edge set number by one.

In the following, we assume $|S| = 2$, that is $S = \{e_1, e_2\}$. First, let S consist of edges incident to v. Since $v_1, v_2, v_3 \in V(G^\bigcirc)$, neither e_1 nor e_2 are bridges.

Towards a contradiction, assume that deleting an edge of S makes the other a bridge. Note that either $S = \{e_1, e_2\}$ in Branching Rule 2 or $S = \{e_2, e_3\}$ in Branching Rule 3. The first case contradicts Observation 3(2), the second case contradicts Observation 3(3).

In the following, we assume that the edges of S are not incident to v, that is, $S = E_i$ for some $1 \leq i \leq 3$. Then, by Definition 1, E_i does not contain B-bridges and by the statements of the branching rules, E_i does not contain A-bridges. Thus, E_i consists of two non-bridges. Hence, $\deg_{G^\bigcirc}(v_i) \geq 3$ and, by Definition 2(iii), there is an edge $e = \{v_i, u\}$ in E_i such that there is a path from u to v that avoids v_i. Let $E_i = \{e, e'\}$. Clearly, deleting e' does not make e a bridge, since, otherwise, all paths from u to v would contain e' and, therefore, also v_i, contradicting the choice of e. Hence, deleting E_i decreases the feedback edge set number by two. $\qquad\square$

Lemma 4. *Reduction Rule 1 is correct and can be applied in linear time.*

Proof (of Lemma 4). First, consider the case that $\deg_{G^\bigcirc}(v) = 2$ and w is dirty. Then, by choice of v, also v is dirty. Hence, by Observation 4(c), $G^*[L]$ is a degree-2 path from u to u. There are no dirty vertices in $L \setminus \{u, v, w\}$, since otherwise, we could apply a branching rule to this vertex. Hence, by reducedness with respect to Path Reduction Rule 4 [14], the 2-claws centered in v and w in G overlap in an edge with maximum distance to u. Deleting an edge from $G[L]$ that does not have maximum distance to u does not destroy both 2-claws centered at v and w. Clearly, if any optimal solution contains two edges of $G[L]$, then replacing these two edges with an edge with maximum distance to u and the B-bridge incident to u yields an optimal solution for G.

In the following, we assume that $\deg\, G^\bigcirc(v) > 2$ or w is not dirty. We prove that there is an optimal solution S for G that contains $\{u, v\}$. We will use the following arguments in the proof.

(1) If there is an optimal solution for G containing $\{u, v\}$, then we are done. Hence, we assume that no optimal solution contains $\{u, v\}$. Then, however, for all solutions S, all 2-claws in $G - ((S \setminus E(G[L]) \cup \{\{u, v\}\})$ have their center in L, that is, "shuffling" edge deletions in $G[L]$ does not create a 2-claw whose center is not in L, as long as $\{u, v\}$ is deleted.
(2) Let S^* be an optimal solution for G and assume that, for some pendant Y-graph Y of a vertex z in $G^* - v$, S^* contains $\{z, \text{center}(Y)\}$. Then, there is an optimal solution for G containing $\{u, v\}$ if and only if there is an optimal solution for $G - \{z, c(Y)\}$ containing $\{u, v\}$. Thus, in the following, we assume that no optimal solution contains an edge of a pendant Y-graph except for the pendant Y-graph of v.
(3) If there is a vertex z with a pendant Y-graph in L, then, by (2), all optimal solutions contain $\deg_G(z) - 1$ edges incident to z. However, deleting $\deg_G(z) - 2$ of these makes the last one a bridge. Hence, the feedback edge set number decreases by only $\deg_G(z) - 2$. Thus, the feedback edge set number of $G[L]$ plus the number of pendant Y-graphs in $G[L]$ is a lower bound for the size of a solution for $G[L]$.

Table 2. Detailed results for running times and search-tree sizes of our $O(3.8^f f^2 + f(n+m))$-time algorithm run on the second batch of tests. The instances consist of a random tree on the vertex set V augmented by $p \cdot |V|$ edges. The fifth column labeled "#" shows the number of instances that were solved within the timelimit of $600s$. For each row, 30 instances were generated. The minimum running times never exceeded $10ms$ and were therefore dropped from the table. Since maxima are not meaningful if the timelimit of $600s$ was hit, maximum running times are omitted and, for maximum search-tree sizes, we just give a lower bound. Herein, $> 2M$ means that a canceled process had explored over 2 million search-tree nodes at the point of termination. Note that averages also loose meaning in this case, but medians do not. For graphs with up to 1000 vertices and $|E|/|V| \leq 1.12$, our algorithm always finishes within half a second. However, for $|E|/|V| = 1.15$, we could not solve all instances containing 600 vertices. Again, the median running times paint a brighter picture. Half of all input instances with 1000 vertices and $|E|/|V| = 1.15$ were solved after about 2 minutes.

| | | | | | running time (s) | | | search-tree size | | | |
$|V|$	p	f	k	#	max	median	avg	min	max	median	avg
100	12%	12	20	30	0.01	0.01	0.01	1	65	5	8
200	12%	24	40	30	0.06	0.01	0.01	1	506	9	50
300	12%	36	61	30	0.24	0.01	0.03	1	2,529	18	173
400	12%	48	81	30	0.06	0.01	0.02	1	327	8	53
500	12%	60	103	30	0.68	0.01	0.04	1	2,569	12	122
600	12%	72	123	30	0.28	0.01	0.03	1	1,234	7	123
700	12%	84	143	30	0.43	0.01	0.04	1	3,882	11	211
800	12%	96	164	30	5.55	0.01	0.37	1	28,684	3	2,013
900	12%	108	185	30	5.30	0.01	0.26	1	23,007	11	1,143
1000	12%	120	206	30	0.49	0.03	0.07	1	1,466	10	172
100	15%	15	22	30	0.06	0.01	0.01	1	873	22	72
200	15%	30	44	30	0.36	0.02	0.06	1	3,249	36	455
300	15%	45	66	30	4.23	0.04	0.43	1	40,287	174	4,475
400	15%	60	88	30	42.05	0.34	3.60	17	255,928	2,862	26,277
500	15%	75	111	30	409.56	0.39	21.38	1	2,774,889	2,592	143,482
600	15%	90	134	27		6.04	32.38	18	> 2M	41,900	194,055
700	15%	105	156	24		21.65	64.31	202	> 2M	112,614	317,224
800	15%	120	180	24		13.24	61.55	3	> 2M	95,136	347,082
900	15%	135	201	22		65.93	73.47	22	> 2M	249,645	368,608
1000	15%	150	225	16		109.78	20.54	104	> 1M	316,186	61,974

(4) u is not dirty, since otherwise, by Observation 4(b) the non-leaf degree of u in $G - b$ is at least three, implying that we could apply a branching rule to u.

Let f_L denote the feedback edge set number in $G[L]$ and let $Z \subseteq E$ denote the set of all relevant A-bridges in $G[L]$. Then, by Observation 4(c and d), $f_L = \deg_{G^*}(v) - 1$. By (3), an optimal solution for $G[L]$ contains at least $|Z| + \deg_{G^*}(v) - 1$ edges. We construct a solution S for $G[L]$ that contains $\{u, v\}$ and matches this lower bound and show that S can be extended to an optimal solution for G.

If $f_L > 2$, then, by Observation 4(c), $\deg_{G^O}(v) > 3$, implying that we could apply a branching rule to v in G. Hence, in the following, we assume that $1 \leq f_L \leq 2$. If $f_L = 1$, let $S := Z \cup \{\{u, v\}\}$ and note that, by definition, w is not dirty. If $f_L = 2$, let $S := Z \cup \{\{u, v\}, \{w, z\}\}$ for some $z \in N_{G^O}(w) \setminus \{u\}$ and note that w is not dirty, since otherwise, we could apply a branching rule to w in $G - \{u, w\}$. Since in both cases $\deg_{G^O - S}(w) < 3$ and $Z \subset S$, we conclude that w has at most two non-leaf neighbors in $G - S$.

We show that S is a solution for $G[N_G[L]]$. If this is not the case, then there is a 2-claw centered at some $x \in L$ in $G[N[L]] - S$. Clearly, $x \in N_{G[L]}[u]$, since, otherwise, we could have applied a reduction rule to x in G. By (4), $x \neq u$. Since $Z \subset S$, we conclude $x \in \{v, w\}$. Since w has at most two non-leaf neighbors in $G - S$, we conclude $x \neq w$ and, hence, $x = v$. However, since $\{u, v\}$ is not in $G[L] - S$, we could apply a reduction rule to v in G, contradicting reducedness of G. Since $|S| = |Z| + f_L$, by (3), S is an *optimal* solution for $G[N[L]]$. Let S^* denote an optimal solution for G and let b denote the B-bridge incident to u. If $|S^* \cap E(G[N[L]])| > |S|$, then $S \cup \{b\}$ can be extended to an optimal solution for G. Otherwise, $G - (S \cup (S^* \setminus E(G[N[L]])))$ contains a 2-claw, which, by (1) and S being an optimal solution for $G[N[L]]$, is centered at u and contains b. However, by (4), the 2-claw contains $\{u, w\}$ and $\{u, v\}$, contradicting $\{u, v\} \in S$.

To prove that Reduction Rule 1 can be applied in linear time, recall that all leaves of the component tree of G can be found in linear time and then solved individually. Clearly, for each leaf L, we can check the conditions in $O(1)$ time and apply the deletion in $O(|L|)$ time, implying linear time overall. $\qquad\square$

In-Out Separation and Column Generation Stabilization by Dual Price Smoothing

Artur Pessoa[1], Ruslan Sadykov[2,3], Eduardo Uchoa[1], and Francois Vanderbeck[3,2]

[1] LOGIS , Universidade Federal Fluminense, Brazil
[2] INRIA Bordeaux, team RealOpt, France
[3] University of Bordeaux, Institut of Mathematics, France

Abstract. Stabilization procedures for column generation can be viewed as cutting plane strategies in the dual. Exploiting the link between in-out separation strategies and dual price smoothing techniques for column generation, we derive a generic bound convergence property for algorithms using a smoothing feature. Such property adds to existing in-out asymptotic convergence results. Beyond theoretically convergence, we describe a proposal for effective finite convergence in practice and we develop a smoothing auto-regulating strategy that makes the need for parameter tuning obsolete. These contributions turn stabilization by smoothing into a general purpose practical scheme that can be used into a generic column generation procedure. We conclude the paper by showing that the approach can be combined with an ascent method, leading to improved performances. Such combination might inspire novel cut separation strategies.

Keywords: Column Generation, Stabilization, Cutting Plane Separation.

Introduction

Separation strategies from the cut generation literature and algorithmic strategies for stabilization in column generation algorithms are dual counterparts. The pricing procedure in column generation is understood as a separation routine for the master dual. Therefore, efficient strategies to define the separation point or select cuts translate into stabilization techniques and column generation strategies, as emphasized in several papers including [2,3,8,9]. In this paper, we specifically formalize the link between in-out separation [2,5] and dual price smoothing techniques whereby the price vector used for column generation is defined as a combination of the optimal solution over the current polyhedral approximation of the master dual (denoted π^{out} hereafter) and a feasible dual solution for the true master (denoted π^{in}). We show that dual price smoothing schemes (such as that of [6,10]) can be understood as an extension of in-out separation, introducing an in-point updating strategy that relies on a valid dual bound computation. Note that dual price smoothing addresses at once the dual oscillations, tailing-off, and degeneracy drawbacks of the column generation procedure. It acts through both smoothing and centralization, and it is simple to implement. Our work brings an additional quality to smoothing. Our proposal for a parameter self-adjusting scheme allows one to avoid the drawback of many alternative stabilization approaches (such as the popular piecewise linear penalty functions [4]) that require fine tuning of several parameters.

More specifically, the contributions of the paper are:

V. Bonifaci et al. (Eds.): SEA 2013, LNCS 7933, pp. 354–365, 2013.

- Establishing the detailed properties of a generic smoothing scheme (that encompasses several variants), including a bound convergence property that has no equivalent in a general in-out procedure. For already existing results, the link with in-out separation has lead to simpler proofs under weaker conditions.
- Proposing a simple scheme for dealing in practice with a sequence of *mis-pricings* (a mis-pricing is a failure to separate π^{out}) that impairs convergence.
- Developing a parameter self-adjusting scheme for automatic tuning that uses gradient information. The scheme is shown to experimentally reproduce the best results obtained by fine tuning the single but critical parameter of the smoothing procedure, essentially making the method parameter-tuning-free. We emphasize that the performance of smoothing techniques, and more generally stabilization techniques, highly depends on proper parameter tuning that is moreover instance dependent. Hence, our automated scheme has practical significance, transforming smoothing into a general purpose technique well suited for a generic branch-and-price solver.
- Extending the smoothing paradigm by combining it with an ascent method that is experimentaly shown to lead to significant improvements.

The paper places dual price smoothing as a key technique in the context of existing column generation stabilization strategies. The features that we introduced in smoothing techniques could inspire dual strategies for cutting plane separation. An extended version of the paper with proposition proofs, illustrative figures and details on experimental test instances is available on the authors' web page.

1 Column Generation

Below we review the main concepts underlying column generation approaches in order to emphasize the properties on which smoothing schemes rely. Consider the integer program:

$$[F] \equiv \min\{cx : x \in X\} \tag{1}$$

where

$$X := Y \cap Z \text{ with } Y := \{x \in \mathbb{R}_+^n : Ax \geq a\} \text{, and } Z := \{x \in \mathbb{N}^n : Bx \geq b, l \leq x \leq u\}.$$

In the decomposition of system X, it is assumed that Z defines a "tractable" subproblem (assumed to be non-empty and bounded to simplify the presentation), but $Ax \geq a$ are "complicating constraints". In other words, we assume that subproblem

$$[SP] \equiv \min\{cx : x \in Z\} \tag{2}$$

is "relatively easy" to solve compared to problem [F]. Then, a natural approach to solve [F], or to estimate its optimal value, is to exploit our ability to optimize over Z. We review this technique below.

Let Q be the enumerated set of subproblem solutions (it is a finite set given the boundedness of Z), i.e. $Q = \{z^1, \ldots, z^{|Q|}\}$ where $z^q \in Z$ is a subproblem solution vector. Abusing notations, $q \in Q$ is used hereafter as a short-cut for $z^q \in Q$. Thus, we can reformulate Z and $conv(Z)$, the convex-hull of the integer solution to Z, as:

$$Z = \{x \in I\!R_+^n : x = \sum_{q \in Q} z^q \lambda_q, \sum_{q \in Q} \lambda_q = 1; \lambda_q \in \{0,1\} \ \forall q \in Q \}, \quad (3)$$

$$conv(Z) = \{x \in I\!R_+^n : x = \sum_{q \in Q} z^q \lambda_q, \sum_{q \in Q} \lambda_q = 1, \lambda_q \geq 0 \ \forall q \in Q \}. \quad (4)$$

Note that $conv(Z)$ defines an ideal formulation for Z. Hence, [SP] can be rewritten as:

$$[\text{SP}] \equiv \min\{cx : x \in Z\} \equiv \min\{cz^q : q \in Q\} \equiv \min\{cx : x \in conv(Z)\}. \quad (5)$$

Exploiting the assumption that the subproblem is tractable, one can derive dual bounds for the original problem [F] by Lagrangian relaxation of the constraints $Ax \geq a$. For any Lagrangian penalty vector $\pi \in I\!R_+^m$, the *Lagrangian function*,

$$L(\pi, x) := \pi \, a + (c - \pi A)x, \quad (6)$$

is optimized over Z to yield a valid dual bound on [F], by solving the *Lagrangian subproblem*:

$$[\text{LSP}(\pi)] \equiv L(\pi) := \min_{x \in Z} L(\pi, x). \quad (7)$$

The *Lagrangian dual function* is defined by $L : \pi \in I\!R_+^m \to L(\pi)$. Maximizing function L leads to the best dual bound that can be derived from the Lagrangian relaxation. The *Lagrangian dual problem* is defined as:

$$[\text{LD}] \equiv \max_{\pi \in I\!R_+^m} L(\pi). \quad (8)$$

The Lagrangian dual problem can be reformulated as a max-min problem, or as a linear program:

$$[\text{LD}] \equiv \max_{\pi \in I\!R_+^m} \min_{x \in Z} \{\pi \, a + (c - \pi A)x\}; \quad (9)$$

$$\equiv \max\{\eta, \quad (10)$$

$$\eta \leq cz^q + \pi(a - Az^q) \quad \forall q \in Q, \quad (11)$$

$$\pi \in I\!R_+^m, \eta \in I\!R^1\}; \quad (12)$$

$$\equiv \min\{\sum_{q \in Q} (c \, z^q) \lambda_q, \quad (13)$$

$$\sum_{q \in Q} (Az^q) \lambda_q \geq a, \quad (14)$$

$$\sum_{q \in Q} \lambda_q = 1, \quad \lambda_q \geq 0 \quad \forall q \in Q\}; \quad (15)$$

$$\equiv \min\{cx : Ax \geq a, \ x \in conv(Z) \}. \quad (16)$$

The *Dantzig-Wolfe reformulation* is a valid reformulation of [F] expressed in terms of variables λ_q that were introduced for the reformulation of Z given in (3). Its linear programming (LP) relaxation, which we denote by [M], is precisely the form (13-15) of [LD].

A *column generation* procedure to solve [M] proceeds as follows. At a stage t, the restriction of [M] to columns defined from $Q^t = \{z^1, \ldots, z^t\}$ is denoted by [M^t]. This *restricted master LP* is:

$$[M^t] \equiv \min\{\sum_{\tau=1}^{t} c\,z^\tau \lambda_\tau : \sum_{\tau=1}^{t} Az^\tau \lambda_\tau \geq a; \sum_{\tau=1}^{t} \lambda_\tau = 1; \lambda_\tau \geq 0, \ \tau = 1, \ldots, t\} \quad (17)$$

Linear program [M^t] is solved to optimality. Let λ^t denote an optimal solution to [M^t]. Its projection in X is:

$$x^t := \sum_{\tau=1}^{t} z^\tau \lambda_\tau^t \,. \quad (18)$$

Let $c\,x^t$ denote its objective value. The linear program dual of [M^t] is:

$$[DM^t] \equiv \max\{\eta : \pi(Az^\tau - a) + \eta \leq cz^\tau, \ \tau = 1, \ldots, t; \pi \in I\!R_+^m; \eta \in I\!R^1\} \quad (19)$$

Let (π^t, η^t) denote an optimal solution to [DM^t]. Using this dual solution, one searches for the most negative reduced cost column, by solving the subproblem:

$$z^{t+1} \leftarrow z_{\pi^t} := argmin_{x \in Z}\{(c - \pi^t A)x\} \,. \quad (20)$$

If $(c - \pi^t A)\,z_{\pi^t} + \pi^t a - \eta^t < 0$, then z_{π^t} defines a negative reduced cost column that is added to the restricted master. Otherwise, the current LP solution is optimal for the unrestricted master program [M].

The above algorithm outputs a sequence of values for the Lagrangian price vector: $\{\pi^t\}_t$, that converges towards an optimal dual price vector, π^*. In the process, one can also derive a sequence of candidate primal solutions, $\{x^t\}_t$, converging towards an optimal solution x^* of problem (16). One can observe the following properties:

Observation 1

(i) The vector x^t defined in (18) is a solution to [M^t] $\equiv \min\{cx : Ax \geq a, \ x \in conv(\{z^1, \ldots, z^t\})\}$.

(ii) The dual solution of [DM^t] is such that $\pi^t = argmax_{\pi \in I\!R_+^m} L^t(\pi)$ where $L^t()$ defines an approximation of the Lagrangian dual function $L()$, considering only the subset of subproblem solutions $\{z^1, \ldots, z^t\}$: i.e.,

$$L^t() : \pi \rightarrow \min_{z \in \{z^1, \ldots, z^t\}} \{\pi a + (c - \pi A)z\} = \min\{L(z^1, \pi) \ldots, L(z^t, \pi)\} \,. \quad (21)$$

Function $L^t()$ is an upper approximation of function $L()$: $L^t(\pi) \geq L(\pi) \ \forall \pi \in I\!R_+^m$. The hypograph of function $L^t()$ defines a polyhedral outer approximation of the master LP dual program (10-12). By duality, $L^t(\pi^t) = \eta^t = c\,x^t$.

(iii) Solving [LSP(π^t)] exactly serves four purposes simultaneously:

(iii.a) it yields the most negative reduced cost column: $z^{t+1} = z_{\pi^t} \in Q \setminus Q^t$ for [M];

(*iii.b*) *it yields the most violated constraint defined by a subproblem solution* $z^q \in Q \setminus Q^t$ *for [DM];*

(*iii.c*) *the constraint violation of the oracle solution* z_{π^t} *defines a sub-gradient of* $L(.)$ *at point* π^t:

$$g^t := (a - A z_{\pi^t}) \ ; \tag{22}$$

(*iii.d*) *the correct value of the Lagrangian function* $L()$ *is now known at point* π^t: $L(\pi^t) = \pi^t a + (c - \pi^t A) z_{\pi^t}$, *and therefore this value remains unchanged in any further approximation of* $L()$, *i.e.,* $L^\tau(\pi^t) = L(\pi^t) \ \forall \tau > t$.

(*iv*) *At stage* t, $\mathrm{conv}(\{(\pi^\tau, L^{\tau+1}(\pi^\tau))\}_{\tau=1,\dots,t})$ *defines an inner approximation of the master LP dual program (10-12). Outer and inner approximation are equal at these points as* $L^{\tau+1}(\pi^\tau) = L(\pi^\tau)$. *One of these points defines the incumbent dual solution* $\hat{\pi} = \mathrm{argmax}_{\tau=1,\dots,t} L^{\tau+1}(\pi^\tau)$ *with value* $\hat{L} = L(\hat{\pi}) = \hat{\eta}$.

(*v*) *If* $L^t(\pi^t) = \hat{L}$, *or equivalently* $cx^t = \hat{L}$, *then the optimal solution is reached, i.e.,* $\eta^* = \hat{L}$.

In the sequel, (π^*, η^*) denotes an optimal solution to the Lagrangian dual, while $\hat{L} = L(\hat{\pi})$ denotes the current best dual (lower) bound on η^*.

2 Stabilization Techniques in Column Generation

The above column generation procedure, also known as Kelley's cutting plane algorithm for the dual master, yields a sequence of dual solution candidates $\{\pi^t\}_t$ converging towards optimal prices, π^*. The sequence of primal solution candidates $\{x^t\}_t$ is a by-product used to prove optimality of the dual solution. Stabilization techniques are devised to accelerate the convergence of the dual sequence $\{\pi^t\}_t$ towards π^* by targetting the following drawbacks, as listed in [9]:

- *Dual oscillations:* Solutions π^t jump erratically. One extreme solution of the restricted dual master (10-12) at iteration t, [DM^t], is followed by a different extreme point of [DM^{t+1}], leading to a behavior often refered to as *"bang-bang"*. Because of these oscillations, it might be that $||\pi^{t+1} - \pi^*|| > ||\pi^t - \pi^*||$. Moreover, the dual bounds $L(\pi^t)$ are converging non monotically, with ups and downs in the value curve (the *yo-yo* phenomenon).
- *The tailing-off effect:* Towards the end of the algorithm, added inequalities in [DM^t] tend to cut only a marginal volume of the dual solution space, making progress very slow.
- *Primal degeneracy and alternative dual optimal solutions:* An extreme point λ of polyhedron [M^t] has typically fewer non zero values than the number of master constraints. The complementary dual solution solves a system with fewer constraints than variables that admits many alternative solutions. As a consequence, the method iterates between alternative dual solutions without making any progress on the objective value.

Techniques to stabilize column generation belongs to one of the three standard families listed in [9]:

Penalty Functions: A penalty is added to the dual objective function to drive the optimization towards dual solutions that are close to a *stability center*, typically defined as the incumbent dual solution $\hat{\pi}$. The dual problem (19) is replaced by

$$\pi^t := argmax_{\pi \in I\!R_+^m}\{L^t(\pi) - \hat{S}(\pi)\}, \tag{23}$$

where the *penalty function*,

$$\hat{S} : \pi \in I\!R_+^m \rightarrow I\!R_+ \,,$$

is typically convex, takes value zero at $\hat{\pi}$, and increases as $||\pi - \hat{\pi}||$ increases. The Bundle method [3] is a special case where $\hat{S}(\pi) = \frac{1}{2\theta}||\pi - \hat{\pi}||^2$. One can also make use of a *piecewise linear penalty function* S (see [4] for instance) in order to ensure that the master problem is still a linear program (with additional artificial variables whose costs and bounds are chosen to model a piecewise linear stabilizing function). Penalty function methods require delicate tuning of several parameters.

Smoothing Techniques: The dual solution π^t used for pricing is "corrected" based on previous dual solutions. In particular, Neame [6] proposes to define smoothed price as:

$$\tilde{\pi}^t = \alpha\tilde{\pi}^{t-1} + (1 - \alpha)\pi^t \,, \tag{24}$$

i.e., $\tilde{\pi}^t$ is a weighted sum of previous iterates: $\tilde{\pi}^t = \sum_{\tau=0}^t (1-\alpha)\alpha^{t-\tau}\pi^\tau$. Wentges [10] proposes another smoothing rule where:

$$\tilde{\pi}^t = \alpha\hat{\pi} + (1 - \alpha)\pi^t \,. \tag{25}$$

i.e., $\tilde{\pi}^t = \hat{\pi} + (1-\alpha)(\pi^t - \hat{\pi})$, which amounts to taking a step of size $(1-\alpha)$ from $\hat{\pi}$ in the direction of π^t. In both rules, $\alpha \in [0,1)$ parameterizes the level of smoothing. The pricing problem is then solved using the smoothed prices, $\tilde{\pi}^t$, instead of π^t:

$$z_{\tilde{\pi}^t} := argmin_{x \in Z}\{(c - \tilde{\pi}^t A)x\} \,. \tag{26}$$

Solving this modified pricing problem might not yield a negative reduced cost column, even when one exists for π^t. This situation is the result of a *mis-pricing*. In such case, applying (24) or (25) with the same π^t solution leads to a new dual price vector that is closer to π^t. Note moreover that the incumbent $\hat{\pi}$ is updated each time the current Lagrangian bound improves over $\hat{L}$.

Centralized Prizes: One makes faster progress in improving the polyhedral outer approximation of the master LP dual program (10-12) when separating a point (π, η_π) in the interior of (10-12) rather than an extreme point. The analytic-center cutting-plane method (ACCPM) defines iterate π^t as the analytic center of the linear program (10-12) augmented with an optimality cut $\eta \geq \hat{L}$ that defines a trust region. Alternatives exist to keep a formulation of the master as a linear program (see references in [9]).

Note that using a smoothed price vector or an interior point for pricing has a drawback. The pricing problem can be harder for some solvers, as there are typically fewer non zero components and less clear dominance that can be exploited in dynamic programming recursions for instance.

3 The Link with In-Out Separation

The above smoothing techniques are related to the in-out separation scheme of [2,5]. The solution over the outer approximation of dual polyhedron (10-12) at iteration t defines an *out-point*, i.e., a point outside polyhedron (10-12):

$$(\pi^{out}, \eta^{out}) := (\pi^t, L^t(\pi^t)) . \tag{27}$$

Symetrically, consider a point inside the inner approximation of polyhedron (10-12). Possible definitions of such *in-point* are provided by the smoothing rules described above:

$$(\pi^{in}, \eta^{in}) := \begin{cases} (\tilde{\pi}^{t-1}, L(\tilde{\pi}^{t-1})) & \text{under rule (24),} \\ (\hat{\pi}, \hat{L}) & \text{under rule (25).} \end{cases} \tag{28}$$

These are in-points, because $L(\tilde{\pi}^{t-1})$ and $L(\hat{\pi})$ have been computed exactly when pricing as noted in Observation 1-$(iii.d)$. On the segment between the in-point and the out-point, one defines a *sep-point* at distance α from the out-point:

$$(\pi^{sep}, \eta^{sep}) := \alpha (\pi^{in}, \eta^{in}) + (1 - \alpha) (\pi^{out}, \eta^{out}) . \tag{29}$$

The in-out separation strategy consists in attempting to cut such sep-point. If an exact separation/pricing oracle fails to yield a separation hyperplan that cuts this sep-point, the point proves to be a valid in-point. Else, the out-point is updated. For standard in-out separation, where either the in-point or the out-point is replaced by the sep-point at each iteration, [2] proves that the distance between them tends to zero during a mis-pricing sequence.

The following proposition formalizes the properties common to Neame's and Wentges' smoothing schemes for column generation. Observe that the smoothing schemes described by rule (24) and (25) differ from the above standard in-out separation by the way in which the component η of the current solution is updated. Indeed, solving the separation/pricing problem yields a supporting hyperplane and a valid Lagrangian bound which is exploited in point (ii) below.

Proposition 1. *Common properties to both Neame's and Wentges' Smoothing Schemes.*
(i) *If the separation point* (π^{sep}, η^{sep}) *is cut by the inequality defined by* $z_{\pi^{sep}}$, *i.e., if* $L(\pi^{sep}) = \pi^{sep} a + (c - \pi^{sep} A) z_{\pi^{sep}} < \eta^{sep}$, *then* (π^{out}, η^{out}) *is cut off and* $z_{\pi^{sep}}$ *defines a negative reduced cost column for* $[M^t]$, *i.e.,* $(c - \pi^{out} A) z_{\pi^{sep}} + \pi^{out} a < \eta^{out}$.
(ii) *In the case* (π^{sep}, η^{sep}) *is not cut, i.e., if* $L(\pi^{sep}) \geq \eta^{sep}$, *then* $(\pi^{sep}, L(\pi^{sep}))$ *defines a new in-point that may be used for the next iteration. Moreover, as* $\eta^{sep} = \alpha \eta^{in} + (1 - \alpha) \eta^{out}$, *the new dual bound,* $L(\pi^{sep})$, *obtained when solving the pricing problem, improves the optimality gap at the smoothed price* $(c x^t - L(\tilde{\pi}^t))$ *by a factor* α:

$$(c x^t - L(\tilde{\pi}^t)) = (\eta^{out} - L(\pi^{sep})) \leq (\eta^{out} - \eta^{sep}) = \alpha (\eta^{out} - \eta^{in}) \leq \alpha (c x^t - L(\tilde{\pi}^{t-1})) . \tag{30}$$

(iii) *The cut defined by* $z_{\pi^{sep}}$ *can cut-off* (π^{out}, η^{out}) *even if it did not cut* (π^{sep}, η^{sep}). *If it does, both the in-point and the out-point can be updated. Otherwise, failing to cut the out-point leads to a mis-pricing. Then,* $(\pi^{out}, \eta^{out}) = (\pi^t, \eta^t)$ *remains solution for*

[DM^{t+1}] defined from [DM^t] by adding generator $z_{\pi^{sep}}$; but, under both rules (24) and (25), the smoothed prices of the next iterate get closer to kelley's prices:

$$\|\bar{\pi}^{t+1} - \pi^{t+1}\| = \alpha \|\bar{\pi}^t - \pi^t\| < \|\bar{\pi}^t - \pi^t\|. \tag{31}$$

Other smoothing schemes that differ by the rules for updating the in-point are possible. Property (30) remains valid provided $\eta^{in} \geq L(\bar{\pi}^{t-1})$.

Hence, the smoothing rules of Neame and Wentges can be understood as a projection in the π-space of the in-out separation procedure of [2], where the in-point is updated even when the sep-point is cut. The update of the η value to a valid dual bound guarantees the feasibility of the udpated in-point. In Wentges'smoothing scheme [10], the in-point is redefined as the dual incumbent at each iterate. Note however that when the separation point cannot be cut, $L(\pi^{sep}) > \hat{L}$ according to Proposition 1-(ii) and $\hat{\pi}$ is updated to π^{sep}. Thus, Wentges'smoothing conforms to the standard in-out paradigm. However, Neame smoothing scheme [6] differs from the standard in-out procedure by the fact that π^{in} is updated to π^{sep} whether or not the sep-point was cut. It can be seen as a valid variant of the in-out procedure as, even if (π^{sep}, η^{sep}) is not an in-point, $(\pi^{sep}, L(\pi^{sep}))$ defines an in-point that can used in the next iteration, as done implicitly in rule (24). In any case, Proposition 1 holds true for Neame smoothing scheme, as well as for Wentges. We emphasize that Proposition 1-(ii) is valid even if there is no mispricing. It has no equivalent for the general in-out procedure of [2] where no special component is associated with the objective value. To the best of our knowledge, such results had not been proven for Neame's smoothing scheme [6]. For Wentges smoothing, Property (iii) was already mentioned in [10], while Property (ii), which then takes the form $(c\,x^t - L(\hat{\pi}^t)) \leq \alpha\,(c\,x^t - L(\hat{\pi}^{t-1}))$, was proven in [7], but in a more intricate manner relying on the concavity of function $L^t()$ defined in (21) and under a mis-pricing assumption.

4 α-Schedule and Convergence

Instead of using the same α for all iterations, one can define iteration-dependent values α_t. We refer to α-schedule as the procedure used to select values of α_t dynamically. Intuitively, a large α can yield deeper cut if no mis-pricing occurs, while a small α can yield large dual bound improvement if a mis-pricing occurs. But a large α resulting in a mis-pricing or a small α with no mis-pricing result in an iterate with little progress being made. The primary concern should be the overall convergence of the method, which can be guaranteed by Proposition 1. If no smoothing is used, i.e., $\alpha_t = 0 \;\forall t$, the procedure is a standard Simplex based column generation for which finite convergence is proven, provided a cycle breaking rule that guarantees that each basis is visited at most once. When smoothing is used on the other hand, the same basis can remain optimal for several iterations in a sequence of mis-pricings. However, Proposition 1-(ii) provides a global convergence measure: the optimality gap $\|c x^t - L(\bar{\pi}^t)\|$ decreases by a factor α in the case of a mis-pricing, hence the total number of mis-pricing iterations is bounded. Alternatively, Proposition 1-(iii) provides a convergence measure local to a mis-pricing sequence: $\|\pi^{sep} - \pi^{out}\|$ decreases during such sequence, thereby bounding the number of mis-pricings for a given LP solution π^{out}. Thus, in the line of the asymptotic convergence proof for in-out separation of [2], one can show that:

Proposition 2. *Finite convergence.*
Applying a Simplex based column generation procedure to (13-15) while pricing on smoothed prices as set in (26), using either Neame (24) or Wentges (25)'s rule, converges to an optimal solution after a finite number of iterations, i.e., for some $t \in \mathbb{N}$, $(\pi^t, \eta^t) = (\pi^*, \eta^*)$, where (π^*, η^*) is an optimal solution to (10-12).

Asymptotically convergent algorithms might not be suitable for practical purposes. For instance, consider setting $\alpha = 0.8$ for all t. Then, the distance reduction in a mis-pricing sequence becomes small very quickly. In practice, it would be better to choose an α-schedule such that $\tilde{\pi}^t = \pi^t$ after a small number of mis-pricing iterations t. Given a static baseline α, we propose as outlined on the left side in Table 1, to adapt α_t during a mis-pricing sequence in such a way that $(1 - \Pi_{\tau=0}^k \alpha_\tau) = k * (1 - \alpha)$. Hence, $\alpha_t = 0$ after $k = \left\lceil \frac{1}{(1-\alpha)} \right\rceil$ mis-pricing iterations, at which point smoothing stops, as $\tilde{\pi}^t = \pi^t$, which forces the end of a mis-pricing sequence.

So far we assumed a static baseline α provided as an input. Let us now consider how the user could be free from having to tune α for his application. In deriving an auto-adaptive α-schedule, one could consider using high α while the out-point is believed to be a bad approximation, and reducing α as the method converges, which is measured by smaller gaps $|\eta^t - \hat{L}|$, and the purpose becomes to prove optimality. Alternatively, one could rely on local information, as we do. We propose to decrease α when the sub-gradient at the sep-point indicates that a larger step from the in-point would further increase the dual bound (i.e., when the angle of the ascent direction, g^{sep}, as defined in (22), and the direction $(\pi^{\text{out}} - \pi^{\text{in}})$ is less than $90°$), and vice versa. We outline this procedure on the right side in Table 1. Functions for increasing and decreasing α are: $f_{\text{incr}}(\alpha_t) = \alpha_t + (1 - \alpha_t) \cdot 0.1$, while $f_{\text{decr}}(\alpha_t) = \alpha_t/1.1$ if $\alpha_t \in [0.5, 1)$, and $f_{\text{decr}}(\alpha_t) = \max\{0, \alpha_t - (1 - \alpha_t) \cdot 0.1\}$, otherwise.

Table 1. α-schedule in a mis-pricing sequence for a given initial α (on the left) and dynamic α-schedule based on sub-gradient information for a given intial α (on the right)

Step 0: $k \leftarrow 1, \pi^0 \leftarrow \pi^{\text{in}}$	Step 0: Let $\alpha_0 \leftarrow \alpha, t \leftarrow 0$.
Step 1: $\tilde{\alpha} \leftarrow [1 - k * (1 - \alpha)]^+$	Step 1: Call pricing on $\pi^{\text{sep}} = \alpha_t \pi^{\text{in}} + (1 - \alpha_t)\pi^{\text{out}}$.
Step 2: $\pi^{\text{sep}} = \tilde{\alpha} \pi^0 + (1 - \tilde{\alpha}) \pi^{\text{out}}$	
Step 3: $k \leftarrow k + 1$	Step 2: If a mispricing occurs, start the mispricing schedule.
Step 4: call the pricing oracle on π^{sep}	Step 3: Else, let g^{sep} be the sub-gradient in sol $z_{\pi^{\text{sep}}}$.
Step 5: if a mis-pricing occurs, goto Step 1;	Step 4: If $g^{\text{sep}}(\pi^{\text{out}} - \pi^{\text{in}}) > 0, \alpha_{t+1} \leftarrow f_{\text{incr}}(\alpha_t)$; otherwise, $\alpha_{t+1} \leftarrow f_{\text{decr}}(\alpha_t)$.
else, let $t \leftarrow t + 1$, solve the master and goto Step 0.	Step 5: Let $t \leftarrow t + 1$, solve the master and goto Step 1.

5 Hybridization with an Ascent Method

With a pure smoothing technique, the price vector is defined by taking a step $(1 - \alpha)$ from the in-point in the direction of the out-point: $\pi^{\text{sep}} = \pi^{\text{in}} + (1 - \alpha)(\pi^{\text{out}} - \pi^{\text{in}})$. Here, we consider modifying the direction $(\pi^{\text{out}} - \pi^{\text{in}})$ by twisting it towards the direction of ascent observed in π^{in}. The resulting method can be viewed as a hybridization of column generation with a sub-gradient method. When Wentges's rule (25) is used, the resulting hybrid method is related to the Volume algorithm [1] where π^t is obtained by taking a step from $\hat{\pi}$ in a direction that combines previous iterate information with the current sub-gradient. However, contrary to the Volume algorithm, our purpose here is not to derive the next π^t iterate, but simply to bring a correction to the price vector that is used in the pricing procedure.

Table 2. Directional smoothing with parameter β

$$\text{Step 1: } \tilde{\pi} = \pi^{\text{in}} + (1 - \alpha)(\pi^{\text{out}} - \pi^{\text{in}})$$
$$\text{Step 2: } \pi^g = \pi^{\text{in}} + \frac{g^{\text{in}}}{\|g^{\text{in}}\|} \|\pi^{\text{out}} - \pi^{\text{in}}\|$$
$$\text{Step 3: } \rho = \beta \pi^g + (1 - \beta)\pi^{\text{out}}$$
$$\text{Step 4: } \pi^{\text{sep}} = \left(\pi^{\text{in}} + \frac{\|\tilde{\pi} - \pi^{\text{in}}\|}{\|\rho - \pi^{\text{in}}\|} (\rho - \pi^{\text{in}})\right)^+$$

The hybrid procedure, that we call *directional smoothing*, is outlined in Table 2. Let g^{in} denote the sub-gradient associated to oracle solution $z_{\pi^{\text{in}}}$. In Step 1, $\tilde{\pi}$ is computed by applying smoothing. In Step 2, π^g is computed as the point located on the steepest ascent direction at a distance from π^{in} equal to the distance to π^{out}. In Step 3, a rotation is performed, defining target ρ as a convex combination between π^g and π^{out}. Then, in Step 4, the sep-point is selected in direction $(\rho - \pi^{\text{in}})$ at the distance from π^{in} equal to $\|\tilde{\pi} - \pi^{\text{in}}\|$ and it is projected on the positive orthant. As is the case with non-directional smoothing, using modified dual prices can result in mis-pricing. When this arises, we switch off directional smoothing by setting $\beta = 0$ in the next iteration. Apart for mis-pricing, directional smoothing can be implemented with a fixed value of parameter β. However, computational experiments showed that the larger the angle γ between vectors $(\pi^{\text{out}} - \pi^{\text{in}})$ and $(\pi^g - \pi^{\text{in}})$, the smaller the value for β should be. Indeed, if the angle γ is large, then twisting the direction is likely to lead to a mis-pricing. Our proposal is to use an adaptive β-schedule by setting $\beta = \cos \gamma$. As γ is always less than $90°$, since vector $(\pi^{\text{out}} - \pi^{\text{in}})$ is an ascent direction, $\beta \in [0, 1]$.

6 Numerical Tests

In the experiments we describe next, we assess numerically the stabilization effect of applying Wentges smoothing with static α-schedule versus auto-adaptive schedule starting with $\alpha_0 = 0.5$. Additionally, we estimate the effect of using directional smoothing, with static and auto-adaptive value of parameter β, in combination with Wentges smoothing. The experiments are conducted on 98 representative instances of

Table 3. Stabilization effect of Wentges smoothing with static versus auto-adaptive α: showing geometric means

	$\alpha = 0$	Best α	$\dfrac{\alpha = 0}{\alpha = best}$		$\dfrac{\alpha = 0}{\alpha = auto}$		$\dfrac{\alpha = best}{\alpha = auto}$	
Problem	time	Range	It	T	It	T	It	T
Generalized Assignment	98	[0.5,0.95]	3.37	4.46	3.36	4.57	1.00	1.03
Lot-Sizing	88	[0.4,0.95]	2.26	3.31	2.51	4.58	1.11	1.38
Machine Scheduling	33	[0.65,0.9]	2.30	3.04	2.29	2.98	1.00	0.98
Bin Packing	7.9	[0.75,0.95]	1.54	1.79	1.49	1.65	0.97	0.92
Vehicle Routing	6.3	[0.2,0.8]	1.32	1.37	1.15	1.28	0.88	0.94

Table 4. Extra stabilization effect when applying directional smoothing

	$\dfrac{\alpha = best, \beta = 0}{\alpha, \beta = best}$		$\dfrac{\alpha = best, \beta = 0}{\alpha = best, \beta = auto}$		$\dfrac{\alpha = best, \beta = 0}{\alpha, \beta = auto}$		$\dfrac{\alpha, \beta = 0}{\alpha, \beta = auto}$	
Problem	It	T	It	T	It	T	It	T
General. Assignment	1.11	1.93	1.35	1.95	1.48	2.25	5.00	10.03
Lot-Sizing	1.17	1.50	1.32	1.61	1.37	1.83	3.09	6.06
Machine Scheduling	0.94	0.91	1.04	1.12	1.10	1.21	2.53	3.68
Bin Packing	0.95	0.94	1.03	0.98	1.04	0.96	1.60	1.72
Vehicle Routing	0.90	0.92	0.94	0.97	0.83	0.92	1.09	1.25

the following problems: Machine Scheduling, Generalized Assignment, Multi-Echelon Small-Bucket Lot-Sizing, Bin Packing, Capacitated Vehicle Routing.

For each instance, we determine experimentally the best static α-value for which the master LP solution time by column generation with Wentges smoothing is minimum, by testing all α values in $\{0.05, 0.1, \ldots, 0.95\}$. The first columns of Table 3 report respectively the geometric mean of CPU time without smoothing on a Dell PowerEdge 1950 (32Go, Intel Xeon X5460, 3.16GHZ) and the range of best α-values that vary a lot from one instance to the next and between applications. In the other columns of Table 3, we compare tuned and self-adjusting smoothing to standard column generation without any smoothing. Next, in Table 4, we compare performance with and without the extra directional feature, using both a static parameter β (the best in the set $\{0.05, 0.1, 0.2, 0.3\}$) and an adaptive β. Thus, in total we compare 6 variants of column generation: (i) without any stabilization ($\alpha = 0$, $\beta = 0$), (ii) with static Wentges stabilization ($\alpha = best$, $\beta = 0$), (iii) with auto-adaptive Wentges stabilization ($\alpha = auto$, $\beta = 0$), (iv) with combined static Wentges and directional stabilization ($\alpha = best$, $\beta = best$), (v) with combined static Wentges and adaptive directional stabilization ($\alpha = best$, $\beta = auto$), and (vi) with combined adaptive Wentges and directional stabilization ($\alpha = auto$, $\beta = auto$). In the tables, we report ratios of geometric means for the following statistics: It is the number of iterations in column generation; T is the solution time. The last columns of Table 4 summarizes the overall performance of smoothing. Note that smoothing improves solution times by a larger factor than the number of iterations, in spite of the potentially harder pricing subproblems.

Conclusion

In this paper, we have specified the link between column generation stabilization by smoothing and in-out separation. We also extended the in-out convergence proof for Neame's and Wentges' smoothing schemes, deriving an extra global convergence property on the optimality gap. These results trivially extend to the case of multiple subproblems. On the practical side, our numerical results confirm the effectiveness of smoothing and show that it can be implemented in a way that does not require parameter tuning. Our hard coded initialization and dynamic auto-adaptive scheme based on local subgradient information experimentally matches or improves the performance of the best user tuning as revealed in Table 3. The extra directional twisting feature is shown in Table 4 to bring further performance improvement in the non-identical subproblem case. When there are identical subproblems, as in Bin Packing and Vehicle Routing, subgradient information is aggregated and hence probably less pertinent. The adaptive setting of parameter β outperforms the static value strategy leading to a generic parameter-tuning-free implementation. This work can hopefully inspire a renewed interest in cut separation strategies, possibly developing in-out separation with a self-adjusting parameter scheme and a gradient direction twist.

Aknowledments. This research was supported by the associated team program of INRIA through the SAMBA project. The remarks of J-P. Richard and anonymous referees were of great value to improve our draft.

References

1. Barahona, F., Anbil, R.: The Volume Algorithm: Producing Primal Solutions with a Subgradient Method. Math. Program. A 87(1), 385–399 (2000)
2. Ben-Ameur, W., Neto, J.: Acceleration of Cutting-Plane and Column Generation Algorithms: Applications to Network Design. Networks 49(1), 3–17 (2007)
3. Briant, O., Lemaréchal, C., Meurdesoif, P., Michel, S., Perrot, N., Vanderbeck, F.: Comparison of Bundle and Classical Column Generation. Math. Program. A 113(2), 299–344 (2008)
4. du Merle, O., Villeneuve, D., Desrosiers, J., Hansen, P.: Stabilized Column Generation. Discrete Math. 194, 229–237 (1999)
5. Fischetti, M., Salvagnin, D.: An In-Out Approach to Disjunctive Optimization. In: Lodi, A., Milano, M., Toth, P. (eds.) CPAIOR 2010. LNCS, vol. 6140, pp. 136–140. Springer, Heidelberg (2010)
6. Neame, P.J.: Nonsmooth Dual Methods in Integer Programming. PhD thesis (1999)
7. Pessoa, A., Uchoa, E., Poggi de Aragão, M., Rodrigues, R.: Exact Algorithm over an Arc-Time-Indexed Formulation for Parallel Machine Scheduling Problems. Math. Prog. Computation 2, 259–290 (2010)
8. Vanderbeck, F., Savelsbergh, M.W.P.: A Generic View of Dantzig-Wolfe Decomposition in Mixed Integer Programming. Operations Research Letters 34(3), 296–306 (2006)
9. Vanderbeck, F.: Implementing Mixed Integer Column Generation. In: Column Generation. Kluwer (2005)
10. Wentges, P.: Weighted Dantzig-Wolfe Decomposition for Linear Mixed-Integer Programming. Int. Trans. O. R. 4, 151–162 (1997)

Energy Minimization via a Primal-Dual Algorithm for a Convex Program

Evripidis Bampis[1,*,**], Vincent Chau[2,*], Dimitrios Letsios[1,2,*],
Giorgio Lucarelli[1,2,*,**], and Ioannis Milis[3,**]

[1] LIP6, Université Pierre et Marie Curie, France
{Evripidis.Bampis,Giorgio.Lucarelli}@lip6.fr
[2] IBISC, Université d'Évry, France
{vincent.chau,dimitris.letsios}@ibisc.univ-evry.fr
[3] Dept. of Informatics, AUEB, Athens, Greece
milis@aueb.gr

Abstract. We present an optimal primal-dual algorithm for the energy minimization preemptive open-shop problem in the speed-scaling setting. Our algorithm uses the approach of Devanur et al. [JACM 2008], by applying the primal-dual method in the setting of convex programs and KKT conditions. We prove that our algorithm converges and that it returns an optimal solution, but we were unable to prove that it converges in polynomial time. For this reason, we conducted a series of experiments showing that the number of iterations of our algorithm increases linearly with the number of jobs, n, when n is greater than the number of machines, m. We also compared the speed of our method with respect to the time spent by a commercial solver to directly solve the corresponding convex program. The computational results give evidence that for $n > m$, our algorithm is clearly faster. However, for the special family of instances where $n = m$, our method is slower.

1 Introduction

The *primal-dual* method has been extensively used for obtaining optimal [7,8,13] and approximate [9,14] solutions for many well known optimization problems. It has been mainly applied for problems formulated as linear programs. Only recently Devanur et al. [6] applied the primal-dual paradigm in the more general setting of convex programming and the Karush-Kuhn-Tucker (KKT) conditions. Our work is in the same vein. We explore the primal-dual paradigm in the context of energy minimization scheduling with respect to the *speed-scaling* model. In this model the speed of each processor can dynamically change at any time,

---

* Partially supported by the French Agency for Research under the DEFIS program TODO, ANR-09-EMER-010, and a PHC CAI YUANPEI France-China bilateral project.
** Partially supported by the project ALGONOW, co-financed by the European Union (European Social Fund - ESF) and Greek national funds, through the Operational Program "Education and Lifelong Learning", under the program THALES.

V. Bonifaci et al. (Eds.): SEA 2013, LNCS 7933, pp. 366–377, 2013.

while the energy consumption is a convex function of the speed. More formally, if a processor runs at speed $s(t)$ at time t, then the power needed is $P(t) = s(t)^\alpha$, where $\alpha > 1$ is a machine-dependent constant.[1] The energy consumption of the processor is equal to the integral of the power over time, i.e., $E = \int P(t)dt$. Moreover, the processor's speed is the rate at which work is executed and, thus, the total of work amount executed by a processor is the integral of its speed, i.e. $w = \int s(t)dt$.

We focus on the speed-scaling preemptive open-shop problem for which there is a natural formulation as a convex program. In the energy minimization speed-scaling preemptive open-shop problem, we are given a set of n jobs $\mathcal{J} = \{J_1, J_2, \ldots, J_n\}$ and a set of m processors $\mathcal{M} = \{M_1, M_2, \ldots, M_m\}$. Each job consists of operations that have to run on different processors. Operations of the same job cannot be executed at the same time. The operation O_{ij} of the job J_j has to be executed on processor M_i and it has an amount of work $w_{ij} \geq 0$. Note that, it is not necessary for each job to have an operation to all processors; in this case $w_{ij} = 0$. The operations may be preempted, that is they can be interrupted and continue their execution later. The goal is to minimize the total energy consumed such that all operations are completed before a given deadline d. Extending the Graham's three-field notation [11] for scheduling problems, we denote our problem by $O|pmtn, d|E$.

Related Work. The preemptive open-shop problem has been extensively studied in the classical setting (without caring about the energy consumption). In this setting, each operation O_{ij} has a processing time p_{ij}, instead of a work. When the goal is the minimization of the length of the schedule (makespan) the problem $O|pmtn|C_{\max}$ can be solved in polynomial time [10]. Several other results for the preemptive open-shop can be found in [5]. The speed-scaling preemptive open-shop problem has beed studied, recently, by Bampis et al. [3], who presented a combinatorial algorithm for $O|pmtn, d|E$ based on a transformation to a convex cost flow problem.

In [12], Gupta et al. apply a primal-dual approach in order to obtain a constant factor competitive algorithm for a speed-scaling problem in the online setting. This algorithm is based on a primal-dual schema using the Lagrangian duality and it is quite different from our primal-dual approach.

There is a lot of work in the speed-scaling area, and the interested reader is referred to the recent reviews [1,2].

Our contribution. In linear programming, the idea of the primal-dual approach is, in general, to modify the dual and the primal variables in turns, based on the complementary slackness conditions. In the convex programming setting, it is not possible to define a dual program as for the linear problems. However, there is always an optimal solution for any convex program in which the primal variables are related with the dual variables through some equality relationships known as the stationarity conditions. Therefore, by modifying the dual variables,

---

[1] In most applications, α is considered to be between two and three.

that correspond to Lagrangian multipliers, there is a direct impact on the values of the primal variables.

In Section 3 we formulate our problem as a convex program. Then, in Section 4, we propose a primal-dual algorithm and we prove that it is optimal and that it converges. Unfortunately, we are unable to prove that it converges in polynomial time. In Section 5, we present a series of experiments showing that the number of iterations of our algorithm increases linearly with the number of jobs when $n > m$. We are also interested in the comparison of the execution time of our method with respect to the time spent by a commercial solver to directly solve the corresponding convex program. The computational results show that for $n > m$, our algorithm is clearly faster. However, for the special family of instances where $n = m$, our method is slower.

2   Preliminaries

In most of the speed-scaling problems, due to the convexity of the speed-to-power function, each job/operation runs at a constant speed during its whole execution in an optimal schedule (see for example [15]). This observation holds also for our problem and its proof directly follows from the convexity of the power function. Note that, given the speed s_{ij} of the operation O_{ij}, the time needed for the execution of O_{ij} is $\frac{w_{ij}}{s_{ij}}$, while the energy consumed during its execution is $\frac{w_{ij}}{s_{ij}} s_{ij}^{\alpha} = w_{ij} s_{ij}^{\alpha-1}$.

In what follows in this paper, we will consider a relaxation Π' of our original problem Π, in which operations of the same job are allowed to be executed simultaneously but the sum of the execution times of the operations of the same job cannot exceed d. A solution of this problem gives the speeds of the operations, without determining their order. Clearly, for an optimal solution E' of Π', it holds that $E' \leq E$, where E is the energy consumption in an optimal solution for Π.

In the following sections we formulate the relaxed problem as a convex program and we propose a primal-dual algorithm to find an optimal solution for it. A solution for Π' determines the speeds of operations, and hence their processing times. Then, we can run a polynomial algorithm for $O|pmtn|C_{\max}$ (see for example [10]) to get a feasible open-shop solution for our problem Π. This procedure is described formally in Algorithm 1.

Algorithm 1.

1: Solve optimally the relaxed problem Π' to define the speeds of operations;
2: For each O_{ij} set processing time $p_{ij} = \frac{w_{ij}}{s_{ij}}$;
3: Run the algorithm proposed in [10] for $O|pmtn|C_{\max}$;

Theorem 1. *Algorithm 1 returns an optimal schedule for $O|pmtn, d|E$.*

Proof. In an optimal solution for Π', all jobs and processors are active for time at most d, while it is easy to see that there is at least one job or processor with processing time exactly d. If not, we can decrease the speeds of all operations and get a feasible schedule for Π' of smaller energy consumption, which is a contradiction as we have considered an optimal schedule. Moreover, it is known (e.g. [5]) that any optimal solution for $O|pmtn|C_{\max}$ has length equal to $\max\{\max_i \sum_{O_{ij} \in M_i} p_{ij}, \max_j \sum_{O_{ij} \in J_j} p_{ij}\}$, which in our case is d. Thus, Algorithm 1 returns a feasible open-shop solution for Π, i.e., a solution where no operations of the same job are executed simultaneously and the completion time of all operations is at most d. In other words, the algorithm for $O|pmtn|C_{\max}$ produces a feasible solution of Π given the speeds obtained by a feasible solution for Π', and hence the energy consumed for Π is equal to E'. As the speeds, and hence the processing times, of operations are selected in Line 1 in such a way that E' is minimized, the theorem holds.   □

3 Convex Programming Formulation and the KKT conditions

In this section we first formulate our relaxed problem as a convex program.

$$\min \sum_{O_{ij} \in J_j} \sum_{O_{ij} \in M_i} w_{ij} s_{ij}^{\alpha-1}$$

$$\sum_{O_{ij} \in M_i} \frac{w_{ij}}{s_{ij}} \leq d \qquad 1 \leq i \leq m \tag{1}$$

$$\sum_{O_{ij} \in J_j} \frac{w_{ij}}{s_{ij}} \leq d \qquad 1 \leq j \leq n \tag{2}$$

$$s_{ij} \geq 0 \qquad O_{ij} \in J_j, O_{ij} \in M_i$$

Constraints (1) and (2) do not allow a job and a processor, respectively, to be active for a time period greater than d.

The Karush-Kuhn-Tucker conditions are necessary and sufficient conditions [4] for a feasible solution of our convex program to be optimal. Note that, the β_i's, $1 \leq i \leq m$, correspond to the Lagrangian multipliers for the Constraints (1) and the γ_j's, $1 \leq j \leq n$, correspond to the Lagrangian multipliers for the Constraints (2).

Stationarity conditions:

$$\nabla \left(\sum_{O_{ij} \in J_j} \sum_{O_{ij} \in M_i} w_{ij} s_{ij}^{\alpha-1} \right) + \sum_{i=1}^{m} \beta_i \cdot \nabla \left(\sum_{O_{ij} \in M_i} \frac{w_{ij}}{s_{ij}} - d \right)$$

$$+ \sum_{j=1}^{n} \gamma_j \cdot \nabla \left(\sum_{O_{ij} \in J_j} \frac{w_{ij}}{s_{ij}} - d \right) + \sum_{O_{ij} \in J_j} \sum_{O_{ij} \in M_i} \delta_{ij} \cdot \nabla(-s_{ij}) = 0$$

or equivalently

$$\sum_{O_{ij} \in J_j} \sum_{O_{ij} \in M_i} \left(w_{ij}(a-1)s_{ij}^{a-2} - (\beta_i + \gamma_j)\frac{w_{ij}}{s_{ij}^2} - \delta_{ij} \right) \nabla s_{ij} = 0 \qquad (3)$$

Complementary slackness conditions:

$$\beta_i \cdot \left(\sum_{O_{ij} \in M_i} \frac{w_{ij}}{s_{ij}} - d \right) = 0 \qquad 1 \le i \le m \qquad (4)$$

$$\gamma_j \cdot \left(\sum_{O_{ij} \in J_j} \frac{w_{ij}}{s_{ij}} - d \right) = 0 \qquad 1 \le j \le n \qquad (5)$$

$$\delta_{ij} \cdot (-s_{ij}) = 0 \qquad O_{ij} \in J_j, O_{ij} \in M_i \qquad (6)$$

Note that operations with work $w_{ij} = 0$ are not counted into the above sums. Then, as each operation has a work $w_{ij} > 0$ to execute, it holds that $s_{ij} > 0$, and hence, by condition (6) we have that $\delta_{ij} = 0$. Thus, Equation (3) can be reformulated as

$$s_{ij}^{\alpha} = \frac{\beta_i + \gamma_j}{\alpha - 1} \qquad O_{ij} \in J_j, O_{ij} \in M_i \qquad (7)$$

As we already mentioned in the introduction, the KKT conditions, and especially the stationarity conditions, give a relation between the primal and the dual variables. Indeed, Equations (7) directly connect our primal variables s_{ij} with our dual variables β_i and γ_j. Intuitively, each dual variable β_i, $1 \le i \le m$, can be considered as the contribution of the processor M_i to the speed of the operations O_{ij}, $1 \le j \le n$. In a similar way, each dual variable γ_j, $1 \le j \le n$, can be considered to be the contribution of the job J_j to the speed of the operations O_{ij}, $1 \le i \le m$.

4   The Primal-Dual Algorithm

In this section we present a combinatorial algorithm based on the primal-dual approach. The main idea of the algorithm is to determine the values of dual variables, β_i and γ_j, and hence the speeds of operations, through a primal-dual scheme. Our algorithm initializes the dual variables according to the following proposition that provides upper and lower bounds for them.

Proposition 1.

(i) *For each β_i, $1 \le i \le m$, it holds that $0 \le \beta_i \le (\alpha - 1)\left(\frac{\sum_{O_{ij} \in M_i} w_{ij}}{d} \right)^{\alpha}$.*

(ii) *For each γ_j, $1 \le j \le n$, it holds that $0 \le \gamma_j \le (\alpha - 1)\left(\frac{\sum_{O_{ij} \in J_j} w_{ij}}{d} \right)^{\alpha}$.*

Proof. The lower bounds follow by the definition of β_i's and γ_j's.

For the upper bound on β_i's, consider a processor M_i, $1 \leq i \leq m$. As we search for an upper bound we can assume that $\beta_i > 0$. Hence, by the complementary slackness conditions (4) and applying the stationarity conditions (7), in an optimal solution it holds that

$$\sum_{O_{ij} \in M_i} \frac{w_{ij}}{s_{ij}} - d = 0 \Leftrightarrow \sum_{O_{ij} \in M_i} \frac{w_{ij}}{\sqrt[\alpha]{\frac{\beta_i + \gamma_j}{\alpha - 1}}} = d$$

To obtain the upper bound on β_i, we can consider that the speeds of all operations $O_{ij} \in M_i$ depend only on the contribution of the processor M_i, that is $\gamma_j = 0$ for all $O_{ij} \in M_i$. Hence, we have that

$$\sum_{O_{ij} \in M_i} \frac{w_{ij}}{\sqrt[\alpha]{\frac{\beta_i}{\alpha - 1}}} \geq d \Leftrightarrow \beta_i \leq (\alpha - 1) \left(\frac{\sum_{O_{ij} \in M_i} w_{ij}}{d} \right)^\alpha$$

The same arguments hold for the upper bounds on γ_j's.   □

Based on the previous proposition, we initialize each dual variable β_i, $1 \leq i \leq m$, to its lower bound and each dual variable γ_j, $1 \leq j \leq n$, to its upper bound. Given these initial values, the obtained schedule may not be feasible. More specifically, the processing time of some processors may be more than d, i.e., $\sum_{O_{ij} \in M_i} \frac{w_{ij}}{\sqrt[\alpha]{\frac{\gamma_j}{\alpha - 1}}} > d$. For such a processor M_i, we increase β_i such that the processing time of M_i is exactly d, i.e., $\sum_{O_{ij} \in M_i} \frac{w_{ij}}{\sqrt[\alpha]{\frac{\beta_i + \gamma_j}{\alpha - 1}}} = d$. We refer to this step as an "infeasible-to-feasible" step. The increasing of β_i's has as a result some jobs to become non-tight, i.e., $\sum_{O_{ij} \in J_j} \frac{w_{ij}}{\sqrt[\alpha]{\frac{\beta_i + \gamma_j}{\alpha - 1}}} < d$. For such a job J_j, we decrease γ_j such that to be equal to the maximum between zero (respecting our definition) and the value of γ_j needed so that J_j becomes tight again, i.e., $\sum_{O_{ij} \in J_j} \frac{w_{ij}}{\sqrt[\alpha]{\frac{\beta_i + \gamma_j}{\alpha - 1}}} = d$. We refer to this step as a "non-tight-to-tight" step. Thus, the decreasing of γ_j's has as a result some processors to become non-feasible, and so on. The criterion to terminate this procedure is when after a "non-tight-to-tight" step all the complementary slackness conditions are satisfied. A formal description of the above procedure is given in Algorithm 2.

Note that, the algorithm modifies a dual variable β_i only if the processor M_i is non-feasible in such a way to make it feasible (and tight). To do this, the speed of each operation $O_{ij} \in M_i$ is increased through the increasing of β_i. By the definition of the algorithm, M_i can be in a feasible and non-tight state only if $\beta_i = 0$. In a similar way the algorithm modifies a dual variable γ_j only if the job J_j is non-tight (and feasible) in such a way to make it tight. To do this, the speed of each operation $O_{ij} \in J_j$ is decreased through the decreasing of γ_j. By the definition of the algorithm, J_j cannot be in an infeasible state. Based on these observations, the following proposition follows.

Algorithm 2.

1: For each i, $1 \leq i \leq m$, set $\beta_i = 0$;

2: For each j, $1 \leq j \leq n$, set $\gamma_j = (\alpha - 1) \left(\frac{\sum_{O_{ij} \in J_j} w_{ij}}{d} \right)^\alpha$;

3: **while** the complementary slackness conditions are not satisfied **do**

4:     **for** each i, $1 \leq i \leq m$, such that the processor M_i is not feasible **do**

5:         Choose β_i such that $\left(\sum_{O_{ij} \in J_j} \frac{w_{ij}}{\sqrt[\alpha]{\frac{\beta_i + \gamma_j}{\alpha - 1}}} - d \right) = 0$;

6:     **for** each j, $1 \leq j \leq j$, such that the job J_j is not tight **do**

7:         Choose the maximum value of γ_j such that $\gamma_j \cdot \left(d - \sum_{O_{ij} \in J_j} \frac{w_{ij}}{\sqrt[\alpha]{\frac{\beta_i + \gamma_j}{\alpha - 1}}} \right) = 0$;

Proposition 2.
(i) For each i, $1 \leq i \leq m$, the value of β_i is always non-decreasing.
(ii) For each j, $1 \leq j \leq n$, the value of γ_j is always non-increasing.

Theorem 2. *Algorithm 2 converges to an optimal solution of the relaxed problem.*

Proof. In each iteration the algorithm modifies at least one dual variable; otherwise the complementary slackness conditions are satisfied and the algorithm terminates. By Proposition 2 the modification of the dual variables is monotone, while by Proposition 1 there are well-defined lower and upper bounds for them. Therefore, the algorithm terminates.

In order to show that the algorithm converges in an optimal solution, we just have to show that the solution obtained satisfies the KKT conditions. The stationarity conditions (7) hold as for any operation O_{ij} the assigned speed by construction can be written as $s_{ij} = \sqrt[\alpha]{\frac{\beta_i + \gamma_j}{\alpha - 1}}$. The complementary slackness conditions (4) hold since after the final "non-tight-to-tight" step any processor M_i is either tight or its $\beta_i = 0$; if not then the algorithm would have executed a new iteration. The complementary slackness conditions (5) hold since in Line 7 we force $\gamma_j \cdot \left(d - \sum_{O_{ij} \in J_j} \frac{w_{ij}}{\sqrt[\alpha]{\frac{\beta_i + \gamma_j}{\alpha - 1}}} \right) = 0$. The complementary slackness conditions (6) hold since for any operation O_{ij} we have set $\delta_{ij} = 0$. □

5   Experimental Results

In this section we experimentally test our primal-dual algorithm towards two directions. The first direction is to observe the behavior of our algorithm when the size of the instance increases. The second direction is to compare the execution time of the primal-dual approach against the execution time of a baseline algorithm, that is a commercial solver that solves directly the corresponding convex program.

5.1   System Specification and Benchmark Generation

Our simulations have been performed on a machine with a CPU Intel Xeon X5650 with 8 cores, running at 2.67GHz. The operating system of the machine is a Linux Debian 6.0. We used Matlab with cvx toolbox. The solver used for the convex program is SeDuMi. For both our algorithm and the convex program, we set $\varepsilon = 10^{-7}$ to be the desired accuracy of the returned solution.

The instance of the problem consists of a matrix $m \times n$ that corresponds to the work of the operations, the value of α and the deadline d. However, we experiment with two more parameters: (i) the density p of the instance, that is the number of non-zero work operations, and (ii) the range $[1, w_{\max}]$ of the values of works.

We have considered several combinations for the parameters m, $1 \le m \le 50$, and n, $1 \le n \le 200$. For each combination, we have first decided randomly with probability p if there is a non-zero work operation in each position of the $m \times n$ matrix. The value of p has been selected to be 0.5 or 0.75 or 1. If the created instance did not correspond to the selected values of m and n, we rejected it and we replaced it by another. In other words, we reject a matrix iff there exists a line or a column in which each value is equal to zero. Then, for each operation with non-zero work, we selected at random an integer in the range of $[1, w_{\max}]$. Note here that $w_{\max}$ and the deadline d are strongly related. Indeed, given a matrix of works and a deadline d, if we increase all works and the deadline by the same factor, then the optimal solutions of the two instances will tend to have very similar (if not the same) speeds and energy consumption. For this reason, we have fixed the value of $d = 1000$ and we examined three different values for $w_{\max}$, i.e., $w_{\max} = 10$, $w_{\max} = 50$ and $w_{\max} = 100$. These values are selected, in general, in the direction of creating instances in which the average speed in the optimal solution is greater than one, almost equal to one and smaller than one, respectively. Finally, as in most applications the value of α is between two and three, we used three different values for it, that is $\alpha = 2$, $\alpha = 2.5$ and $\alpha = 3$.

For each combination of parameters we have repeated the experiments with 30 different matrices. All results we present below, concern the average of these 30 instances.

The benchmark as well as the code we used in our experiments are freely available at http://todo.lamsade.dauphine.fr/spip.php?article85.

5.2   Results

The main goal of our experiments is to study the behavior of the primal-dual algorithm when the size of the instance increases. However, during our experiments we noticed that the speed of convergence strongly depends on the relation between the number of jobs n and the number of processors m.

In Table 1, we show how the size of the instance affects the number of modifications of the dual variables made by the primal-dual algorithm. We observe that, if $n > m$ then the number of modifications increases linearly with the size of the instance (see also Fig. 1). Moreover, the parameters α, $w_{\max}$ and p do not play any role to the number of modifications.

Table 1. The number of modifications of the dual variables done by the primal-dual algorithm. The values of the table correspond to $\alpha = 2$, $w_{\max} = 10$, $p = 1$. Each entry of the table is the average over 30 instances. The empty entries correspond to cases with $m = n$ and take time longer than 30 minutes each and are interrupted.

n	$m = 5$	$m = 10$	$m = 15$	$m = 20$	$m = 25$	$m = 30$	$m = 40$	$m = 50$
5	40101	1	2	2	2	2	2	2
10	151	279611	3	4	3	4	4	4
20	255	295	384	–	34	7	7	10
30	355	410	443	500	593	–	12	15
40	455	510	565	572	640	756	–	32
50	555	610	665	720	768	755	947	–
60	655	710	765	820	872	864	1040	1294
70	755	810	865	920	975	1030	1034	1250
100	1055	1110	1165	1220	1275	1330	1440	1495
150	1555	1610	1665	1720	1775	1830	1940	2050
200	2055	2110	2165	2220	2275	2330	2440	2550

Note also that if $n < m$ then the number of modifications increases linearly with the size of the instance. In fact the two cases $n > m$ and $n < m$ should be symmetric. However, the initialization step of our algorithm breaks this symmetry. Recall that the algorithm initializes the dual variables that correspond to processors (β_i's) to zero and the dual variables that correspond to jobs (γ_j's) to their upper bounds. In the case where $n < m$, we expect to have all jobs tight and most of the processors non-tight in the optimal schedule of a random instance. Hence, the number of non-zero β_i's is expected to be very small. The initialization step helps in this direction, and this is the reason why the number of modifications is very small if $n < m$.

However, if $n = m$ the behavior of our algorithm completely changes. For example, for $m = 10$ and $n = 10$ we need 279611 modifications, while for $m = 10$ and $n = 20$ we need only 295. Even more, for $m = n = 20$ the primal-dual algorithm does not even converge in 30 minutes. Furthermore, if $m = n$ then the parameters α, $w_{\max}$ and p affect the convergence of the algorithm. For example, in the case where $m = 10$ and $n = 10$, then the following table shows the number of modifications of the dual variables performed by our algorithm when we fix the two of the three parameters. Note that in the last line of the table, the algorithm did not terminate within the time threshold.

Parameters		Modifications
$\alpha = 2$ $w_{\max} = 10$	$p = 0.5$	344
	$p = 0.75$	23915
	$p = 1$	179611
$w_{\max} = 10$ $p = 1$	$\alpha = 2$	279611
	$\alpha = 2.5$	59785
	$\alpha = 3$	10716
$\alpha = 2$ $p = 1$	$w_{\max} = 10$	279611
	$w_{\max} = 50$	406608
	$w_{\max} = 100$	–

In Table 2 we give a comparison of the execution time of the primal-dual algorithm with the execution time of solving directly the convex program using the SeDuMi solver in Matlab. We observe again the difference between $n \neq m$ and $n = m$. In the first case, our algorithm highly outperforms the solver (see Fig. 1). In the second case, our algorithm does not even terminate within 30 minutes if $n = m = 20$, while the execution time of the solver is not affected. Note also that the solver's execution time as well as the execution time of the primal-dual algorithm when $n = m$ depend on the parameters α, w_{max} and p.

Table 2. A comparison of the execution time of the primal-dual approach (PD) with the execution time of the SeDuMi solver for convex programs (CP). The execution times are computed in *seconds*. The values of the table correspond to $\alpha = 2$, $w_{max} = 10$, $p = 1$. Each entry of the table is the average over 30 instances. The empty entries correspond to cases with $m = n$ and take time longer than 30 minutes each and are interrupted.

n	$m = 10$		$m = 20$		$m = 30$		$m = 40$		$m = 50$	
	CP	PD	CP	PD	CP	PD	CP	PD	CP	PD
5	0.59	0.00	0.99	0.00	1.41	0.01	1.83	0.01	2.11	0.01
10	1.22	147.93	1.26	0.01	1.81	0.01	2.42	0.01	2.59	0.01
20	1.25	0.06	3.12	–	2.57	0.02	3.11	0.02	3.92	0.03
30	1.72	0.08	2.58	0.12	5.57	–	4.36	0.03	5.30	0.04
40	2.17	0.10	3.28	0.13	4.38	0.21	8.31	–	6.48	0.05
50	2.67	0.12	4.00	0.16	5.19	0.19	6.72	0.33	11.49	–
60	3.47	0.15	4.96	0.18	6.72	0.23	8.39	0.29	9.87	0.47
70	3.86	0.16	5.99	0.21	7.73	0.26	9.84	0.28	11.42	0.40
100	5.85	0.22	8.62	0.27	11.85	0.32	13.86	0.38	17.56	0.42
150	9.31	0.31	14.34	0.38	19.30	0.47	24.66	0.52	31.10	0.56
200	12.89	0.42	19.87	0.51	28.78	0.59	36.83	0.68	46.31	0.74

The results presented above motivated us to further explore the case $n = m$. For this reason, we performed more experiments for $m = 10, 20, 30, 40, 50$ and $n = m - 5, m - 4, \ldots, m + 4, m + 5$. The results of these experiments are shown in Fig. 2. The horizontal axis corresponds to the difference $m - n$, while the vertical axis corresponds to the logarithm of the modifications of the dual variables made by our algorithm.

We observe that the behavior of the primal-dual algorithm dramatically changes when $n = m$, while there is a much smaller perturbation when $n = m \pm 1$ and $n = m \pm 2$. In all other cases the number of modifications seems to increase linearly with the size of the instance. The problem with the case where $n = m$ probably occurs because in an optimal solution of a random instance almost all processors and jobs are tight, that is the total execution time of each processor and each job is equal to the deadline d. In other words, all β_i's and γ_j's are expected to be non-zero. The primal-dual algorithm, in each iteration "corrects" first the values of β_i's and then the values of γ_j's. As all of them are expected to be non-zero in the optimal solution the required precision plays a significant role to the speed of the convergence of the algorithm.

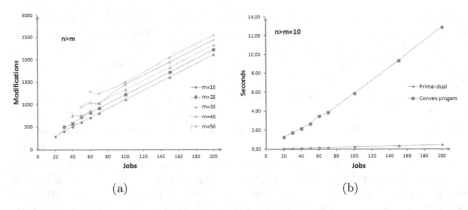

Fig. 1. Parameters: $\alpha = 2$, $w_{\max} = 10$, $p = 1$. (a) The number of modifications of the dual variables made by the primal-dual algorithm if $n > m$. (b) A comparison of the execution times of the primal-dual algorithm and the SeDuMi solver for convex programs if $n > m$ ($m = 10$).

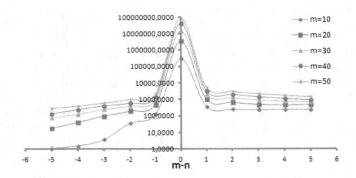

Fig. 2. Parameters: $\alpha = 2$, $w_{\max} = 10$, $p = 1$. The vertical axis represent the logarithm of the modifications of the dual variables made by the primal-dual algorithm.

6   Conclusions

We have presented a primal-dual algorithm in the general setting of convex programming and KKT conditions and we have proved that it converges to an optimal solution. In the direction of exploring the complexity of our algorithm, we performed simulations to observe its behavior when the size of the instance increases. Our experiments highlight the case in which the primal-dual algorithm has a problematic behavior, i.e., when $n = m$. In all other cases, and more interestingly in the case where $n > m$ that is closer to applications, the complexity of the algorithm seems to depend linearly on the size of the instance. An interesting open question remaining is whether our algorithm has a polynomial-time complexity. If not, the design of another algorithm based on the primal-dual paradigm that runs in polynomial time would be a challenging direction.

References

1. Albers, S.: Energy-efficient algorithms. Communications of ACM 53, 86–96 (2010)
2. Albers, S.: Algorithms for dynamic speed scaling. In: STACS 2011, pp. 1–11 (2011)
3. Bampis, E., Letsios, D., Lucarelli, G.: Green scheduling, flows and matchings. In: Chao, K.-M., Hsu, T.-S., Lee, D.-T. (eds.) ISAAC 2012. LNCS, vol. 7676, pp. 106–115. Springer, Heidelberg (2012)
4. Boyd, S., Vandenberghe, L.: Convex Optimization. Cambridge Un. Press (2004)
5. Brucker, P.: Scheduling algorithms, 4th edn. Springer (2004)
6. Devanur, N.R., Papadimitriou, C.H., Saberi, A., Vazirani, V.V.: Market equilibrium via a primal-dual algorithm for a convex program. Journal of the ACM 55(5) (2008)
7. Edmonds, J.: Maximum matching and a polyhedron with 0, 1-vertices. Journal of Research of the National Bureau of Standards, Section B 69, 125–130 (1965)
8. Ford, L.R., Fulkerson, D.R.: Maximal flow through a network. Canadian Journal of Mathematics 8, 399–404 (1956)
9. Goemans, M.X., Williamson, D.P.: The primal-dual method for approximation algorithms and its application to network design problems, ch. 4, pp. 144–191. PWS Publishing Company (1997)
10. Gonzalez, T.: A note on open shop preemptive schedules. IEEE Transactions on Computers C-28, 782–786 (1979)
11. Graham, R.L., Lawler, E.L., Lenstra, J.K., Rinnooy Kan, A.H.G.: Optimization and approximation in deterministic sequencing and scheduling. Annals of Discrete Mathematics 5, 287–326 (1979)
12. Gupta, A., Krishnaswamy, R., Pruhs, K.: Online primal-dual for non-linear optimization with applications to speed scaling. CoRR, abs/1109.5931 (2011)
13. Kuhn, H.W.: The hungarian method for the assignment problem. Naval Research Logistics Quarterly 2, 83–97 (1955)
14. Vazirani, V.V.: Approximation algorithms, ch. 12. Springer (2001)
15. Yao, F., Demers, A., Shenker, S.: A scheduling model for reduced CPU energy. In: FOCS 1995, pp. 374–382 (1995)

Reoptimization in Branch-and-Bound Algorithms with an Application to Elevator Control

Benjamin Hiller, Torsten Klug, and Jakob Witzig

Zuse Institute Berlin
Takustraße 7
D–14195 Berlin, Germany
{hiller,klug,witzig}@zib.de

Abstract. We consider reoptimization (i.e., the solution of a problem based on information available from solving a similar problem) for branch-and-bound algorithms and propose a generic framework to construct a re-optimizing branch-and-bound algorithm. We apply this to an elevator scheduling algorithm solving similar subproblems to generate columns using branch-and-bound. Our results indicate that reoptimization techniques can substantially reduce the running times of the overall algorithm.

1 Introduction

Many powerful solution methods for hard optimization problems, e.g., Lagrangian relaxation and column generation, are based on decomposing a problem into a master problem and one or more subproblems. The subproblems are then repeatedly solved to update the master problem that will eventually be solved to optimality. Usually, the subproblems solved in successive rounds are rather similar; typically, only the cost vector changes reflecting updated information from the master problem (i.e., Lagrangian multipliers in the case of Lagrangian relaxation approaches and dual prices in column generation methods). It is obvious that this similarity in subproblems should be exploited by "warmstarting" the solving process of a subproblem using information from the last round in order to reduce the running time. Methods to achieve this are known as *reoptimization techniques* and have been investigated in the context of decomposition methods for some time now, see e.g., [12] as an example for reoptimization in the context of Lagrangian methods and [2] as an example for column generation methods.

There is much literature on reoptimization of polynomially solvable optimization problems like the (standard) shortest path problems, see e.g. [13] and the references therein. However, for theoretical reasons the decomposition is usually done such that the resulting subproblems are NP-hard, though solvable effectively in practice. Recently, there is also growing theoretical interest in reoptimization methods focusing on the case that only the optimal solution from the last problem is known, see e.g. [1] for a survey.

V. Bonifaci et al. (Eds.): SEA 2013, LNCS 7933, pp. 378–389, 2013.

Usually, specialized combinatorial algorithms are used to solve the subproblems, but there are also cases in which the subproblems may be solved using branch-and-bound algorithms, see e.g., [11,4,10]. Moreover, there is a recent interest to automatically reformulate mixed-integer programs (MIPs) and then apply decomposition techniques [6,7,8,5,3]. The subproblems in these reformulated models are solved by standard MIP solvers, which are very sophisticated branch-and-bound algorithms. Thus there is a need for reoptimization techniques for branch-and-bound-type algorithms.

In this paper, we propose a generic way to implement a reoptimizing branch-and-bound algorithm using the typical ingredients of a branch-and-bound algorithm. The essential idea is to "continue" the branch-and-bound search; in order to do that correctly, we do not only have to keep the search frontier, but also the set of pruned nodes and all solutions found so far. The intuition is that once the cost vector has converged to some extent, the branch-and-bound trees generated in successive rounds are basically the same in the higher levels. Thus the effort for creating this part again may be saved by reoptimization. In contrast to e.g., [2], our approach does not require any assumption on the structure of the change of the cost function, although we propose a way to exploit a special common structure. The details of our approach are presented in Section 2.

The main part of the paper is devoted to an application to elevator control, where reoptimization allows to substantially improve the running times of the column generation algorithm EXACTREPLAN presented in [9,10]. In addition to the extensions suggested by the generic reoptimization scheme, we also adjust the branching rule to facilitate reoptimization. This is necessary to take advantage of additional pruning possibilities that arise from properties of the lower bound used in EXACTREPLAN.

2 Construction of a Reoptimizing Branch-and-Bound Algorithm

To formally introduce our concept of a reoptimizing branch-and-bound algorithm, we consider the following abstract setting. The aim is to solve the combinatorial optimization problem

$$\min\{c(x) \mid x \in \mathcal{S}\}, \tag{1}$$

given by a finite set of feasible solutions $\mathcal{S}$ and a cost function $c \colon \mathcal{S} \to \mathbb{R}$, successively for a sequence of cost functions $(c_i \colon \mathcal{S} \to \mathbb{R})_{i \in I}$, $I \subseteq \mathbb{N}$. For the applications we have in mind, c_{i+1} depends on the solutions obtained for c_i, but this is not used in the following.

Consider a branch-and-bound algorithm $\mathcal{A}$ that solves (1). We could just invoke $\mathcal{A}$ as-is once for every c_i to solve the sequence of optimization problems. We now design a branch-and-bound algorithm $\mathcal{A}'$ that allows to benefit from computations done for c_i when solving (1) with cost function c_{i+1}. To do that, we denote by v a node in $\mathcal{A}$'s search tree corresponding to a subproblem of (1) and think of $\mathcal{A}$ as being specified by the following subalgorithms:

lb(v) A function to compute a lower bound for node v in the search tree.

branch(v) A branching rule that partitions the search space corresponding to v into smaller regions, creating nodes $v_1, \ldots, v_k$, $k \geq 2$, of the search tree.

heu(v) A function to compute a heuristic solution for node v in the search tree; this function may also fail in the sense that no solution is produced. If, however, v corresponds to a single solution x (i.e., v is a leaf in the search tree), heu(v) returns x.

feas(v) A function to check whether the search tree rooted at node v still contains feasible solutions.

Using these subalgorithms, $\mathcal{A}$ basically divides the search space $\mathcal{S}$ into

- the set $\Sigma \subseteq \mathcal{S}$ of feasible solutions found so far,
- a set of subproblems $\mathcal{P}_f$ that have been pruned since the corresponding regions do not contain elements of $\mathcal{S}$ (i.e., for $v \subseteq \mathcal{S}$ feas(v) is false),
- a set of subproblems $\mathcal{P}_c$ that have been pruned since the corresponding regions do not contain elements with cost less than some upper bound U (i.e., for $v \subseteq \mathcal{S}$ we have lb(v) $> U$),
- and a set $\mathcal{O}$ of as-yet unexplored subproblems,

such that we have $\mathcal{S} \subseteq \Sigma \cup \mathcal{S}(\mathcal{P}_c) \cup \mathcal{S}(\mathcal{O})$, where $\mathcal{S}(V) \subseteq \mathcal{S}$ denotes the solutions represented by the search tree nodes V.

As an example, consider the well-known LP-based branch-and-bound algorithm to solve mixed-integer programs. In this case, lb(v) is just the value of the LP relaxation at node v and branch(v) is a (possibly quite sophisticated) branching rule that selects a fractional variable and creates two child nodes with integer bounds for that variable excluding its current fractional value. heu(v) corresponds to the set of heuristics applied at v, including the trivial heuristic that returns the current LP solution if it is integer. Finally, feas(v) is the check whether the LP relaxation at node v is still feasible.

Assuming that $(\Sigma, \mathcal{P}_f, \mathcal{P}_c, \mathcal{O})$ are the corresponding sets after running $\mathcal{A}$ for cost function c_i, we may observe the following:

- The nodes in $\mathcal{P}_f$ do not have to be considered for cost function c_{i+1}.
- An optimal solution for cost function c_{i+1} is contained in $\Sigma \cup \mathcal{S}(\mathcal{P}_c) \cup \mathcal{S}(\mathcal{O})$.
- Any solution $x \in \Sigma$ may be optimal for c_{i+1}.
- A node $v \in \mathcal{P}_c$ might be attractive for cost function c_{i+1}, i.e., lb(v) $\leq U$ for some given upper bound U.
- A node $v \in \mathcal{O}$ might be pruned due to cost for cost function c_{i+1}, i.e., lb(v) $> U$.

Based on these observations it is straightforward to construct a branch-and-bound algorithm $\mathcal{A}'$ that uses the subalgorithms of $\mathcal{A}$ and exploits the computations done in the last round. To this end, $\mathcal{A}'$ maintains the sets Σ and $\mathcal{P}_c$ in addition to $\mathcal{O}^1$ for use in the next round and initializes them properly based on the sets from the last round. A pseudo-code for this reoptimizing branch-and-bound algorithm is shown

---

[1] $\mathcal{O}$ may not be empty in case the branch-and-bound search is stopped early, which is useful when applied to subproblems as a part of a decomposition scheme.

Input: cost function $c\colon S \to \mathbb{R}$; upper bound U; sets Σ', $\mathcal{P}'_c$, $\mathcal{O}'$
Output: sets Σ, $\mathcal{P}_c$, $\mathcal{O}$; set of solutions Σ^* costing at most U
1: $\Sigma \leftarrow \Sigma'$, $\mathcal{P}_c \leftarrow \emptyset$, $\mathcal{O} \leftarrow \emptyset$, $\Sigma^* \leftarrow \emptyset$ ▷ Initialization
2: **for all** $x \in \Sigma$ **do**
3:     Put x in Σ^* if $c(x) \leq U$.
4: **for all** $v \in \mathcal{P}'_c \cup \mathcal{O}'$ **do**
5:     Put v in $\mathcal{O}$ if $\mathsf{lb}(v) \leq U$ and in $\mathcal{P}_c$ otherwise.
6: **while** $\mathcal{O} \neq \emptyset$ **do**   ▷ Standard branch-and-bound
7:     Choose $v \in \mathcal{O}$.
8:     **if** $\mathsf{heu}(v)$ is successful and returns solution x **then**
9:         Put x in Σ.
10:        Put x in Σ^* if $c(x) \leq U$.
11:     $v_1, \ldots, v_k \leftarrow \mathsf{branch}(v)$
12:     **for** $i = 1, \ldots, k$ **do**
13:        **if** $\mathsf{feas}(v)$ **then**
14:           Put v in $\mathcal{O}$ if $\mathsf{lb}(v) \leq U$ and in $\mathcal{P}_c$ otherwise.

Fig. 1. Pseudocode for reoptimizing branch-and-bound algorithm $\mathcal{A}'$

in Figure 1. This version computes the set Σ^* of all feasible solutions with cost at most that of a given upper bound U. This formulation of the optimization is due to our application, where (1) corresponds to a pricing problem in a column generation context and U is some small negative constant, i.e., we look for any columns with negative reduced cost. We may as well consider all truely optimal solutions only.

From the preceding discussion and the logic of standard branch-and-bound, we have the following result.

Theorem 1. *Assuming that* $S = \Sigma' \cup S(\mathcal{P}'_c) \cup S(\mathcal{O}')$, *algorithm* $\mathcal{A}'$ *defined in Figure 1 correctly computes the set* Σ^* *and upon termination we have* $S = \Sigma \cup S(\mathcal{P}_c) \cup S(\mathcal{O})$.

Proof. Consider a solution $x \in S$. In the case that x is in Σ', by Step 1 it will be in Σ, too. Moreover, it will also be in Σ^* iff its cost are at most U. If x is represented by a search tree node $v' \in \mathcal{P}'_c \cup \mathcal{O}'$, i.e., $x \in S(\mathcal{P}'_c) \cup S(\mathcal{O}') = S(\mathcal{P}'_c \cup \mathcal{O}')$, after Step 4 node v' is either in $\mathcal{O}$ (if $\mathsf{lb}(v') \leq U$) or in $\mathcal{P}_c$ (if $c(x) \geq \mathsf{lb}(v') > U$).

Assume now that $v' \in \mathcal{O}$ after the initialization phase. The remaining steps of $\mathcal{A}'$ maintain the invariant $x \in \Sigma \cup S(\mathcal{P}_c \cup \mathcal{O})$. To see this, let v be the node representing x at the beginning of the while loop. In case x is found by $\mathsf{heu}(v)$, it is put in Σ (and in Σ^* if necessary). Otherwise, it will be represented by at least one of the nodes $v_1, \ldots, v_k$ created by $\mathsf{branch}(v)$, say v_1. By definition, $\mathsf{feas}(v_1)$ is true, so v_1 is either put in $\mathcal{O}$ or $\mathcal{P}_c$. Moreover, v_1 is only put in $\mathcal{P}_c$ if $c(x) \geq \mathsf{lb}(v_1) > U$. Thus in case $c(x) \leq U$ it will eventually be found by $\mathsf{heu}()$, and thus be contained in Σ^*.

As for the running time, initializing Σ^* in Step 2 takes $|\Sigma'|$ evaluations of c, which is usually cheap. To initialize $\mathcal{O}$ and $\mathcal{P}_c$ in Step 4 requires $|\mathcal{P}'_c \cup \mathcal{O}|$ evaluations of $\mathsf{lb}()$, which may be rather expensive. It is, however, possible to avoid the recomputation of the lower bounds if the cost functions c_i and c_{i+1} and the lower bound method exhibit special structure. Assume that the cost functions c_i and c_{i+1} have the same (separable) structure of the form

$$c(x) = c_0(x) + \sum_{j=1}^{m} c_j(x)\pi_j, \tag{2}$$

where only the coefficients π_j change from iteration i to iteration $i + 1$. Denote by $c(v)$ the minimum cost of a solution represented by node v. If $\mathsf{lb}(v)$ actually provides (additionally) lower bounds $\underline{c}_j(v)$ for $c_j(x)$, $0 \le j \le m$, for any $x \in \mathcal{S}$ represented by v, we can compute a lower bound for v as

$$c(v) \ge \underline{c}_0(v) + \sum_{j=1}^{m} \underline{c}_j(v)\pi_j, \tag{3}$$

which takes time $O(m)$ for each node v if $\underline{c}_j(v)$, $0 \le j \le m$, are stored with v. A cost structure like (2) arises in the contexts of column generation and Lagrangian relaxation, where the π_j are dual prices or Lagrangian multipliers, respectively.

3   Elevator Control as an Application of a Reoptimizing Branch-and-Bound Algorithm

We now apply our framework for reoptimizing branch-and-bound algorithms to the column-generation-based elevator scheduling algorithm EXACTREPLAN from [9,10]. The EXACTREPLAN algorithm is designed to schedule elevators in destination call systems, where a passenger registers his destination floor upon his arrival at the start floor. Let $\mathcal{E}$ be the set of elevators. A (destination) *call* is a triple of the release time, the start floor and the destination floor corresponding to this registration. Note that the elevator control knows only calls, not about passengers. At any point in time we can build a *snapshot problem* describing the current system state. EXACTREPLAN determines an optimal solution for each snapshot problem, giving the schedule to follow until new information becomes available. In a snapshot problem, the calls are grouped to *requests* according to certain rules reflecting the communication between the passenger and the elevator control. A request has a start floor and a set of destinations floors. We distinguish between *assigned requests* $\mathcal{R}(e)$ for each elevator, for which it has already been decided that elevator e is going to serve them, and *unassigned requests* $\mathcal{R}_u$, which still may be assigned to any elevator. In fact, determining the elevator serving each request $\rho \in \mathcal{R}_u$ is the main task of an elevator control algorithm. Solving a snapshot problem requires to schedule the elevators such that each request is *served*, i.e., there is an elevator traveling to the corresponding start floor (to pick up the calls/passengers) and visiting its destination floors (to drop the calls/passengers) afterwards. In particular, a feasible schedule for elevator e needs to serve all assigned requests $\mathcal{R}(e)$ and may serve any subset of the unassigned requests $\mathcal{R}_u$. We call a selection of feasible schedules, one for each elevator, that together serve all requests, a *dispatch*.

3.1   The Original ExactReplan Algorithm

Let $\mathcal{S}(e)$ be the set of all feasible schedules for elevator e and define $\mathcal{S} := \bigcup_{e \in \mathcal{E}} \mathcal{S}(e)$. For each $S \in \mathcal{S}$ we introduce a decision variable $x_S \in \{0,1\}$ for including a schedule in the current dispatch or not. Denoting by $c(S)$ the cost of schedule S, the following set partitioning model describes the problem:

$$\min \sum_{S \in \mathcal{S}} c(S) x_S \tag{4}$$

$$\text{s.t.} \sum_{S \in \mathcal{S} : \rho \in S} x_S = 1 \quad \forall \rho \in \mathcal{R}_u \tag{5}$$

$$\sum_{S \in \mathcal{S}(e)} x_S = 1 \quad \forall e \in \mathcal{E} \tag{6}$$

$$x_S \in \{0,1\} \quad \forall S \in \mathcal{S} \tag{7}$$

Equations (5) and (6) ensure that each request is served by exactly one elevator and each elevator has exactly one schedule, respectively. Note that the model only decides assignment for the unassigned requests and the assigned requests are treated implicitly by the sets $\mathcal{S}(e)$. The number of variables of this Integer Programming (IP) problem is very large, because each permutation serving a request subset $R \subseteq \mathcal{R}_u$ corresponds to a feasible schedule. We therefore use column generation to solve the LP relaxation of the model above using a branch-and-bound algorithm to solve the following pricing problem.

For all requests $\rho \in \mathcal{R}$ and $e \in \mathcal{E}$ we denote the dual prices associated with constraints (5) and (6) by π_ρ and π_e, respectively. Moreover, let $\mathcal{R}_u(S)$ be the unassigned requests served by schedule S. For each elevator e we have to find $S \in \mathcal{S}(e)$ with negative reduced cost

$$\tilde{c}(S) := c(S) - \sum_{\rho \in \mathcal{R}_u(S)} \pi_\rho - \pi_e \tag{8}$$

or to decide that no such schedule exists. The cost of S is the sum of the cost $c(\rho)$ for serving each request ρ, i.e.,

$$c(S) = \sum_{\rho \in \mathcal{R}(e) \cup \mathcal{R}_u(S)} c(\rho). \tag{9}$$

Pricing via Branch-and-Bound. A schedule S is a sequence of stops $(s_0, \ldots, s_k)$ describing future visits to floors. We enumerate all feasible schedules for elevator e by constructing a schedule stop by stop, branching if there is more than one possibility for the next stop. Thus each search tree node v corresponds to a feasible schedule S_v and one of its stops s_v; the schedule up to s_v is fixed and the later stops correspond to dropping off passengers. Moreover, we maintain for each v the set $A_v \subseteq \mathcal{R}(e)$ of not yet picked up assigned requests and the set $O_v \subseteq \mathcal{R}_u$ of not yet picked up optional requests. At v there are the following branching possibilities: Either the next stop is the one following s_v (if there is one) or the next stop is at a starting floor of a request in $A_v \cup O_v$, which is then picked up there. We create a child node for any of these possibilities.

Our branch-and-bound pricing algorithm computes for each node v a lower bound of the reduced costs by

$$\tilde{\underline{c}}(v) = c(S_v) + \sum_{\rho \in A_v} \underline{c}(\rho) + \sum_{\rho \in O_v\,:\,\underline{c}(\rho) - \pi_\rho < 0} (\underline{c}(\rho) - \pi_\rho) \quad - \pi_e, \qquad (10)$$

where $\underline{c}(\rho)$ is a lower bound on the primal-costs of requests ρ. An important observation is that we can prune all optional requests with $\underline{c}(\rho) - \pi_\rho \geq 0$, leading to a much smaller search tree.

Proposition 1. *Consider a node v of the search tree corresponding to an elevator e and dual prices $(\pi_\rho)_{\rho \in O_v}$. If the search tree rooted at v contains a schedule with negative reduced cost, then it also contains one with negative reduced cost that does not serve the requests in $O_v^{\geq} := \{\rho \in O_v \mid \underline{c}(\rho) - \pi_\rho \geq 0\}$.*

An important feature of our pricing algorithm is that we do not solve the pricing problem to optimality, but stop as soon as k schedules with negative reduced cost are found. These schedules are then added to the set partitioning master problem, whose LP relaxation is then resolved to obtain new dual prices. The rationale for this is to avoid to spend too much time due to bad dual prices.

Similarity of Search Trees. In our computations we observed that the sets of generated nodes in successively generated search trees get more and more similar. To quantify this, we use the following similarity measure for rooted trees [14]. We denote by $\mathfrak{T}$ the set of all rooted trees and define for two rooted trees $\mathcal{T}, \mathcal{T}' \in \mathfrak{T}$ the number

$$\alpha(\mathcal{T}, \mathcal{T}') := |\{v \in \mathcal{T} \mid \text{the unique } (r, v)\text{-path in } \mathcal{T} \text{ is contained in } \mathcal{T}'\}|,$$

where r is the root of $\mathcal{T}$. The similarity $\Lambda(\mathcal{T}, \mathcal{T}') \in [0, 1]$ between $\mathcal{T}$ and $\mathcal{T}'$ is then given by

$$\Lambda : \mathfrak{T} \times \mathfrak{T} \to [0, 1], (\mathcal{T}, \mathcal{T}') \mapsto \begin{cases} \frac{\alpha(\mathcal{T}, \mathcal{T}')}{|V| + |V'| - \alpha(\mathcal{T}, \mathcal{T}')}, & V \neq \emptyset, V' \neq \emptyset \\ 0, & \text{otherwise.} \end{cases} \qquad (11)$$

An example of the evolution of this similarity measure from pricing round to pricing round is shown in Table 1.

3.2   The Reoptimizing ExactReplan Algorithm

In Section 2 we presented a straightforward way to transform a standard branch-and-bound algorithm to a reoptimizing one. Now we aim to apply this scheme to the ExactReplan algorithm. Recall that an upper bound is given by U, the set

Table 1. Example for similarity of rooted trees. Entry $(i,j), i \geq j$, represents the similarity between the rooted search tree $\mathcal{T}_i$ at the end of pricing round i and the rooted search tree $\mathcal{T}_j$ at the end of pricing round j. For instance, 91.87 % of all nodes from the rooted search trees $\mathcal{T}_5$ and $\mathcal{T}_6$ are part of both trees.

$\mathcal{T}_i$	1	2	3	4	5	6	7
1	1.0000	0.2581	0.1048	0.1044	0.1069	0.1030	0.1039
2		1.0000	0.2591	0.2637	0.2658	0.2604	0.2626
3			1.0000	0.5706	0.5105	0.5537	0.5629
4				1.0000	0.8484	0.9098	0.9286
5					1.0000	0.9187	0.8965
6						1.0000	0.9778
7							1.0000

of solutions found so far is denoted by Σ, $\mathcal{P}_c$ is the set of nodes v that have been pruned since $\tilde{c}(v) > U$ and finally, $\mathcal{O}$ is the the set of as-yet unexplored nodes. The implementation of these basic structures and the initialization procedure in Algorithm 1 is straightforward.

We already mentioned that, assuming that the cost function has the structure (2) and the lower bound method "is compatible" with this structure, the updates of the lower bounds for all nodes $v \in \mathcal{P}_c \cup \mathcal{O}$ can be done in time $O(m)$. Observe that the schedule cost function (9) is exactly of type (2) and also the lower bound (10) matches this structure. We can thus use Formula (3) to update the lower bounds for each node v, which only requires storing $\underline{c}_0(v) := c(S_v) + \sum_{\rho \in A_v} \underline{c}(\rho)$ and $\underline{c}(\rho)$ for $\rho \in O_v$ with v. Our computational experiments show [14] that using this fast update of the lower bounds reduces the time spent in the initialization phase by 60–85 %.

A disadvantage of the straightforward reoptimizing branch-and-bound algorithm is that we cannot exploit Proposition 1: It might happen that a request ρ with $\underline{c}(\rho) - \pi_\rho \geq 0$ at iteration i will have $\underline{c}(\rho) - \pi_\rho < 0$ at iteration $j > i$, which we would not detect if we just remove ρ from O_v in iteration i. An immediate consequence is an unnecessarily high number of generated nodes in the reoptimizing branch-and-bound algorithm. To avoid that, we use a different branching procedure when reoptimizing that records pruning due to Proposition 1 explicitly. It is thus equivalent to the original one in the sense that it generates the same search tree when used without reoptimization. To describe this, we introduce the following notation.

- The set of all floors which are branching possibilities at node v is denoted by $\mathcal{B}(v)$.
- $O_v^< := \{\rho \in O_v \mid$ the start floor of ρ is in $\mathcal{B}(v)$ and $\underline{c}(\rho) - \pi_\rho < 0\} \subseteq O_v$
- $O_v^\geq := \{\rho \in O_v \mid$ the start floor of ρ is in $\mathcal{B}(v)$ and $\underline{c}(\rho) - \pi_\rho \geq 0\} \subseteq O_v$
- A node v is called *branched*, if $O_v^\geq = O_v^< = \emptyset$.
- A node v is called *pseudo-branched*, if $O_v^\geq \neq \emptyset$ and $O_v^< = \emptyset$.

In each branching step we branch only on the start floors corresponding to optional requests in $O_v^<$. If $O_v^\geq = \emptyset$, v is branched and can be deleted as in

Table 2. Building A, computational results. The second and fourth column shows the average of generated nodes per snapshot problem and per elevator. The total time in the third and sixth column is the sum over all snapshot problems. Analog the time needed for the initialization at the beginning of each pricing round (fifth column). All times are represented in seconds. The time ratio is the quotient of (total time EXACTREPLAN-REOPT)/(total time EXACTREPLAN).

traffic pattern	#snapshots	EXACTREPLAN		EXACTREPLAN-REOPT			
name		Ø gen. nodes	total time	Ø gen. nodes	Σ init. time	total time	time ratio
Down Peak 80 %	963	39.57	68	**17.13**	2	**44**	0.642
Down Peak 100 %	1133	73.92	134	**32.45**	5	**86**	0.640
Down Peak 144 %	1544	391.18	824	**168.13**	64	**514**	0.624
Interfloor 80 %	980	71.14	117	**30.35**	5	**76**	0.650
Interfloor 100 %	1174	296.45	515	**129.78**	43	**328**	0.637
Interfloor 144 %	1571	10426.06	26621	**2772.96**	3110	**12124**	0.455
Lunch Peak 80 %	980	26.28	52	**11.33**	1	**36**	0.690
Lunch Peak 100 %	1172	75.47	146	**31.92**	6	**94**	0.641
Lunch Peak 144 %	1568	2774.93	6764	**748.98**	598	**2827**	0.418
Real Down Peak 80 %	979	27.77	55	**12.06**	1	**38**	0.689
Real Down Peak 100 %	1171	61.80	119	**27.75**	4	**79**	0.669
Real Down Peak 144 %	1565	337.59	738	**146.38**	61	**471**	0.639
Real Up Peak 80 %	964	125.26	203	**30.67**	6	**74**	0.364
Real Up Peak 100 %	1164	308.59	585	**73.65**	19	**199**	0.340
Real Up Peak 144 %	1714	3914.48	12690	**553.04**	392	**2401**	0.189
Up Peak 80 %	966	65.54	105	**16.91**	2	**45**	0.426
Up Peak 100 %	1159	1305.92	2702	**259.06**	70	**670**	0.248
Up Peak 144 %	1656	29064.20	106597	**3564.73**	2652	**16004**	0.150

Table 3. Building B, computational results. The second and fourth column shows the average of generated nodes per snapshot problem and per elevator. The total time in the third and sixth column is the sum over all snapshot problems. Analog the time needed for the initialization at the beginning of each pricing round (fifth column). All times are represented in seconds. The time ratio is the quotient of (total time ExactReplan-Reopt)/(total time ExactReplan).

| traffic pattern | | ExactReplan | | ExactReplan-Reopt | | | |
name	#snapshots	Ø gen. nodes	total time	Ø gen. nodes	Σ init. time	total time	time ratio
Down Peak 80 %	1830	19.11	105	8.36	1	67	0.641
Down Peak 100 %	2131	25.27	147	10.97	3	92	0.626
Down Peak 144 %	2871	42.22	293	18.35	8	181	0.618
Interfloor 80 %	1868	100.27	403	48.58	23	286	0.710
Interfloor 100 %	2182	246.01	1078	136.23	106	887	0.823
Interfloor 144 %	2752	4637.07	26507	1681.62	3432	16225	0.612
Lunch Peak 80 %	1844	48.40	217	23.09	7	147	0.680
Lunch Peak 100 %	2183	247.52	1087	133.77	105	875	0.805
Lunch Peak 144 %	2772	441.79	2417	249.33	267	1991	0.824
Real Down Peak 80 %	1868	21.29	116	9.16	2	74	0.642
Real Down Peak 100 %	2182	34.22	195	14.77	5	125	0.640
Real Down Peak 144 %	2809	81.17	499	35.93	20	317	0.637
Real Up Peak 80 %	1893	46.19	211	14.86	4	110	0.522
Real Up Peak 100 %	2226	89.98	447	26.14	11	199	0.445
Real Up Peak 144 %	11658	1218.45	37309	257.86	1180	10539	0.282
Up Peak 80 %	1829	46.68	201	14.28	3	99	0.495
Up Peak 100 %	2264	81.75	427	22.01	8	171	0.402
Up Peak 144 %	13268	614.23	27857	121.03	467	6834	0.245

the non-reoptimizing branch-and-bound algorithm. Otherwise we store v in $\mathcal{P}_c$. Additionally, we extend the initialization phase: If a node v from $\mathcal{P}_c$ is not yet branched, we compute the new sets $O_{\bar{v}}^{\geq}$ and $O_v^{<}$ from the set $O_{\bar{v}}^{\geq}$ of the last iteration, creating child nodes for each start floor of an request in $O_v^{<}$. These child nodes are stored in $\mathcal{O}$ for further processing. Moreover, the requests in $O_v^{<}$ are removed from $O_{\bar{v}}^{\geq}$, recording the fact that the corresponding branches have been created. We call this modified branching method PSEUDO-BRANCHING.

The PSEUDO-BRANCHING technique has a positive side effect, namely a reduction of schedules which have to be stored, because we are generating fewer nodes. Thus the time needed for the initialization phases decreases, too. Our computational experiments show that on average, the number of generated nodes decreases by 60 %, the initialization time by 25 % and the number of stored schedules by 24 % due to PSEUDO-BRANCHING.

3.3   Computational Results

In our simulations we consider two buildings and six traffic patterns with three different traffic intensities [10]. Building A has a population of 1400 people, 23 floors and 6 elevators; building B has a population of 3300 people, 12 floors and 8 elevators. The traffic patterns are standard for assessing elevator control algorithms and mimic traffic arising in a typical office building. In the morning, passengers enter the building from the ground floor, causing *up peak traffic*. Then there is some *interfloor traffic* where the passengers travel roughly evenly between the floors. During *lunch traffic*, people leave and reenter the building via the ground floor. Finally, there is *down peak traffic* when people leave the building in the afternoon. In addition, we also consider *real up peak traffic* and *real down peak traffic*, which mix the up peak and down peak traffic which 5% of interfloor and 5% of down peak and up peak traffic, respectively. These two patterns are supposed to model the real traffic conditions more closely than the pure ones. One hour of each traffic pattern is simulated for three different intensities: 80 %, 100 % and 144 % of the population arriving in one hour.

We compare the original EXACTREPLAN algorithm to its reoptimizing version EXACTREPLAN-REOPT with fast updating of lower bounds and PSEUDO-BRANCHING. Both variants of the EXACTREPLAN algorithm solve the LP relaxation of any snapshot problem in the root node to optimality and then solve the resulting IP to optimality without generating further columns. All computations ran under Linux on a system with an Intel Core 2 Extreme CPU X9650 with 3.0 GHz and 16 GB of RAM. We did not use the 64bit facility on this machine. Results are shown in Tables 2 and 3. The number of generated nodes is at least halved on the average. A comparison between the *total time* and the *initialization time* shows that the initialization of the reoptimizing branch-and-bound algorithm is cheap compared to the branching procedure. Moreover, the column *time ratio* shows that it is possible to save up to 85 % of computation time (Up Peak 144 % on building A) when using the reoptimizing variant. Since the branching rules are equivalent w. r. t. generated search trees without reoptimization, the speedup is entirely due to our reoptimization techniques.

4   Conclusion

We proposed a general scheme to use reoptimization in a branch-and-bound algorithm and applied this scheme to the elevator scheduling algorithm ExactReplan based on column generation. Moreover, we adjusted the branching rule of our reoptimizing version of ExactReplan to take advantage of additional pruning possibilities also when using reoptimization. This reoptimizing version of ExactReplan outperforms ExactReplan substanstially up to a factor of 6. As a next step, we want to employ reoptimization also to the branch-and-price version of ExactReplan, which also uses column generation to solve the LP relaxation of nodes below the root. Moreover, we will study reoptimization for LP-based branch-and-bound used in state-of-the-art MIP solvers to improve the performance of automatic decomposition frameworks like GCG [8].

References

1. Ausiello, G., Bonifaci, V., Escoffier, B.: Complexity and Approximation in Reoptimization. In: Computability in Context: Computation and Logic in the Real World. Imperial College Press/World Scientific (2011)
2. Desrochers, M., Soumis, F.: A reoptimization algorithm for the shortest path problem with time windows. European J. Oper. Res. 35, 242–254 (1988)
3. DIP – Decomposition for Integer Programming, https://projects.coin-or.org/Dip
4. Friese, P., Rambau, J.: Online-optimization of a multi-elevator transport system with reoptimization algorithms based on set-partitioning models. Discrete Appl. Math. 154(13), 1908–1931 (2006)
5. Galati, M.: Decomposition in Integer Linear Programming. PhD thesis, Lehigh University (2009)
6. Gamrath, G.: Generic branch-cut-and-price. Diploma thesis, TU Berlin (2010)
7. Gamrath, G., Lübbecke, M.E.: Experiments with a generic Dantzig-Wolfe decomposition for integer programs. In: Festa, P. (ed.) SEA 2010. LNCS, vol. 6049, pp. 239–252. Springer, Heidelberg (2010)
8. GCG – Generic Column Generation, http://www.or.rwth-aachen.de/gcg/
9. Hiller, B.: Online Optimization: Probabilistic Analysis and Algorithm Engineering. PhD thesis, TU Berlin (2009)
10. Hiller, B., Klug, T., Tuchscherer, A.: An exact reoptimization algorithm for the scheduling of elevator groups. Flexible Services and Manufacturing Journal (to appear)
11. Krumke, S.O., Rambau, J., Torres, L.M.: Real-time dispatching of guided and unguided automobile service units with soft time windows. In: Möhring, R.H., Raman, R. (eds.) ESA 2002. LNCS, vol. 2461, pp. 637–648. Springer, Heidelberg (2002)
12. Létocart, L., Nagih, A., Plateau, G.: Reoptimization in Lagrangian methods for the quadratic knapsack problem. Comput. Oper. Res. 39(1), 12–18 (2012)
13. Miller-Hooks, E., Yang, B.: Updating paths in time-varying networks with arc weight changes. Transportation Sci. 39(4), 451–464 (2005)
14. Witzig, J.: Effiziente Reoptimierung in Branch&Bound-Verfahren für die Steuerung von Aufzügen. Bachelor thesis, TU Berlin (2013)

Cluster-Based Heuristics for the Team Orienteering Problem with Time Windows

Damianos Gavalas[1], Charalampos Konstantopoulos[2], Konstantinos Mastakas[3], Grammati Pantziou[4], and Yiannis Tasoulas[2]

[1] Department of Cultural Technology and Communication,
University of the Aegean, Mytilene, Greece
dgavalas@aegean.gr
[2] Department of Informatics,
University of Piraeus, Piraeus, Greece
{konstant,jtas}@unipi.gr
[3] Department of Mathematics, University of Athens, Athens, Greece
kmast@math.uoa.gr
[4] Department of Informatics,
Technological Educational Institution of Athens, Athens, Greece
pantziou@teiath.gr

Abstract. The Team Orienteering Problem with Time Windows (TOPTW) deals with deriving a number of tours comprising a subset of candidate nodes (each associated with a "profit" value and a visiting time window) so as to maximize the overall "profit", while respecting a specified time span. TOPTW has been used as a reference model for the Tourist Trip Design Problem (TTDP) in order to derive near-optimal multiple-day tours for tourists visiting a destination featuring several points of interest (POIs), taking into account a multitude of POI attributes. TOPTW is an NP-hard problem and the most efficient known heuristic is based on Iterated Local Search (ILS). However, ILS treats each POI separately; hence it tends to overlook highly profitable areas of POIs situated far from the current location, considering them too time-expensive to visit. We propose two cluster-based extensions to ILS addressing the aforementioned weakness by grouping POIs on disjoint clusters (based on geographical criteria), thereby making visits to such POIs more attractive. Our approaches improve on ILS with respect to solutions quality, while executing at comparable time and reducing the frequency of overly long transfers among POIs.

Keywords: Tourist Trip Design Problem, Point of Interest, Team Orienteering Problem with Time Windows, Iterated Local Search, Clustering.

1 Introduction

A TTDP [15] refers to a route-planning problem for tourists interested in visiting multiple points of interest (POIs). The objective of the TTDP is to select POIs that match tourist preferences, thereby maximizing tourist satisfaction, while

V. Bonifaci et al. (Eds.): SEA 2013, LNCS 7933, pp. 390–401, 2013.

taking into account a multitude of parameters and constraints (e.g., distances among POIs, visiting time required for each POI, POIs visiting hours, entrance fees) and respecting the time available for sightseeing in daily basis. Different versions of TTDP have been studied in the literature. Herein, we deal with a version of TTDP that considers the following input data: (a) a set of candidate POIs, each associated with the following attributes: a location (i.e. geographical coordinates), time windows(TW) (i.e. opening hours), a "profit" value, calculated as a weighted function of the objective and subjective importance of the POI (subjectivity refers to the users' individual preferences and interests on specific POI categories) and a visiting time (i.e. the anticipated duration of visit of a user at the POI), (b) the travel time among POIs, based on the topological distance between a pair of POIs, (c) the number k of routes that must be generated, based upon the period of stay (number of days) of the tourist at the destination, and (d) the daily time budget B that a tourist wishes to spend on visiting sights; the overall daily route duration (i.e. the sum of visiting times plus the overall travel time among visited POIs) should be kept below B.

By solving the TTDP we expect to derive k routes (typically starting and ending at the tourist's accommodation location) each of length at most B, that maximize the overall collected profit. A well-known optimization problem that may formulate this version of TTDP is the team orienteering problem with time windows (TOPTW) [13]. TOPTW is NP-hard (e.g. see [3], [6]). Hence, exact solutions for TOPTW are feasible for instances with very restricted number of locations (e.g. see the work of Li and Hu [7], which is tested on networks of up to 30 nodes). Note that since the TTDP is typically dealt with online web and mobile applications with strict execution time restrictions, only highly efficient heuristic approaches are eligible for solving it. The most efficient known heuristic for TOPTW is based on Iterated Local Search (ILS) [14], offering a fair compromise with respect to execution time versus deriving routes of reasonable quality. However, ILS treats each POI separately, thereby commonly overlooking highly profitable areas of POIs situated far from current location considering them too time-expensive to visit. ILS is also often trapped in areas with isolated high-profit POIs, possibly leaving considerable amount of the overall time budget unused.

Herein, we introduce CSCRatio and CSCRoutes, two cluster-based algorithmic approaches to the TTDP, which address the shortcomings of ILS. The main incentive behind our approaches is to motivate visits to topology areas featuring high density of 'good' candidate nodes (such areas are identified by a geographical clustering method performed offline); the aim is to improve the quality of derived solutions while not sacrificing time efficiency. Furthermore, both our algorithms favor solutions with reduced number of overly long transfers among nodes, which typically require public transportation rides (such transfers are costly and usually less attractive to tourists than short walking transfers). The remainder of this article is organized as follows: Section 2 overviews TOPTW heuristics. Section 3 presents our novel cluster-based heuristics, while Section 4 discusses the experimental results. Section 5 concludes the paper.

2   Related Work

Labadi et al. [4] proposed a local search heuristic algorithm for TOPTW based on a variable neighborhood structure. In the local search routine the algorithm tries to replace a segment of a path by nodes offering more profit. For that, an assignment problem related to the TOPTW is solved and based on that solution the algorithm decides which arcs to select. Lin et al. [9] proposed a heuristic algorithm for TOPTW based on simulated annealing. On each iteration a neighbouring solution is obtained from the current solution by applying one of the moves swap, insertion or inversion, with equal probability. A new solution is adopted provided that it is more profitable than the current one; otherwise, the new solution might again replace the current one with a probability inversely proportional to their difference in profits. After applying the above procedure for a certain number of iterations the best solution found so far is further improved by applying local search.

The Iterated Local Search (ILS) heuristic proposed by Vansteenwegen et al. [14] is the fastest known algorithm proposed for TOPTW [13]. The algorithm is discussed in the following section. Montemanni and Gambardella proposed an ant colony system (ACS) algorithm [10] to derive solutions for a hierarchical generalization of TOPTW, wherein more than the k required routes are constructed. At the expense of the additional overhead, those additional fragments are used to perform exchanges/insertions so as to improve the quality of the k tours. ACS has been shown to obtain high quality results (that is, low average gap to the best known solution) at the expense of prolonged execution time, practically prohibitive for online applications. In [2] a modified ACS framework (Enhanced ACS) is presented and implemented for the TOPTW to improve the results of ACS.

Labadi et al. [5] recently proposed a method that combines the greedy randomized adaptive search procedure (GRASP) with the evolutionary local search (ELS). Their approach derives solutions of comparable quality and significantly less computational effort to ACS. Compared to ILS, GRASP-ELS gives better quality solutions at the expense of increased computational effort [5]. Tricoire et al. [12] deal with the Multi-Period Orienteering Problem with Multiple Time Windows (MuPOPTW), a generalization of TOPTW, wherein each node may be assigned more than one time window on a given day, while time windows may differ on different days. Both mandatory and optional visits are considered. The authors developed two heuristic algorithms for the MuPOPTW: a deterministic constructive heuristic which provides a starting solution, and a stochastic local search algorithm, the Variable Neighbourhood Search (VNS), which considers random exchanges between chains of nodes. Vansteenwegen et al. [13] argue that a detailed comparison of ILS, ACS and the algorithm of Tricoire et al. [12], is impossible since the respective authors have used (slightly) different benchmark instances. Nevertheless, it can be concluded that ILS has the advantage of being very fast (its execution time is no longer than a few seconds), while ACS, Enhanced ACS and the approach of Tricoire et al. have the advantage of obtaining high quality solutions.

3   Cluster-Based Heuristics

In TOPTW we are given a directed graph $G = (V, A)$ where $V = \{1, ..., N\}$ is the set of nodes (POIs) and A is the set of links, an integer k, and a time budget B. The main attributes of each node are: the service or visiting time (visit$_i$), the profit gained by visiting i (profit$_i$), and each day's time window ([open$_{im}$, close$_{im}$], $m = 1, 2, \ldots, k$) (a POI may have different time windows per day). Every link $(i, j) \in A$ denotes the transportation link from i to j and is assigned a travel cost travel$_{ij}$. The objective is to find k disjoint routes starting from 1 and ending at N, each with overall duration limited by the time budget B, that maximize the overall profit collected by visited POIs in all routes.

The ILS heuristic proposed by Vansteenwegen et al. [14] defines an "insertion" and a "shake" step. At each insertion step (**ILS_Insert**) a node is inserted in a route, ensuring that all following nodes in the route remain feasible to visit, i.e. their time window constraints are satisfied and the time budget is not violated. ILS modeling involves two additional variables for each node i: (a) wait$_i$ defined as the waiting time in case the arrival at i takes place before i's opening time, and (b) maxShift$_i$ defined as the maximum time the start of the visit of i can be delayed without making any visit of a POI in the route infeasible. If a node p is inserted in a route t between i and j, let shift$_p$ = travel$_{ip}$ + wait$_p$ + visit$_p$ + travel$_{pj}$ − travel$_{ij}$ denote the time cost added to the overall route time due to the insertion of p. The node p can be inserted in a route t between i and j if and only if start$_{it}$ +visit$_i$ + travel$_{ip}$ ≤ close$_{pt}$ and at the same time shift$_p$ ≤ wait$_j$ + maxShift$_j$. For each node p not included in a route, its best possible insert position is determined by computing the lowest insertion time cost (shift$_p$). For each of these possible insertions the heuristic calculates the ratio ratio$_p = \dfrac{\text{profit}^2}{\text{shift}_p}$, which represents a measure of how profitable is to visit p versus the time delay this visit incurs. Among all candidate nodes, the heuristic selects for insertion the one with the highest ratio .

At the shake step (**Shake**) the algorithm tries to escape from local optimum by removing a number of nodes in each route of the current solution, in search of non-included nodes that may either decrease the route time length or increase the overall collected profit. The shake step takes as input two integers: (a) the removeNumber that determines the number of the consecutive nodes to be removed from each route and (b) the startNumber that indicates where to start removing nodes on each route of the current solution. If throughout the process, the end location is reached, then the removal continues with the nodes following the start location.

To the best of our knowledge, ILS is the fastest known algorithm for solving the TOPTW offering a fair compromise in terms of speed versus deriving routes of reasonable quality. However, it presents the following weaknesses: (i) During the insertion step, ILS may rule out candidate nodes with high profit value because they are relatively time-expensive to reach (from nodes already included in routes). This is also the case even when whole groups of high profit nodes are located within a restricted area of the plane but far from the current route instance. In case that the route instance gradually grows and converges towards the high profit nodes, those may be no longer feasible to insert due to overall route time constraints. (ii) In the insertion step, ILS may be attracted and include into the solution some high-score nodes isolated from high-density topology areas. This may trap ILS and make it infeasible to visit far located areas with "good" candidate nodes due to prohibitively large traveling time (possibly leaving considerable amount of the overall time budget unused).

Herein, we propose two heuristic algorithms, Cluster Search Cluster Ratio (CSCRatio) and Cluster Search Cluster Routes (CSCRoutes), which address the aforementioned weaknesses of the ILS algorithm. Both algorithms employ clustering to organize POIs into groups (clusters) based on topological distance criteria. POIs at the same cluster are close to each other e.g., they are within walking distance or they belong to the same area of the city. Having visited a high-profit POI that belongs to a certain cluster, our algorithms encourage visits to other POIs at the same cluster because such visits reduce (a) the duration of the routes and (b) the number of transfers among clusters. Note that a tourist apart from maximizing the total profit, may also prefer to minimize inter-cluster tranfers as those are typically long and require usage of public transportation, this may incur a considerable budget cost, while walking is usually a preferred option than using the public transportation.

Both CSCRatio and CSCRoutes employ the global k-means algorithm [8] to build a clustering structure consisting of an appropriate (based on the network topology) number of clusters (numberOfClusters). Once the clusters of POIs have been formed during a preprocessing (clustering) phase, a route initialization phase **RouteInitPhase** starts. During this phase one POI is inserted into each of the k initially empty routes. Each of the k inserted POIs comes from a different cluster, i.e. no two inserted POIs belong to the same cluster. Since the number of clusters is usually larger that k we need to decide which k clusters will be chosen in the route initialization phase. Different approaches may be followed such as choosing the k clusters with the highest total profit, or trying different sets of k high-profit clusters and run CSCRatio and CSCRoutes algorithms for each such set searching for the best possible solution. Following the second approach, we consider a listOfClusterSets list containing a specific number of different sets of k high-profit clusters. The list may contain all k-combinations of the elements of a small set S with the most profitable clusters. **RouteInitPhase** takes as argument a set of k clusters from listOfClusterSet and proceeds as follows: for each cluster C_i in the set, it finds the POI $p \in C_i$ with the highest $ratio_p$ and inserts it into one of the empty routes. By initializing each one of the k routes of the TOPTW solution with a POI from different clusters the algorithms encourage searching different areas of the network and avoid getting trapped at specific high-scored nodes. Then the algorithms combine an insertion step and a shake step to escape from local optima.

Cluster Search Cluster Ratio Algorithm. The CSCRatio algorithm introduces an insertion step **CSCRatio_Insert** which takes into account the clustering of the POIs by using a parameter clusterParameter ≥ 1. The higher the value of clusterParameter, the more the insertion of a node p before or after a node that belongs to the same cluster with p is favored. Specifically, the parameter clusterParameter is used to increase the likelihood of inserting p between i and j if p belongs to the same cluster with either i or j. For that, CSCRatio considers the variable shiftCluster$_p$ defined as the ratio $\frac{\text{shift}_p}{\text{clusterParameter}}$ in the case that cluster(p) coincides with cluster(i) or cluster(j) (cluster(l) denotes the cluster where a node l belongs to). Otherwise, shiftCluster$_p$ = shift$_p$. Then the lowest insertion time cost (shiftCluster$_p$) i.e. the best possible insert position for p, is determined. For each of those best possible insertions, the heuristic calculates ratio$_p$ = $\frac{\text{profit}_p^2}{\text{shiftCluster}_p}$. CSCRatio initializes the clusterParameter with the value of 1.3 in order to initially encourage visits to be within the same clusters and decreases the value of clusterParameter by 0.1 every a quarter of maxIterations. At the last quarter the **CSCRatio_Insert** step becomes the same as **ILS_Insert**. Thus, routes with a lot of POIs belonging to the same cluster are initially favored, while as the number of iterations without improvement increases, the diversification given by ILS is obtained.

The maximum value of the parameter removeNumber used in the shake step is allowed to be half of the size of the largest route (currentSolution.maxSize) in the current solution and not $\frac{N}{3k}$ as in ILS [14]. In this way, execution time is saved, since local optimum is reached in short time, if a small portion of the solution has been removed. As a result, the number of iterations of CSCRatio can be larger than the number of iterations of ILS [14] without increasing the overall algorithm's execution time.

CSCRatio loops for a number of times equal to the size of the listOfClusterSets. Within the loop, firstly all POIs included into the current solution's routes are removed and the route initialization phase is executed with argument a set of high-profit clusters taken (pop operation) from the listOfClusterSets list. Secondly, the algorithm initializes the parameters startNumber and removeNumber of **Shake** to 1 and the parameter clusterParameter of **CSCRatio_Insert** as discussed above, and executes an inner loop until there is no improvement of the best solution for maxIterations successive iterations. The insertion step is iteratively applied within this loop until a local optimum is reached. Lastly, the shake step is applied. The pseudo code of CSCRatio algorithm is listed below (Algorithm 1). In order to reduce the search space (therefore, the execution time) of **CSCRatio_Insert**, in case that a non-included POI p is found infeasible to insert in any route, it is removed from the list of candidate POIs and added back, only after **Shake** has been applied.

Cluster Search Cluster Routes Algorithm. Given a route t of a TOPTW solution, any maximal sub-route in t comprising a sequence of nodes within the same cluster C is defined as a *Cluster Route (CR)* of t *associated with cluster* C and denoted as CR_C^t. The length of CR_C^t may be any number between 1 and $|C|$. Note that a route t of a TOPTW solution constructed by the ILS or CSCRatio algorithm may include more than one cluster route CR_C^t for the same cluster C, i.e., a tour t may visit and leave cluster C more than once. CSCRoutes

run the global k-means algorithm with k=numberOfClusters
construct the list listOfClusterSets
$it1 \leftarrow \frac{maxIterations}{4}$; $it2 \leftarrow \frac{2 \cdot maxIterations}{4}$; $it3 \leftarrow \frac{3 \cdot maxIterations}{4}$
while listOfClusterSets is not empty **do**
    remove all POIs visited in the currentSolution
    theClusterSetIdToInsert $\leftarrow$ listOfClusterSets.pop
    RouteInitPhase(theClusterSetIdToInsert)
    startNumber $\leftarrow 1$; removeNumber $\leftarrow 1$; notImproved $\leftarrow 0$
    while notImproved < maxIterations **do**
        if notImproved < it2 **then**
            if notImproved < it1 **then** clusterParameter $\leftarrow 1.3$
            else clusterParameter $\leftarrow 1.2$
            end if
        else
            if notImproved < it3 **then** clusterParameter $\leftarrow 1.1$
            else clusterParameter $\leftarrow 1.0$
            end if
        end if
        while not local optimum **do**
            CSCRatio_Insert(clusterParameter)
        end while
        if currentSolution.profit > bestSolution.profit **then**
            bestSolution $\leftarrow$ currentSolution ; removeNumber $\leftarrow 1$; notImproved $\leftarrow 0$
        else increase notImproved by 1
        end if
        if removeNumber > $\frac{currentSolution.maxSize}{2}$ **then** removeNumber $\leftarrow 1$
        end if
        Shake(removeNumber,startNumber)
        increase startNumber by removeNumber
        increase removeNumber by 1
        if startNumber $\geq$ currentSolution.sizeOfSmallestTour **then**
            decrease startNumber by currentSolution.sizeOfSmallestTour
        end if
    end while
end while
return bestSolution

Algorithm 1. CSCRatio(numberOfClusters,maxIterations)

algorithm is designed to construct routes that visit each cluster at most once, i.e. if a cluster C has been visited in a route t it cannot be revisited in the same route and therefore, for each cluster C there is only one cluster route in any route t associated with C. The only exception allowed is when the start and the end node of a route t belong to the same cluster C'. In this case, a route t may start and end with nodes of cluster C', i.e. C' may be visited twice in the route t and therefore, for a route t there might be two cluster routes $CR_{C'}^{t}$.

The insertion step **CSCRoutes_Insert** of CSCRoutes does not allow the insertion of a node p in a route t, if this insertion creates more than one cluster routes CR_C^{t} for some cluster C. Therefore, a POI cannot be inserted at any position in the route t. In the sequel, the description of insertion step **CSCRoutes_Insert** is given, based on the following assumptions. Consider w.l.o.g. that the start and end nodes in the TOTPW coincide (*depot*). If a route t contains two CRs associated with the cluster of the *depot*, then let CR_f^t be the first cluster route (starts at the *depot*) in t, and CR_l^t be the last cluster route (ends at the *depot*) in t. Also, assume that for each POI p ratio$_p$ is calculated as in ILS algorithm. Finally, consider for each route t, the list listOfClusters(t) containing any cluster C for which there is a nonempty CR_C^t. Given a candidate for insertion node p and a route t, **CSCRoutes_Insert** distinguishes among the following cases:

- cluster(p)=cluster(*depot*) and listOfClusters(t) contains only the cluster (*depot*). Then p can be inserted anywhere in the route, since the insertion would not violate the CR constraints.
- cluster(p)=cluster(*depot*) and listOfClusters(t) contains more than one cluster. Then p can be inserted anywhere in CR_f^t and in CR_l^t.
- cluster(p)$\neq$cluster(*depot*) and listOfClusters(t) contains only cluster(*depot*), then the insertion is feasible anywhere in t. If the insertion occurs, then a new CR will be created with p as its only POI.
- cluster(p)$\neq$cluster(*depot*) and listOfClusters(t) contains two or more clusters but not cluster(p). Then p can be inserted after the end of every CR in t. If the insertion occurs, then a new CR will be created with p as its only POI.
- cluster(p)$\neq$cluster(*depot*) and listOfClusters(t) contains two or more clusters and also includes cluster(p). Then p can be inserted anywhere in $CR_{cluster(p)}^t$.

The CSCRoutes algorithm is likely to create solutions of lower quality (w.r.t. overall profit), especially in instances featuring tight time windows. However, it significantly reduces the number of transfers among clusters and therefore it favors routes that include POIs of the same cluster. Thus, walking transfers are preferred while overly long travel distances are minimized. In effect, CSCRoutes is expected to perform better than ILS and CSCRatio with respect to execution time, since **CSCRoutes_Insert** is faster than **ILS_Insert** and **CSCRatio_Insert** (this is because the number of possible insertion positions for any candidate node is much lower).

4 Experimental Results

Test Instances. Montemanni and Gambardella [10] designed TOPTW instances based on previous OPTW instances of Solomon [11] and Cordeau et al. [1] (data sets for vehicle routing problems with time windows). Solomons instances comprise 100 nodes, with c1*, r1* and rc1* featuring much shorter time budget and tighter time windows than c2*, r2* and rc2* instances. Likewise, Cordeau et al. instances feature 48-288 nodes, constant time budget (=1000 min) and average time windows equal to 135 min and 269 min for pr01-10 and pr11-20 instances, respectively. All the aforementioned instances involve one, two, three and four tours.

The aforementioned instances, though, are not suitable for real-life TTDP problems, wherein: (a) POIs are typically associated with much wider, overlapping, multiple time windows; (b) POIs are densely located at certain areas, while isolated POIs are few; (c) visiting time at a POI is typically correlated with its profit value ; (d) the daily time budget available for sightseeing is typically in the order of a few hours per day (in contrast, most existing instances define unrealistically long time budgets). Along this line, we have created 100 new TOPTW instances (t*) with the following characteristics: the number of tours is 1-3; the number of nodes is 100-200; 80% of the nodes are located around 1-10 zones; the visiting time at any vertex is 1-120 min and proportional to the profit; regarding time windows, we assume that 50% of the nodes are open in 24h basis, while the remaining are closed either for one or two days per week (during their opening days, the latter are open 08:30-17:00); the daily time budget is set to 10h in t1* and 5h in t2* instances, respectively. The instances of Montemanni and Gambardella are available in http://www.mech.kuleuven.be/en/cib/op/, while the t* instances in http://www2.aegean.gr/dgavalas/public/op_instances/.

Results. All computations were carried out on a personal computer Intel Core i5 with 2.50 GHz processor and 4 GB RAM. Our tests compared our proposed algorithms against the best known real-time TOPTW approach (ILS). Clearly, mostly preferred solutions are those associated with high profit values, low number of transfers and reduced execution time. CSCRatio and CSCRoutes set the value of maxIterations equal to $\frac{400}{|\text{listOfClusterSets}|} \cdot \frac{k+1}{2 \cdot k}$. ListOfClusterSets is implemented by adding $\lceil \text{numberOfClusters}/k \rceil$ disjoint sets of k clusters which are randomly selected from the set of the clusters. The value of numberOfClusters is set to $N/10$.

Table 1 illustrates the average gaps among CSCRatio and ILS over all the existing and new test instances, with respect to profit, number of transfers and execution time; the existing instances have been tested on 1 to 4 tours. Positive gaps denote prevalence of our algorithm against ILS (the opposite is signified by negative gap values). CSCRatio yields significantly higher profit values, especially for instances with tight daily time budget and small number of tours (e.g. 0.79 in r1* and 2.04 in rc1*, for one tour). This is because ILS is commonly trapped in isolated areas with few high profit nodes, failing to explore remote areas with considerable numbers of fairly profited candidate nodes, As regards the number of transfers, CSCRatio clearly prevails, mainly when the time budget is prolonged (e.g. in c2*, r2* and rc2* instances), as it prioritizes the successive placement of nodes assigned to the same cluster into the tours. ILS and CSCRatio attain similar execution times in most cases, however the former clearly executes faster when examining instances with both long time budget and wide time windows. With regards to our new benchmark instances (i.e. t1* and t2*) CSCRatio achieves considerably higher profit gaps than ILS, especially when considering instances featuring tight time budgets (t2* instances). This improvement is attributed to the RouteInitPhase incorporated into both our proposed algorithms, which increases the probability of initially inserting high-profit nodes located on far-reached clusters (such nodes are typically overlooked by ILS itineraries due to the high travel time, hence, low insertion ratio). On the other hand, ILS performs better as regards the number of transfers yield on t2* instances (CSCRatio commonly explores areas far located from the depot, hence, it is forced to perform a number of inter-cluster transfers to connect those areas to the depot). Last, the two algorithms present comparable execution times.

Table 1. Average gaps between ILS and CSCRatio for Solomon, Cordeau et al., t* instances

Name	Profit Gap(%)				Transfers Gap (%)				Time Gap (%)			
	1	2	3	4	1	2	3	4	1	2	3	4
c1*	0.21	0.32	0.53	0.68	-0.2	-0.01	3.12	6.21	-40.8	8.45	35	24.9
c2*	0.84	0.79	0.29	0	19.1	12.6	12.3	20	-4.29	-18.2	-101	-398
r1*	0.79	0.91	-0.57	0.33	4.96	4.55	-0.11	2.71	-20.7	18.1	39.3	21.9
r2*	0.11	0.47	0.03	0	9.78	9.86	10.8	14	-4.88	-120	-305	-608
rc1*	2.04	0.87	0.81	-0.47	9.17	3.75	4.81	4.94	-1.07	36.3	34.7	44.7
rc2*	0.45	-0.34	0.32	0	5.49	1.83	0.48	5.49	11.9	-38.4	-197	-416
pr*	1.46	-0.02	0.4	0.9	-0.72	-9.99	4.5	4.62	29.1	27.4	7.44	-27.4
t1*	0.28				2.19				-5.27			
t2*	2				-13.2				8.33			

Table 2 illustrates the average gaps among CSCRoutes and ILS. The results indicate a trade-off between the profit and the number of transfers. In particular, ILS yields improved quality solutions as it inserts best candidate nodes freely, irrespective of their cluster assignment. This is especially true when considering instances which combine long time budgets with tight time windows (e.g. r2*), whereby CSCRoutes fails to use the time budget effectively, as it might get trapped within clusters, spending considerable amounts of time waiting for the nodes opening time, while not allowed to escape by visiting neighbor cluster nodes. This disadvantage is mitigated when the number of tours increases, as high-profit nodes are then more likely to be selected. On the other hand, CSCRoutes clearly improves on ILS with respect to the number of transfers due to its focal design objective to prohibit inter-cluster transfers. CSCRoutes also attains shorter execution times (excluding the c2*, r2* and rc2* instances for 4 tours), as it significantly reduces the search space on its insertion phase (i.e. in order to insert a new vertex between a pair of nodes that belongs to the same cluster, it only examines nodes assigned to the same cluster). As regards the new benchmark instances, ILS yields higher profit values than CSCRoutes in t1*, however, the performance gap is decreased compared to the results reported on previous instances. This is due to the wider and overlapping time windows chosen in t* instances, which diminishes the wait time (until opening) and allows more effective use of the budget time by CSCRoutes. CSCRoutes performs much better with respect to number of transfers and execution time. Interestingly, the results differ significantly on t2* instances, with CSCRoutes deriving solutions of considerably higher quality at the expense of increased number of transfers. This is mainly due to some outlier values, which largely affect the average value. In those instances, CSCRoutes is initialized inserting a far-located high-profit vertex and is forced to traverse a number of intermediate clusters in order to

Table 2. Average gaps between ILS and CSCRoutes for Solomon, Cordeau et al., t* instances

Name	Profit Gap(%)				Transfers Gap (%)				Time Gap (%)			
	1	2	3	4	1	2	3	4	1	2	3	4
c1*	-1.65	-3.59	-1.03	-1.36	19.2	22.1	23.8	20.2	-21.1	29	38.2	31.2
c2*	-0.82	0.79	0.14	0	36	30.5	25.7	37.7	65.5	57.9	10.4	-170
r1*	-1.2	-1.27	-2.37	-2.15	23	21.7	16.6	22.2	0.1	36.5	49.4	38.6
r2*	-15.5	-10.3	-3.79	-1.21	56.3	55.4	52.6	46.4	76.7	25.1	-77.7	-284
rc1*	1.06	-1.8	-1.26	-1.86	16.7	11.9	14.8	14.4	10.7	50.3	42.9	52.9
rc2*	-9.5	-12.5	-8.21	-2.63	39.7	42.6	44.5	45.2	76	51.1	-40	-203
pr*	-8.11	-8.11	-5.44	-4.8	35.5	34.1	32.6	32.4	62.2	62.9	42.9	14.6
t1*	-0.52				5.31				22.2			
t2*	1.91				-4.5				4.59			

connect it to the depot vertex. It is noted that CSCRoutes retains lead over ILS with regard to the execution time on t2* instances.

5   Conclusions

The comparison of CSCRatio over ILS demonstrated that CSCRatio achieves higher quality solutions in comparable execution time (especially when considering limited itinerary time budget), while also reducing the average number of transfers. As regards the CSCRoutes-ILS comparison, the former clearly prevails in situations where the reduction of inter-cluster transfers is of critical importance. The transfers gap, though, is achieved at the expense of slightly lower quality solutions. CSCRoutes achieves the best performance results with respect to execution time, compared to ILS and CSCRatio. Notably, the performance gap of our algorithms over ILS increases when tested on realistic TTDP instances, wherein nodes are located nearby each other and feature wide, overlapping time windows, while the daily time budget is 5-10h. We argue that our two cluster-based heuristics may be thought of as complementary TTDP algorithmic options. The choice among CSCRatio and CSCRoutes (when considering real-world online TTDP applications) should be determined by user-stated preferences. For instance, a user willing to partially trade the quality of derived solutions with itineraries more meaningful to most tourists (i.e. mostly walking between successive POI visits, rather than public transportation transfers) should opt for the CSCRoutes algorithm.

Acknowledgement. This work was supported by the EU FP7/2007-2013 (DG CONNECT.H5-Smart Cities and Sustainability), under grant agreement no. 288094 (project eCOMPASS).

References

1. Cordeau, J.-F., Gendreau, M., Laporte, G.: A tabu search heuristic for periodic and multi-depot vehicle routing problems. Networks 30, 105–119 (1997)
2. Gambardella, L.M., Montemanni, R., Weyland, D.: Coupling ant colony systems with strong local searches. European Journal of Operational Research 220(3), 831–843 (2012)

3. Golden, B.L., Levy, L., Vohra, R.: The orienteering problem. Naval Research Logistics (NRL) 34(3), 307–318 (1987)
4. Labadi, N., Mansini, R., Melechovský, J., Wolfler Calvo, R.: The team orienteering problem with time windows: An lp-based granular variable neighborhood search. European Journal of Operational Research 220(1), 15–27 (2012)
5. Labadi, N., Melechovský, J., Wolfler Calvo, R.: Hybridized evolutionary local search algorithm for the team orienteering problem with time windows. Journal of Heuristics 17, 729–753 (2011)
6. Laporte, G., Martello, S.: The selective travelling salesman problem. Discrete Applied Mathematics 26(2-3), 193–207 (1990)
7. Li, Z., Hu, X.: The team orienteering problem with capacity constraint and time window. In: The Tenth International Symposium on Operations Research and Its Applications (ISORA 2011), pp. 157–163 (August 2011)
8. Likas, A., Vlassis, N., Verbeek, J.: The global k-means clustering algorithm. Pattern Recognition 36(2), 451–461 (2003)
9. Lin, S.-W., Yu, V.F.: A simulated annealing heuristic for the team orienteering problem with time windows. European Journal of Operational Research 217(1), 94–107 (2012)
10. Montemanni, R., Gambardella, L.M.: An ant colony system for team orienteering problems with time windows. Foundations of Computing and Decision Sciences 34(4), 287–306 (2009)
11. Solomon, M.: Algorithms for the Vehicle Routing and Scheduling Problems with Time Window Constraints. Operations Research 35, 254–265 (1987)
12. Tricoire, F., Romauch, M., Doerner, K.F., Hartl, R.F.: Heuristics for the multiperiod orienteering problem with multiple time windows. Computers & Operations Research 37(2), 351–367 (2010)
13. Vansteenwegen, P., Souffriau, W., Van Oudheusden, D.: The orienteering problem: A survey. European Journal of Operational Research 209(1), 1–10 (2011)
14. Vansteenwegen, P., Souffriau, W., Vanden Berghe, G., Van Oudheusden, D.: Iterated local search for the team orienteering problem with time windows. Comput. Oper. Res. 36, 3281–3290 (2009)
15. Vansteenwegen, P., Van Oudheusden, D.: The mobile tourist guide: An or opportunity. Operational Research Insight 20(3), 21–27 (2007)

Finding Robust Solutions for the Stochastic Job Shop Scheduling Problem by Including Simulation in Local Search

Marjan van den Akker, Kevin van Blokland, and Han Hoogeveen

Department for Information and Computing Sciences
Utrecht University
P.O. Box 80089, 3508 TB Utrecht, The Netherlands
{J.M.vandenAkker,C.H.M.vanBlokland,J.A.Hoogeveen}@uu.nl

Abstract. Although combinatorial algorithms have been designed for problems with given, deterministic data, they are often used to find good, approximate solutions for practical problems in which the input data are stochastic variables. To compensate for the stochasticity, in many cases the stochastic data are replaced, either by some percentile of the distribution, or by the expected value multiplied by a 'robustness' factor; the resulting, deterministic instance is then solved, and this solution is run in practice. We apply a different approach based on a combination of local search and simulation. In the local search, the comparison between the current solution and a neighbor is based on simulating both solutions a number of times. Because of the flexibility of simulation, each stochastic variable can have its own probability distribution, and the variables do not have to be independent. We have applied this method to the job shop scheduling problem, where we used simulated annealing as our local search method. It turned out that this method clearly outperformed the traditional rule-of-thumb methods.

Keywords: Stochastic variables, simulation, local search, job shop scheduling, simulated annealing, simulation optimization.

1 Introduction

One of the standard assumptions in traditional machine scheduling theory is that the processing times are given, deterministic values. In practice, however, this assumption is violated once in a while, and more and more attention is spent on problems with stochastic processing times. In this paper, we look at the job shop scheduling problem where (some of the) processing times are stochastic variables. We assume that all characteristics are known at time zero, that is, we are scheduling a *batch of stochastic jobs*, according to the division by Niño-Mora (2008) of stochastic scheduling problems. Moreover, we assume that we have to come up with a *nonpreemptive static list policy* in the terminology of Pinedo (2005), which implies that it is not possible (desired) to adjust the solution when information becomes available about the realized processing times.

V. Bonifaci et al. (Eds.): SEA 2013, LNCS 7933, pp. 402–413, 2013.
© Springer-Verlag Berlin Heidelberg 2013

The theoretical results obtained for stochastic scheduling problems are quite different from the ones obtained for their deterministic counterparts. Most importantly, only problems in which all processing times follow a similar distribution with nice characteristics are analyzable. Pinedo (2005) describes a number of such problems in which an optimal nonpreemptive static list policy can be constructed. In most of these cases, a similar rule can be applied to solve both the stochastic problem and its deterministic counterpart. For example, Rothkopf (1966) showed that the stochastic problem of minimizing the expected total weighted completion time can be tackled by solving the deterministic problem that is obtained by taking the expected processing time; hence, the single-machine problem can be solved using Smith's rule (Smith, 1956), but the problem with two or more parallel, identical machines is $\mathcal{NP}$-hard. On the contrary, if the processing times are exponentially distributed, then the problem of minimizing the expected maximum completion time in case of two parallel, identical machines is solvable through the list scheduling rule where the jobs are added in order of nonincreasing expected processing time (Pinedo and Weiss, 1979), whereas its deterministic counterpart is $\mathcal{NP}$-hard. Moreover, in some cases the problem has to be adjusted to make it meaningful. For example, since the completion times become stochastic variables as well, hard deadlines do not make sense anymore and have to be reformulated. Van den Akker and Hoogeveen (2008) and Trietsch and Baker (2008) use a relaxation of these 'hard' deadlines by issuing *chance constraints*, which state that the probability that a deadline is missed should remain below a given upper bound. Using this concept, they show that, when the processing times are drawn from a probability distribution that possesses a number of properties, then the problem of minimizing the number of tardy jobs on one machine is solvable by Moore-Hodgson's rule (Moore, 1968).

In this paper, we look at the job shop scheduling problem, for which no clear-cut solution method exists; Williamson et al. (1997) have shown that already the problem of deciding whether there exists a feasible schedule of length 4 is $\mathcal{NP}$-hard in the strong sense. Moreover, the job shop scheduling problem is much more affected by the presence of stochasticity in the processing times than for example single machine problems, because of the possibility of delay propagation between machines due to the precedence constraints. To counter this, we apply a local search algorithm that has been designed for the deterministic case, which we make applicable to deal with the stochastic processing times. Our goal is to find a good, robust solution, where the quality of the solution is measured by the expected value. The only assumption we need here is that we, either know for each random variable the probability distribution that it follows, or that we can generate it from historic data; we do not require that each random variables comes from the same probability distribution. Furthermore, we do not require that the stochastic variables are independent, as long as we know their covariance matrix. We compare several methods to adapt the local search algorithm to the stochastic processing times, some of which are based on discrete-event simulation (Law, 2007). We have conducted extensive computational experiments to test the resulting algorithms. It turns out that incorporating simulations in the local search clearly outperforms the traditional methods.

The outline of our paper is as follows. In Section 2 we give an introduction to the job shop scheduling problem. In Section 3 we describe the local search algorithm that we want to apply. In Section 4 we discuss the possible adaptations of the local search algorithm that are needed to make these applicable to the case with stochastic processing times. In Section 5 we discuss the computational experiments, and in Section 6 we draw some conclusions.

2   The Job Shop Scheduling Problem

Our method to deal with stochastic processing times is based on local search. We illustrate our approach on the *job shop scheduling problem*, which is defined as follows. In a job shop scheduling problem we have m machines, which have to carry out n jobs. Each job j ($j = 1, \ldots, n$) consists of a chain of operations, which must be executed without interruption in the given order: operation $i+1$ of job j can only start when operation i of job j has been completed. For each operation, we know the machine that has to execute it, and we know the characteristics of the random variable that represents its processing time. We assume that each machine is constantly available from time zero onwards, and that each machine can handle only one operation at a time. The goal is to find for each machine an order in which the operations should be executed, such that the expected makespan, which is defined as the time by which all jobs have been completed, is minimum. We assume that all relevant data are known at time zero. We follow the common assumption that each job should visit each machine exactly once; hence, each job consists of m operations. Formally, we use J_j ($j = 1, \ldots, n$) to denote job j, and we use O_{ij} ($i = 1, \ldots, m; j = 1, \ldots, n$) to denote the ith operation of J_j. Extending the three-field notation scheme introduced by Graham et al (1979), we denote our problem by $J|\text{stoch } p_{ij}|E(C_{\max})$.

Our approach can easily be made suitable for any kind of job shop scheduling problem, as long as the starting times and completion times are not bounded from above, which implies that we cannot handle deadlines and precedence constraints that incur exact and/or maximum delays between operations. Extensions that can be handled are minimum delays (which decree that at least a given amount of time should elapse between the execution of two operations in the same job), release dates, and more complex precedence constraints between the operations. In case of deadlines and/or maximum delays, we can use our approach, if these deadlines and/or maximum delays are stated in the form of chance constraints.

Since the deterministic job shop scheduling problem is $\mathcal{NP}$-hard and, more importantly, since it is very hard from a computational point of view, many researchers have studied local search methods, like for example Tabu Search based algorithms (Taillard, 1994; Nowicki and Smutnicki, 1996; Ten Eikelder et al., 1999), Simulated Annealing based algorithms (Yamada and Nakano, 1996) and, more recently, the hybrid Genetic Algorithms by: Wang and Zheng (2001), who combine a GA with Simulated Annealing; Gonçalves et al. (2005), who use a GA in combination with Iterative Improvement; and Moraglio et al. (2005), who combine their GA with Tabu Search. All of these studies report that good results are obtained. We have decided to use Simulated Annealing, since it requires some effort to compare two solutions, which seems to make Tabu Search less attractive. We have not applied any enhancements of Simulated Annealing, like the commonalities used by Kammer et al. (2011), which approach is based on the thesis by Schilham (2001).

Since a local search move can cause the schedule to change drastically, there is no algebraic way to compute the difference in solution value; therefore, we use simulation as a tool to evaluate each neighbor. The combination of local search with simulation has been applied before. We refer to the survey by Bianchi et al. (2009).

3   Local Search

All local search methods for the job shop scheduling problem are based on the *disjunctive graph* model that was introduced by Roy and Sussman (1964). This graph is constructed as follows. For each operation O_{ij} we introduce a vertex v_{ij}, which gets a weight equal to its processing time. Furthermore, there are two dummy vertices v_{start} and v_{end} with weight zero. The precedence constraints between operations O_{ij} and $O_{i+1,j}$ $(i = 1, \ldots, m-1)$ within a job j $(j = 1, \ldots, n)$ are modeled by including an arc from v_{ij} to $v_{i+1,j}$. Furthermore, we include an edge between each pair of vertices that correspond to two operations that must be executed by the same machine. All arcs and edges get weight zero. Finally, we add arcs from v_{start} to v_{1j}, for $j = 1, \ldots, n$, and we add arcs from vertices v_{mj} $(j = 1, \ldots, n)$ to v_{end}. Since a schedule is fully specified when the order of the operations on the machines is given, we have to direct the edges such that an acyclic graph remains. After the edges have been oriented, we call them *machine arcs*; to distinguish these from the original arcs in the graph, the latter ones are called *job arcs*.

Given the directed graph, we can compute the starting time of each operation as the length of the longest path in the graph from v_{start} to the vertex corresponding to this operation. Hence, the makespan is equal to the length of the longest path to v_{end}. Adams et al. (1988) have shown that the calculation of the longest path on a directed acyclic graph can be done in linear time.

A longest path in the directed acyclic graph is also called a *critical path*; the critical path does not have to be unique. We can decompose a critical path into *critical blocks*, where each critical block consists of one or more operations that are carried out contiguously on the same machine; at the end of a critical block, the critical path jumps to another machine.

In this paper we focus on a neighborhood that consists of two different parts, which both come down to swapping, that is, reversing the order of a machine arc.

- Critical path block swap (Nowicki and Smutnicki, 1996);
- Waiting left shift.

The makespan can only be decreased by removing the currently longest path from the disjunctive graph. Therefore, a common neighborhood is to reverse the order of a machine arc on the critical path; Van Laarhoven et al. (1992) have shown that such a reversal will never lead to a cycle in the disjunctive graph. Moreover, Nowicki and Smutnicki (1996) have shown that we can restrict ourselves to reversing machine arcs that are, either between the first pair of operations, or between the last pair of operations in a critical block. This is the first part of our neighborhood.

The second part of our neighborhood is the so-called *waiting left shift*. When optimizing the job shop scheduling problem, we see that in order to improve the objective value, waiting times need to be reduced. Bad schedules come from operations that cause a significant amount of waiting time on machines. By reducing these waiting times, the length of the schedule may become shorter. The idea of the waiting left shift neighborhood is to reduce the waiting time of the operation with the largest waiting time; suppose this is operation O_{ab}. During this waiting time, the machine waits until the predecessor $O_{a-1,b}$ of O_{ab} gets finished. Therefore, we want to make changes in the execution of the operations belonging to J_b. Our neighborhood consists of swapping a machine arc between an operation O_{ib} and its immediate predecessor on this machine; here O_{ib} can be any operation in J_b, as long as it is not the first operation on its machine. If everything works out as we hope, then all operations of that job shift left. Since the relative positions of jobs are changed, this neighborhood is able to modify the structure in such a way that new solutions can be explored to get out of a local optimum. A negative side-effect is that it is possible to create a cycle in the disjunctive graph. When this is detected, we discard this neighbor.

In our local search, we select the part of the neighborhood that we use deterministically. Given an initial solution, which is determined by scheduling all jobs sequentially, we twice apply a sequence of critical path block swaps and then once a sequence of waiting left shift swaps, after which we continue with a new series of sequences. We quit the current sequence and start the next one as soon as we encounter a series of X iterations in which we have not seen an improvement of the best solution found so far.

We use the traditional idea of Simulated Annealing that we always accept improvements, and that we accept equal or worse solutions according to a probability scheme. In contrast to traditional Simulated Annealing, the probability of accepting a worse solution does not depend on the amount of deterioration Δ. We apply a *deterministic* cooling scheme, that is, an equal or worse solution is accepted with probability T, which we call the *temperature*. The value of T depends on the number of iterations that we have had so far in the local search: after each series of Q iterations, we multiply the current value of T by some given value $\alpha < 1$. We stop the Simulated Annealing after a fixed number of iterations.

We have chosen to apply this probability mechanism, because the difference in objective function value Δ between the two solutions will be a stochastic value, as we are working with stochastic processing times. Therefore, if we would apply the classical method, the behavior of accepting solutions will be unstable. By using the temperature T as a probability number, we make sure we have a stable acceptance probability that is not influenced by stochastic realizations.

4 Dealing with Stochastic Processing Times

In this section we discuss a number of possible approaches to deal with the complications caused by the stochastic processing times. We distinguish between *classical methods*, which are often used in practice as rule-of-thumb, and *simulation-based methods*, which we have applied in this research. At the end of this section we describe the adaptations to the local search algorithm needed in our simulation-based methods and the simulation model that we have used.

4.1 Classical Methods

The classical methods try to find a deterministic value for a stochastic variable; this deterministic value is then used instead. Usually the derived value is chosen in such a way that a small amount of 'slack' is included. This is done by making sure that the derived value is (slightly) larger than the mean of the underlying distribution. The quality of the solutions that are produced by these classical methods highly depends on whether the correct amount of slack is included. A successful solution will contain just enough slack to cope with a reasonable amount of disturbances. The derived value must also somehow resemble the underlying distribution's shape to be effective.

Essentially, there are two classical methods, which in some cases are identical. The first one is to use a *robustness factor*: the deterministic value that is plugged in for the stochastic value is then equal to the expected value times the robustness factor. The second classical method works with *robustness percentiles*: the stochastic variable is replaced by a deterministic value, which is put equal to a given percentile of the probability distribution. For some distributions, these methods boil down to the same. For example, the 70% percentile of the exponential distribution is equal to 1.2 times the mean.

4.2 Simulation-Based Methods

The core of this paper concerns our derivation of more sophisticated methods based on including simulation in the local search. We first discuss this type of method, and after that discuss the adaptations that we have to make to the local search algorithm.

Result Sampling. The disadvantage of the classical methods is that it is not possible to have very small and very large values anymore for the processing times: all values are average or average-plus. In practice, only a portion of the operations will receive an average processing time while other operations will have smaller or larger processing times. Taking average or average-plus process-ing times for all operations will remove this interaction. To overcome this, we apply simulation. Given the order in which each machine executes its operations, we apply a single run of discrete event simulation to find a realization of this schedule, which is given to the local search method. Since all operations in the schedule are influenced by stochastic processing times the makespan can behave very erratically. This behavior allows us to make decisions based on situations that could occur in real life. However, a large disadvantage of the *single result sampling* method is that no two realizations will look the same. Even in differ-ent iterations of the local search, a schedule that first might have been a good schedule, can turn out to be very bad in later iterations.

To limit this effect we extended the heuristic to *five times result sampling* and *ten times result sampling*. Instead of running the simulation once, the simulation will be run either five or ten times, depending on the heuristic. The results of executing the schedule are then averaged. Again in each individual simulation run only one sample for each processing time is obtained. As the collaboration of operations determines in the end how a schedule performs, and since each individual simulation obtains one sample from each processing time variable, this interaction is preserved.

Finally, we have applied *Cutoff Sampling* to better guide the local search. The reasoning behind is that for a 'wild' probability like the exponential distribution it is nearly impossible to be robust against realizations of the processing times that are either far above or far below the mean. To counter this, we do not use the results of all ten runs of the simulation, but just the ones that appear in the middle to construct the neighborhood of the local search. We sort the obtained makespans and disregard the three schedules with smallest makespan and the two schedules with largest makespan, and proceed as before with the remaining five schedules.

4.3   Necessary Adaptations

To make the local search work, we have to make a number of adaptations. First of all, we have to find a way to compare the current schedule with the selected neighbor. Fortunately, since we have the outcomes of the simulation, we can just compute the value of the selected neighbor as the average of the realized makespans in the five or ten runs. To have a fair comparison, we apply *Common Random Numbers* (Law, 2007), that is, we use the same realizations of the processing times generated in the simulations to compute the value of the current solution. In this way we avoid that a solution gets an advantage from a fortunate set of processing times. If the neighbor is better than the current solution, then we test it against the best solution known so far as well.

Second, we have to define the moves that form the neighborhood. If only one solution has been generated according to a given schedule, like in the case of the classical methods and the single result sampling, then we follow the standard local search as described in Section 3. If on the other hand five or ten solutions have been generated, then we compute for each solution the neighborhood as described. These moves are then combined, after which a move is selected from this set randomly. In this way, moves that occur in multiple solutions have a better chance of getting selected.

Third, we have to apply the discrete event simulation to compute each of the solutions needed in the one/five/ten result sampling. This is a discrete event simulation as generally applied to a queuing system, but now the servers are the machines and the customers are the operations. However, the machines do not execute the operations in a first-come-first-served order, but in the order fixed by the schedule. Hence, the machine can have operations in the queue, but still it has to wait, if the next operation in the schedule is not available yet. Next to the makespan, we store the critical path and the waiting times of the jobs, such that we can compute the neighborhood of the solution.

5   Computational Results

To test our methods, we have conducted a large number of computational experiments. Due to a lack of space, we only present a few of the results; we refer to Van Blokland (2012) for the full set of results. The instances used in the experiments can also be downloaded from this website.

Set Up Local Search. To avoid any bias, we have kept the local search settings the same in all experiments. We apply

- At most 10 restarts
- At most 25000 iterations per restart
- The initial temperature is equal to 0.8
- The current temperature is multiplied with 0.95 after 400 iterations
- A new sequence of neighborhood search is started after 15 successive failures to improve the best solution known so far.

When a restart takes place, the local search is repeated starting with the best solution known so far as initial solution. The final solution will be simulated 1000 times to find the expected makespan and the corresponding variance in the outcomes. Furthermore, each specific experiment is repeated 5 times, resulting in 5 possibly different solutions to the problem instance. The instances we used in our experiments are randomly generated since there are no benchmarks instances known for the stochastic job shop scheduling problem in the literature.

Instance Generation. We first have created an initial set of job shop scheduling instances with 10 machines and 10 jobs and moderate processing times; later on, we have considered bigger instances with larger processing times as well. Given the number of machines, we generate for each job a random order of the indices $1, \ldots, m$ to find the order in which the machines are visited. Then, we determine for each operation the *base processing time* by drawing a random integer from $U[10, 20]$; the base processing time corresponds to the parameter in the probability distribution. Next, we select the probability distribution that the processing time will follow. In general, we assume independent distributions, and we assume that the random variables, either follow the same type of probability distribution, or are deterministic. Given a base processing time value p, the realization of the processing time is, either equal to p, or drawn from one of the following distributions:

- The uniform distribution $U[0.8 \times p, 1.2 \times p]$;
- The exponential distribution $Exp(p)$;
- The 4-Erlang distribution $Gamma(4, p/4)$;
- The LogNormal distribution $LogNormal(ln(p) - \frac{ln(2)}{2}, ln(2))$.

For each probability distribution, we apply 7 different scenarios, which indicate how many operations will get a stochastic processing time. Depending on the scenario, we decree that, either all operations within a job, or all operations on the same machine get a stochastic processing time. In the scenarios 30%, 50%, 70%, 100% of the machines or jobs yield stochastic processing times, respectively.

Experiments and Initial Conclusions. We have experimented in three phases. In the first phase, we have applied all possible methods to just one instance to get an idea of which methods work and which ones do not work. We have chosen to use an instance with 10 machines and 10 jobs. The processing times of the operations of 5 jobs are deterministic; the other operations have processing times that follow an exponential distribution. The base processing times are drawn from $U[10, 20]$. There are four series of experiments (always on the same, single instance):

1. Robustness factor 1.2; percentiles 60, 70, 80, 90; single sampling.
2. Multiple $(1, 5, 10)$ times sampling; neighborhood based on the first sample; no dominance test;
3. Multiple $(5, 10)$ times sampling; neighborhood based on all samples; dominance test; cut-off sampling.

The remark *with dominance test* indicates that we check, each time that the neighbor is better than the current solution, whether it also dominates the best solution known so far by running 5 or 10 simulations. It turns out that this dominance test is crucial. The best solutions are found when sampling 10 times with the dominance test; this is about 2 % better than the same experiment with 5 times sampling. After that, trailing the ten times sampling by 8% is the experiment with percentile 70. The solutions without dominance test are far away.

Next, we have executed Phase 2 of the experiments. For each combination of a scenario (out of the 7 mentioned above) and a probability distribution (out of the 4 mentioned above) we have generated an instance with 10 jobs and 10 machines, where all base processing times are random integers drawn from $U[10, 20]$. To each of these instances, we have tested the following methods:

- Robustness factors 1.1, 1.2, 1.3;
- Percentiles 60, 70, 80, 90;
- Local search with 1 time sampling;
- Local search with 5 times sampling, neighborhood based on all samples, dominance test, and cut-off sampling (*LS5*);
- Local search with 10 times sampling, neighborhood based on all samples, dominance test, and cut-off sampling (*LS10*).

It turns out that for most of the simulation results, the difference in performance between the *LS5* and *LS10* methods is relatively small, and sometimes *LS5* outperforms *LS10*, which suggests that taking only five samples is sometimes sufficient for the search process.

Remarkably enough, for two out of the 28 experiments the sampling methods are beaten (albeit by at most one percent) by one of the classical methods. In both cases, this occurs when 70% of the machines or jobs yields stochastic processing times that follow a LogNormal probability distribution. Surprisingly, the winners then are the methods with Robustness Factors 1.1 and 1.3 respectively. Overall, the sampling methods yield a consistently better and more stable performance.

Finally, we have executed Phase 3 of the experiments. Here we increase the number of jobs and machines to 20 jobs and 20 machines at most with a base processing time drawn from $U[10, 30]$. Moreover, we have generated a number of 10 job, 10 machine instances with a base processing time drawn from $U[10, 40]$, $U[10, 80]$, and $U[10, 120]$, respectively. In all instances, we have that all operations have stochastic processing times that, either all come from the exponential distribution, or all come from the 4-Erlang distribution. For these instances we have only applied the Percentile Robustness method with percentile 70 and the multiple times sampling (either 5 or 10 samples) with a neighborhood based on all samples, including a dominance test and cut-off sampling. In all our experiments, the 10 times sampling method outperformed the other methods with a typical gap of 3% with the 5 times sampling method and a large gap with the 70 percentile method, which was more than 50% for the largest instance.

6   Conclusions and Future Research

We have presented a simulation based method as an alternative for the classical methods of using a robustness factor or a robustness percentile. Our experimental results, depicted in the figure below clearly show that our new method dominates the classical ones.

So far, we have looked at the stochastic job shop scheduling problem with stochastic processing times. Obviously, there are more sources of stochasticity available, like for example

- Machine failures;
- Uncertain changeover times or set-up times;
- Travel times between machines that are stochastic.

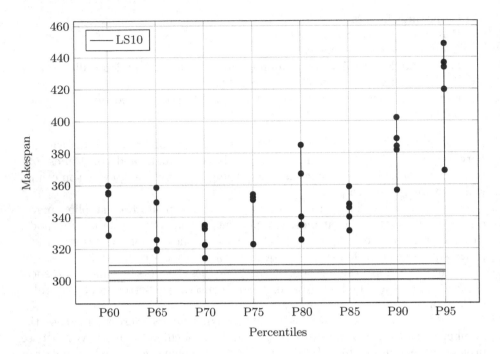

Fig. 1. Overview of the Percentile Robustness method compared to the Ten Times Sampling method. The data represented in this figure is based on the mean makespan of 1000 simulations. Each method is executed five times, resulting in five schedules.

It is an interesting question whether a simulation-based method can provide good solutions for these situations as well.

Another question is whether it is possible to combine a simulation-based method with Tabu Search. Based on our experiments, you need to sample at least five times to evaluate each neighbor, which makes the method computationally unattractive.

Finally, we have used cut-off sampling to look at the average solutions in our local search. It might be an interesting idea to use a similar technique to find solutions that are both good and stable, that is, with a small variance in outcome value.

References

1. Adams, J., Balas, E., Zawack, D.: The shifting bottleneck procedure for job shop scheduling. Management Science 34, 391–401 (1988)
2. van den Akker, M., Hoogeveen, H.: Minimizing the number of late jobs in a stochastic setting using a chance constraint. Journal of Scheduling 11, 59–69 (2008)
3. Bianchi, L., Dorigo, M., Gambardella, L.M., Gutjahr, W.J.: A survey on metaheuristics for stochastical combinatorial optimization. Natural Computing 8, 239–287 (2007)

4. van Blokland, C.H.M.: Solution Approaches for Solving Stochastic Job Shop and Blocking Job Shop Problems (2012), http://www.staff.science.uu.nl/~3235777/thesis.pdf
5. ten Eikelder, H.M.M., Aarts, B.J.M., Verhoeven, M.G.A., Aarts, E.H.L.: Sequential and parallel local search algorithms for job shop scheduling. In: Voss, S., Martello, S., Osman, I.H., Roucairol, C. (eds.) Meta-Heuristics, Advances and Trends in Local Search Paradigms for Optimization, pp. 359–371. Kluwer (1999)
6. Gonçalves, J.F., de Magalhães Mendes, J.J., Resende, M.G.C.: A hybrid genetic algorithm for the job shop scheduling problem. European Journal of Operational Research 167, 77–95 (2005)
7. Graham, R.L., Lawler, E.L., Lenstra, J.K., Rinnooy Kan, A.H.G.: Optimization and Approximation in Deterministic Sequencing and Scheduling: a Survey. Annals of Discrete Mathematics 5, 287–326 (1979)
8. Kammer, M.L., Hoogeveen, J.A., van den Akker, J.M.: Identifying and exploiting commonalities for the job-shop scheduling problem. Computers & Operations Research 38, 1556–1561 (2011)
9. van Laarhoven, P.J.M., Aarts, E.H.L., Lenstra, J.K.: Job shop scheduling by simulated annealing. Operations Research 40, 113–125 (1992)
10. Law, A.M.: Simulation Modeling and Analysis. Mc Graw-Hill (2007)
11. Moore, J.M.: An n job, one machine sequencing algorithm for minimizing the number of late jobs. Management Science 15, 102–109 (1968)
12. Moraglio, A., ten Eikelder, H., Tadei, R.: Genetic Local Search for Job Shop Scheduling Problem, Technical Report CSM-435, University of Essex, UK (2005)
13. Niño-Mora, J.: Stochastic scheduling. In: Floudas, C.A., Pardalos, P.M. (eds.) Encyclopidea of Optimization, pp. 3818–3824 (2008), http://halweb.uc3m.es/jnino/eng/pubs/EoO08.pdf
14. Nowicki, E., Smutnicki, C.: A fast taboo search algorithm for the job shop problem. Management Science 42, 797–813 (1996)
15. Pinedo, M.L.: Scheduling Theory, Algorithms and Systems. Prentice Hall (2002)
16. Pinedo, M.L., Weiss, G.: The largest variance first policy in some stochastic scheduling problems. Operations Research 35, 884–891 (1979)
17. Rothkopf, M.H.: Scheduling with random service times. Management Science 12, 707–713 (1966)
18. Roy, B., Sussmann, B.: Les Problèmes d'ordonnancement avec contraintes disjonctives, Note DS no. 9 bis, SEMA, Montrouge (1964)
19. Schilham, R.M.F.: Commonalities in local search. PhD Thesis, Eindhove University of Technology, The Netherlands (2001)
20. Smith, W.E.: Various optimizers for single-stage production. Naval Research Logistics Quarterly 3, 59–66 (1956)
21. Taillard, E.D.: Parallel taboo search techniques for the job shop scheduling problem. ORSA Journal on Computing 6, 108–117 (1994)
22. Trietsch, D., Baker, K.R.: Minimizing the number of tardy jobs with stochastically-ordered processing times. Journal of Scheduling 11, 71–73 (2008)
23. Wang, L., Zheng, D.: An effective hybrid optimisation strategy for job-shop scheduling problems. Computers and Operations Research 28, 585–596 (2001)
24. Williamson, D.P., Hall, L.A., Hoogeveen, J.A., Hurkens, C.A.J., Lenstra, J.K., Sevast'janov, S.V., Shmoys, D.B.: Short shop schedules. Operations Research 45, 288–294 (1997)

Author Index